WAVES AND RAYS IN ELASTIC CONTINUA

Fourth Edition

Other Related Titles from World Scientific

Waves and Rays in Seismology: Answers to Unasked Questions
Third Edition
by Michael A. Slawinski
ISBN: 978-981-122-643-4

Wavefronts and Rays as Characteristics and Asymptotics
Third Edition
by Andrej Bóna and Michael A. Slawinski
ISBN: 978-981-122-646-5

On Foundations of Seismology: Bringing Idealizations Down to Earth
by James Robert Brown and Michael A. Slawinski
ISBN: 978-981-4329-49-1

WAVES AND RAYS IN ELASTIC CONTINUA

Fourth Edition

Michael A. Slawinski

World Scientific

NEW JERSEY • LONDON • SINGAPORE • BEIJING • SHANGHAI • HONG KONG • TAIPEI • CHENNAI

Published by

World Scientific Publishing Co. Pte. Ltd.
5 Toh Tuck Link, Singapore 596224
USA office: 27 Warren Street, Suite 401-402, Hackensack, NJ 07601
UK office: 57 Shelton Street, Covent Garden, London WC2H 9HE

Library of Congress Control Number: 2020945151

British Library Cataloguing-in-Publication Data
A catalogue record for this book is available from the British Library.

WAVES AND RAYS IN ELASTIC CONTINUA
Fourth Edition

ISBN 978-981-122-640-3 (hardcover)
ISBN 978-981-122-641-0 (ebook for institutions)
ISBN 978-981-122-642-7 (ebook for individuals)

For any available supplementary material, please visit
https://www.worldscientific.com/worldscibooks/10.1142/11994#t=suppl

Desk Editor: Ng Kah Fee

Typeset by Stallion Press
Email: enquiries@stallionpress.com

Portrait by Ida Czechowska

Dedication

This book is dedicated to the scientific spirit and accomplishments of Maurycy Pius Rudzki, Chair of Geophysics at the Jagiellonian University, where, in 1895, the first Institute of Geophysics in the world was created.

In order to interpret the wealth of detail contained in the seismographic record of a distant earthquake, we must know the path or trajectory of each ray that leaves the focus or origin in any given direction. An indirect solution of this problem was attempted by several earlier investigators, prominent among whom were Rudzki and Benndorf.

James B. Macelwane (1936)
Introduction to theoretical seismology: Geodynamics

Seismological studies appear to have stimulated Rudzki to make the first quantitative calculations on elastic waves.

Michael J.P. Musgrave (1970)
Crystal acoustics: Introduction to the study of elastic waves and vibrations in crystals

In the first decade of the [twentieth] century M.P. Rudzki in Cracow began to investigate the consequences of anisotropy in the earth for seismic waves. As far as I can make out, he was the first to determine the wave surface for elastic waves in an anisotropic solid.

Klaus Helbig (1994)
Foundations of anisotropy for exploration seismics

Acknowledgments

The author wishes to acknowledge the substantial improvements of this book that have resulted from numerous collaborations.

The manuscript of this book originated as the notes for a graduate course. The quality of these notes was improved by collaborations with Rachid Ait-Haddou, Dan Calistrate, Nilanjan Ganguly, Hugh Geiger, Bill Goodway, Jeff Grossman, Marcelo Epstein, Xinxiang Li, John Parkin and Paul Webster.

The manuscript of this book has reached its final form as a result of collaborations with Andrej Bóna, Len Bos, Nelu Bucataru, David Dalton, Peter Gibson, Andrzej Hanyga, Klaus Helbig, Misha Kochetov, Ed Krebes, Michael Rochester, Yves Rogister, Raphaël Slawinski, Jędrzej Śniatycki and Chad Wheaton.

The second edition of this book benefited from further collaborations with several people stated above, namely, Andrej Bóna, Len Bos, Nelu Bucataru, David Dalton, Klaus Helbig, Misha Kochetov, Michael Rochester and Yves Rogister, as well as Çağrı Diner, Garry Quinlan, Peter Smith and Jason Soo. Norm Bleistein's review of the first edition provided feedback that benefited the second one. Furthermore, reviews of manuscripts of the second edition by Vassily M. Babich, Ed Krebes and Małgorzata Seredyńska—as well as insightful questions from Andreas Atle, Ayiaz Kaderali and Fabrice Oehler—have improved the final version.

Editorial revisions of the text with Cathy Beveridge, from the original sketch to the final drafts for both editions, enhanced the structure and coherence of this book.

The third edition benefited from discussions with Michael Rochester, computational results of Tomasz Danek, refinements of figures by Elena Patarini and proofreading of David Dalton.

The fourth edition benefited from collaborations with Len Bos, Sandra Forte and Maurizio Vianello, as well as from comments of Md Abu Sayed, Andrej Bóna, Yuriy Ivanov, Thomas Meehan, Andrey Melnikov, Michael Rochester and Theodore Stanoev. It also benefitted from the graphic support of Elena Patarini as well as proofreading of Filip Adamus, David Dalton and Theodore Stanoev. Kah-Fee Ng ensured the adjustments of final details prior to printing.

Preface to First Edition

> Il ne suffit pas d'observer, il faut se servir de ses observations, et pour cela il faut généraliser. [...] Le savant doit ordonner; on fait la science avec des faits comme une maison avec des pierres; mais une accumulation de faits n'est pas plus une science qu'un tas de pierres n'est une maison.[1]
>
> *Henri Poincaré (1902) La science et l'hypothèse*

Theoretical formulations of applied seismology are substantiated by observable phenomena. Reciprocally, our perception and understanding of these phenomena necessitate rigorous descriptions of physical behaviours.

1

> It is not enough to observe. One must use these observations, and for this purpose one must generalize. [...] The scientist must organize [knowledge]; science is composed of facts as a house is composed of bricks; but an accumulation of facts is no more a science than a pile of bricks is a house.

To emphasize this statement of Poincaré, let us also consider the following quotation.

> As the biggest library if it is in disorder is not as useful as a small but well-arranged one, so you may accumulate a vast amount of knowledge but it will be of far less value to you than a much smaller amount if you have not thought it over for yourself; because only through ordering what you know by comparing every truth with every other truth can you take complete possession of your knowledge and get it into your power.
>
> *Arthur Schopenhauer (1851) Parerga and Paralipomena, Volume 2*

Let us also consider the distinction between the deductive formulation of a theory, implied by the statement of Poincaré, and the process of development of science *sensu lato.*

> The development of science does not in the least resemble the building of a house, or of a palace, or of a planned garden-suburb. There is nothing in its development comparable to a single foundation upon which, once well and securely laid, the building may be erected by the labours of many workers, with or without the control of a master-builder.
>
> *L. Susan Stebbing (1937/1958) Philosophy and the physicists*

As stated by Bunge in his book on "Philosophy of Science, Vol. I: From problem to theory",

> A nice illustration of the intertwining of empirical and theoretical events in the actual practice of science is offered by seismology, the study of elastic disturbances of Terra. [...] In conclusion, in order to "read" a seismogram so that it may become a set of data regarding an event (e.g., an earthquake) or an evidence relevant to a theory (e.g., about the inner structure of our planet), the seismologist employs elasticity theory and all the theories that may enter the design and interpretation of the seismograph.

The present book emphasizes the interdependence of mathematical formulation and physical meaning in the description of seismic phenomena. Herein, we use aspects of continuum mechanics, wave theory and ray theory to explain phenomena resulting from the propagation of seismic waves.

The book is divided into three main parts: *Elastic continua*, *Waves and rays* and *Variational formulation of rays*. There is also a fourth part, which consists of *Appendices*. In *Part 1*, we use continuum mechanics to describe the material through which seismic waves propagate, and to formulate a system of equations to study the behaviour of such a material. In *Part 2*, we use these equations to identify the types of body waves propagating in elastic continua as well as to express their velocities and displacements in terms of the properties of these continua. To solve the equations of motion in anisotropic inhomogeneous continua, we use the high-frequency approximation and, hence, establish the concept of a ray. In *Part 3*, we show that, in elastic continua, a ray is tantamount to a trajectory along which a seismic signal propagates in accordance with the variational principle of stationary traveltime. Consequently, many seismic problems in elastic continua can be conveniently formulated and solved using the calculus of variations. In *Part 4*, we describe two mathematical concepts that are used in the book; namely, homogeneity of a function and Legendre's transformation. This part also contains a *List of symbols*.

The book contains an *Index* that focuses on technical terms. The purpose of this index is to contribute to the coherence of the book and to facilitate its use as a study manual and a reference text. Numerous terms are grouped to indicate the relations among their meanings and nomenclatures. Some references to selected pages are marked in bold font. These pages contain a defining statement of a given term.

This book is intended for senior undergraduate and graduate students as well as scientists interested in quantitative seismology. We assume that the reader is familiar with linear algebra, differential and integral calculus, vector calculus, tensor analysis, as well as ordinary and partial differential equations. The chapters of this book are intended to be studied in sequence. In that manner, the entire book can be used as a manual for a single course. If the variational formulation of ray theory is not to be included in such a course, the entire *Part 3* can be omitted.

Each part begins with an *Introduction*, which situates the topics discussed therein in the overall context of the book as well as in a broader scientific context. Each chapter begins with *Preliminary remarks*, which state the motivation for the specific concepts discussed therein, outline the structure of the chapter and provide links to other chapters in the book. Each chapter ends with *Closing remarks*, which specify the limitations of the concepts discussed and direct the reader to related chapters. Each chapter is followed by *Exercises* and their solutions. While some exercises extend the topics covered, others are referred to in the main text. Reciprocally, the footnotes attached to these latter exercises refer the reader to the sections in the main text, where a given exercise is mentioned. Often, the exercises referred to in the main text supply steps that are omitted from the exposition in the text. Also, throughout the book, footnotes refer the reader to specific sources included in the *Bibliography*.

"Seismic waves and rays in elastic media" strives to respect the scientific spirit of Rudzki, described in the following statement[2] of Marian Smoluchowski, Rudzki's colleague and friend.

> Tematyka geofizyczna musiała nęcić Rudzkiego, tak wielkiego, fantastycznego miłośnika przyrody, z drugiej zaś strony ta właśnie nauka odpowiadała najwybitniejszej właściwości

[2] Smoluchowski, M., (1916) Maurycy Rudzki jako geofizyk / Maurycy Rudzki as a geophysicist: Kosmos, **41**, 105–119.

umysłu Rudzkiego, jego dążeniu do matematycznej ścisłości w rozumowaniu.[3]

[3] The subject of geophysics must have attracted Rudzki, a great lover of nature. Also, this very science accommodated the most outstanding quality of Rudzki's mind, his striving for mathematical rigour in reasoning.

Changes from First Edition

> Mathematics gives to science the power of abstraction and generalization, and a symbolism that says what it has to say with the greatest possible clarity and economy.
>
> *John Lighton Synge and Byron A. Griffith (1949)*
> *Principles of mechanics*

The conviction expressed by the above quote has inspired both the first and second editions of this book. Significant additions and modifications have been included in the second edition. These changes came about as a result of teaching from this text, readers's enquiries, and research into concepts discussed in the first edition. A summary of these changes and the reason for their inclusion are given below.

Part 1

Chapter 1: To familiarize the reader with the fundamentals that underlie the seismic theory, we introduce in Section 1.2 the three rudimentary concepts of continuum mechanics, namely, material body, manifold of experience and system of forces.

Chapter 2: To deepen the reader's understanding of the arguments used to formulate the balance principles of continuum mechanics, we include in Sections 2.4 and 2.7 the particle-mechanics motivation of the balance of linear momentum and the balance of angular momentum, respectively. To make the reader aware of possible extensions within continuum mechanics, we discuss in Section 2.7 the distinction between the strong and weak forms of Newton's third law of motion. As a result, we show that the symmetry of the stress tensor is a consequence of the strong form, and not a fundamental physical law.

Chapter 3: To deepen the reader's understanding of the fundamentals that underlie the formulation of seismic theory, we introduce in Section 3.1 the three principles of constitutive equations, namely, determinism, local action and objectivity. To provide the background that allows us to emphasize the meaning of elasticity, we introduce in Section 3.4 constitutive equations of anelastic continua.

Chapter 5: Following a demonstration that only eight classes of the elasticity tensor exist,[4] all these classes are included in this edition. The addition consists of Sections 5.8 and 5.11, where we study trigonal and cubic continua.

Part 2

Chapter 6: Since this chapter deals with the wave equation, which plays an important role in seismology, we wish to deepen the reader's understanding of different aspects of this equation. In Section 6.5, we discuss well-posedness of the wave equation with its initial conditions. To provide further insight into the properties of the wave equation, we compare it in Section 6.7 to the evolution equation, which is ill-posed. Another significant addition consists of the formulation and study of solutions of the wave equation in two and three spatial dimensions, as presented in Section 6.6. Therein, we discuss also the range of influence and the domain of dependence of these solutions. To familiarize the reader with the concepts of reflection and transmission in the context of the wave equation, we study the solutions of one-dimensional scattering, which is discussed in Section 6.8. To familiarize the reader with the several-century-long debate concerning the applicability of the wave equation, which is a differential equation, to study nondifferentiable solutions, we present the theory of distributions and its application to the wave equation. The crux of this addition is Section 6.9. Furthermore, in Section 6.10, we elaborate on the concept of the reduced wave equation and Fourier's transform of the wave equation, which is used in the following chapter to discuss trial solutions of the equations of motion.

Chapter 7: Since this chapter deals with the equations of motion whose trial solutions necessitate asymptotic methods, we wish to familiarize the reader with the concept of the asymptotic series in the context of ray theory,

[4] Readers interested in the existence of the eight symmetry classes might refer to Bóna, A., Bucataru, I., Slawinski, M.A., (2005) Characterization of elasticity-tensor symmetries using SU(2). Journal of Elasticity **75**(3), 267–289, and to Bóna, A., Bucataru, I., Slawinski, M.A., (2004) Material symmetries of elasticity tensor. The Quarterly Journal of Mechanics and Applied Mathematics **57**(4), 583–598.

which is commonly referred to as the asymptotic ray theory. This addition is contained in Sections 7.2.3 and 7.2.4.

Chapter 8: Since Hamilton's ray equations are the basis of ray theory, we wish to deepen the reader's understanding of their meaning, as well as the meaning of their solutions. To do so, we present an analytic solution of Hamilton's ray equations, which allows us to gain both mathematical and physical insights into Hamilton's equations. Furthermore, an equivalent solution is presented using Lagrange's ray equations. Hence, this analytic solution allows us to relate the Hamiltonian and Lagrangian formulations of ray theory—two distinct approaches. The crux of this addition consists of Section 8.5.

Chapter 9: The convexity and detachment of the innermost phase-slowness surface are commonly invoked in seismology. In this edition, we show particular cases for which this surface is not detached, and we comment on the fact that detachment is not necessary to prove the convexity of the innermost surface.

Chapter 10: In this edition, we use Newton's third law of motion—rather than the second one used in the first edition—to derive in Section 10.2.1, the dynamic boundary conditions on the interface between two media. Since we are dealing with two media acting on one another, the third law lends itself more naturally than the second one for which we had to treat the interface as a layer with a vanishing thickness.

Part 3

Chapter 14: In this edition, the traveltime expression for a signal in a continuum exhibiting an elliptical velocity dependence with direction and a linear dependence with depth is valid for the entire ray: both the downward and upward segments. The expression presented in the first edition is valid for a downward segment only.

In this edition, the chapter on Lagrange's ray equations is found at the end of *Part 2*, rather than immediately after the chapter on Hamilton's ray equations. This placement of the chapter emphasizes the physical motivation; namely, we can study the entire ray theory in the context of Hamilton's ray equations. Lagrange's ray equations, on the other hand, open a new path of study to which we devote *Part 3*.

As a result of teaching from the first edition, the second edition acquired more exercises and figures. There is also a *List of Figures* included.

With the above additions, the title of the book has been modified to emphasize the generality of the approach and the importance of the concept of continuum. Thus, "Seismic waves and rays in elastic media" became "Waves and rays in elastic continua".

Changes from Second Edition

> Every partially true theory is as alive as the physicists who care to work it out, apply and analyze it: even classical theories continue to evolve and deserve being overhauled both for their own sake and as models of theory construction.
>
> *Mario Bunge (1967)*
> *Foundations of physics*

A signifiant addition to the third edition is Section 5.9.2, where we discuss different coordinate expressions of the same tensor, which we illustrate in Figure 5.9.1 and Exercises 5.5 and 5.6. Another addition in the third edition is Section 6.4, in which we formulate the wave equations in terms of the displacement vector, not in terms of its derivatives. Therein, we address also the necessary and sufficient conditions for existence of P and S waves in isotropic Hookean solids.

Also, in the third edition, there are several signifiant modifications with additions. In Section 3.4.3, we examine in detail the issues of viscoelasticity, with additions of Figures 3.4.2, 3.4.3 and Exercise 3.6. Also, Section 3.4.4 and Figure 3.4.4 are added. In Section 5.3, we discuss relations between rotations in three dimensions and their expressions in terms of 6×6 matrices. Also, Exercises 5.9 and 5.10 are added. In Section 6.1.4, an application of the Jacobian matrix in the context of the wave equations for P and S waves allows us to explain the role of the divergence and the curl in formulating these equations. To set the stage for this application, the last paragraph of Section 1.5 is modified. Another modification is at the end of Section 6.5.1 in which d'Alembert's approach is extended to the higher spatial dimension for waves exhibiting spherical symmetries; this modification is followed by the addition of Exercises 6.3 and 6.4. In Section 6.10.2, certain properties of Fourier's transform are clarified. The last modification is in Exercise 9.12, where the physical meaning of equality among the wavefront orientation and the ray and polarization directions is emphasized.

The third edition benefits from inclusion of references to Bóna, A. and Slawinski, M.A., (2011) Wavefronts and rays as characteristics and asymptotics: World Scientific—a book that can be viewed as a follow up of subjects discussed in Chapters 6–8, and which was published after the second edition. The figures have been refined and captions accommodated accordingly. Also, references to recent papers and books have been included and several index entries added.

Changes from Third Edition

> The aphorism that nature abhors a vacuum was the accepted wisdom for 2,000 years, well into the Middle Ages, because it was the simplest, seemingly obvious, explanation of a whole range of everyday phenomena. [...] As Pascal himself noted, nature does not abhor a vacuum less on the top of a mountain than in a valley [...] it is the weight of the air that causes all the phenomena that the philosophers had attributed to an 'imaginary cause'.

Frank Close (2009)
Nothing: A Very Short Introduction

Changes in the fourth edition continue to be a result of teaching from this text, readers's enquiries, and research into concepts discussed in the second edition. A summary of these changes, and the reason for their inclusion, are given below.

A significant addition to the third edition is in Chapter 6. Therein, Section 6.4 is now divided into three subsections: *Introductory comments*, *Sufficient conditions for P and S waves* and *Necessary conditions for P and S waves*. As a result of this addition, the derivation of the P and S waves in terms of displacements—originally presented by Rochester (2010)—is strengthened and yields more detailed information.

Throughout the book, several new figures are included. An extension of Figure 2.5.2 and its explanations in the main text emphasize the distinction between the surface forces acting on the coordinate planes and the corresponding average force per unit area. To offer an insight into the theory of viscoelasticity, a mechanical illustration of the Kelvin-Voigt model is added as Figure 3.4.1. An examination of Figure B.1.1 allows us to discuss subtleties of Legendre's transformation, which plays an important role in relating *Part 2* and *Part 3* of the book. An ambiguity of this transformation is avoided by a piecewise operation.

In a new example, Example B.2.2, we discuss the details of the aforementioned piecewise Legendre's transformation.

There are two new exercises with detailed solutions. Both emphasize mathematical subtleties that are crucial for a profound understanding of the subjects discussed in the book. In Exercise 1.8, we show that, fundamentally, the result of the cross product is not a vector but a second-rank tensor. In Exercise 9.1, we examine operations on tensors that entail properties of transformations of Christoffel's equations, which underly ray theory discussed in this book.

Essential aspects of the changes in this edition are contained in new remarks, whose purpose is to extend and clarify subjects discussed in the previous editions. In Remark 1.3.1, we discuss the concept of the material time derivative, and its reduction to a partial derivative. In Remark 2.1.1, we clarify subtleties of a derivation of a differential equation from an integral equation, and, in Remark 2.9.1, we illustrate the fact that the equality of definite integrals does not imply the equality of their integrands.

An important subject of this book is the material symmetry, which is discussed in Chapter 5. Therein, in Remark 5.12.1, we show an uncommon parameterization of the elasticity matrix for an isotropic continuum, which allows for a consistency with a common parameterization of the elasticity matrix for transverse isotropy. In Remark 5.12.2, we consider a positive semidefinite elasticity matrix for an isotropic continuum, which results in a distinction between a fluid and a solid.

Chapter 8 is devoted to Hamilton's ray equations. Therein, in Remark 8.5.1, we examine singularities of the traveltime expressions in anisotropic inhomogeneous media that exhibit a linear velocity dependance with depth and an elliptical velocity dependance with direction. This remark includes a detailed analysis of several averages for a linear velocity dependance.

In Remark B.1.2, in a manner similar to the aforementioned Figure B.1.1 and Example B.2.2, we discuss the invertibility of Legendre's transformation, which requires the relation between the function and its slope to be one-to-one.

Continuous work with, and lecturing from, this book results in many new index entries, which facilitate the use of this large document as a reference for a specific topic. Also, there are references to new publications and books, including the more recently published volumes of this series, within the World Scientific, namely, Slawinski, M.A. (2021) Waves and rays in seismology: Answers to unasked questions (*3rd* edition), and Brown, J.R. and Slawinski, M.A., (2017) On foundations of seismology: Bringing idealizations down to Earth.

Contents

List of Figures

Part 1

Elastic continua

Introduction to Part 1

> One conceives the causes of all natural effects in terms of mechanical motion. This, in my opinion, we must necessarily do, or else renounce all hopes of ever comprehending anything in Physics.*
>
> *Christian Huygens (1690)*
> *Treatise on light: In which are explained the causes of that which occurs in reflection and refraction*

Our focus in this book is the description of seismic phenomena in elastic media. The physical basis of seismic wave propagation lies in interactions among the grains that form materials within which disturbances propagate. It is difficult to describe quantitatively these interactions. However, since our experimental data are the result of a large number of such interactions, we can consider these interactions as an ensemble and describe seismic wave propagation through a granular material in terms of wave propagation

* Readers interested in the modern view of this statement, in the context of analytical mechanics, might refer to Born, M., and Wolf, E., (1999) Principles of optics (*7th* edition): Cambridge University Press, p. xxix. Readers interested in this statement in the context of variational mechanics, which we discuss in Section 13.2, might refer to Poincaré, H., (1902/1968) La science et l'hypothèse: Flammarion, pp. 219–225.

Also, readers might refer to Einstein, A., and Infeld, L., (1938) Evolution of physics from early concepts to relativity and quanta: Simon & Schuster, p. 125:

> During the second half of the nineteenth century new and revolutionary ideas were introduced into physics; they opened the way to a new philosphical view, differing from the mechanical one.

through a medium that is continuous. We refer to such a medium as a continuum.

Consequently, in this book, we follow the concepts of continuum mechanics where any material is described by a continuum. A continuum is formulated mathematically in terms of continuous functions representing the average properties of many microscopic objects forming the actual material. In this context, all the associated quantities become scalar, vector or tensor fields, and the formulated problems are governed by differential equations.

Using the methods of continuum mechanics, we adhere to the following statement of Kennett from his book "The seismic wavefields".

> We adopt a viewpoint in which the details of the microscopic structure of the medium through which seismic waves propagate is ignored. The material is supposed to comprise a continuum of which every subdivision possesses the macroscopic properties.

At the beginning of *Part 1*, we formulate the methods for describing deformations of continua and we introduce the concept of strain. This is followed by a description of forces acting within the continuum and the introduction of the concept of stress. We also derive the fundamental equations; namely, the equation of continuity and the equations of motion, which result from the conservation of mass and the balance of linear momentum, respectively.

To supplement these equations and, hence, to formulate a determined system that governs the behaviour of a continuum, we consider a particular class of continua that is, however, general enough to be of significance in applied seismology. Our attention focuses on elastic continua. Any continuum is characterized by its deformation in response to applied loads. In this book, we assume that this response can be adequately described by linear stress-strain equations. Also, we assume that all the energy expended on deformation is transformed into potential energy, which is stored in the deformed continuum. Consequently, upon the removal of the load, the stored energy—to which we refer to as the strain energy—allows this continuum to return to its undeformed state.

The original formulation of the theory of continuum mechanics can be dated to the second half of the eighteenth century and is associated with the work of Leonhard Euler. At the beginning of the nineteenth century, further development was achieved by Augustin-Louis Cauchy and George Green, as well as several other European scientists. The modern development of the theory of continuum mechanics is mainly associated with the work of American scientists, in particular, the work of Walter Noll, Ronald Rivlin and Clifford Ambrose Truesdell III, in the second half of the twentieth century.

We should also note that too literal an interpretation of the concept of continuum can lead to inaccurate conclusions. This can be illustrated by an example given by Schrödinger in his book entitled "Nature and the Greeks".

> Let a cone be cut in two by a plane parallel to its base; are the two circles, produced by the cut on the two parts equal or unequal? If unequal, then, since this would hold for any such a cut, the ascending part of the cone's surface would not be smooth but covered with indentations; if you say equal, then for the same reason, would it not mean that all these parallel sections are equal and thus the cone is a cylinder?

Also, in view of the abstract nature of continuum mechanics, we must carefully consider the definition of exactness of a solution. While exact mathematical solutions to the equations formulated in continuum mechanics exist, the equations themselves are not exact representations of nature since they rely on abstract formulations. Hardy expresses a similar thought in his book entitled "A mathematician's apology".

> It is quite common for a physicist to claim that he has found a 'mathematical proof' that the physical universe must behave in a particular way. All such claims, if interpreted literally, are strictly nonsense. It cannot be possible to prove mathematically that there will be an eclipse tomorrow, because eclipses, and other physical phenomena, do not form part of the abstract world of mathematics.

Nevertheless, the notion of continuum, as it pertains to the theory of elasticity, is particularly useful for seismological purposes because it permits convenient mathematical analysis that gives rise to scientific theory validated by experimental data.

CHAPTER 1

Deformations

> … au lieu de considérer la masse donnée comme un assemblage d'une infinité de points contigus, il faudra, suivant l'esprit du calcul infinitésimal, la considérer plutôt comme composée d'éléments infiniment petits, qui soient du même ordre de dimension que la masse entière;[1]
>
> *Joseph-Louis Lagrange (1788)*
> *Mécanique Analytique*

Preliminary Remarks

We begin our study of seismic wave propagation by considering the materials through which these waves propagate. Physical materials are composed of atoms and, hence, the fundamental treatment of this propagation would require the study of interactions among the atoms. At present, such an approach is impractical and, perhaps, impossible with the available mathematical tools. Consequently, we seek a more convenient approach. An alternative approach is offered by continuum mechanics, which allows us to obtain results consistent with observable phenomena without dealing directly with the discrete properties of the materials through which seismic waves propagate.

1 … instead of considering a given mass as an assembly of an infinity of neighbouring points, one shall—following the spirit of calculus—consider rather the mass as composed of infinitely small elements, which would be of the same dimension as the entire body;

As all mathematical physics, continuum mechanics utilizes abstract concepts to model physical reality.[2] In a seismological context, the Earth is regarded as a continuum that transmits mechanical disturbances. The notion of continuum allows us to describe the deformations and forces experienced by a deformable body in terms of strains and stresses within a continuum.

We begin this chapter with an explanation of the notion of continuum followed by a description of deformations within it. In particular, we derive the strain tensor, which allows us to describe both a relative change in volume and a change in shape within the continuum.

1.1. Notion of Continuum

In continuum mechanics, we choose to disregard the atomic structure of matter and the explicit interactions among particles. The notion of continuum is justified by the assumption that a material is composed of sufficiently closely spaced particles, so that its descriptive functions can be considered to be continuous. In other words, the infinitesimal elements of the material are assumed to possess the same physical properties as the properties observed in macroscopic studies. Although the microscopic structure of real materials is not consistent with the concept of continuum, this idealization provides a useful platform for mathematical analysis, which in turn permits us to model physical reality using abstract concepts.[3]

In the context of the philosophy of science, continuum mechanics is associated with the concept of emergence, which is also called methodological holism. In physics, emergence is used to describe properties, laws or phenomena that occur at macroscopic scales but not at microscopic ones, in spite of the fact that a macroscopic system can be viewed as an ensemble of microscopic ones. As an example, let us consider colour. Elementary particles, such as protons or electrons, have no colour. Colour emerges if

[2] Readers interested in the concept of models and physical understanding might refer to Weinert, F., (2005) The scientist as philosopher: Springer-Verlag, pp. 45–47.

[3] Readers interested in rigorous mathematical foundations of elasticity might refer to Marsden, J.E., and Hughes, T.J.R., (1983/1994) Mathematical foundations of elasticity: Dover. For general aspects of continuum-mechanics formulations, readers might refer to Malvern, L.E., (1969) Introduction to the mechanics of a continuous medium: Prentice-Hall.

these particles are arranged in atoms, which absorb or emit specific wavelengths of light and can thus be said to have colour. Emergent concepts in continuum mechanics are elasticity, rigidity, viscosity, friction, and so on. The approach in which we invoke elementary particles to describe properties of matter belongs to the field of condensed-matter physics, and philosophically is associated also with the concept of reductionism and so-called methodological individualism.

The concept of continuum allows us to consider materials in such a way that their descriptive functions are continuous and differentiable. In particular, we can define stress at a given point, thereby enabling us to apply calculus to the study of forces within a continuum. This definition and the subsequent application of calculus is associated with the work of Augustin-Louis Cauchy in the first half of the nineteenth century. Instead of studying atomic forces among individual particles, he introduced the notions of stress and strain in a continuum, which resulted in the equations associated with the theory of elasticity.

Using a continuum-mechanics approach to describe seismic wave propagation raises some concerns. In continuum mechanics, the behaviour of a multitude of grains in a portion of a material is discussed by studying the behaviour of the ensemble. Consequently, information relating to the grains themselves is lost in the averaging process. In other words, the application of continuum mechanics raises the question whether the loss of information about the granular structures of the material allows us to represent properly the macroscopic behaviour of that material. To answer this question, we state that our ability to formulate a coherent theory that accurately describes and predicts observable seismic phenomena is a key criterion to justify our usage of the notion of continuum. To gain further insight into our statement, let us consider the following quote from "La science et l'hypothèse" of Poincaré.

> Dans la plupart des questions, l'analyste suppose, au début de son calcul, soit que la matière est continue, soit, inversement, qu'elle est formée d'atomes. Il aurait fait le

> contraire que ses résultats n'en auraient pas été changés; il aurait eu plus de peine à les obtenir, voilà tout.[4]

1.2. Rudiments of Continuum Mechanics

1.2.1. Axiomatic format

Modern continuum mechanics is a physical theory that adopts an axiomatic format rather than a historical exposition or a heuristic approach. In this format, the structure of the theory is a hypotheticodeductive system. In other words, it is a system that starts from a set of hypotheses and proceeds deductively; hence, conclusions are of no greater generality than premises, and conclusions are as certain as the premises.[5]

To axiomatize a theory, we need to lay down a set of primitive concepts such that there is a sufficient characterization of basic ideas and a platform for the subsequent statements of the theory. To do so, we need to use the language of mathematics and the principles of logic. In other words, an axiomatic theory of physics presupposes both logic and mathematics; it requires them as a formal apparatus of description. However, this apparatus does not suffice to construct a physical theory because it is devoid of intrinsic physical meaning. Herein, we might recall that the purpose of physics is to accurately describe physical phenomena, while the purpose of logic and mathematics is to consistently define abstract concepts. To axiomatize a physical theory is to articulate its explicitness. Indeed, only the theories with explicitly articulated physical meaning can be tested.

Among the advantages of an axiomatic format are the recognition of presuppositions and assumptions contained within the theory, as well as the rigour and consistency, which provide clarity and foster coherent developments of the theory. Among the inconveniences of an axiomatic approach are the propensity for formalism at the expense of intuition, as well as the propensity for generality at the expense of concreteness. Since the

---

[4] For most problems, the researcher assumes, at the beginning of his calculations, that either the matter is continuous or, otherwise, it is composed of atoms. If he makes the other assumption, his results do not change; they will be just more difficult to obtain.

[5] If we consider physical interpretations of these conclusions, we might refer to continuum mechanics as a hypotheticoinferential system, whose conclusions do not inherit the certainty of premises (Brown, J.R. and Slawinski, M.A., (2017) On foundations of seismology: Bringing idealizations down to Earth: World Scientific, pp. 57, 122).

purpose of this book is concrete—namely, the description of wave phenomena in elastic continua—we trust that our presentation might enjoy the advantages of the axiomatic format and avoid its inconveniences.

1.2.2. Primitive concepts of continuum mechanics

Introductory comments

The primitive basis of a physical theory is a set of formal concepts that are assigned a physical meaning. These concepts cannot be proven within a given theory but are assumed to be true.[6] The three primitive concepts upon which the continuum mechanics is formed are the material body, the manifold of physical experience, and the system of internal forces.[7] These concepts are consistent with each other since the set they form is free of contradictions. They exhibit a weak deductive completeness since all known statements within the theory can be formulated in the context of the three primitive concepts, but not every statement is entailed by these three concepts—hence, the development of the theory can continue to accommodate new ideas and observations.

Material body

The first primitive concept is the material body, $\mathscr{B}$. It is a Euclidean three-dimensional smooth manifold composed of material points $\mathbf{X}$, where we define a material point as an infinitesimal element of volume that possesses the same physical properties as the properties observed in macroscopic studies. This element of volume is sufficiently large that it contains enough discrete particles of matter to allow us to establish a concept of continuum, while it is sufficiently small to be perceived as a mathematical point.[8]

---

[6] Note that, as stated on page 5, only abstract statements can be proven. Herein, the primitive concepts, although abstract, cannot be proven within the theory whose foundation they form. If they could be proven, they would be a part of the theory and not its primitive basis.
[7] Readers interested in more detail and an elegant exposition might refer to Truesdell, C., (1966) Six lectures on modern natural philosophy: Springer-Verlag, pp. 2–3 and pp. 96–97, and to Bunge, M., (1967) Foundations of physics: Springer-Verlag, pp. 143–157.
[8] According to Coirier, J., (1997) Mécanique des milieux continues: Concepts de base: Cours et exercices corrigés: Dunod, pp. 4–5, the size of such an element of volume is of the order of $10^{-15}\ m^3$ and contains of the order of 10^{10} molecules. According to Truesdell, C.,

The body manifold possesses the following properties. Every sufficiently smooth portion of a body is a body. Also, there is a measure called mass, m, which is a nonnegative scalar quantity such that

$$m(\mathscr{B}_1 \cup \mathscr{B}_2) = m(\mathscr{B}_1) + m(\mathscr{B}_2),$$

where $\mathscr{B}_1$ and $\mathscr{B}_2$ are disjoint subsets of $\mathscr{B}$. In other words, the mass of a body is the sum of the masses of its parts. Furthermore, mass of a material body occupying volume V is

$$m(\mathscr{B}) = \iiint_V \rho \, \mathrm{d}V,$$

where ρ is the mass density of the material composing $\mathscr{B}$.

As an abstract entity, $\mathscr{B}$ is not accessible to direct observations. We encounter its representations at particular times, t, and in particular spatial locations, $\mathbf{x}$. Points t and $\mathbf{x}$ constitute a manifold of physical experience, which is the second primitive concept of the theory of continuum mechanics.

Manifold of physical experience

The second primitive concept is the manifold of physical experience. It is a Euclidean space-time, $\mathbb{E}^3 \times t$, where $\mathbb{E}^3$ is composed of points $\mathbf{x}$ and is endowed with the Euclidean metric, while t denotes time. Symbol $\times$ stands for the direct product, which is a set of all possible ordered pairs $(\mathbf{x}, t)$, which we can write explicitly as (x_1, x_2, x_3, t). We assume that a given time interval is the same for all observers—the time is absolute and universal. Also, we assume that the distance between two given locations is the same for all observers—the space is absolute. These assumptions are tantamount to limiting our study to nonrelativistic continuum mechanics.

---

and Toupin, R., in Flügge, S., (1960) Handbuch der Physik: Springer-Verlag, Vol. III/1, p. 227:

> The corpuscular theories and the field theories are mutually contradictory as direct models of nature. [...] To speak of an element of volume [...] as "a region large enough to contain many molecules but small enough to be used as an element of integration" is not only loose but also needless and bottomless.

In the encounters with $\mathscr{B}$, we consider a sequence of configurations given by

$$\mathbf{x} = \chi(\mathbf{X}, t), \tag{1.2.1}$$

where χ is the motion of $\mathscr{B}$ given for a fixed t by an isomorphism: a structure-preserving continuous map between topological spaces that is injective, which means that to every element of one set there corresponds at most one element of the other set, and surjective, which means that to every element of one set there corresponds at least one element of the other set; thus the mapping between $\mathbf{X}$ and $\mathbf{x}$ is one-to-one. Hence, two distinct elements of $\mathscr{B}$ cannot occupy the same point on the manifold of physical experience, and no single element of $\mathscr{B}$ can occupy two distinct points. Since χ is an isomorphism, its inverse is continuous. Thus, we can write

$$\mathbf{X} = \chi^{-1}(\mathbf{x}, t),$$

which allows us to consider the spatial description, as opposed to the material one given by equation (1.2.1). In a material description, the observer identifies the location of points $\mathbf{X} \in \mathscr{B}$ that are immersed in $\mathbb{E}^3$; material point (X_1, X_2, X_3) and time t are the independent variables of this description. In a spatial description, the observer cannot identify particular material points but only spatial locations $\mathbf{x} \in \mathbb{E}^3$; in other words, the observer is unable to follow the displacement of a given material point. In this description, location (x_1, x_2, x_3) and time t are the independent variables. We discuss these two descriptions in more detail in Section 1.3.

Note that the goal of theoretical sciences is to give the best possible conceptual representation of a given system. The best representation is as close as possible to an isomorphism, which is attainable only in mathematics. Isomorphism is a perfect formal analogy; herein, whatever happens in $\mathbf{X}$ has its isomorphic image in $\mathbf{x}$, and vice versa. Commonly, the same system can be represented in a variety of ways.

System of internal forces

The third primitive concept is the system of internal forces within body $\mathscr{B}$. These forces are described by establishing Cauchy's stress principle, which can be stated in the following way.

Cauchy's stress principle: The action of the material occupying a portion within the body that is exterior to the closed surface on the material within this surface is represented by vector field $\mathbf{T}$.

Vector $\mathbf{T}$ is called traction. It acts on a surface whose outward normal unit vector is $\mathbf{n}$. $\mathbf{T}$ is assumed to depend continuously on $\mathbf{n}$. Its physical dimensions are force per unit area. The concept of the stress tensor that is invoked by the system of internal forces is central to the theory of continuum mechanics. We discuss it in detail in Section 2.3.

1.3. Material and Spatial Descriptions

1.3.1. Fundamental concepts

While using the concept of continuum, which does not involve any discrete particles, we must carefully consider methods that allow us to describe the displacement of material points within the continuum. In continuum mechanics, we can describe such a displacement in at least two ways; namely, by studying material and spatial descriptions.[9] We can observe the displacement either by following a given material point—in other words, following an infinitesimal element of the continuum, which is analogous to following a particle in particle mechanics—or by studying the flow of the continuum across a fixed position, which does not have an analogue in particle mechanics. As stated in Section 1.2.2, the first approach is called the material description of motion while the second one is called the spatial description of motion. These approaches are also known as the Lagrangian description and the Eulerian description, respectively.[10]

In global geodynamics, the fundamental laws that govern deformations of the Earth necessitate the distinction between the equations derived using the material and the spatial formulations. However, in applied seismology, we can often accurately analyze observable phenomena while ignoring the

[9] Material and spatial descriptions correspond to the referential and spatial descriptions of Malvern, L.E., (1969) Introduction to the mechanics of a continuous medium: Prentice-Hall, p. 138, where the relative description is also discussed.

[10] Readers interested in detailed descriptions of these approaches and their consequences might refer to Malvern, L.E., (1969) Introduction to the mechanics of a continuous medium: Prentice-Hall, pp. 138 – 145. Readers interested in heuristic descriptions might refer to Snieder, R., (2004) A guided tour of mathematical methods for the physical sciences: Cambridge University Press, pp. 57–61.

distinction between the material and the spatial descriptions. For instance, examining a recorded signal, we consider the spatial location of the seismic receiver, even though such a receiver, which is attached to the material that moves due to deformation, is an example of the material description.

To gain insight into the meaning of the material and spatial descriptions, consider a moving continuum and let the observer focus the attention on a given material point within the continuum. Suppose the position of a material point at initial time t_0 is given by vector $\mathbf{X}$. Although position vector $\mathbf{X}$ is not a material point, we refer to a given material point as "material point $\mathbf{X}$", which is a concise way of referring to a material point that at time t_0 occupied position $\mathbf{X}$, as shown in Remark 1.6.1, which follows Exercise 1.1. At a later time t, the position vector of the material point, $\mathbf{X}$, is given by $\mathbf{x}$. With a certain abuse of notation, we write expression (1.2.1) as $\mathbf{x}(\mathbf{X},t)$; this mapping gives position $\mathbf{x}$ of material point $\mathbf{X}$ at time t. This is the material description, where variable $\mathbf{X}$ identifies the material point. We assumed that, for a given time t, this mapping is one-to-one and continuous, as well as possessing the continuous inverse. Also, we have to assume that this mapping and its inverse have continuous partial derivatives to whatever order is required. Since we assume that the transition of the material point from the initial position to the present one occurs in a smooth fashion, vector $\mathbf{x}$ is a continuous function of time and, by symmetry, its inverse is also continuous. Again, with a certain abuse of notation, this inverse can be written as $\mathbf{X}(\mathbf{x},t)$, which fixes our attention on a given region in space and takes position, $\mathbf{x}$, and time, t, as independent variables. To introduce the material and spatial coordinates, consider an orthonormal coordinate system, where

$$x_i = x_i(X_1, X_2, X_3, t), \qquad i \in \{1,2,3\},$$

and

$$X_i = X_i(x_1, x_2, x_3, t), \qquad i \in \{1,2,3\},$$

with the components x_i and X_i being the spatial and material coordinates, respectively.

In general, a physical quantity that characterizes a continuum can be described by a function $f(\mathbf{x},t)$, which is a spatial description of this quantity, or by a function $F(\mathbf{X},t)$, which is a material description of this quantity. The material and spatial descriptions are consistent with one another.

The relation between f and F is given by $f(\mathbf{x}(\mathbf{X},t),t) = F(\mathbf{X},t)$, or by $f(\mathbf{x},t) = F(\mathbf{X}(\mathbf{x},t),t)$.

1.3.2. Material time derivative

In view of the previous section, we see that either the material or the spatial description can be used to describe the temporal variation of a given physical quantity. Let us consider time derivatives in the context of either description.

The material description consists of fixing our attention on a given material point, $\mathbf{X}$, and observing the variation of the quantity F with time. The time derivative associated with this viewpoint can be written as

$$\frac{\mathrm{d}F}{\mathrm{d}t} = \left.\frac{\mathrm{d}F(\mathbf{X},t)}{\mathrm{d}t}\right|_{\mathbf{X}}, \tag{1.3.1}$$

where symbol $|_{\mathbf{X}}$ means that the derivative is evaluated at $\mathbf{X}$.

The spatial description consists of fixing our attention on a given spatial location, $\mathbf{x}$, and observing the variation of quantity f with time. The time derivative associated with this viewpoint can be written as

$$\frac{\partial f}{\partial t} = \left.\frac{\partial f(\mathbf{x},t)}{\partial t}\right|_{\mathbf{x}}, \tag{1.3.2}$$

where symbol $|_{\mathbf{x}}$ means that the derivative is evaluated at $\mathbf{x}$.

The material and spatial descriptions are related by the chain rule of differentiation. To see this relation, consider a three-dimensional continuum and explicitly write

$$F(X_1,X_2,X_3,t) = f(x_1(X_1,X_2,X_3,t), x_2(X_1,X_2,X_3,t), x_3(X_1,X_2,X_3,t), t). \tag{1.3.3}$$

Taking the time derivative of both sides, we get

$$\frac{\mathrm{d}F}{\mathrm{d}t} = \left.\left[\frac{\partial f}{\partial t} + \frac{\partial f}{\partial x_1}\frac{\partial x_1}{\partial t} + \frac{\partial f}{\partial x_2}\frac{\partial x_2}{\partial t} + \frac{\partial f}{\partial x_3}\frac{\partial x_3}{\partial t}\right]\right|_{\mathbf{X}}.$$

Since the derivative is evaluated for a given material point, $\mathbf{X}$, it implies that $\partial x_i/\partial t$ are the components of velocity of this point moving in space;

we write these components as v_i. Thus,

$$\frac{\mathrm{d}F}{\mathrm{d}t} = \frac{\partial f}{\partial t} + \frac{\partial f}{\partial x_1} v_1 + \frac{\partial f}{\partial x_2} v_2 + \frac{\partial f}{\partial x_3} v_3 . \tag{1.3.4}$$

As shown in expression (1.3.1), $\mathrm{d}F/\mathrm{d}t$ describes the temporal variation of a given quantity for a particular material point within the continuum, and as shown in expression (1.3.2), $\partial f/\partial t$ describes the temporal variation of this quantity at a particular point in space. To discuss the relation between $\mathrm{d}F/\mathrm{d}t$ and $\partial f/\partial t$, we write the last three terms on the right-hand side of the above equation as a scalar product to get

$$\frac{\mathrm{d}F}{\mathrm{d}t} = \frac{\partial f}{\partial t} + [v_1, v_2, v_3] \cdot \left[\frac{\partial f}{\partial x_1}, \frac{\partial f}{\partial x_2}, \frac{\partial f}{\partial x_3} \right] .$$

Recognizing that the second term in brackets is the gradient of function f, and denoting the velocity vector by $\mathbf{v} = [v_1, v_2, v_3]$, we write

$$\frac{\mathrm{d}F}{\mathrm{d}t} = \frac{\partial f}{\partial t} + \mathbf{v} \cdot \nabla f,$$

which is the relation between the time derivatives of F and f.

We can formally rewrite the right-hand side of the above equation as

$$\frac{\mathrm{d}F(\mathbf{X},t)}{\mathrm{d}t} = \left(\frac{\partial}{\partial t} + \mathbf{v} \cdot \nabla \right) f(\mathbf{x},t) , \tag{1.3.5}$$

where the term in parentheses is an operator acting on function f, and the material and spatial coordinates are related by

$$\mathbf{x} = \mathbf{x}(\mathbf{X},t) . \tag{1.3.6}$$

The term in parentheses on the right-hand side of equation (1.3.5) is called the material time-derivative operator. It can be applied to a scalar, to a vector, or to a tensor function of position and time coordinates. It is common to denote this operator by $\mathrm{D}/\mathrm{D}t$ so as to concisely write equation (1.3.5) as

$$\frac{\mathrm{d}F(\mathbf{X},t)}{\mathrm{d}t} = \frac{\mathrm{D}f(\mathbf{x},t)}{\mathrm{D}t} , \tag{1.3.7}$$

where $\mathrm{D}/\mathrm{D}t := \partial/\partial t + \mathbf{v} \cdot \nabla$.

Examining the left-hand side of equation (1.3.5) in view of expression (1.3.1), we conclude that the material time derivative is a rate of change associated with a particular element of the continuum. In other words, it is measured by an observer travelling with this element. Mathematically, the material time derivative is the time derivative with material coordinates held constant. To consider spatial coordinates in this context, let us examine the right-hand side of equation (1.3.5). The first term of $(\partial/\partial t + \mathbf{v}\cdot\nabla)$ describes the time rate of change at the location $\mathbf{x}$, while the second term describes the rate of change associated with the motion of material points; more explicitly, the second term describes the spatial rate of change of material point $\mathbf{X}$ moving with velocity $\mathbf{v}$. Recalling expression (1.3.3), we can write expression (1.3.4) with a certain abuse of notation as

$$\frac{\mathrm{d}F}{\mathrm{d}t} = \left.\frac{\partial f}{\partial t}\right|_{\mathbf{x}} + \sum_{i=1}^{3} \left.\frac{\partial f}{\partial x_i}\right|_{t} v_i|_{\mathbf{X}},$$

where $\partial f/\partial t$ is evaluated at location $\mathbf{x}$, $\partial f/\partial x_i$ is evaluated at instant t for a given material point, $\mathbf{X}$, moving with velocity $\mathbf{v}$.

As we see from the formulation presented above, in general, $\mathrm{d}F/\mathrm{d}t$ differs from $\partial f/\partial t$ by term $\mathbf{v}\cdot\nabla f$.[11] This term vanishes in the absence of motion, $\mathbf{v} = \mathbf{0}$, or if f does not vary spatially, $\nabla f = \mathbf{0}$. Also, if this term is negligible, we need not distinguish between the material and spatial descriptions. This is an important concept that often allows us simplify our approach; we discuss it in the next section.

REMARK 1.3.1. To gain further insight into the concept of the material time derivative, let us show that it is a time derivative, $\partial/\partial t$, of quantity f, transported by the flow of a vector field, $\mathbf{W}$. To do so, we consider $\mathbf{W}(\mathbf{c}(\mathbf{x},t)) = \partial\mathbf{c}/\partial t$,[12] where position $\mathbf{x}$ and time t specify the location of a material point, $\mathbf{X}$, along a curve, $\mathbf{c}$, in the vector field, $\mathbf{W}$. Curves $\mathbf{c}$ share their domain with the field itself in such a manner that their tangents, $\partial\mathbf{c}/\partial t$, coincide with the field lines and, herein, have the units of velocity. Following the chain rule—with a certain abuse of notation—we

[11] Readers interested in this term and its meaning as the convective derivative might refer to Sedov, L.I., (1971) A course in continuum mechanics: Wolters-Noordhoff Publishing, Vol. I, pp. 31–32.

[12] Readers interested in this formulation and its meaning might refer to Marsden, J.E., and Tromba, A., (1981) Vector calculus (*2nd* edition): W.H. Freeman & Co., pp. 176–177 and p. 551, as well as to Marsden, J.E., and Tromba, A., (2012) Vector calculus (*6th* edition): W.H. Freeman & Co., pp. 241–242.

write

$$\frac{\mathrm{d}f(\mathbf{c}(\mathbf{x},t),t)}{\mathrm{d}t} = \frac{\partial f}{\partial t} + \sum_{i=1}^{3} \frac{\partial f}{\partial x_i}\frac{\partial c_i}{\partial t} \equiv \frac{\partial f}{\partial t} + \nabla f \cdot \mathbf{W}(\mathbf{c}(\mathbf{x},t)),$$

which is the required result. Also, we see that the material time derivative is reduced to a partial derivative if $\nabla f \perp \mathbf{W}(\mathbf{c}(\mathbf{x},t))$, which is tantamount to the spatial variation of a given quantity being perpendicular to the velocity of a material point.

1.3.3. Conditions of linearized theory

In general, equations governing wave phenomena in elastic media are nonlinear. However, seismic experiments indicate that important aspects of wave propagation can be adequately described by linear equations, which greatly simplify mathematical formulations. The process of going from nonlinear equations to linear ones is called the linearization process and the resulting theory is the linearized theory. This linearization is achieved by the fact that, under certain assumptions that appear to be satisfied for many seismological studies, the material and spatial descriptions are equivalent to one another.

The linearization allows us to formulate mathematical statements of seismic wave phenomena in a form that is simpler than it would be otherwise possible. In this section, we briefly discuss the conditions that allow us to use linearization. A more detailed description of the linearization process is beyond the scope of this book.[13]

In applied seismology, we often assume that the displacements of material elements resulting from the propagation of seismic waves can be considered as infinitesimal. Such an assumption is used in this entire book. As a consequence of this assumption and in view of the material time derivative, discussed in Section 1.3.2, we can conclude that, while considering

[13] Readers interested in a thorough analysis of physical quantities in the material and spatial descriptions, and the subsequent linearization might refer to Achenbach, J.D., (1973) Wave propagation in elastic solids: North Holland, pp. 11–21 and 46–47, to Malvern, L.E., (1969) Introduction to the mechanics of a continuous medium: Prentice-Hall, pp. 497–565, and to Marsden, J.E., and Hughes, T.J.R., (1983/1994) Mathematical foundations of elasticity: Dover, pp. 9–10 and 226 – 246.

displacements, it is unnecessary to distinguish between the material and spatial descriptions.

To arrive at this conclusion, let us consider the notion of displacement using both the material and spatial descriptions. Displacement is the difference between the final position and the initial position. Using the material description, we can write the displacement vector as

$$\mathbf{U}(\mathbf{X},t) = \mathbf{x}(\mathbf{X},t) - \mathbf{X}, \tag{1.3.8}$$

while using the spatial description, we note that the displacement vector is

$$\mathbf{u}(\mathbf{x},t) = \mathbf{x} - \mathbf{X}(\mathbf{x},t). \tag{1.3.9}$$

Note that at the initial time, $\mathbf{x} = \mathbf{X}$.

Since the same quantity is given by expressions (1.3.8) and (1.3.9), we can write

$$\mathbf{U}(\mathbf{X},t) = \mathbf{u}(\mathbf{x},t), \tag{1.3.10}$$

where the material and spatial coordinates are related by equation (1.3.6).

We can develop each component of $\mathbf{U}(\mathbf{X},t)$ into Taylor's series about $\mathbf{x}$ to obtain

$$U_i(\mathbf{X},t) = U_i(\mathbf{X},t)|_{\mathbf{X}=\mathbf{x}} + \left[\left.\frac{\partial U_i(\mathbf{X},t)}{\partial X_1}\right|_{\mathbf{X}=\mathbf{x}}, \left.\frac{\partial U_i(\mathbf{X},t)}{\partial X_2}\right|_{\mathbf{X}=\mathbf{x}}, \left.\frac{\partial U_i(\mathbf{X},t)}{\partial X_3}\right|_{\mathbf{X}=\mathbf{x}} \right] \cdot (\mathbf{X} - \mathbf{x}) + \ldots,$$

where $i \in \{1,2,3\}$. Assuming that the gradient of the displacement, which is shown in brackets, is vanishingly small, we can consider only the first term of the series. Thus, we can write

$$\mathbf{U}(\mathbf{X},t) \approx \mathbf{U}(\mathbf{x},t). \tag{1.3.11}$$

Hence, expression (1.3.8) can be written as

$$\mathbf{U}(\mathbf{x},t) \approx \mathbf{x}(\mathbf{X},t) - \mathbf{X}. \tag{1.3.12}$$

Since in expression (1.3.12) $\mathbf{U}$ is a function of $\mathbf{x}$, we rewrite—using expression (1.3.9)—the displacement as a function of $\mathbf{x}$ to get

$$\mathbf{U}(\mathbf{x},t) \approx \mathbf{x} - \mathbf{X}(\mathbf{x},t). \tag{1.3.13}$$

Comparing expressions (1.3.9) and (1.3.13), we see that

$$\mathbf{U}(\mathbf{x},t) \approx \mathbf{u}(\mathbf{x},t) .$$

Thus, in view of expression (1.3.11), we conclude that—for infinitesimal displacements—we can write

$$\mathbf{U}(\mathbf{X},t) \approx \mathbf{u}(\mathbf{x},t) . \tag{1.3.14}$$

To gain insight into the meaning of this result, we examine equations (1.3.10) and (1.3.14). Equation (1.3.10) states that $\mathbf{U} = \mathbf{u}$, with $\mathbf{x}$ related to $\mathbf{X}$ by equation (1.3.6). Equation (1.3.14) states that $\mathbf{U} \approx \mathbf{u}$, where we can simply replace $\mathbf{x}$ by $\mathbf{X}$, without invoking equation (1.3.6). This approximation is illustrated in Exercise 1.2. Now, let us consider the velocity using both the material and spatial descriptions. To do so, let the physical quantity considered in the material time derivative be given by displacement. In such a case, expression (1.3.5) becomes

$$\frac{\mathrm{d}\mathbf{U}}{\mathrm{d}t} = \frac{\partial \mathbf{u}}{\partial t} + (\mathbf{v} \cdot \nabla)\,\mathbf{u} .$$

If both the gradient of the displacement $\mathbf{u}$ and the velocity $\mathbf{v}$ are infinitesimal, we can ignore the second term on the right-hand side to obtain

$$\frac{\mathrm{d}\mathbf{U}}{\mathrm{d}t} \approx \frac{\partial \mathbf{u}}{\partial t} .$$

Also, let us consider the acceleration using both the material and spatial descriptions. To do so, let the physical quantity considered in the material time derivative be given by velocity. In such a case, expression (1.3.5) becomes

$$\frac{\mathrm{d}^2\mathbf{U}}{\mathrm{d}t^2} = \frac{\partial^2 \mathbf{u}}{\partial t^2} + (\mathbf{v} \cdot \nabla)\,\frac{\partial \mathbf{u}}{\partial t} .$$

If both the gradient of $\partial \mathbf{u}/\partial t$ and the velocity $\mathbf{v}$ are infinitesimal, we can ignore the second term on the right-hand side to obtain

$$\frac{\mathrm{d}^2\mathbf{U}}{\mathrm{d}t^2} \approx \frac{\partial^2 \mathbf{u}}{\partial t^2} .$$

As stated by Ruggeri (2007, Section 7.4),

> se i gradienti sono piccoli ed anche la velocità è piccola,
> derivata materiale e derivata parziale rispetto al tempo

> coincidono così come sono circa eguali le derivate spaziali rispetto a $\mathbf{X}$ e rispetto a $\mathbf{x}$ insieme alle densità.[14]

This property of the time derivative of displacement, which results from the linearized theory, is used, for instance, in the derivation of equations of motion (2.6.3).

Thus, we can conclude that, under the assumption of infinitesimal displacements of a given element of the continuum, we do not need to distinguish between either the material and spatial coordinates or the material and spatial descriptions of displacements. In other words, $\mathbf{X} \approx \mathbf{x}$ and $\mathbf{U} \approx \mathbf{u}$. Furthermore, if we also assume that the velocities of these displacements are infinitesimal, that the gradients of these displacements are infinitesimal, and that the gradients of these velocities are also infinitesimal, there is no need to distinguish between the material and spatial descriptions while studying velocities and accelerations. In other words, $\mathrm{d}\mathbf{U}/\mathrm{d}t \approx \partial \mathbf{u}/\partial t$ and $\mathrm{d}^2\mathbf{U}/\mathrm{d}t^2 \approx \partial^2 \mathbf{u}/\partial t^2$, respectively.

It is important to note that the assumptions about the properties of the displacements, gradients of displacements, velocities and gradients of velocities are independent of each other. They result from the physical context in which we consider a given mathematical formulation. For instance, in the context of applied seismology, we assume that the displacement amplitude of a material point is small compared to the wavelength. Also, we assume that the velocity of this displacement is small compared to the wave propagation velocity.[15]

In this section, we attempt to justify the linearization by *a priori* arguments. Let us also mention another approach that plays an important role in continuum mechanics and consists of an *a posteriori* justification; in other words, the results justify the assumptions. Herein, the initial approach could be to try the linear formulation by ignoring the nonlinear terms without invoking any physical reason; it is common to begin with a linear formulation and to investigate its applicability. Having developed a theory based on such a formulation, we would compare the predictions of

[14] If the gradients are small and also the velocity is small, the material derivative and partial derivative with respect to time coincide, just as the spatial derivatives with respect to $\mathbf{X}$ and $\mathbf{x}$ are nearly equal to one another, and so is the density in terms of $\mathbf{X}$ and $\mathbf{x}$.

[15] Readers interested in details of this linearization might refer to Achenbach, J.D., (1973) Wave propagation in elastic solids: North Holland, pp. 17–21.

this theory with experimental results. If the agreement between the theory and experiments is satisfactory, we could accept the assumptions. In our context, many seismological measurements agree with theoretical prediction of a linearized theory, thus providing an *a posteriori* justification.

Following our decision to make no distinction between the material and spatial descriptions, we follow the customary notation to describe the coordinates as well as the displacements of a given element of the continuum using lower-case letters. Also, to avoid any confusion, we note that the velocities denoted by v and V, in Parts 2 and 3 of the book, refer to the propagation velocities; namely, the phase velocity and the ray velocity, respectively. They are not directly associated with the velocities of displacements of a given element of the continuum, which we discuss herein.

1.4. Strain

1.4.1. Introductory comments

[16] Seismic waves consist of the propagation of deformations through a material. To study these waves, we wish to describe the associated deformations of the continuum in the context of infinitesimal displacements.

Deformation of a continuum is a change of positions of points within it relative to each other. If such a change occurs, a continuum is said to be strained. This strain is accompanied by stress. The produced stress resists deformation and attempts to restore the continuum to its unstrained state. The resistance of a continuum to the deformation and the continuum's tendency to restore itself to its undeformed state account for the propagation of seismic waves.

The relation between stress and strain is one of mutual dependence and is an intrinsic concept of elasticity theory. In this theory, applied forces are formulated in terms of a stress tensor, discussed in Chapter 2, while the associated deformations are formulated in terms of a strain tensor, discussed below.

---

[16] Readers interested in a thorough description of strain and deformation might refer to Malvern, L.E., (1969) Introduction to the mechanics of a continuous medium: Prentice-Hall, Chapter 4: Strain and Deformation.

1.4.2. Derivation of strain tensor

In this section, we formulate the strain tensor, which is a mathematical entity that allows us to quantify the deformation within the continuum.

The concept of deformation—which is the essence of continuum mechanics as a study of deformable bodies—implies that distances among points within the continuum change. To derive the strain tensor in a three-dimensional continuum, consider two infinitesimally close points therein whose coordinates are (x, y, z) and $(x + \mathrm{d}x, y + \mathrm{d}y, z + \mathrm{d}z)$. The square of the distance between these points is

$$\mathrm{d}s^2 = \mathrm{d}x^2 + \mathrm{d}y^2 + \mathrm{d}z^2 . \tag{1.4.1}$$

Let the continuum be subject to a deformation described by a vector field composed of the displacement vectors,

$$\mathbf{u} = [u_x (x, y, z), u_y (x, y, z), u_z (x, y, z)] .$$

After deformation, the coordinates of the first point are

$$(x + u_x , y + u_y , z + u_z)|_{x,y,z} , \tag{1.4.2}$$

and the coordinates of the second point are

$$(x + \mathrm{d}x + u_x , y + \mathrm{d}y + u_y , z + \mathrm{d}z + u_z)|_{x+\mathrm{d}x,y+\mathrm{d}y,z+\mathrm{d}z} , \tag{1.4.3}$$

where the subscripts are values at which these coordinates are evaluated. Subtracting the coordinates in expression (1.4.2) from the corresponding coordinates in expression (1.4.3), we obtain the differences between the coordinates of these points after the deformation,

$$\begin{aligned}\Big(&\mathrm{d}x + u_x|_{x+\mathrm{d}x,y+\mathrm{d}y,z+\mathrm{d}z} - u_x|_{x,y,z} , \\ &\mathrm{d}y + u_y|_{x+\mathrm{d}x,y+\mathrm{d}y,z+\mathrm{d}z} - u_y|_{x,y,z} , \\ &\mathrm{d}z + u_z|_{x+\mathrm{d}x,y+\mathrm{d}y,z+\mathrm{d}z} - u_z|_{x,y,z}\Big) .\end{aligned} \tag{1.4.4}$$

Each component—u_x, u_y, u_z—is a function of x, y and z. Thus, the length and orientation of a given displacement vector depends on its position, and—in general—each vector within the field has different length and orientation, since $\mathbf{u}|_{x+\mathrm{d}x,y+\mathrm{d}y,z+\mathrm{d}z} \neq \mathbf{u}|_{x,y,z}$.

It is the dependence of the field on x, y, z that results in a relative change of position of points with respect to each other and, hence, in deformation within the continuum; prior to deformation, their coordinates are separated by $\mathrm{d}x$, $\mathrm{d}y$ and $\mathrm{d}z$; after the deformation, they are separated by the values shown in expression (1.4.4).

Let us consider a special case. If $\mathbf{u}$ does not depend on position, then $\mathbf{u}|_{x+\mathrm{d}x,y+\mathrm{d}y,z+\mathrm{d}z} = \mathbf{u}|_{x,y,z}$, and expression (1.4.4) reduces to $(\mathrm{d}x, \mathrm{d}y, \mathrm{d}z)$, which means that the effect of $\mathbf{u}$ does not result in change of position of points relative to each other; it results in a translation without deformation. Other special cases are discussed in Section 1.4.3.

Let us return to our derivation. To avoid $\mathbf{u}$ being evaluated both at $(x+\mathrm{d}x,\ y+\mathrm{d}y,\ z+\mathrm{d}z)$ and at $(x,\ y,\ z)$ in expression (1.4.4), we assume infinitesimal displacements and a smooth change of properties of $\mathbf{u}$ between two adjacent points. Hence, the components evaluated at $(x+\mathrm{d}x,\ y+\mathrm{d}y,\ z+\mathrm{d}z)$ can be approximated by the first two terms of Taylor's series about $(x,\ y,\ z)$,

$$u_x|_{x+\mathrm{d}x,y+\mathrm{d}y,z+\mathrm{d}z} \approx u_x|_{x,y,z} + \left.\frac{\partial u_x}{\partial x}\right|_{x,y,z} \mathrm{d}x + \left.\frac{\partial u_x}{\partial y}\right|_{x,y,z} \mathrm{d}y + \left.\frac{\partial u_x}{\partial z}\right|_{x,y,z} \mathrm{d}z,$$

$$u_y|_{x+\mathrm{d}x,y+\mathrm{d}y,z+\mathrm{d}z} \approx u_y|_{x,y,z} + \left.\frac{\partial u_y}{\partial x}\right|_{x,y,z} \mathrm{d}x + \left.\frac{\partial u_y}{\partial y}\right|_{x,y,z} \mathrm{d}y + \left.\frac{\partial u_y}{\partial z}\right|_{x,y,z} \mathrm{d}z,$$

$$u_z|_{x+\mathrm{d}x,y+\mathrm{d}y,z+\mathrm{d}z} \approx u_z|_{x,y,z} + \left.\frac{\partial u_z}{\partial x}\right|_{x,y,z} \mathrm{d}x + \left.\frac{\partial u_z}{\partial y}\right|_{x,y,z} \mathrm{d}y + \left.\frac{\partial u_z}{\partial z}\right|_{x,y,z} \mathrm{d}z.$$

Inserting these terms into expression (1.4.4), we obtain the approximation for the difference of the corresponding coordinates of the two points after the deformation,

$$\left[\mathrm{d}x + \left.\frac{\partial u_x}{\partial x}\right|_{x,y,z} \mathrm{d}x + \left.\frac{\partial u_x}{\partial y}\right|_{x,y,z} \mathrm{d}y + \left.\frac{\partial u_x}{\partial z}\right|_{x,y,z} \mathrm{d}z,\right.$$
$$\mathrm{d}y + \left.\frac{\partial u_y}{\partial x}\right|_{x,y,z} \mathrm{d}x + \left.\frac{\partial u_y}{\partial y}\right|_{x,y,z} \mathrm{d}y + \left.\frac{\partial u_y}{\partial z}\right|_{x,y,z} \mathrm{d}z,$$
$$\left.\mathrm{d}z + \left.\frac{\partial u_z}{\partial x}\right|_{x,y,z} \mathrm{d}x + \left.\frac{\partial u_z}{\partial y}\right|_{x,y,z} \mathrm{d}y + \left.\frac{\partial u_z}{\partial z}\right|_{x,y,z} \mathrm{d}z\right].$$

To compare this result with expression (1.4.1), we write the square of the distance between the two points after the deformation,

$$\mathrm{d}\breve{s}^2 \approx \left(\mathrm{d}x + \left.\frac{\partial u_x}{\partial x}\right|_{x,y,z} \mathrm{d}x + \left.\frac{\partial u_x}{\partial y}\right|_{x,y,z} \mathrm{d}y + \left.\frac{\partial u_x}{\partial z}\right|_{x,y,z} \mathrm{d}z\right)^2 + \left(\mathrm{d}y + \left.\frac{\partial u_y}{\partial x}\right|_{x,y,z} \mathrm{d}x + \left.\frac{\partial u_y}{\partial y}\right|_{x,y,z} \mathrm{d}y + \left.\frac{\partial u_y}{\partial z}\right|_{x,y,z} \mathrm{d}z\right)^2 + \left(\mathrm{d}z + \left.\frac{\partial u_z}{\partial x}\right|_{x,y,z} \mathrm{d}x + \left.\frac{\partial u_z}{\partial y}\right|_{x,y,z} \mathrm{d}y + \left.\frac{\partial u_z}{\partial z}\right|_{x,y,z} \mathrm{d}z\right)^2 .$$

Squaring the expressions in parentheses and—in view of infinitesimal gradients of the displacement—neglecting the terms that contain the products of two derivatives, we obtain

$$\begin{aligned}\mathrm{d}\breve{s}^2 \approx{} & \mathrm{d}x^2 + \mathrm{d}y^2 + \mathrm{d}z^2 \\ & + 2\left(\left.\frac{\partial u_x}{\partial x}\right|_{x,y,z} \mathrm{d}x^2 + \left.\frac{\partial u_y}{\partial y}\right|_{x,y,z} \mathrm{d}y^2 + \left.\frac{\partial u_z}{\partial z}\right|_{x,y,z} \mathrm{d}z^2\right. \\ & + \left.\frac{\partial u_x}{\partial y}\right|_{x,y,z} \mathrm{d}x\,\mathrm{d}y + \left.\frac{\partial u_x}{\partial z}\right|_{x,y,z} \mathrm{d}x\,\mathrm{d}z + \left.\frac{\partial u_y}{\partial x}\right|_{x,y,z} \mathrm{d}x\,\mathrm{d}y \\ & \left. + \left.\frac{\partial u_y}{\partial z}\right|_{x,y,z} \mathrm{d}y\,\mathrm{d}z + \left.\frac{\partial u_z}{\partial x}\right|_{x,y,z} \mathrm{d}x\,\mathrm{d}z + \left.\frac{\partial u_z}{\partial y}\right|_{x,y,z} \mathrm{d}y\,\mathrm{d}z\right) .\end{aligned} \tag{1.4.5}$$

Using expressions (1.4.1) and (1.4.5), we obtain the difference in the square of the distance between the two points that results from the deformation,

$$\begin{aligned}\mathrm{d}\breve{s}^2 - \mathrm{d}s^2 \approx 2 \Bigg[& \left.\frac{\partial u_x}{\partial x}\right|_{x,y,z} \mathrm{d}x^2 + \left.\frac{\partial u_y}{\partial y}\right|_{x,y,z} \mathrm{d}y^2 + \left.\frac{\partial u_z}{\partial z}\right|_{x,y,z} \mathrm{d}z^2 \\ & + \left.\left(\frac{\partial u_x}{\partial y} + \frac{\partial u_y}{\partial x}\right)\right|_{x,y,z} \mathrm{d}x\,\mathrm{d}y + \left.\left(\frac{\partial u_y}{\partial z} + \frac{\partial u_z}{\partial y}\right)\right|_{x,y,z} \mathrm{d}y\,\mathrm{d}z \\ & + \left.\left(\frac{\partial u_x}{\partial z} + \frac{\partial u_z}{\partial x}\right)\right|_{x,y,z} \mathrm{d}x\,\mathrm{d}z \Bigg] .\end{aligned}$$

For convenience, letting $x_1 := x$, $x_2 := y$, $x_3 := z$ and $u_1 := u_x$, $u_2 := u_y$ and $u_3 := u_z$,[17] we write this expression concisely as

$$\mathrm{d}\breve{s}^2 - \mathrm{d}s^2 \approx \sum_{i=1}^{3}\sum_{j=1}^{3}\left(\frac{\partial u_i}{\partial x_j} + \frac{\partial u_j}{\partial x_i}\right)\Bigg|_{\mathbf{x}} \mathrm{d}x_i \mathrm{d}x_j \,,$$

where derivatives are evaluated at $\mathbf{x} := (x_1, x_2, x_3)$. Since the left-hand side is a scalar, so is the right-hand side. Therein, the term in parentheses is a component of a second-rank tensor,[18] as shown in Exercise 1.4.

In elasticity theory, the term in parentheses is the definition of the strain tensor for infinitesimal displacements,

$$\varepsilon_{ij} := \frac{1}{2}\left(\frac{\partial u_i}{\partial x_j} + \frac{\partial u_j}{\partial x_i}\right), \qquad i, j \in \{1, 2, 3\}\,, \tag{1.4.6}$$

expressed in Cartesian coordinates.[19] The strain tensor allows us to quantify the concept of deformation. For instance, examining expression (1.4.6), we see that in the special case discussed on page 25, where $\mathbf{u}$ is constant, $\varepsilon_{ij} = 0$ for all i and j; there is no deformation, as expected.

In view of its definition, the strain tensor is symmetric, $\varepsilon_{ij} = \varepsilon_{ji}$; consequently, in a three-dimensional continuum, there are only six independent components. Also, in view of its definition, the strain tensor is dimensionless.

Note an analogy between vector calculus and tensor calculus: the gradient operator applied to the scalar field, f, which is a tensor field of rank

[17] In this book, symbol $:=$ implies a definition; "$:=$" is read as "defines/denotes" and "$=:$" as "is defined/denoted by".

[18] Both terms "rank" and "order" are commonly used to describe the number of indices of a tensor. In this book, we use the former term since it does not appear in any other context, while the latter term is used in the context of differential equations. Note that although the term "rank" also has a specific meaning in matrix algebra, we do not use it in such a context in this book.

[19] Readers interested in formulation of the strain tensor leading to its form that is valid for curvilinear coordinates might refer to Synge, J.L., and Schild, A., (1949/1978) Tensor calculus: Dover, pp. 202–205. Readers interested in a concise formulation of the strain tensor that invokes tensorial properties might refer to Semay, C., and Silvestre-Brac, B., (2007) Introduction au calcul tensoriel: Applications à la physique: Dunod, pp. 203–204.

zero, results in a vector field,

$$\nabla f(x_1,x_2,x_3) := \left[\frac{\partial f}{\partial x_1}, \frac{\partial f}{\partial x_2}, \frac{\partial f}{\partial x_3}\right],$$

which is a tensor field of rank one, and—as shown in the derivation of expression (1.4.6)—the gradient operator applied to the vector field, $\mathbf{u}$, results in a tensor field of rank two.

1.4.3. Physical meaning of strain tensor

Introductory comments

The strain tensor describes two types of deformation. Firstly, the sides of a volume element within a continuum can change in length. This can result in a change of volume without necessarily a change in shape. Components of the strain tensor, which we use to describe such deformations, are dimensionless quantities given by a change in length per unit length. Secondly, the sides of an element within a continuum can change orientation with respect to each other. This results in a change of shape without necessarily a change in volume. Components of the corresponding strain tensor are measured in radians and describe the change in angles before and after the deformation. Thus, the strain tensor describes relative linear displacement and relative angular displacement.

Relative change in length

To illustrate a length change expressed by a strain tensor, we revisit the derivation shown in Section 1.4.2 and consider the one-dimensional case.

Let $\mathbf{x} = [x_1, 0, 0]$ and $\mathbf{x} + \mathrm{d}\mathbf{x} = [x_1 + \mathrm{d}x_1, 0, 0]$ be two close points on the x_1-axis prior to deformation. During deformation, these points may be removed from the x_1-axis, however, their coordinates along this axis after the deformation are

$$\breve{x}_1 = x_1 + u_1|_{x_1,0,0}, \tag{1.4.7}$$

and

$$\breve{x}_1 + \mathrm{d}\breve{x}_1 = x_1 + \mathrm{d}x_1 + u_1|_{x_1+\mathrm{d}x_1,0,0}. \tag{1.4.8}$$

The distance between their components along the x_1-axis after the deformation is given by the difference between expressions (1.4.7) and (1.4.8);

namely,

$$\mathrm{d}\breve{x}_1 = \mathrm{d}x_1 + u_1|_{x_1+\mathrm{d}x_1,0,0} - u_1|_{x_1,0,0} \,. \tag{1.4.9}$$

Taylor's series of the middle term on the right-hand side can be written as

$$u_1|_{x_1+\mathrm{d}x_1,0,0} = u_1|_{x_1,0,0} + \left.\frac{\partial u_1}{\partial x_1}\right|_{x_1,0,0} \mathrm{d}x_1 + \frac{1}{2}\left.\frac{\partial^2 u_1}{\partial x_1^2}\right|_{x_1,0,0} (\mathrm{d}x_1)^2 + \dots .$$

Using the approximation consisting of the first two terms, we can write expression (1.4.9) as

$$\mathrm{d}\breve{x}_1 \approx \mathrm{d}x_1 + \left.\frac{\partial u_1}{\partial x_1}\right|_{x_1,0,0} \mathrm{d}x_1 \,,$$

which can be restated as

$$\mathrm{d}\breve{x}_1 \approx \left(1 + \frac{\partial u_1}{\partial x_1}\right) \mathrm{d}x_1 \,.$$

Hence, in view of definition (1.4.6), we can write the distance between the two points after deformation as

$$\mathrm{d}\breve{x}_1 \approx (1 + \varepsilon_{11})\, \mathrm{d}x_1 \,, \tag{1.4.10}$$

where $\mathrm{d}x_1$ is the distance between these two points prior to deformation.

Thus, ε_{11} is a relative elongation or contraction along the x_1-axis. Similarly, $\partial u_2/\partial x_2 = \varepsilon_{22}$ and $\partial u_3/\partial x_3 = \varepsilon_{33}$ correspond to relative elongations or contractions along the x_2-axis and the x_3-axis, respectively.

To see the meaning of ε_{ii}, where $i \in \{1,2,3\}$, consider Figure 1.4.1 with axes defined in terms of the material coordinates that correspond to the configuration of the element of the continuum before deformation. The relative elongation along the X_1-axis can be written as

$$\frac{\Delta X_1 + \Delta u_1}{\Delta X_1} = 1 + \frac{\Delta u_1}{\Delta X_1} \,. \tag{1.4.11}$$

Considering infinitesimal gradients of the displacement, discussed in Section 1.3.3, and in view of Exercise 1.5, we can restate expression (1.4.11) as

$$1 + \frac{\partial u_1}{\partial x_1} \,. \tag{1.4.12}$$

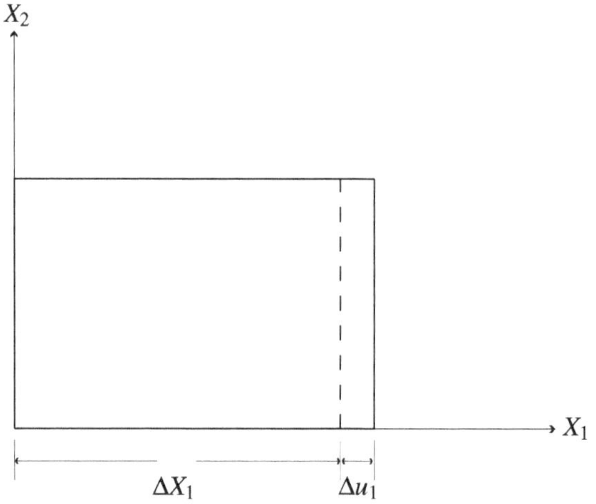

Fig. 1.4.1 Deformation resulting in a change of length. Uniaxial extension in the X_1-axis direction. Segment ΔX_1, before deformation, becomes $\Delta X_1 + \Delta u_1$, after the deformation.

Expression (1.4.12) is a relative change in length due to deformation. Now, recall equation (1.4.10), which we can restate as

$$\frac{\mathrm{d}\breve{x}_1}{\mathrm{d}x_1} \approx 1 + \varepsilon_{11}, \tag{1.4.13}$$

to describe a relative change in length due to deformation. Hence, examining expressions (1.4.12) and (1.4.13), we conclude that $\varepsilon_{11} \equiv \partial u_1 / \partial x_1$, as expected.

Relative change in volume

Having formulated the relative change in length, we can express a relative change in volume.

Consider a rectangular box with edge lengths Δx_1, Δx_2, and Δx_3, along the x_1-axis, the x_2-axis and the x_3-axis, respectively. Its volume is

$$V = \Delta x_1 \Delta x_2 \Delta x_3 . \tag{1.4.14}$$

After the deformation, following expression (1.4.10), the edge lengths become $(1 + \varepsilon_{11}) \Delta x_1$, $(1 + \varepsilon_{22}) \Delta x_2$ and $(1 + \varepsilon_{33}) \Delta x_3$, respectively, and, the volume becomes

$$\breve{V} = (1 + \varepsilon_{11})(1 + \varepsilon_{22})(1 + \varepsilon_{33})V . \tag{1.4.15}$$

Note that to state $\breve{V}$ as written in expression (1.4.15), we require that after the deformation the original rectangular box remains rectangular. In the context of our study, where the deformations are assumed to be small, the errors resulting from departing from this requirement are considered to be negligible. In other words, we use expression (1.4.15) even if Δx_1 , Δx_2 and Δx_3 are no longer parallel to the corresponding axes.

Assuming small deformations and, consequently, retaining only first-order strain-component terms resulting from the triple product, the volume of the deformed rectangular box can be written as

$$\breve{V} \approx (1 + \varepsilon_{11} + \varepsilon_{22} + \varepsilon_{33})V . \tag{1.4.16}$$

Thus, using expressions (1.4.14) and (1.4.16), we can state the relative change in volume as

$$\frac{\breve{V} - V}{V} \approx \varepsilon_{11} + \varepsilon_{22} + \varepsilon_{33} =: \varphi . \tag{1.4.17}$$

We refer to φ as dilatation.

Using vector calculus, we can conveniently state the relative change in volume in terms of the displacement vector, $\mathbf{u}$. In view of definition (1.4.6), expression (1.4.17) can be stated as divergence, since we can write

$$\varphi = \frac{\partial u_1}{\partial x_1} + \frac{\partial u_2}{\partial x_2} + \frac{\partial u_3}{\partial x_3} = \nabla \cdot \mathbf{u} . \tag{1.4.18}$$

To gain insight into the physical meaning of φ , we can revisit a special case discussed on page 25. If $\mathbf{u}$ is given by constants, there is no deformation and, hence, no change in volume, as expected. Also, if $\mathbf{u} = [u_1(x_2,x_3), u_2(x_1,x_3), u_3(x_1,x_2)]$, then $\varphi = 0$. Such a displacement vector causes an infinitesimal deformation that results in change of shape, as discussed in the next section, but not in change of volume.

The dilatation appears in stress-strain equations (5.12.5). It appears also in the wave equation for P waves given in expression (6.1.12). Since dilatation is associated with a change in volume, P waves can be viewed as the propagation of compression within the continuum.

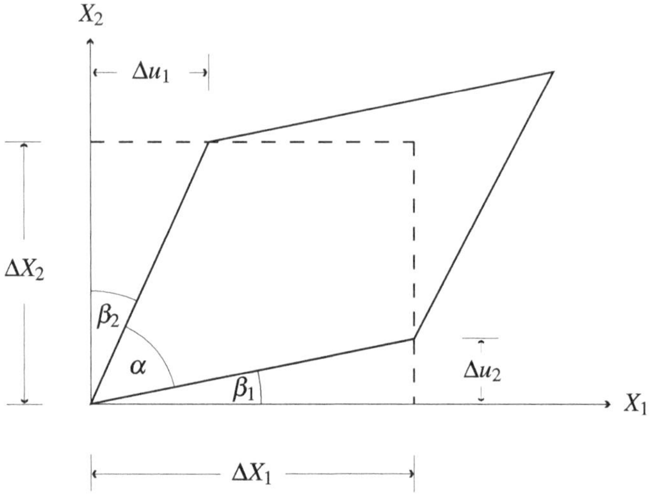

Fig. 1.4.2 Deformation resulting in a change of shape. A rectangle with sides ΔX_1 and ΔX_2 becomes a parallelogram. If $\beta_1 \neq \beta_2$, the object is also rotated; otherwise, the orientation of the diagonal remains unchanged: there is no rotation.

Note that, in terms of tensor algebra, expression (1.4.17) is the trace of the strain tensor, $tr(\varepsilon_{ij})$; namely, the sum of the diagonal terms. The trace of a second-rank tensor is a scalar; hence, it is invariant under the coordinate transformations, as proved in Exercise 1.6. Thus, as expected, the description of the change in volume is independent of the choice of the coordinate system. Relative change in volume in the context of material properties is shown in Exercise 5.14.

Change in shape

The strain tensor also describes deformations leading to a change in shape. Consider Figure 1.4.2 with axes defined in terms of the material coordinates that correspond to the configuration of the element of the continuum before deformation. A rectangular element of the continuum is deformed into a parallelogram. In other words, the original right angle is reduced to angle α. We can write this reduction as

$$\frac{\pi}{2} - \alpha = \beta_1 + \beta_2 ,$$

where β_1 and β_2 are the angles measured with respect to the X_1-axis and the X_2-axis, respectively. Assuming that angles β_1 and β_2 are small and measured in radians, we can approximate them by the corresponding tangents. Hence, examining Figure 1.4.2, we can write

$$\beta_1 + \beta_2 \approx \frac{\Delta u_2}{\Delta X_1} + \frac{\Delta u_1}{\Delta X_2}. \qquad (1.4.19)$$

Considering infinitesimal displacements, discussed in Section 1.3.3, and in view of Exercise 1.5, we can write equation (1.4.19) as

$$\beta_1 + \beta_2 \approx \frac{\partial u_2}{\partial x_1} + \frac{\partial u_1}{\partial x_2} = 2\varepsilon_{21} = 2\varepsilon_{12}, \qquad (1.4.20)$$

where we assume the equivalence between X_i and x_i. In other words, a function of coordinates that is evaluated at a point corresponding to the original configuration is approximately equal to this function evaluated at a point corresponding to the final position.[20]

Examining Figure 1.4.2, we see that equation (1.4.20) implies that the original segments are deviated by small angles β_1 and β_2 that can be stated as $\partial u_2/\partial x_1$ and $\partial u_1/\partial x_2$, respectively. Consequently, the initial right angle between segments, coinciding with the two axes, is changed by the sum of these two angles.[21]

1.5. Rotation Tensor and Rotation Vector

In Section 1.4.3, we defined dilatation, φ, which allows us to describe a relative change in volume using the divergence operator and the displacement vector, as shown in expression (1.4.18). In this section, we associate a change in shape with the displacement vector by using the curl operator.

Let us define a tensor given by

$$\xi_{ij} := \frac{1}{2}\left(\frac{\partial u_i}{\partial x_j} - \frac{\partial u_j}{\partial x_i}\right), \qquad i,j \in \{1,2,3\}. \qquad (1.5.1)$$

---

[20] Readers interested in more details associated with the strain tensor in the context of the material and the spatial coordinates might refer to Malvern, L.E., (1969) Introduction to the mechanics of a continuous medium: Prentice-Hall, pp. 120–135.

[21] Readers interested in a geometrical interpretation of the strain-tensor components might refer to Fung, Y.C., (1977) A first course in continuum mechanics: Prentice-Hall, Inc., pp. 129–130.

In view of definition (1.5.1), $\xi_{11} = \xi_{22} = \xi_{33} = 0$, and tensor ξ_{ij} has only three independent components; namely, $\xi_{23} = -\xi_{32}$, $\xi_{13} = -\xi_{31}$ and $\xi_{12} = -\xi_{21}$. Thus, ξ_{ij} is an antisymmetric tensor. We refer to ξ_{ij} as the rotation tensor. As discussed in Section 1.4.3 and illustrated in Figure 1.4.2, the quantities $\partial u_i/\partial x_j$, where $i \neq j$, are tantamount to the small deviation angles. Following the properties of the curl operator, we can associate tensor (1.5.1) with a vector given by

$$\Psi = \nabla \times \mathbf{u}, \tag{1.5.2}$$

as shown in Exercise 1.7 and further discussed in Exercise 1.8. We refer to Ψ as the rotation vector.[22]

Rotation vector (1.5.2) is used in formulating the wave equation involving S waves, as shown in expression (6.1.16). In other words, S waves can be viewed as the propagation of rotation within the continuum.

Note that we can use tensor calculus to relate the components of the strain tensor, the components of the rotation tensor and the components of the gradient of the displacement vector. Using expressions (1.4.6) and (1.5.1), we can write the partial derivative of a component of displacement as

$$\frac{\partial u_i}{\partial x_j} = \varepsilon_{ij} + \xi_{ij}, \qquad i,j \in \{1,2,3\}. \tag{1.5.3}$$

Equation (1.5.3) illustrates the fact that any second-rank tensor can be written as a sum of symmetric and antisymmetric tensors; herein, the strain and the rotation tensors. The left-hand side is associated with the deformation gradient, which in the Cartesian coordinates can be written as $F_{ij} := \delta_{ij} + \partial u_i/\partial x_j$, where δ_{ij} is Kronecker's delta; in matrix notation, we

[22] Readers interested in a relation between the rotation tensor and rotation vector might also refer to Fung, Y.C., (1977) A first course in continuum mechanics: Prentice-Hall, Inc., pp. 130–132.

write

$$\mathbf{F} := \begin{bmatrix} 1+\dfrac{\partial u_1}{\partial x_1} & \dfrac{\partial u_1}{\partial x_2} & \dfrac{\partial u_1}{\partial x_3} \\ \\ \dfrac{\partial u_2}{\partial x_1} & 1+\dfrac{\partial u_2}{\partial x_2} & \dfrac{\partial u_2}{\partial x_3} \\ \\ \dfrac{\partial u_3}{\partial x_1} & \dfrac{\partial u_3}{\partial x_2} & 1+\dfrac{\partial u_3}{\partial x_3} \end{bmatrix} .$$

Application of this gradient to an infinitesimal segment, $[\mathrm{d}x_1, \mathrm{d}x_2, \mathrm{d}x_3]$, leaves it unchanged if $\partial u_i/\partial x_j = 0$ for all i and j—otherwise, it subjects it to stretch and rotation.

Closing Remarks

Formulations of continuum mechanics allow us to describe deformation in a three-dimensional continuum. Let us emphasize this concept by defining continuum mechanics as a study of deformable bodies. In that manner, continuum mechanics extends beyond the subject of rigid-body mechanics.

In subsequent chapters, these formulations allow us to study and describe phenomena associated with wave propagation. In this study, we use the linearized theory of elasticity. Although linearization results in a loss of subtle details, the agreement between the theory and experiments is satisfactory for our purposes.

Also, the linearity is associated with several convenient properties, notably, with the principle of superposition. According to this principle, the net effect of several causes acting together is equivalent to the sum of the effects of each of these causes acting separately.

1.6. Exercises

EXERCISE 1.1. [23]Given a material description of motion,

$$\mathbf{x}(\mathbf{X},t) = \begin{cases} x_1 = X_1 e^t + X_3 \left(e^t - 1\right) \\ x_2 = X_2 + X_3 \left(e^t - e^{-t}\right) \\ x_3 = X_3 \end{cases}, \tag{1.6.1}$$

verify that the transformation between the material, $\mathbf{X}$, and spatial, $\mathbf{x}$, coordinates exists, and find the spatial description of this motion.[24]

SOLUTION 1.1. The transformation between the material and spatial coordinates exists if and only if the Jacobian, which is given by

$$J := \det \begin{bmatrix} \frac{\partial x_1}{\partial X_1} & \frac{\partial x_1}{\partial X_2} & \frac{\partial x_1}{\partial X_3} \\ \frac{\partial x_2}{\partial X_1} & \frac{\partial x_2}{\partial X_2} & \frac{\partial x_2}{\partial X_3} \\ \frac{\partial x_3}{\partial X_1} & \frac{\partial x_3}{\partial X_2} & \frac{\partial x_3}{\partial X_3} \end{bmatrix}, \tag{1.6.2}$$

does not vanish. Using equations (1.6.1), we obtain

$$J = \det \begin{bmatrix} e^t & 0 & e^t - 1 \\ 0 & 1 & e^t - e^{-t} \\ 0 & 0 & 1 \end{bmatrix} = e^t \neq 0 .$$

Thus, the transformation exists. Since, in this exercise, mapping $\mathbf{x} = \mathbf{x}(\mathbf{X},t)$ is linear, we can write it using matrix notation $\mathbf{x} = \mathbf{A}\mathbf{X}$. We can explicitly write,

$$\begin{bmatrix} x_1 \\ x_2 \\ x_3 \end{bmatrix} = \begin{bmatrix} e^t & 0 & e^t - 1 \\ 0 & 1 & e^t - e^{-t} \\ 0 & 0 & 1 \end{bmatrix} \begin{bmatrix} X_1 \\ X_2 \\ X_3 \end{bmatrix},$$

where $\mathbf{A}$ is the transformation matrix. Since $\det \mathbf{A} = J \neq 0$, transformation matrix $\mathbf{A}$ has an inverse. Thus, the spatial description of motion, namely,

[23] See also Section 1.3.1.

[24] In this book, $e^{(\cdot)}$ and $\exp(\cdot)$ are used as synonymous notations.

$\mathbf{X} = \mathbf{X}(\mathbf{x},t)$, is $\mathbf{X} = \mathbf{A}^{-1}\mathbf{x}$. In other words,

$$\begin{bmatrix} X_1 \\ X_2 \\ X_3 \end{bmatrix} = \begin{bmatrix} e^{-t} & 0 & e^{-t}(1-e^t) \\ 0 & 1 & e^{-t}\left(1-e^{2t}\right) \\ 0 & 0 & 1 \end{bmatrix} \begin{bmatrix} x_1 \\ x_2 \\ x_3 \end{bmatrix}.$$

REMARK 1.6.1. Note that at $t = 0$, $\mathbf{A} = \mathbf{A}^{-1} = \mathbf{I}$; hence, $\mathbf{X}(0) = \mathbf{x}(0)$. In other words, at the initial time, both material and spatial descriptions of motion coincide. At a later time, the material point that occupied position $\mathbf{X}$ at time $t = 0$, occupies position $\mathbf{x}$.

EXERCISE 1.2. [25]Consider

$$F(X) = a\sin\frac{X}{b}, \tag{1.6.3}$$

where a and b are constants. Let the change of variables be given by $X = x - u(x)$. Show that if both a and $u(x)$ are infinitesimal while b is finite, we obtain

$$F(X) = F(x).$$

SOLUTION 1.2. Considering the given change of variables, we can write expression (1.6.3) as

$$\begin{aligned} F(X(x)) &= a\sin\frac{x-u(x)}{b} \\ &= a\left(\sin\frac{x}{b}\cos\frac{u(x)}{b} - \sin\frac{u(x)}{b}\cos\frac{x}{b}\right). \end{aligned}$$

Since $u(x)$ is an infinitesimal quantity and b is finite

$$\lim_{(u/b)\to 0} \cos\frac{u(x)}{b} = 1$$

and

$$\lim_{(u/b)\to 0} \sin\frac{u(x)}{b} = 0.$$

Thus, we can write

$$F(X(x)) \approx a\sin\frac{x}{b} = F(x),$$

as required.

[25] See also Section 1.3.3.

REMARK 1.6.2. The result of Exercise 1.2, as well as the equivalence of the material and spatial coordinates for the infinitesimal displacements, is quite intuitive. In other words, considering the change of variables given by $X = x - u(x)$, we get $X \approx x$, for infinitesimal values of $u(x)$.

EXERCISE 1.3. A bar of length l would have an elongation, u_1, due to strain ε_1, that is, $u_1 = \varepsilon_1 l$. The same bar would have another elongation u_2 due to strain ε_2, that is, $u_2 = \varepsilon_2 l$. Show that considering only linear terms, under the assumption of small strains, the total elongation due to both strains is equal to the sum of both elongations.

SOLUTION 1.3. Assume that ε_1 is applied first. This results in the elongation,

$$u_1 = \varepsilon_1 l.$$

Hence, the new length of the bar is

$$l + u_1 = l + \varepsilon_1 l = l(1 + \varepsilon_1).$$

Subsequently, applying strain, ε_2, we obtain the final elongation,

$$u_f = u_1 + \varepsilon_2 l(1 + \varepsilon_1) = u_1 + \varepsilon_2 l + \varepsilon_1 \varepsilon_2 l = u_1 + u_2 + \varepsilon_1 \varepsilon_2 l.$$

Assuming that the value of the product, $\varepsilon_1 \varepsilon_2 l$, is small compared with the values of both $\varepsilon_1 l$ and $\varepsilon_2 l$—in other words, both ε_1 and ε_2 are much smaller than unity—we obtain

$$u_f \approx u_1 + u_2.$$

REMARK 1.6.3. The same result is obtained if the order is reversed, or if ε_1 and ε_2 are applied simultaneously. This is the illustration of the fact that the principle of superposition is applicable to linear systems—a commonly used property in mathematical physics. [26]

EXERCISE 1.4. [27]Using definition (1.4.6) and considering orthonormal coordinate systems, show that strain, ε_{ij}, which is given in terms of first partial derivatives of a vector, is a second-rank tensor.

[26] Readers interested in the epistemological justification of linearity and the principle of superposition might refer to Steiner, M., (1998) The applicability of mathematics as a philosophical problem: Harvard University Press , pp. 30–32.

[27] See also Sections 1.4.2, 5.1.1 and 5.2.3.

NOTATION 1.6.4. The repeated-index summation notation is used in this solution. Any term in which an index appears twice stands for the sum of all such terms as the index assumes values 1 , 2 and 3 .

SOLUTION 1.4. Following definition (1.4.6), consider $\partial\hat{u}_i/\partial\hat{x}_j$,where $\hat{u}_i$ are the components of the displacement vector, $\mathbf{u}$, in the transformed coordinates $\hat{x}_j$. The transformation rule of the coordinate points is given by

$$\hat{x}_j = a_{j\ell}x_\ell\,, \qquad j \in \{1,2,3\}\,, \tag{1.6.4}$$

where the entries of matrix $\mathbf{a}$ are the projections between the transformed and original axes. Matrix $\mathbf{a}$ is an orthogonal matrix, which means that its inverse is equal to its transpose. Hence,

$$x_j = a_{\ell j}\hat{x}_\ell\,, \qquad j \in \{1,2,3\}\,.$$

Consequently, we obtain

$$\frac{\partial x_j}{\partial \hat{x}_\ell} = a_{\ell j}\,. \tag{1.6.5}$$

Since $\mathbf{u}$ is a vector, its components transform according to the rule

$$\hat{u}_i = a_{ik}u_k\,, \qquad i \in \{1,2,3\}\,.$$

Thus, we can write

$$\frac{\partial \hat{u}_i}{\partial \hat{x}_\ell} = a_{ik}\frac{\partial u_k}{\partial \hat{x}_\ell}\,, \qquad i,l \in \{1,2,3\}\,,$$

which can be restated as

$$\frac{\partial \hat{u}_i}{\partial \hat{x}_\ell} = a_{ik}\frac{\partial u_k}{\partial x_j}\frac{\partial x_j}{\partial \hat{x}_\ell}\,, \qquad i,l \in \{1,2,3\}\,.$$

Hence, in view of equation (1.6.5), we can write

$$\frac{\partial \hat{u}_i}{\partial \hat{x}_\ell} = a_{ik}a_{\ell j}\frac{\partial u_k}{\partial x_j}\,, \qquad i,l \in \{1,2,3\}\,, \tag{1.6.6}$$

which is a transformation rule for the second-rank tensors. Consequently, since the sum of second-rank tensors is a second-rank tensor, an entity given by $\varepsilon_{ij} := (\partial u_i/\partial x_j + \partial u_j/\partial x_i)/2$ is a second-rank Cartesian tensor.

EXERCISE 1.5. [28]Considering the one-dimensional case and assuming infinitesimal displacement gradients, in view of expressions (1.3.8) and (1.3.9), show that

$$\frac{\partial u}{\partial x} \approx \frac{\partial U}{\partial X}. \tag{1.6.7}$$

SOLUTION 1.5. Consider the one-dimensional case of expressions (1.3.8) and (1.3.9), namely

$$\begin{cases} U(X,t) = x(X,t) - X \\ u(x,t) = x - X(x,t) \end{cases}.$$

Taking partial derivatives with respect to the first arguments, we obtain

$$\begin{cases} \dfrac{\partial U}{\partial X} = \dfrac{\partial x}{\partial X} - 1 \\ \\ \dfrac{\partial u}{\partial x} = 1 - \dfrac{\partial X}{\partial x} \end{cases}. \tag{1.6.8}$$

Since $x(X,t)$ and $X(x,t)$ are inverses of one another, we use the properties of the derivative of an inverse to obtain

$$\frac{\partial x}{\partial X} = \frac{1}{\dfrac{\partial X}{\partial x}}.$$

Hence, we can write expression (1.6.8) as

$$\begin{cases} \dfrac{\partial U}{\partial X} = \dfrac{1}{\dfrac{\partial X}{\partial x}} - 1 \\ \\ \dfrac{\partial u}{\partial x} = 1 - \dfrac{\partial X}{\partial x} \end{cases}.$$

[28] See also Section 1.4.3.

Solving both equations for $\partial X/\partial x$, we obtain

$$\begin{cases} \dfrac{\partial X}{\partial x} = \dfrac{1}{\dfrac{\partial U}{\partial X}+1} \\ \\ \dfrac{\partial X}{\partial x} = 1 - \dfrac{\partial u}{\partial x} \end{cases} .$$

Equating the right-hand sides and solving for $\partial u/\partial x$, we get

$$\frac{\partial u}{\partial x} = \frac{\dfrac{\partial U}{\partial X}}{\dfrac{\partial U}{\partial X}+1} . \tag{1.6.9}$$

Examining equation (1.6.9), we notice that for the infinitesimal displacement gradients, namely, $\partial U/\partial X \ll 1$, we can write $\partial u/\partial x \approx \partial U/\partial X$, which is expression (1.6.7), as required.

EXERCISE 1.6. [29]Prove the following theorem.

THEOREM 1.6.5. *The sum of diagonal elements of a second-rank tensor is invariant under orthogonal transformations of the coordinate system.*

SOLUTION 1.6. Let us proceed with the following proof.

PROOF. By definition, the components of the second-rank tensor $\varepsilon_{\ell m}$ transform to the components $\hat{\varepsilon}_{ik}$, which are expressed in another coordinate system, according to the rule given by

$$\hat{\varepsilon}_{ik} = \sum_{\ell=1}^{3}\sum_{m=1}^{3} b_{i\ell} b_{km} \varepsilon_{\ell m} , \qquad i,k \in \{1,2,3\} ,$$

where $\mathbf{b}$ is an orthogonal transformation matrix. Setting $k = i$, we obtain the sum of the components along the main diagonal; namely,

$$\sum_{i=1}^{3} \hat{\varepsilon}_{ii} = \sum_{i=1}^{3}\sum_{\ell=1}^{3}\sum_{m=1}^{3} b_{i\ell} b_{im} \varepsilon_{\ell m} .$$

[29]See also Section 1.4.3.

Hence, by orthogonality of $\mathbf{b}$, we have

$$\sum_{i=1}^{3} b_{i\ell} b_{im} = \delta_{\ell m}, \qquad \ell, m \in \{1,2,3\}.$$

Thus, we can write

$$\sum_{i=1}^{3} \hat{\varepsilon}_{ii} = \sum_{\ell=1}^{3} \sum_{m=1}^{3} \delta_{\ell m} \varepsilon_{\ell m} = \sum_{m=1}^{3} \varepsilon_{mm}.$$

Since both i and m are the summation indices, we are allowed to write

$$\sum_{j=1}^{3} \hat{\varepsilon}_{jj} = \sum_{j=1}^{3} \varepsilon_{jj}.$$

This implies that the value of the sum of the diagonal elements is invariant under orthogonal transformations of the coordinate system. The sum of the diagonal elements of a second-rank tensor is a scalar. □

REMARK 1.6.6. Following Exercise 1.4, we can see that dilatation, φ, defined by expression (1.4.18) is the sum of diagonal elements of the second-rank tensor, namely, the trace of the strain tensor,

$$\varepsilon_{ij} := \left(\partial u_i / \partial x_j + \partial u_j / \partial x_i\right) / 2.$$

Consequently, as shown in Exercise 1.6, we can prove that dilatation is a scalar quantity. This is expected because of the physical meaning of dilatation. In other words, the change of volume must be independent of the coordinate system.

EXERCISE 1.7. [30]In view of the properties of vector operators, show that the components of the second-rank tensor, given by expression (1.5.1), namely,

$$\xi_{ij} := \frac{1}{2}\left(\frac{\partial u_i}{\partial x_j} - \frac{\partial u_j}{\partial x_i}\right), \qquad i, j \in \{1,2,3\},$$

are associated with rotation vector (1.5.2).

[30] See also Section 1.5.

SOLUTION 1.7. Consider the displacement vector $\mathbf{u} = [u_1, u_2, u_3]$. We can write its curl as

$$\nabla \times \mathbf{u} = \left(\frac{\partial u_3}{\partial x_2} - \frac{\partial u_2}{\partial x_3}\right)\mathbf{e}_1 + \left(\frac{\partial u_1}{\partial x_3} - \frac{\partial u_3}{\partial x_1}\right)\mathbf{e}_2 + \left(\frac{\partial u_2}{\partial x_1} - \frac{\partial u_1}{\partial x_2}\right)\mathbf{e}_3 ,$$

where $\mathbf{e}_i$ denotes a unit vector along the x_i-axis. Following expression (1.5.1), we can rewrite the curl as

$$\nabla \times \mathbf{u} = [2\xi_{32}, 2\xi_{13}, 2\xi_{21}] .$$

Thus, ξ_{ij} can be viewed as the components of the vector that results from the rotation of $\mathbf{u}/2$. Denoting $\Psi = [2\xi_{32}, 2\xi_{13}, 2\xi_{21}]$, we obtain definition (1.5.2).

REMARK 1.6.7. As illustrated in Exercise 1.8, below, the association between the components of the second-rank tensor, ξ_{ij}, and the components of vector Ψ is due to the antisymmetry of this tensor that results in only three independent components.

EXERCISE 1.8. [31] Using the permutation symbol, ε_{ijk}, illustrate that the result of the cross product is not a vector but a second-rank tensor, which—in three dimensions—has three independent components.

SOLUTION 1.8. Consider an antisymmetric second-rank tensor, whose components in $\mathbb{R}^n$ can be written as

$$\begin{bmatrix} 0 & \ldots & a_1 b_n - a_n b_1 \\ \vdots & \ddots & \vdots \\ a_n b_1 - a_1 b_n & \ldots & 0 \end{bmatrix},$$

where a_j and b_k are components of two first-rank tensors. This second-rank tensor has $n(n-1)/2$ independent components, c_{jk}. Consequently, since in $\mathbb{R}^3$, it has only three independent components, it can be viewed—in these dimensions only—as a first-rank tensor.

To reduce c_{jk} to a first-rank tensor in $\mathbb{R}^3$, we write

$$\frac{1}{2}\sum_{j=1}^{3}\sum_{k=1}^{3} c_{jk}\,\varepsilon_{ijk} = \frac{1}{2}\sum_{j=1}^{3}\sum_{k=1}^{3}(a_j b_k - a_k b_j)\,\varepsilon_{ijk} =: (\mathbf{a}\times\mathbf{b})_i , \quad i \in \{1,2,3\};$$

[31] See also Section 1.5.

explicitly, $(\mathbf{a}\times\mathbf{b})_1 = a_2 b_3 - a_3 b_2$, $(\mathbf{a}\times\mathbf{b})_2 = a_3 b_1 - a_1 b_3$ and $(\mathbf{a}\times\mathbf{b})_3 = a_1 b_2 - a_2 b_1$, which can be viewed as the components of the cross product of vectors $\mathbf{a}$ and $\mathbf{b}$, in $\mathbb{R}^3$.

In the reduction process, we use—together with the properties of the permutation symbol—the linearity of the operations and the antisymmetry of the tensor to obtain

$$
\begin{aligned}
(\mathbf{a}\times\mathbf{b})_i &= \frac{1}{2}(\mathbf{a}\times\mathbf{b})_i + \frac{1}{2}(\mathbf{a}\times\mathbf{b})_i = \frac{1}{2}(\mathbf{a}\times\mathbf{b})_i - \frac{1}{2}(\mathbf{b}\times\mathbf{a})_i \\
&= \frac{1}{2}\sum_{j=1}^{3}\sum_{k=1}^{3}(a_j b_k \varepsilon_{ijk} - b_j a_k \varepsilon_{ijk}) = \frac{1}{2}\sum_{j=1}^{3}\sum_{k=1}^{3}(a_j b_k - a_k b_j)\,\varepsilon_{ijk} \\
&= \frac{1}{2}\sum_{j=1}^{3}\sum_{k=1}^{3} c_{jk}\,\varepsilon_{ijk}, \quad i \in \{1,2,3\}.
\end{aligned}
$$

The cross product of vectors in $\mathbb{R}^3$ behaves as a pseudovector; it transforms as a vector under the rotations of the coordinate system but not under its reflections. In physics, both the magnetic field and the angular momentum, discussed in Section 2.7, are pseudovectors.

CHAPTER 2

Forces and Balance Principles

> It is as necessary to science as to pure mathematics that the fundamental principles should be clearly stated and that the conclusions shall follow from them. But in science it is also necessary that the principles taken as fundamental should be as closely related to observation as possible.
>
> *Harold Jeffreys and Bertha Jeffreys (1946)*
> *Methods of mathematical physics*

Preliminary Remarks

In the context of continuum mechanics, seismic waves are deformations that propagate in a continuum. These deformations are associated with forces. To describe the propagation of deformations, we now seek to formulate the equations that relate these deformations to forces acting within the continuum.

In general, the fundamental principles of continuum mechanics consist of the conservation of mass, balance of linear momentum, balance of angular momentum, balance of energy, balance of electric charge, balance of magnetic flux, and the principle of irreversibility: entropy. In this book, in which we work within the theory of linear elasticity, we need to invoke explicitly only a few of these principles.

We begin this chapter with the study of the conservation of mass, which is associated with the motion of mass within the continuum. Using the conservation of mass, we derive the equation of continuity. Then we formulate the balance of linear momentum. Subsequently, to take into account the forces acting within the continuum, we formulate the stress tensor. Using the balance of linear momentum and the concept of stress, we derive

Cauchy's equations of motion. To obtain all fundamental equations that relate the unknowns that appear in the equation of continuity and in Cauchy's equations of motion, we also invoke the balance of angular momentum. These three balance principles lead to a system of equations that is associated with the propagation of deformations in an elastic continuum.

2.1. Conservation of Mass

2.1.1. Introductory comments

A fundamental principle in which our description of continuum mechanics must be rooted is the conservation of mass. We use this principle to derive an equation that relates mass density, ρ, and displacement vector, $\mathbf{u}$.

Note that, in general, conservation principles are special cases of the corresponding balance principles. Herein, discussing the balance of mass, we wish to emphasize that we do not consider production or destruction of mass and, hence, the total amount of mass is conserved. Discussing the balance of linear momentum and the balance of angular momentum in Sections 2.4 and 2.7, respectively, we wish to emphasize that—for a given portion of a continuum—the total amount of these momenta changes and these changes are balanced by forces acting within the continuum.

2.1.2. Integral equation

The amount of mass, m, occupying a fixed volume, V, at an instant of time is given by

$$m(t) = \iiint_V \rho(\mathbf{x}, t)\, \mathrm{d}V, \tag{2.1.1}$$

where ρ denotes mass density. The rate of change of mass contained in this volume is given by the differentiation of equation (2.1.1) with respect to time; namely,

$$\frac{\mathrm{d}}{\mathrm{d}t} m(t) = \frac{\mathrm{d}}{\mathrm{d}t} \iiint_V \rho(\mathbf{x}, t)\, \mathrm{d}V. \tag{2.1.2}$$

Furthermore, for an arbitrary volume that is fixed, V, we can rewrite expression (2.1.2) as

$$\frac{\mathrm{d}}{\mathrm{d}t}m(t)=\iiint\limits_V \frac{\partial\rho(\mathbf{x},t)}{\partial t}\,\mathrm{d}V\,. \tag{2.1.3}$$

We can also express $\mathrm{d}m/\mathrm{d}t$ in a different way. Since we assume that there is no production or destruction of mass, the rate of change of mass contained in a fixed volume is only a function of the mass flowing through this volume. In other words, the rate of change of mass contained in volume V per unit time is equal to the amount of mass that passes through the surface, S, bounding this volume. This can be written as

$$\frac{\mathrm{d}}{\mathrm{d}t}m(t)=-\iint\limits_S \rho(\mathbf{x},t)\,\mathbf{v}\cdot\mathbf{n}\,\mathrm{d}S\,, \tag{2.1.4}$$

where $\mathbf{v}$ represents the velocity of a portion of mass that passes through this surface, and where $\mathbf{n}$ denotes an outward unit normal vector to this surface. The negative sign results from the fact that the vector normal to the surface points away from volume V. Herein, we assume the surface element, dS, to be sufficiently small that it might be considered as a plane and to have the same mass flow across all its points.

Expressions (2.1.3) and (2.1.4) describe the same quantity; namely,

$$\iiint\limits_V \frac{\partial\rho(\mathbf{x},t)}{\partial t}\,\mathrm{d}V=-\iint\limits_S \rho(\mathbf{x},t)\,\mathbf{v}\cdot\mathbf{n}\,\mathrm{d}S\,,$$

which means that the rate of change of the amount of material inside a closed surface is equal to the net rate with which this material flows through this surface. To combine expressions (2.1.3) and (2.1.4), we express the right-hand side of equation (2.1.4) as a volume integral. Invoking the divergence theorem—according to which the surface integral of vector field $\rho(\mathbf{x},t)\,\mathbf{v}(\mathbf{x})$ over a closed surface can be expressed as the volume integral of the divergence of this vector field integrated over the volume enclosed by this surface—we write

$$\iint\limits_S \rho\mathbf{v}\cdot\mathbf{n}\,\mathrm{d}S=\iiint\limits_V \nabla\cdot(\rho\mathbf{v})\,\mathrm{d}V\,. \tag{2.1.5}$$

Equating the right-hand sides of equations (2.1.3) and (2.1.4), and using expression (2.1.5) for the latter one, we obtain

$$\iiint\limits_V \frac{\partial \rho}{\partial t}\,\mathrm{d}V = -\iiint\limits_V \nabla \cdot (\rho \mathbf{v})\,\mathrm{d}V\,. \tag{2.1.6}$$

Combining the two volume integrals, we write

$$\iiint\limits_V \left[\frac{\partial \rho}{\partial t} + \nabla \cdot (\rho \mathbf{v})\right] \mathrm{d}V = 0\,, \tag{2.1.7}$$

which is a statement of the conservation of mass for a fixed volume V.

Note that in this derivation of equation (2.1.7), the change in the amount of mass in a volume at any instant is balanced by the mass flowing through the surface that encloses this volume. Consequently, considering a given volume, one could refer to equation (2.1.7) as a balance-of-mass equation rather than a conservation-of-mass equation. However, as stated above, we choose to use only the latter term. Our choice is also justified by the fact that discussing the balance of linear momentum and the balance of angular momentum in Sections 2.4 and 2.7, respectively, we consider a moving volume that consistently contains the same portion of the continuum, as discussed in Section 2.2; in such a case, there is no mass flowing through the surface that encloses this moving volume. Recalling our discussion in Section 1.3, we might note that for conservation of mass we use the spatial point of view, while for balances of linear and angular momenta we use the material point of view.

2.1.3. Equation of continuity

The equation of continuity is a differential equation expressing the conservation of mass within the continuum. To derive the equation of continuity, consider integral equation (2.1.7). For this equation to be true for an arbitrary fixed volume, the integrand must be identically zero. Thus, we require

$$\frac{\partial \rho}{\partial t} + \nabla \cdot (\rho \mathbf{v}) = 0\,. \tag{2.1.8}$$

Note that if there were a point where the integrand were nonzero, we could consider a sufficiently small volume around that point. This would result in a nonzero value of the integral, as illustrated in Exercise 2.1.

REMARK 2.1.1. We could proceed directly from equation (2.1.6) to equation (2.1.8) by requiring the volume, V, to be arbitrary and to be the same on both sides of equation (2.1.6); herein, the second condition is necessarily satisfied, since equation (2.1.8) results from applying the divergence theorem. Without stating these conditions, it might appear that the equality of integrands is due, in general, to the equality of integrals, which is not the case, as illustrated in Remark 2.9.1 on page 79, below. The step between equations (2.1.6) and (2.1.8) can be viewed as a proof of equality stated by the latter; this proof implicitly invokes the linearity of the integral operator.

Since $\mathbf{v} = \partial \mathbf{u}/\partial t$, where $\mathbf{u}$ denotes the displacement vector, we can write equation (2.1.8) as

$$\frac{\partial \rho}{\partial t} + \nabla \cdot \left(\rho \frac{\partial \mathbf{u}}{\partial t} \right) = 0. \tag{2.1.9}$$

This is the equation of continuity, which relates mass density and displacement.

Note that we could rewrite equation (2.1.9) as

$$\left[\frac{\partial}{\partial t}, \frac{\partial}{\partial x_1}, \frac{\partial}{\partial x_2}, \frac{\partial}{\partial x_3} \right] \cdot \left[\rho, \rho \frac{\partial u_1}{\partial t}, \rho \frac{\partial u_2}{\partial t}, \rho \frac{\partial u_3}{\partial t}, \right] = 0,$$

where the differential operator is a divergence in the four-dimensional spacetime. This is a typical form of a fundamental conservation principle.[1]

2.2. Time Derivative of Volume Integral

To derive the remaining two balance principles, namely, the balance of linear momentum and the balance of angular momentum, we use the concept of the time derivative of a moving-volume integral, which is associated with the conservation of mass. For this purpose, let us consider a moving volume that consistently contains the same portion of the continuum. In other words, there is no mass transport through the surface encompassing this volume. In such a case, the portion of the continuum possessing a

[1] Readers interested in fundamentals of conservation principles might refer to Bunge, M., (1967) Foundations of physics: Springer-Verlag, pp. 47–51.

given velocity and acceleration is identifiable. Hence, such a description lends itself to a convenient extension of particle mechanics and, therefore, allows us to use Newton's laws of motion.

To consider the temporal variation of a physical quantity enclosed in a moving volume, we study

$$\frac{\mathrm{d}}{\mathrm{d}t}\iiint\limits_{V(t)} \rho\left(\mathbf{x},t\right)\mathscr{A}\left(\mathbf{x},t\right)\mathrm{d}V\,, \tag{2.2.1}$$

where ρ is mass density, $\mathscr{A}$ is a scalar, a vector or a tensor that describes a physical quantity of interest, while $V\left(t\right)$ is a volume that varies with time but always contains the same portion of the continuum. The temporal variation of $\mathscr{A}$ in $V\left(t\right)$ can be also expressed in a different way; namely,

$$\iiint\limits_{V(t)} \frac{\partial}{\partial t}\left(\rho\mathscr{A}\right)\mathrm{d}V + \iint\limits_{S(t)} \rho\mathscr{A}\sum_{j=1}^{3} v_j n_j \mathrm{d}S. \tag{2.2.2}$$

Herein, the volume integral describes the change of the amount of $\mathscr{A}$ due to its creation or destruction within the volume, and the surface integral describes the net change of the amount of $\mathscr{A}$ due to its flow across the surface; v_j are the components of the velocity vector with which $\mathscr{A}$ flows across the surface whose normal is $\mathbf{n} = \left[n_1, n_2, n_3\right]$. Since expressions (2.2.1) and (2.2.2) describe the same entity, we can equate them to write

$$\frac{\mathrm{d}}{\mathrm{d}t}\iiint\limits_{V(t)} \rho\left(\mathbf{x},t\right)\mathscr{A}\left(\mathbf{x},t\right)\mathrm{d}V = \iiint\limits_{V(t)} \frac{\partial}{\partial t}\left(\rho\mathscr{A}\right)\mathrm{d}V + \iint\limits_{S(t)} \rho\mathscr{A}\sum_{j=1}^{3} v_j n_j \mathrm{d}S, \tag{2.2.3}$$

and proceed to express the right-hand side in terms of volume integrals.

Note that the above formulation is analogous to the one presented in Section 2.1. Expressions (2.2.1) and (2.2.2) are analogous to (2.1.2) and (2.1.4), respectively. The key difference between expressions (2.1.4) and (2.2.2) is the appearance of $\iiint_{V(t)} \partial\left(\rho\mathscr{A}\right)/\partial t\,\mathrm{d}V$ in the latter one, which is due to the fact that, although mass cannot be created or destroyed in classical physics, no such restriction governs quantity $\mathscr{A}$.

To express the right-hand side in terms of volume integrals, we invoke the divergence theorem to write equation (2.2.3) as

$$\frac{\mathrm{d}}{\mathrm{d}t}\iiint\limits_{V(t)} \rho\mathscr{A}\,\mathrm{d}V = \iiint\limits_{V(t)} \frac{\partial}{\partial t}(\rho\mathscr{A})\,\mathrm{d}V + \iiint\limits_{V(t)} \sum_{j=1}^{3} \frac{\partial(\rho\mathscr{A}v_j)}{\partial x_j}\mathrm{d}V\,.$$

Differentiating and rearranging, we get

$$\frac{\mathrm{d}}{\mathrm{d}t}\iiint\limits_{V(t)} \rho\mathscr{A}\,\mathrm{d}V = \iiint\limits_{V(t)} \left\{\rho\left(\frac{\partial\mathscr{A}}{\partial t} + \mathbf{v}\cdot\nabla\mathscr{A}\right) + \mathscr{A}\left[\frac{\partial\rho}{\partial t} + \nabla\cdot(\rho\mathbf{v})\right]\right\}\mathrm{d}V\,.$$

We note that the term in brackets vanishes due to equation of continuity (2.1.8). Thus, we write

$$\frac{\mathrm{d}}{\mathrm{d}t}\iiint\limits_{V(t)} \rho\mathscr{A}\,\mathrm{d}V = \iiint\limits_{V(t)} \rho\left(\frac{\partial\mathscr{A}}{\partial t} + \mathbf{v}\cdot\nabla\mathscr{A}\right)\mathrm{d}V\,,$$

which can be restated as

$$\frac{\mathrm{d}}{\mathrm{d}t}\iiint\limits_{V(t)} \rho\mathscr{A}\,\mathrm{d}V = \iiint\limits_{V(t)} \rho\left(\frac{\partial}{\partial t} + \mathbf{v}\cdot\nabla\right)\mathscr{A}\,\mathrm{d}V\,. \qquad (2.2.4)$$

In view of expression (1.3.5), we see that the term in parentheses is the material time-derivative operator acting on $\mathscr{A}(\mathbf{x},t)$, which is equivalent to $\mathrm{d}\mathfrak{A}(\mathbf{X},t)/\mathrm{d}t$, where $\mathfrak{A}$ is the material description of $\mathscr{A}$. We can rewrite expression (2.2.4) as

$$\frac{\mathrm{d}}{\mathrm{d}t}\iiint\limits_{V(t)} \rho(\mathbf{x},t)\,\mathscr{A}(\mathbf{x},t)\,\mathrm{d}V = \iiint\limits_{V(t)} \rho(\mathbf{x},t)\,\frac{\mathrm{D}\mathscr{A}(\mathbf{x},t)}{\mathrm{D}t}\mathrm{d}V, \qquad (2.2.5)$$

which is the desired result. As defined in expression (1.3.7), $\mathrm{D}/\mathrm{D}t$ denotes the material time-derivative operator. Also, we can view the time derivative of a moving-volume integral as a case of Leibniz's integration rule, as illustrated in Exercises 2.3 and 2.4.

2.3. Stress

2.3.1. Stress as description of surface forces

We wish to analyze the internal forces acting among the adjacent material elements within the continuum. For this purpose, we introduce the concept of stress.

Together with the material and spatial descriptions and with the material time derivative, the concept of stress distinguishes continuum mechanics from particle mechanics. Stress, as a mathematical entity, was introduced by Cauchy in 1827 to express the interaction of a material with the surrounding material in terms of surface forces.[2]

When a material is subject to loads, internal forces are induced within it. Deformation of this material is a function of the distribution of these forces. In a continuum, stress is associated with internal surface forces that an element of the continuum exerts on another element of the continuum across an imaginary surface that separates them. Stress is a system of surface forces producing strain within a continuum. Owing to the mutual dependence of stress and strain, strains cannot be produced without inducing stresses, and stresses cannot be induced without producing, or tending to produce, strains. This interrelation between stress and strain is an intrinsic property of the elasticity theory.

2.3.2. Traction

As a result of forces being transmitted within the continuum, the portion of the continuum enclosed by an imaginary surface interacts with the portion of the continuum outside of this surface, as discussed on page 14. Let $\Delta\mathbf{F}$ be the force exerted on the surface element, ΔS, by the continuum on either side of this surface. The average force per unit area can be written in terms of the ratio given by

$$\bar{\mathbf{T}} = \frac{\Delta\mathbf{F}}{\Delta S}. \tag{2.3.1}$$

Cauchy's stress principle—the fundamental principle of continuum mechanics stated on page 14—implies that as $\Delta S \to 0$, ratio (2.3.1) tends to a

[2] Interested readers might refer to Cauchy, A. L., (1827) De la pression ou tension dans un corps solide: Ex. de Math, **2**, pp. 42 – 56.

finite limit.[3] The resulting traction vector is given by

$$\mathbf{T}^{(\mathbf{n})} = \lim_{\Delta S \to 0} \frac{\Delta \mathbf{F}}{\Delta S} = \frac{\mathrm{d}\mathbf{F}}{\mathrm{d}S}, \tag{2.3.2}$$

where the superscript, $\mathbf{n}$, specifies the surface element, ΔS, upon which the traction is acting, by stating unit vector $\mathbf{n}$ that is normal to this surface element. Thus, traction is a vector that describes the contact force with which the elements at each side of an internal surface within the continuum act upon each other.

Since the value of traction is finite even if the element of the surface area becomes infinitesimal, we can describe forces at any given point within the continuum. Also, since the traction is explicitly dependent on the orientation of the surface element, as indicated by vector $\mathbf{n}$, we can describe forces in any given direction within the continuum. Consequently, we can study both inhomogeneity and anisotropy of the continuum.

2.4. Balance of Linear Momentum

In general, the forces acting within a continuum are classified as either surface forces or body forces according to their mode of application. Surface forces are transmitted by direct mechanical contacts across imaginary surfaces separating given portions of the continuum. Body forces, such as gravitational force, are assumed to behave as an action-at-a-distance, which is the instantaneous action between two bodies in spatial separation. In general, this is not the case, as illustrated by forces used in the theories of electromagnetism or general relativity.

Consider a portion of a continuum contained in volume V and subjected to time-varying and space-varying forces. The surface forces are given as the traction vector shown in expression (2.3.2); namely,

$$\mathbf{T} = \frac{\mathrm{d}\mathbf{F}}{\mathrm{d}S},$$

and the body forces are given by

$$\mathbf{f} = \mathbf{f}(\mathbf{x}, t).$$

[3] Interested readers might refer to Malvern, L.E., (1969) Introduction to the mechanics of a continuous medium: Prentice-Hall, p. 70.

Consequently, the total force is

$$\mathbf{F}_T = \iint\limits_S \mathbf{T}\,\mathrm{d}S + \iiint\limits_V \mathbf{f}\,\mathrm{d}V\,, \tag{2.4.1}$$

where S is the surface enclosing volume V.

Note that $\mathbf{T}$ and $\mathbf{f}$ have units of force per area and force per volume, respectively. Hence, both integrals have units of force, as required.

To study the effect of this force, we choose to consider a moving volume that consistently contains the same portion of the continuum. Hence, invoking Newton's second law of motion, we can write

$$\frac{\mathrm{d}}{\mathrm{d}t}\iiint\limits_{V(t)} \rho \frac{\mathrm{d}\mathbf{u}}{\mathrm{d}t}\,\mathrm{d}V = \iint\limits_{S(t)} \mathbf{T}\,\mathrm{d}S + \iiint\limits_{V(t)} \mathbf{f}\,\mathrm{d}V\,, \tag{2.4.2}$$

where the displacement,

$$\mathbf{u} = \left[u_1(\mathbf{x},t)\,, u_2(\mathbf{x},t)\,, u_3(\mathbf{x},t)\right]\,, \tag{2.4.3}$$

is a function of both space and time.

In view of the definition of linear momentum, which is the product of mass and velocity, the integrand on the left-hand side of equation (2.4.2) stands for $\rho\,\mathbf{v}$. In expressing $\mathbf{v}$ as $\mathrm{d}\mathbf{u}/\mathrm{d}t$, we write the standard symbol to denote generically a derivative, rather than the symbol of the material time derivative, introduced in Section 1.3.2 and used for linearization discussed in Section 1.3.3, which contains the partial derivative.

This integral equation in expression (2.4.2) states the balance of linear momentum: The rate of change of the linear momentum of an element within the continuum is equal to the sum of the external forces acting upon this element. This statement is analogous to Newton's second law of motion in particle mechanics.

To see this analogy, let us consider a system of n particles whose masses are m_i. For such a system, we can write Newton's second law of motion as

$$\frac{\mathrm{d}}{\mathrm{d}t}\sum_{i=1}^{n} m_i \frac{\mathrm{d}\mathbf{x}_i}{\mathrm{d}t} = \sum_{i=1}^{n} \mathbf{F}_i + \sum_{i=1}^{n}\sum_{j=1}^{n} \mathbf{F}_{ji}\,. \tag{2.4.4}$$

The sum on the left-hand side is the linear momentum of the system of particles. The first sum on the right-hand side denotes the external forces acting on the particles of the system. The second sum on the right-hand side denotes the internal forces among the particles of the system; $\mathbf{F}_{ji}$ is the force on the ith particle due to the jth particle. Since i and j are summation indices, we can rewrite the second sum on the right-hand side as

$$\frac{1}{2}\sum_{i=1}^{n}\sum_{j=1}^{n}\left(\mathbf{F}_{ji}+\mathbf{F}_{ij}\right).$$

Now, let us invoke Newton's third law of motion: Forces between particles are equal in magnitude and opposite in direction, $\mathbf{F}_{ji}=-\mathbf{F}_{ij}$. Also, in classical mechanics, it is assumed—though seldom explicitly stated—that a point particle exerts no force on itself, only on other particles; thus, $\mathbf{F}_{ii}=\mathbf{0}$. Consequently, the double sum vanishes and, hence, we can write Newton's second law of motion for a system of particles as

$$\frac{\mathrm{d}}{\mathrm{d}t}\sum_{i=1}^{n} m_i\frac{\mathrm{d}\mathbf{x}_i}{\mathrm{d}t}=\sum_{i=1}^{n}\mathbf{F}_i.$$

In other words, only the external forces affect the change of linear momentum. This means that the system itself cannot change its own linear momentum. In equation (2.4.2), the external forces are $\mathbf{T}$ and $\mathbf{f}$; these forces are external to the portion of a continuum contained in volume V.

Also, in applying Newton's third law of motion, we assume that the two particles remain in constant contact.[4] This assumption allows us to avoid the physically undesirable concept of action-at-a-distance.

Before returning to equation (2.4.2), let us emphasize that in considering the analogy of Newton's laws of motion we must be aware that particle and continuum mechanics are two distinct physical theories. Even though one might start with particle mechanics for pedagogical reasons, yet—both methodologically and logically—continuum mechanics comes before it: methodologically, because we experience bulk matter; logically,

[4] Readers interested in an insightful explanation for the requirement of the constant contact might refer to Schutz, B., (2003) Gravity from the ground up: Cambridge University Press, p. 12.

because by suitably weakening continuum mechanics we can obtain particle mechanics.[5]

Now, let us return to equation (2.4.2). Invoking expression (2.2.5) and letting $\mathscr{A} = \mathrm{d}\mathbf{u}/\mathrm{d}t$, we bring the derivative inside the volume integral to write equation (2.4.2) as

$$\iiint\limits_{V(t)} \rho \frac{\mathrm{d}^2\mathbf{u}}{\mathrm{d}t^2}\mathrm{d}V = \iint\limits_{S(t)} \mathbf{T}\,\mathrm{d}S + \iiint\limits_{V(t)} \mathbf{f}\,\mathrm{d}V\,. \tag{2.4.5}$$

Herein—with a certain abuse of notation in view of expression (2.2.5)—$\mathrm{d}^2/\mathrm{d}t^2$ contains the material time-derivative operator, introduced in Section 1.3.2, which in turn contains partial derivatives. One could write equation (2.4.5) distinguishing among d/dt, D/Dt and $\partial/\partial t$. In particular, as a consequence of linearization discussed in Section 1.3.3, $\mathbf{v} = \partial\mathbf{u}/\partial t$, and as a consequence of bringing the derivative inside the integral, discussed in Section 2.2, $\mathrm{d}^2\mathbf{u}/\mathrm{d}t^2 \equiv \mathrm{D}\mathbf{v}/\mathrm{Dt} := \partial\mathbf{v}/\partial\mathrm{t} + \mathbf{v}\cdot\nabla\mathbf{v}$.

In Section 2.5.2, we use the balance of linear momentum to formulate the stress tensor. In Section 2.6, following the formulation of the stress tensor, we use equation (2.4.5) to derive Cauchy's equations of motion.

2.5. Stress Tensor

2.5.1. Traction on coordinate planes

We wish to describe the state of stress at a given point in a continuum. At an arbitrary point within a continuum, Cauchy's stress principle associates a traction and a unit normal of a surface element on which this vector is acting. Consider a fixed coordinate system with the orthonormal vectors given by $\mathbf{e}_1$, $\mathbf{e}_2$ and $\mathbf{e}_3$. The traction acting on the ith coordinate plane is represented by a vector, which can be written as

$$\mathbf{T}^{(\mathbf{e}_i)} = T_1^{(\mathbf{e}_i)}\mathbf{e}_1 + T_2^{(\mathbf{e}_i)}\mathbf{e}_2 + T_3^{(\mathbf{e}_i)}\mathbf{e}_3\,,$$

where $T_j^{(\mathbf{e}_i)}$ are the components of this vector along the x_j-axis. At a given point, the three tractions associated with the three mutually orthogonal

[5] Readers interested in a discussion of the relation between continuum and particle mechanics might refer to Bunge, M., (1967) Foundations of physics: Springer-Verlag, p. 143 and p. 155.

planes can be explicitly written as three vectors given by

$$\begin{bmatrix} \mathbf{T}^{(\mathbf{e}_1)} \\ \mathbf{T}^{(\mathbf{e}_2)} \\ \mathbf{T}^{(\mathbf{e}_3)} \end{bmatrix} = \begin{bmatrix} T_1^{(\mathbf{e}_1)} & T_2^{(\mathbf{e}_1)} & T_3^{(\mathbf{e}_1)} \\ T_1^{(\mathbf{e}_2)} & T_2^{(\mathbf{e}_2)} & T_3^{(\mathbf{e}_2)} \\ T_1^{(\mathbf{e}_3)} & T_2^{(\mathbf{e}_3)} & T_3^{(\mathbf{e}_3)} \end{bmatrix} \begin{bmatrix} \mathbf{e}_1 \\ \mathbf{e}_2 \\ \mathbf{e}_3 \end{bmatrix}. \tag{2.5.1}$$

Considering the traction components shown in the 3×3 matrix, we see that the subscript refers to the component of a given traction, while the superscript identifies the plane on which this traction is acting. For instance, $T_2^{(\mathbf{e}_1)}$ is the x_2-component of a traction acting on the plane normal to the x_1-axis. For convenience, we write the square matrix in equations (2.5.1) as

$$\sigma = \begin{bmatrix} \sigma_{11} & \sigma_{12} & \sigma_{13} \\ \sigma_{21} & \sigma_{22} & \sigma_{23} \\ \sigma_{31} & \sigma_{32} & \sigma_{33} \end{bmatrix}. \tag{2.5.2}$$

By examining expressions (2.5.1) and (2.5.2), we immediately see that σ_{ij} represents the jth component of the surface force acting on the surface whose normal is parallel to the x_i-axis. This index convention, which allows us to describe the direction of the force and the orientation of the surface on which it is acting, is illustrated in Figure 2.5.1.[6] Also, we wish to distinguish between tension and compression for the traction components normal to a given face, as well as denote the direction of the traction components tangential to a given face. For this purpose, we adopt the following sign convention. On a surface whose outward normal points in the positive direction of the corresponding coordinate axis, all traction components that act in the positive direction of a given axis are positive. On a surface whose outward normal points in the negative direction of the corresponding coordinate axis, all traction components that act in the negative direction of a given axis are positive. This convention applies to both the normal and the tangential components. Examining Figure 2.5.1, we see

---

[6] This index convention is consistent with Malvern, L.E., (1969) Introduction to the mechanics of a continuous medium: Prentice-Hall, p. 80, and with Aki, K., and Richards, P.G., (2002) Quantitative seismology (2*nd* edition): University Science Books, pp. 17–18.

One can also use the opposite convention; for instance, Kolsky, H., (1953/1963) Stress waves in solids: Dover, p. 5, and Sheriff, R.E., and Geldart, L.P., (1982) Exploration Seismology: Cambridge University Press, Vol. I, p. 33.

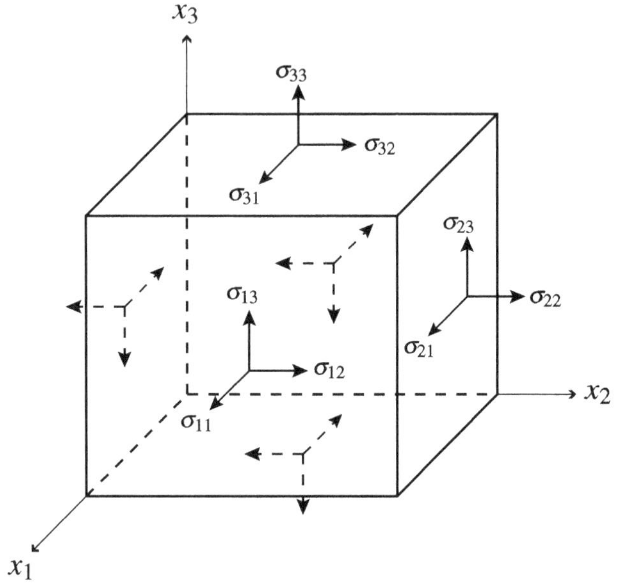

Fig. 2.5.1 Index convention for the stress-tensor components. Symbol σ_{ij} represents the jth component of the surface force acting on the surface whose normal is parallel to the x_i-axis. All components shown herein are positive.

that all the traction components on each of the six faces illustrated therein are positive. In the context of the normal components, our sign convention implies that tension is positive while compression is negative.[7]

Note that, if we wished, we could reverse our sign convention without affecting Newton's third law of motion. In other words,

$$\mathbf{T}^{(\mathbf{n})} = -\mathbf{T}^{(-\mathbf{n})} \tag{2.5.3}$$

is always true; herein, $\mathbf{T}^{(\mathbf{n})}$ is due to the force acting on the plane whose outward normal is $\mathbf{n}$, and $-\mathbf{T}^{(-\mathbf{n})}$ is due to the force acting on the other side of this plane.

As formulated herein, the entries of matrix (2.5.2) determine the stress state within a continuum at a given point with respect to the coordinate

[7] This sign convention is consistent with Aki, K. and Richards, P.G., (2002) Quantitative seismology (2*nd* edition): University Science Books, p. 15.

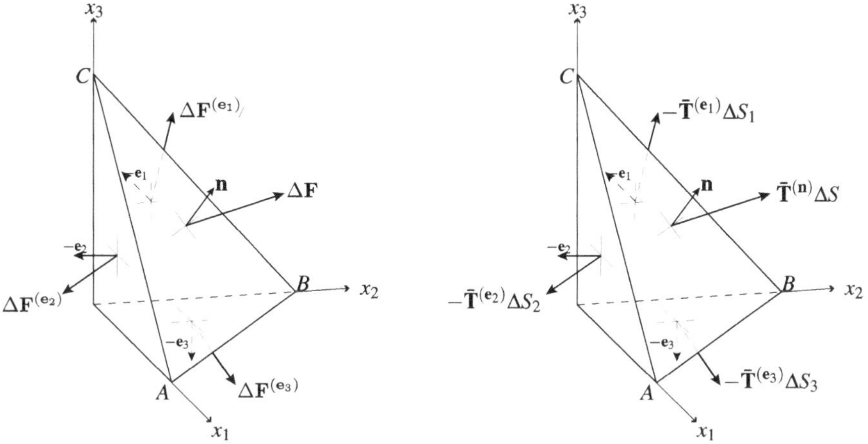

Fig. 2.5.2 Cauchy's tetrahedron. Three of the faces are orthogonal to each other and coincide with the coordinate planes; their unit normals are $-\mathbf{e}_i$, where $i \in \{1,2,3\}$. The fourth face has an arbitrary orientation described by the unit vector, $\mathbf{n}$, normal to it. Each face of the left-hand-side tetrahedron is subject to a force, which—for the right-hand-side tetrahedron—is expressed by the product of the corresponding traction and surface area, using the sign convention introduced on page 57.

planes. As shown in Section 2.5.2, these entries can also be used to describe the stress state with respect to an arbitrary plane within the continuum.

2.5.2. Traction on arbitrary planes

To study forces within the continuum, we wish to describe them with respect to a plane of arbitrary orientation. For this purpose, consider an element of a continuum in the form of a tetrahedron; such a construction is also called Cauchy's tetrahedron. Let the tetrahedron be spanned by four points $O(0,0,0)$, $A(a,0,0)$, $B(0,b,0)$ and $C(0,0,c)$, shown in Figure 2.5.2. Thus, the four faces of the tetrahedron consist of the oblique face, namely, ABC, and of three orthogonal faces, namely, OAB, OBC and OAC.

Given forces acting on the coordinate planes, as illustrated on the left-hand side of Figure 2.5.2, we seek to determine the force, $\Delta\mathbf{F}$, acting on the oblique face whose area is ΔS and whose unit normal is $\mathbf{n}$. The key statement of this derivation relies on the balance of linear momentum, discussed in Section 2.4, and the fact that the tetrahedron is subjected to both surface and body forces. In view of equation (2.4.5), for a finite-size tetrahedron, we can write

$$\Delta\mathbf{F}+\Delta\mathbf{F}^{(\mathbf{e}_1)}+\Delta\mathbf{F}^{(\mathbf{e}_2)}+\Delta\mathbf{F}^{(\mathbf{e}_3)}+\bar{\mathbf{f}}\Delta V=\bar{\rho}\Delta V\frac{\mathrm{d}\bar{\mathbf{v}}}{\mathrm{d}t}, \tag{2.5.4}$$

where $\Delta\mathbf{F}$ is the surface force acting on the oblique face, $\Delta\mathbf{F}^{(\mathbf{e}_i)}$ is the surface force acting on the orthogonal face normal to the x_i-axis, and $\bar{\mathbf{f}}$ refers to the body force acting on the tetrahedron with volume ΔV and mass density $\bar{\rho}$. Thus, the left-hand side of equation (2.5.4) gives the sum of forces, while the right-hand side gives the rate of change of linear momentum with $\bar{\mathbf{v}}$ denoting velocity. The bars above a given symbol denote the average value of the corresponding quantity for this finite-size tetrahedron.

In view of expression (2.3.1), we can write

$$\Delta\mathbf{F}=\bar{\mathbf{T}}^{(\mathbf{n})}\Delta S. \tag{2.5.5}$$

Using expressions (2.5.3) and (2.5.5), as illustrated on the right-hand side of Figure 2.5.2, we can rewrite equation (2.5.4) as

$$\bar{\mathbf{T}}^{(\mathbf{n})}\Delta S-\bar{\mathbf{T}}^{(\mathbf{e}_1)}\Delta S_1-\bar{\mathbf{T}}^{(\mathbf{e}_2)}\Delta S_2-\bar{\mathbf{T}}^{(\mathbf{e}_3)}\Delta S_3+\bar{\mathbf{f}}\Delta V=\bar{\rho}\Delta V\frac{\mathrm{d}\bar{\mathbf{v}}}{\mathrm{d}t}, \tag{2.5.6}$$

where ΔS is the area of the oblique face and ΔS_i is the area of an orthogonal face normal to the x_i-axis. In equation (2.5.6), $\bar{\mathbf{T}}^{(\cdot)}$ is a resultant traction that corresponds to a given face. Note that the orthogonal faces have unit outward normals parallel and opposite in sign to the unit vectors of the coordinate axes, $\mathbf{e}_i$. Hence, in view of Newton's third law, we introduced the negative signs in the summation.

The surface forces, which are used in equation (2.5.6), are illustrated in Figure 2.5.2.

To study equation (2.5.6), we wish to relate geometrically the surface areas of the tetrahedron, ΔS and ΔS_i, where $i\in\{1,2,3\}$, and its volume, V. The areas of the orthogonal faces are

$$\Delta S_i=n_i\Delta S, \qquad i\in\{1,2,3\}, \tag{2.5.7}$$

where n_i are the components of the unit vector, $\mathbf{n}$, which is normal to the oblique face. Using expression (2.5.7), we can rewrite equation (2.5.6) as

$$\bar{\mathbf{T}}^{(\mathbf{n})}\Delta S - \bar{\mathbf{T}}^{(\mathbf{e}_1)}n_1\Delta S - \bar{\mathbf{T}}^{(\mathbf{e}_2)}n_2\Delta S - \bar{\mathbf{T}}^{(\mathbf{e}_3)}n_3\Delta S + \bar{\mathbf{f}}\Delta V = \bar{\rho}\Delta V\frac{\mathrm{d}\bar{\mathbf{v}}}{\mathrm{d}t}. \quad (2.5.8)$$

Now, we wish to relate the volume, ΔV, to the area of the oblique face, ΔS. Considering the oblique face as the base of the tetrahedron, we can state its volume as

$$\Delta V = \frac{h}{3}\Delta S, \quad (2.5.9)$$

where h is the height of the tetrahedron. Hence, using expression (2.5.9), we can rewrite equation (2.5.8) as

$$\bar{\mathbf{T}}^{(\mathbf{n})}\Delta S - \bar{\mathbf{T}}^{(\mathbf{e}_1)}n_1\Delta S - \bar{\mathbf{T}}^{(\mathbf{e}_2)}n_2\Delta S - \bar{\mathbf{T}}^{(\mathbf{e}_3)}n_3\Delta S + \frac{\bar{\mathbf{f}}h}{3}\Delta S = \frac{\bar{\rho}h}{3}\Delta S\frac{\mathrm{d}\bar{\mathbf{v}}}{\mathrm{d}t}. \quad (2.5.10)$$

Dividing both sides of equation (2.5.10) by ΔS, we obtain

$$\bar{\mathbf{T}}^{(\mathbf{n})} - \bar{\mathbf{T}}^{(\mathbf{e}_1)}n_1 - \bar{\mathbf{T}}^{(\mathbf{e}_2)}n_2 - \bar{\mathbf{T}}^{(\mathbf{e}_3)}n_3 + \frac{\bar{\mathbf{f}}h}{3} = \frac{\bar{\rho}h}{3}\frac{\mathrm{d}\bar{\mathbf{v}}}{\mathrm{d}t}. \quad (2.5.11)$$

To describe the state of stress at a point within the continuum, we let $h \to 0$ in such a way that the areas of all faces simultaneously approach zero, the orientation of the height, h, does not change, and the origin of the coordinate system does not move. In other words, the finite-size tetrahedron reduces to an infinitesimal tetrahedron at point $O(0,0,0)$. Thus, we obtain

$$\mathbf{T}^{(\mathbf{n})} = \mathbf{T}^{(\mathbf{e}_1)}n_1 + \mathbf{T}^{(\mathbf{e}_2)}n_2 + \mathbf{T}^{(\mathbf{e}_3)}n_3. \quad (2.5.12)$$

For an infinitesimal tetrahedron, where h tends to zero, the body-force term and the change-of-momentum term vanish due to the disappearance of a volume and, hence, mass. Thus, according to expression (2.5.12), given tractions acting on infinitesimal surface elements parallel to the coordinate planes, we obtain the traction in an arbitrary direction.

Note that in equation (2.5.12), the tractions no longer correspond to the average values but to the local values at point $O(0,0,0)$. This also implies that equation (2.5.12) is valid for any coordinate system.

Equation (2.5.12) can be viewed as an equilibrium equation of an infinitesimal element within the continuum. Note, however, that the derivation of this equation stems from the balance of linear momentum without *a priori* assuming such an equilibrium.[8]

Expressing the orthogonal-face tractions in terms of their components, equation (2.5.12) can be explicitly written as

$$\mathbf{T}^{(\mathbf{n})} = \begin{bmatrix} T_1^{(\mathbf{e}_1)} \\ T_2^{(\mathbf{e}_1)} \\ T_3^{(\mathbf{e}_1)} \end{bmatrix} n_1 + \begin{bmatrix} T_1^{(\mathbf{e}_2)} \\ T_2^{(\mathbf{e}_2)} \\ T_3^{(\mathbf{e}_2)} \end{bmatrix} n_2 + \begin{bmatrix} T_1^{(\mathbf{e}_3)} \\ T_2^{(\mathbf{e}_3)} \\ T_3^{(\mathbf{e}_3)} \end{bmatrix} n_3$$

$$= \begin{bmatrix} T_1^{(\mathbf{e}_1)} & T_1^{(\mathbf{e}_2)} & T_1^{(\mathbf{e}_3)} \\ T_2^{(\mathbf{e}_1)} & T_2^{(\mathbf{e}_2)} & T_2^{(\mathbf{e}_3)} \\ T_3^{(\mathbf{e}_1)} & T_3^{(\mathbf{e}_2)} & T_3^{(\mathbf{e}_3)} \end{bmatrix} \begin{bmatrix} n_1 \\ n_2 \\ n_3 \end{bmatrix}. \tag{2.5.13}$$

Equation (2.5.13) states explicitly that, at a given point, we can determine traction $\mathbf{T}^{(\mathbf{n})}$ that acts on an arbitrary plane through that point, provided we know the tractions at this point that act on the three mutually orthogonal planes. Examining expressions (2.5.1) and (2.5.13), we conclude that

$$\mathbf{T}^{(\mathbf{n})} = \sigma^T \mathbf{n}, \tag{2.5.14}$$

where σ is given in expression (2.5.2) and ^{T} denotes transpose. In the context of an arbitrary plane, we see that the entries of matrix σ are the components of a second-rank tensor. The fact that σ is a second-rank tensor is shown in Exercise 2.5.

Tensor σ_{ij} is called the stress tensor, and expression (2.5.14) is the statement of Cauchy's stress theorem, even though Cauchy proposed it as a hypothesis only. This tensor is also known as Cauchy's stress tensor. The stress tensor allows us to determine the stress state associated with an infinitesimal plane of arbitrary orientation. The stress tensor takes into account both the direction of the traction and the orientation of the surface upon which the traction is acting.

---

[8] Readers interested in the theorem relating the stress tensor and the balance of linear momentum might refer to Marsden, J.E., and Hughes, T.J.R., (1983/1994) Mathematical foundations of elasticity: Dover, pp. 132–135.

In view of expression (2.5.2), we can rewrite equation (2.5.14) as

$$\mathbf{T}^{(\mathbf{n})} = \begin{bmatrix} \sigma_{11} & \sigma_{21} & \sigma_{31} \\ \sigma_{12} & \sigma_{22} & \sigma_{32} \\ \sigma_{13} & \sigma_{23} & \sigma_{33} \end{bmatrix} \begin{bmatrix} n_1 \\ n_2 \\ n_3 \end{bmatrix},$$

which can be concisely stated as

$$T_i^{(\mathbf{n})} = \sum_{j=1}^{3} \sigma_{ji} n_j, \qquad i \in \{1,2,3\}. \tag{2.5.15}$$

Expression (2.5.15) is an important statement of continuum mechanics. It relates the components of forces acting within the continuum to the orientation of the plane upon which the forces are acting. In other words, two vectorial properties, namely, traction, $\mathbf{T}^{(\mathbf{n})}$, and surface-normal vector, $\mathbf{n}$, are uniquely related by the stress tensor, σ_{ij}.[9] The derivation performed in this section shows that to describe a traction related to an arbitrary plane, it is enough to consider tractions on three planes with linearly independent normals. The main-diagonal entries describe the components acting along the surface-normal vector. Hence, in view of the sign convention stated on page 57, if σ_{ii} is positive, it is a tensile component; if it is negative, it is a compressional component. The nondiagonal entries describe shear components.[10]

Also, expression (2.5.15) allows us to invoke the divergence theorem to express the surface integral in expression (2.4.5) as a volume integral. Hence, it allows us to proceed with the derivation of Cauchy's equations of motion.

---

[9] Readers interested in a discussion of the stress tensor using modern mathematical language might refer to Segev, R. and Rodnay, G., (1999) Cauchy's theorem on manifolds: Journal of Elasticity, **56**, 129–144.

[10] Readers interested in formulation of the stress tensor as a generalization of the concept of hydrostatic pressure might refer to Synge, J.L. and Schild, A., (1949/1978) Tensor calculus: Dover, pp. 205–208.

2.6. Cauchy's Equations of Motion

2.6.1. General formulation

To formulate the equations of motion, we consider the balance of linear momentum and the concept of the stress tensor. In view of expression (2.5.15), we can write the balance of linear momentum, stated in equation (2.4.5), in terms of components, as

$$\iiint\limits_{V(t)} \rho \frac{\mathrm{d}^2 u_i}{\mathrm{d}t^2} \mathrm{d}V = \iint\limits_{S(t)} \sum_{j=1}^{3} \sigma_{ji} n_j \mathrm{d}S + \iiint\limits_{V(t)} f_i \mathrm{d}V, \qquad i \in \{1,2,3\},$$

where, as discussed for equation (2.4.5), the time derivative contains the material time-derivative operator.

We wish to express all integrals as volume integrals. Hence, invoking the divergence theorem, we can write

$$\iiint\limits_{V(t)} \rho \frac{\mathrm{d}^2 u_i}{\mathrm{d}t^2} \mathrm{d}V = \iiint\limits_{V(t)} \sum_{j=1}^{3} \frac{\partial \sigma_{ji}}{\partial x_j} \mathrm{d}V + \iiint\limits_{V(t)} f_i \mathrm{d}V, \qquad i \in \{1,2,3\}.$$

Using the linearity of the integral operator, we can rewrite this equation as

$$\iiint\limits_{V(t)} \left(\sum_{j=1}^{3} \frac{\partial \sigma_{ji}}{\partial x_j} + f_i - \rho \frac{\mathrm{d}^2 u_i}{\mathrm{d}t^2} \right) \mathrm{d}V = 0, \qquad i \in \{1,2,3\}, \qquad (2.6.1)$$

which states the balance of linear momentum, as long as the portion of the continuum contained in volume $V(t)$ remains the same.

To derive Cauchy's equations of motion, consider equation (2.6.1). For this integral equation to be satisfied for an arbitrary volume that contains the same portion of the continuum, the integrand must be identically zero, as illustrated in Exercise 2.1. Thus,

$$\sum_{j=1}^{3} \frac{\partial \sigma_{ji}}{\partial x_j} + f_i = \rho \frac{\mathrm{d}^2 u_i}{\mathrm{d}t^2}, \qquad i \in \{1,2,3\}. \qquad (2.6.2)$$

Even though, in principle, $\mathrm{d}^2/\mathrm{d}t^2$ contains the material time-derivative operator—in view of the linearized formulation discussed in Section 1.3.3—we write equations (2.6.2) as[11]

$$\sum_{j=1}^{3}\frac{\partial\sigma_{ji}}{\partial x_j}+f_i=\rho\frac{\partial^2 u_i}{\partial t^2}, \qquad i\in\{1,2,3\}. \tag{2.6.3}$$

These are Cauchy's equations of motion. As shown in Exercise 2.7, the *SI* units of Cauchy's equations of motion are N/m^3—force per volume.[12] These equations are also known as Cauchy's first law of motion.[13]

Cauchy's equations of motion relate two vectorial quantities, namely, the surface force—which corresponds to the summation term defining the divergence of tensor σ_{ji}—and the body force, to the acceleration vector. In other words, Cauchy's equations of motion state that the acceleration of an element within a continuum results from the application of surface and body forces.

If the acceleration term vanishes in equations of motion (2.6.3), we obtain the equations of static equilibrium,

$$\sum_{j=1}^{3}\frac{\partial\sigma_{ji}}{\partial x_j}+f_i=0, \qquad i\in\{1,2,3\}. \tag{2.6.4}$$

These equations describe the equilibrium state of an element of the continuum arising from the application of forces whose resultant is zero. Equations (2.6.4) are used to illustrate the symmetry of the stress tensor, as shown in Exercise 2.8. Equations (2.6.4) are also valid for rectilinear, constant-velocity motion.

Consider a system composed of equation of continuity (2.1.9) and of Cauchy's equations of motion (2.6.3) in a three-dimensional continuum.

---

[11]Readers interested in this approximation might refer to Grant, F.S., and West, G.F., (1965) Interpretation theory in applied geophysics: McGraw-Hill Inc., pp. 28–29, to Graff, K.F., (1975/1991) Wave motion in elastic solids: Dover, pp. 586–587, and to Bóna, A. and Slawinski, M.A., (2021) Wavefronts and rays as characteristics and asymptotics (*3rd* edition): World Scientific, Appendix B.1.

[12]Readers interested in formulation of Cauchy's equations of motion as a generalization of equations of motion for a perfect fluid might refer to Synge, J.L., and Schild, A., (1949/1978) Tensor calculus: Dover, p. 208.

[13]Readers interested in an insightful interpretation of Cauchy's first law of motion might refer to Bunge, M., (1967) Foundations of physics: Springer-Verlag, pp. 152–154.

This system contains four equations and sixteen unknowns, namely, mass density, ρ, stress-tensor components, σ_{11}, σ_{12}, σ_{13}, σ_{21}, σ_{22}, σ_{23}, σ_{31}, σ_{32}, σ_{33}, body-force components, f_1, f_2, f_3, and displacement-vector components, u_1, u_2, u_3.

Note that if we consider conservative systems, the three body-force components are derived from a single scalar function. In other words, $\mathbf{f} = \nabla U(\mathbf{x})$.

In our subsequent studies, we reduce the discrepancy between the number of equations and the number of unknowns. In Section 2.7, we show that the stress tensor is symmetric, which results in only six independent stress-tensor components. Also, we ignore the body force, $\mathbf{f} = [f_1, f_2, f_3]$, for the following two reasons.

The body force is irrelevant if we consider an infinitesimal element of the continuum, as we do in *Part 1*. We can see this in view of the tetrahedron argument, discussed in Section 2.5.2; in particular, by examining the step between equations (2.5.11) and (2.5.12).

The effects of the body forces are negligible as compared to the effects of the surface forces if we consider sufficiently high frequencies, which is the case of applied seismology discussed in *Part* 2 and *Part* 3. In other words, the effects of gravitation are negligible as compared to the effects of elasticity.[14]

2.6.2. Surface-forces formulation

To gain insight into the equations of motion without body forces, we rederive equations (2.6.3) without using the divergence theorem, which relates surface and volume integrals.

Consider forces acting in the positive direction of the x_1-axis on each coordinate plane. In view of definition (2.3.2), we can write the resultant force acting along this axis as

$$T_1^{(\mathbf{e}_1)}\,\mathrm{d}x_2\mathrm{d}x_3 + T_1^{(\mathbf{e}_2)}\,\mathrm{d}x_1\mathrm{d}x_3 + T_1^{(\mathbf{e}_3)}\,\mathrm{d}x_1\mathrm{d}x_2\,, \tag{2.6.5}$$

[14]Readers interested in the effect of gravitation on seismic wave propagation might refer to Udías, A., (1999) Principles of seismology: Cambridge University Press, pp. 39–40.

Let us note here that term 'gravity' is often used carelessly. One should use 'gravitation' to refer to the universal attraction between masses, and 'gravity' to refer to the downward acceleration experienced by a mass at rest in a reference frame rotating with the Earth; gravity is the vector resultant of gravitation and centrifugal force.

where $\mathbf{e}_i$ denotes the unit normal of the coordinate plane on which $T_1^{(\mathbf{e}_i)}$ is acting, and $\mathrm{d}x_j \mathrm{d}x_k$ is the surface area of the planar element. Following expression (2.5.2), expression (2.6.5) can be rewritten as

$$\sigma_{11}\,\mathrm{d}x_2\mathrm{d}x_3 + \sigma_{21}\,\mathrm{d}x_1\mathrm{d}x_3 + \sigma_{31}\,\mathrm{d}x_1\mathrm{d}x_2 . \tag{2.6.6}$$

Consider a rectangular box with its sides parallel to the coordinate planes. Let this box be spanned by $\mathrm{d}x_1$, $\mathrm{d}x_2$ and $\mathrm{d}x_3$. Consider forces acting in the positive direction of the x_1-axis on each face of the box. The resultant force along this axis is the sum of forces acting on the three sets of the parallel faces. Within each set, the two parallel faces are separated by a distance $\mathrm{d}x_i$. By convention, stated in Section 2.5.1, a stress component is positive if it acts in the positive direction of the coordinate axis and on the plane whose outward normal points in the positive coordinate direction. For each set of the two parallel faces of the aforementioned rectangular box, one face exhibits an outward normal that points in the positive coordinate direction and the other face exhibits an outward normal that points in the negative coordinate direction. Thus, in view of expression (2.6.6), we can write the resultant force along the x_1-axis as

$$\begin{aligned}\mathrm{d}F_1 &= \left[\left(\sigma_{11} + \frac{\partial\sigma_{11}}{\partial x_1}\,\mathrm{d}x_1\right)\mathrm{d}x_2\mathrm{d}x_3 + \left(-\sigma_{11}\,\mathrm{d}x_2\mathrm{d}x_3\right)\right] \\ &+ \left[\left(\sigma_{21} + \frac{\partial\sigma_{21}}{\partial x_2}\,\mathrm{d}x_2\right)\mathrm{d}x_1\mathrm{d}x_3 + \left(-\sigma_{21}\,\mathrm{d}x_1\mathrm{d}x_3\right)\right] \\ &+ \left[\left(\sigma_{31} + \frac{\partial\sigma_{31}}{\partial x_3}\,\mathrm{d}x_3\right)\mathrm{d}x_1\mathrm{d}x_2 + \left(-\sigma_{31}\,\mathrm{d}x_1\mathrm{d}x_2\right)\right],\end{aligned} \tag{2.6.7}$$

which, for a given direction, contains six forces; each force acting on one face of the box. Each expression in brackets corresponds to the sum of the two forces along the x_1-axis acting on faces orthogonal to the x_1-axis, the x_2-axis, and the x_3-axis, respectively. In other words, the first bracket denotes a sum of the two forces acting along the x_1-axis on the faces normal to it, as shown in Figure 2.6.1; the second and the third brackets denote the sums of forces acting along the x_1-axis on the faces parallel to it.

Note that, in view of terms $\sigma_{i1} + (\partial\sigma_{i1}/\partial x_i)\,\mathrm{d}x_i$, expression (2.6.7) is a first-order approximation. This approximation is consistent with our study in the context of linearized theory.

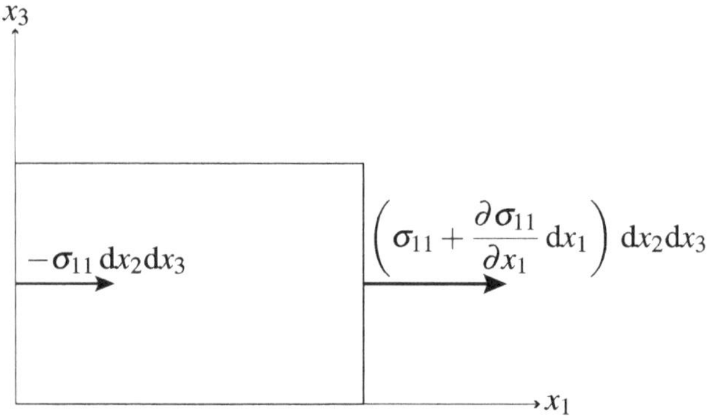

Fig. 2.6.1 Two forces acting along the x_1-axis on faces that are normal to it, as denoted by the first bracket in expression (2.6.7). The two forces have opposite signs since they act in the same direction on the planes whose outward normals are in the opposite directions. Also, in general, the magnitudes of these forces are different from one another, unless $\partial\sigma_{11}/\partial x_1 = 0$, which means that the stress is the same at all points. Consequently, in general, the sum of the two forces is nonzero, which results in a deformation, as illustrated in Figure 1.4.1. Note that applying a force in a given direction to the outside of a face has the same effect as applying it to the inside, as a consequence of Newton's third law; in other words, we could place the arrowheads where the tails are, and $-\sigma_{11}\,dx_2dx_3$ would remain a compression and $(\sigma_{11}+\partial\sigma_{11}/\partial x_1\,dx_1)\,dx_2dx_3$ a tension—the concept of push versus pull is inconsequential. Such dynamic boundary conditions are discussed in detail on page 446.

Upon simplification, we write expression (2.6.7) as

$$dF_1 = \left(\frac{\partial\sigma_{11}}{\partial x_1}+\frac{\partial\sigma_{21}}{\partial x_2}+\frac{\partial\sigma_{31}}{\partial x_3}\right)dx_1dx_2dx_3\,. \tag{2.6.8}$$

Invoking Newton's second law of motion in the form given by

$$dF_1 = \rho\,dx_1dx_2dx_3\frac{d^2u_1}{dt^2}\,,$$

where ρ is the mass density of the small rectangular box and u_1 is the displacement in the x_1-direction, we can write expression (2.6.8) as

$$\rho\frac{d^2u_1}{dt^2} = \frac{\partial\sigma_{11}}{\partial x_1}+\frac{\partial\sigma_{21}}{\partial x_2}+\frac{\partial\sigma_{31}}{\partial x_3}\,. \tag{2.6.9}$$

Analogously, for the displacement-vector component along the x_2-axis and the displacement-vector component along the x_3-axis, we can write

$$\rho\frac{\mathrm{d}^2u_2}{\mathrm{d}t^2}=\frac{\partial\sigma_{12}}{\partial x_1}+\frac{\partial\sigma_{22}}{\partial x_2}+\frac{\partial\sigma_{32}}{\partial x_3} \tag{2.6.10}$$

and

$$\rho\frac{\mathrm{d}^2u_3}{\mathrm{d}t^2}=\frac{\partial\sigma_{13}}{\partial x_1}+\frac{\partial\sigma_{23}}{\partial x_2}+\frac{\partial\sigma_{33}}{\partial x_3}, \tag{2.6.11}$$

respectively.

In view of linearization discussed in Section 1.3.3, total derivatives with respect to time are equivalent to partial derivatives. Consequently, expressions (2.6.9), (2.6.10) and (2.6.11) are equivalent to Cauchy's equations of motion (2.6.3) with no body forces.

2.7. Balance of Angular Momentum

2.7.1. Introductory comments

Above we postulated that the continuum obeys the conservation of mass and the balance of linear momentum. In this section, we formulate the balance of angular momentum in the context of the conservation of mass, the balance of linear momentum and Newton's third law of motion. To obtain the balance of angular momentum, we use the fact that the time rate of change of angular momentum for a given system is equal to the vector sum of the torques due to the external forces acting on that system. This property allows us to study the behaviour of a system without investigating the details of its internal behaviour. In the context of particle mechanics, we can formulate such a system by assuming that each two particles act on one another with forces of equal magnitude that are collinear and have opposite directions; such forces are called the central forces. In view of this assumption, all the internal forces within the system cancel out, which means that the system cannot change its own angular momentum.

To see this formulation in the context of particle mechanics, let us return to Newton's second law of motion for a system of particles as stated by equation (2.4.4). Multiplying both sides by vector $\mathbf{r}_i$, which is the vector between a reference point and the ith particle, we get

$$\frac{\mathrm{d}}{\mathrm{d}t}\sum_{i=1}^{n}\mathbf{r}_i\times m_i\frac{\mathrm{d}\mathbf{x}_i}{\mathrm{d}t}=\sum_{i=1}^{n}\mathbf{r}_i\times\mathbf{F}_i+\sum_{i=1}^{n}\sum_{j=1}^{n}\mathbf{r}_i\times\mathbf{F}_{ji}.$$

The sum on the left-hand side is the angular momentum of the system of n particles. The first sum on the right-hand side is the torque on the system due to external forces. The second sum on the right-hand side is the torque due to the forces within the system. In a manner analogous to the one used in Section 2.4, we wish to eliminate the second sum. Since i and j are summation indices, we can rewrite the double sum as

$$\sum_{i=1}^{n}\sum_{j=1}^{n}\mathbf{r}_i\times\mathbf{F}_{ji}=\frac{1}{2}\sum_{i=1}^{n}\sum_{j=1}^{n}\left(\mathbf{r}_i\times\mathbf{F}_{ji}+\mathbf{r}_j\times\mathbf{F}_{ij}\right).$$

Invoking Newton's third law of motion, $\mathbf{F}_{ji}=-\mathbf{F}_{ij}$, together with $\mathbf{F}_{ii}=\mathbf{0}$, we can rewrite this sum as

$$\sum_{i=1}^{n}\sum_{j=1}^{n}\mathbf{r}_i\times\mathbf{F}_{ji}=\frac{1}{2}\sum_{i=1}^{n}\sum_{j=1}^{n}\left[\left(\mathbf{r}_i-\mathbf{r}_j\right)\times\mathbf{F}_{ji}\right].$$

Since $\mathbf{r}_i$ is the vector between the reference point and the ith particle while $\mathbf{r}_j$ is the vector between the reference point and the jth particle, $\mathbf{r}_i-\mathbf{r}_j$ is the vector between the two particles. For this sum to vanish, we require that $\mathbf{r}_i-\mathbf{r}_j$ be parallel to $\mathbf{F}_{ji}$. This means that we require the forces between the particles to be central forces.[15] This formulation relies on the strong form of Newton's third law of motion: Forces between particles are equal in magnitude and opposite in direction as well as acting along the line joining the particles. Notably, the particle-mechanics analogue that we used in

[15] Readers interested in more details of such formulations in the context of particle mechanics might refer to Doran, C., and Lasenby, A., (2003) Geometric algebra for physicists, Cambridge University Press, pp. 57–61, Feynman, R.P., Leighton, R.B., and Sands, M., (1963/1989) Feynman's lectures on physics: Addison-Wesley Publishing Co. Vol. I, p. 18-7, to Goldstein, H., (1950/1980) Classical mechanics: Addison-Wesley Publishing Co. pp. 5–9, and to Radin, S.H., and Folk, R.T., (1982) Physics for scientists and engineers: Prentice-Hall, Inc., pp. 156–159, 164.

Readers interested in a formulation that invokes Lagrange's equations of motion, which we discuss in Section 13.2, might refer to Landau, L.D., and Lifshitz, E.M., (1969) Course of theoretical physics, Vol. 1: Mechanics: Elsevier, pp. 18–19. Therein, the balance of angular momentum is derived as a fundamental consequence of the inertial frame of reference.

Section 2.4 for the balance of linear momentum required only the weak form of Newton's third law of motion; the forces are not required to act along the line joining the particles. Herein, in view of the strong form of Newton's third law of motion, we write the balance of angular momentum for the system of particles as

$$\frac{\mathrm{d}}{\mathrm{d}t}\sum_{i=1}^{n}\mathbf{r}_i \times m_i \frac{\mathrm{d}\mathbf{x}_i}{\mathrm{d}t} = \sum_{i=1}^{n}\mathbf{r}_i \times \mathbf{F}_i . \qquad (2.7.1)$$

We also use the strong form of Newton's third law of motion for a continuum. This means that for our formulation we consider only central forces; we do not consider couple stresses within the continuum.[16]

Similarly to the comment on page 55 about the relation between particle and continuum mechanics, we must be aware of the logical weakness of the analogy that invokes Newton's laws of motion: as a pedagogical concession, we can use particle mechanics to explain certain aspects of continuum mechanics but we cannot use it to justify them logically; a logical formulation calls for postulating torques that include couples.[17]

2.7.2. Integral equation

Considering only the central forces acting within the continuum, we can state the balance of angular momentum as

$$\frac{\mathrm{d}}{\mathrm{d}t}\iiint\limits_{V(t)}\left(\mathbf{x}\times\rho\frac{\mathrm{d}\mathbf{u}}{\mathrm{d}t}\right)\mathrm{d}V = \iint\limits_{S(t)}(\mathbf{x}\times\mathbf{T})\,\mathrm{d}S + \iiint\limits_{V(t)}(\mathbf{x}\times\mathbf{f})\,\mathrm{d}V , \qquad (2.7.2)$$

where $V(t)$ is a volume that moves while always containing the same portion of the continuum and $S(t)$ is the surface enclosing this volume.[18] In other words, the rate of change of the angular momentum of an element within the continuum is equal to the sum of torques acting upon this element. The integrand on the left-hand side is the angular momentum,

---

[16] Readers interested in a description of couple stresses might refer to Malvern, L.E., (1969) Introduction to the mechanics of a continuous medium: Prentice-Hall, pp. 215–220.

[17] Readers interested in a description of torques that contain couples might refer to Truesdell, C., (1966) Six lectures on modern natural philosophy: Springer-Verlag, Lecture II.

[18] Readers interested in the balance of angular momentum that includes couple stresses might refer to Malvern, L.E., (1969) Introduction to the mechanics of a continuous medium: Prentice-Hall, pp. 218–219.

namely, the vector product of the distance, $\mathbf{x}$, between a reference point and the element of the continuum with the linear-momentum density, $\rho\,\mathrm{d}\mathbf{u}/\mathrm{d}t$. The first integrand on the right-hand side is the vector product of this distance and force per unit area, $\mathbf{T}$, associated with this element, while the second integrand on the right-hand side is the vector product of that distance and force per unit volume, $\mathbf{f}$, associated with this element.

This integral equation states the balance of angular momentum: The rate of change of the angular momentum of an element within the continuum is equal to the sum of torques due to the external forces acting upon this element. $\mathbf{x}\times\mathbf{T}$ and $\mathbf{x}\times\mathbf{f}$ in equation (2.7.2) are torque densities per unit area and unit volume, respectively; the forces, $\mathbf{T}$ and $\mathbf{f}$, causing the torques are external to the portion of the continuum contained in volume V. The sum of the two integrals on the right-hand side of equation (2.7.2) is analogous to the sum on the right-hand side in equation (2.7.1).

The balance of angular momentum stated in equation (2.7.2) is valid only for the strong form of Newton's third law of motion.

Note that we have arrived at the balance of angular momentum as a consequence of the conservation of mass, balance of linear momentum and the strong form of Newton's third law of motion. It is worth noting that one could adopt the balance of angular momentum as an independent postulate.

2.7.3. Symmetry of stress tensor

Considering the conservation of mass and the balance of linear momentum—together with the balance of angular momentum under the assumption of only central forces acting within the continuum—we obtain an important consequence of these laws: The stress tensor is symmetric. In deriving the equations resulting from the balance of angular momentum, we implicitly use the conservation of mass—by invoking time derivative of volume integral—and explicitly use the balance of linear momentum.

THEOREM 2.7.1. *Consider a linearized formulation of displacement and its associated quantities in a three-dimensional continuum. Let the principles of the conservation of mass and the balance of linear momentum hold. Then, the balance of angular momentum stated in expression (2.7.2) holds if*

$$\sigma_{ij}=\sigma_{ji}\,,\qquad i,j\in\{1,2,3\}\,.$$

NOTATION 2.7.2. The repeated-index summation notation is used in this proof. Any term in which an index appears twice stands for the sum of all such terms as the index assumes values 1, 2 and 3.

The time-derivative operators, $\mathrm{d}/\mathrm{d}t$, inside the volume integrals are the material time-derivative operators.

PROOF. Since ρ is a scalar, we can rewrite expression (2.7.2) as

$$\frac{\mathrm{d}}{\mathrm{d}t}\iiint\limits_{V(t)}\left[\rho\left(\mathbf{x}\times\frac{\mathrm{d}\mathbf{u}}{\mathrm{d}t}\right)\right]\mathrm{d}V=\iint\limits_{S(t)}(\mathbf{x}\times\mathbf{T})\,\mathrm{d}S+\iiint\limits_{V(t)}(\mathbf{x}\times\mathbf{f})\,\mathrm{d}V\,. \quad (2.7.3)$$

Invoking the time derivative of a moving-volume integral, which is given by expression (2.2.5) and where we let $\mathscr{A}=\mathbf{x}\times\mathrm{d}\mathbf{u}/\mathrm{d}t$, we can restate expression (2.7.3) as

$$\iiint\limits_{V(t)}\rho\left(\mathbf{x}\times\frac{\mathrm{d}^2\mathbf{u}}{\mathrm{d}t^2}\right)\mathrm{d}V=\iint\limits_{S(t)}(\mathbf{x}\times\mathbf{T})\,\mathrm{d}S+\iiint\limits_{V(t)}(\mathbf{x}\times\mathbf{f})\,\mathrm{d}V\,. \quad (2.7.4)$$

In view of the linearized formulation discussed in Section 1.3.3, we can rewrite expression (2.7.4) as

$$\iiint\limits_{V(t)}\rho\left(\mathbf{x}\times\frac{\partial^2\mathbf{u}}{\partial t^2}\right)\mathrm{d}V=\iint\limits_{S(t)}(\mathbf{x}\times\mathbf{T})\,\mathrm{d}S+\iiint\limits_{V(t)}(\mathbf{x}\times\mathbf{f})\,\mathrm{d}V\,. \quad (2.7.5)$$

Using the stress tensor given in expression (2.5.14), invoking the divergence theorem, and requiring expression (2.7.5) to be valid for an arbitrary volume that consistently contains the same portion of the continuum, we obtain the differential equation given by

$$\rho\left(\mathbf{x}\times\frac{\partial^2\mathbf{u}}{\partial t^2}\right)=\nabla\cdot\left(\mathbf{x}\times\sigma^T\right)+\mathbf{x}\times\mathbf{f}, \quad (2.7.6)$$

where ^{T} denotes transpose. Consider the first term on the right-hand side in equation (2.7.6). The expression in parentheses is a second-rank tensor whose ilth component can be written as

$$\left(\mathbf{x}\times\sigma^T\right)_{i\ell}=\varepsilon_{ijk}x_j\sigma_{\ell k}\,, \qquad i,l\in\{1,2,3\}\,,$$

where ε_{ijk} is the permutation symbol and where we reversed the subscripts of the stress tensor to consider its transpose. Taking the ith component of the divergence and using the product rule, we obtain

$$
\begin{aligned}
\left[\nabla \cdot \left(\mathbf{x} \times \sigma^T\right)\right]_i &= \frac{\partial}{\partial x_\ell}\left(\varepsilon_{ijk} x_j \sigma_{\ell k}\right) \\
&= \left(\mathbf{x} \times \nabla \cdot \sigma^T\right)_i + \varepsilon_{ijk}\delta_{j\ell}\sigma_{\ell k}, \qquad i \in \{1,2,3\}. \qquad (2.7.7)
\end{aligned}
$$

Substituting expression (2.7.7) into equation (2.7.6), we obtain

$$
\rho\left(\mathbf{x} \times \frac{\partial^2 \mathbf{u}}{\partial t^2}\right)_i = \left(\mathbf{x} \times \nabla \cdot \sigma^T\right)_i + \varepsilon_{ijk}\delta_{j\ell}\sigma_{\ell k} + (\mathbf{x} \times \mathbf{f})_i, \qquad i \in \{1,2,3\}.
$$

Using the linearity of the cross-product operator, we can rearrange the above equation and write

$$
\begin{aligned}
\left[\mathbf{x} \times \left(\rho \frac{\partial^2 \mathbf{u}}{\partial t^2}\right)\right]_i - \left(\mathbf{x} \times \nabla \cdot \sigma^T\right)_i - (\mathbf{x} \times \mathbf{f})_i &= \left[\mathbf{x} \times \left(\rho \frac{\partial^2 \mathbf{u}}{\partial t^2} - \nabla \cdot \sigma^T - \mathbf{f}\right)\right]_i \\
&= \varepsilon_{ijk}\delta_{j\ell}\sigma_{\ell k}, \qquad i \in \{1,2,3\}. \\
&\qquad (2.7.8)
\end{aligned}
$$

Now, we invoke Cauchy's equations of motion (2.6.3), which can be written as

$$
\rho \frac{\partial^2 \mathbf{u}}{\partial t^2} - \nabla \cdot \sigma^T - \mathbf{f} = 0;
$$

hence, the term in brackets in equation (2.7.8) vanishes. This implies

$$
\varepsilon_{ijk}\delta_{j\ell}\sigma_{\ell k} = 0, \qquad i \in \{1,2,3\}.
$$

Using the properties of Kronecker's delta, we can rewrite this equation as

$$
\varepsilon_{ijk}\sigma_{jk} = 0, \qquad i \in \{1,2,3\},
$$

which, in view of the properties of the permutation symbol, represents the equation given by

$$
\sigma_{jk} = \sigma_{kj},
$$

as required. □

REMARK 2.7.3. $\varepsilon_{ijk}\sigma_{jk} = 0$ is a summation of terms for a given i. For instance, for $i = 1$, the summation can be written as

$$
\sum_{j=1}^{3}\sum_{k=1}^{3} \varepsilon_{1jk}\sigma_{jk} = 0.
$$

By the properties of the permutation symbol, ε_{ijk}, only two terms are nonzero; they are σ_{23} and σ_{32}. Also by the properties of the permutation symbol, these terms exhibit opposite signs. Thus we obtain

$$\sigma_{23} - \sigma_{32} = 0\,.$$

Thus, in view of the balance of angular momentum given by equation (2.7.2), we see that the stress tensor is symmetric. Herein, this symmetry is established without explicitly assuming the equilibrium of an element within the continuum; this assumption is explicitly invoked in Exercise 2.8. We conclude that the symmetry of the stress tensor is a general property for any continuum as long as we impose the strong form of Newton's third law of motion; otherwise, the stress tensor is not symmetric.

Herein, due to its symmetry, the stress tensor has only six independent components and, in view of Section 2.5.2, these components are sufficient to determine the state of stress at any given point within a continuum. Thus, the constraints imposed by the balance of angular momentum reduced the number of unknowns.

The symmetry of the stress tensor is also known as Cauchy's second law of motion. It is one of the key tenets of classical continuum mechanics. [19]There are generalizations of the classical theory that depart from Cauchy's second law of motion; these generalizations are not discussed in this book.[20]

Now, considering the system of fundamental equations and not including the body forces, we have four equations and ten unknowns.

---

[19]Readers interested in an insightful interpretation of Cauchy's second law of motion within classical continuum mechanics might refer to Bunge, M., (1967) Foundations of physics: Springer-Verlag, pp. 153–154. Readers interested in the symmetry of the stress tensor in the context of classical field theories might refer to Landau, L.D., and Lifshitz, E.M., Course of theoretical physics: Elsevier, Vol. 7 (1986): Theory of elasticity, p. 7, and Vol. 2 (1975): The classical theory of fields, pp. 82–85.

[20]Readers interested in the generalizations of classical continuum mechanics might refer to Malvern, L.E., (1969) Introduction to the mechanics of a continuous medium: Prentice-Hall, pp. 217–220, to Rymarz, C., (1993) Mechanika ośrodkow ciągłych: PWN, pp. 109–112, to Truesdell, C., (1966) Six lectures on modern natural philosophy: Springer-Verlag, Lecture II: Polar and oriented media, and study Eringen, A.C., (1999) Microcontinuum field theories I: Foundations and solids: Springer-Verlag, and Forest, S., (2006) Milieux continus généralisés et matériaux hétérogènes: Mines Paris, as well as the books referenced therein.

2.8. Fundamental Equations

The conservation of mass, the balance of linear momentum and the balance of angular momentum are the only three fundamental principles that relate the unknowns in our system. No other balance principles furnish us with additional constraints. For instance, the balance of energy, which deals with thermodynamic processes, does not add another fundamental equation or reduce the number of unknowns since we assume that the heat generated by the deformation is negligible and does not affect the process of deformation. The balance of energy does, however, play a key role in the formulation of the constitutive equations, which are discussed in Chapters 3 and 4.

Let us summarize the fundamental equations that describe the motion within a continuum.

In view of the symmetry of the stress tensor, the system of equations formed by expressions (2.1.9), (2.6.9), (2.6.10) and (2.6.11) consists of four equations, namely, the equation of continuity,

$$\frac{\partial \rho}{\partial t}+\sum_{i=1}^{3}\frac{\partial}{\partial x_i}\left(\rho\frac{\partial u_i}{\partial t}\right)=0,$$

and the three Cauchy's equations of motion with no body forces,

$$\sum_{j=1}^{3}\frac{\partial \sigma_{ij}}{\partial x_j}=\rho\frac{\partial^2 u_i}{\partial t^2}, \qquad i\in\{1,2,3\}, \tag{2.8.1}$$

where $\sigma_{ij}=\sigma_{ji}$, with $i,j\in\{1,2,3\}$. Explicitly, these equations can be written as

$$\frac{\partial \rho}{\partial t}+\frac{\partial}{\partial x_1}\left(\rho\frac{\partial u_1}{\partial t}\right)+\frac{\partial}{\partial x_2}\left(\rho\frac{\partial u_2}{\partial t}\right)+\frac{\partial}{\partial x_3}\left(\rho\frac{\partial u_3}{\partial t}\right)=0, \tag{2.8.2}$$

and

$$\frac{\partial \sigma_{11}}{\partial x_1}+\frac{\partial \sigma_{12}}{\partial x_2}+\frac{\partial \sigma_{13}}{\partial x_3}-\rho\frac{\partial^2 u_1}{\partial t^2}=0, \tag{2.8.3}$$

$$\frac{\partial \sigma_{12}}{\partial x_1}+\frac{\partial \sigma_{22}}{\partial x_2}+\frac{\partial \sigma_{23}}{\partial x_3}-\rho\frac{\partial^2 u_2}{\partial t^2}=0, \tag{2.8.4}$$

$$\frac{\partial \sigma_{13}}{\partial x_1}+\frac{\partial \sigma_{23}}{\partial x_2}+\frac{\partial \sigma_{33}}{\partial x_3}-\rho\frac{\partial^2 u_3}{\partial t^2}=0. \tag{2.8.5}$$

The resulting system of four equations contains ten unknowns; namely, ρ , u_1 , u_2 , u_3 , σ_{11} , σ_{12} , σ_{13} , σ_{22} , σ_{23} and σ_{33} . This system of equations is underdetermined; there are not enough equations to uniquely determine the behaviour of the continuum. To render the system determined, we turn to constitutive equations, which we discuss in Chapter 3.

To motivate further the formulation of constitutive equations, let us examine again equations (2.8.1). We see that—given σ and ρ —we can obtain $\partial^2\mathbf{u}/\partial t^2$. However, given the mass density and acceleration, we can obtain only the divergence of the stress tensor and not the stress tensor itself, which would allow us to describe the internal forces. Notably, the same accelerations associated with two different materials of the same mass density might result in two different responses. Hence, the general principles do not suffice to determine internal forces.

Closing Remarks

In Chapters 1 and 2, two important entities that underly quantitative seismology are introduced: the strain and stress tensors. In Chapter 3, we introduce the elasticity tensor to form the linear relation between the stress and strain tensors: the constitutive equations of an elastic continuum. These equations complete the system containing Cauchy's equations of motion and the equation of continuity. Also, these equations allow us to associate the fundamental principles discussed in Chapter 2 with observable properties of materials. Notably, the wave equation and the eikonal equation—used in examining seismic phenomena and discussed throughout the book—are rooted in Cauchy's equations of motion and the constitutive equations for elastic continua.

Concluding Chapters 1 and 2, let us comment on a difference between formulations of the strain and stress tensors. The former, which allows us to consider deformations within continua, is formulated without invoking any entities beyond the displacement vector. The latter, which allows us to examine forces within continua, is intrinsically linked to fundamental principles. The validity of the stress tensor in considering tractions on planes with arbitrary orientations requires the balance of linear momentum. In other words, we invoke this principle to prove Cauchy's stress theorem. Also, in contrast to the strain tensor, the stress tensor appears in derivations of fundamental equations. In the derivation of Cauchy's equations of motion from the balance of linear momentum, it is used to apply the divergence theorem.

Furthermore, the term that results in the definition of the strain tensor, which is stated in expression (1.4.6), is a result of neglecting higher-order terms. On the other hand, the relation between $\mathbf{T}$ and $\mathbf{n}$, as stated in expression (2.5.15), is an intrinsic property: at a given instant and a given point within a continuum, the traction acting on a surface depends only on the surface normal, $\mathbf{T} = \mathbf{T}(\mathbf{n})$, not on, say, its curvature. It is a theorem proven by Noll a century-and-a-half after being stated as a hypothesis by Cauchy.[21]

2.9. Exercises

EXERCISE 2.1. [22] Several physical laws discussed in this book are stated as the vanishing of a definite integral, which—under certain conditions—results in the vanishing of the integrand. Using a one-dimensional case, illustrate the condition of arbitrary integration limits required for this result.

SOLUTION 2.1. Consider an integral equation given by

$$\int_A^B f(x)\,\mathrm{d}x = 0 .$$

Let $f(x)$ be a continuous function in the interval $[A,B]$ and let $f(x_0) \neq 0$, for $x_0 \in [A,B]$. Because of the continuity, $f(x) \neq 0$ also in the neighbourhood of x_0, and, hence, the integral taken over this neighbourhood does not vanish. Since we require $\int_A^B f(x)\,\mathrm{d}x = 0$, for arbitrary limits of integration, we must require that $f(x) = 0$, for all $x \in [A,B]$. Also, $f(x) = 0$ implies $\int_A^B f(x)\,\mathrm{d}x = 0$, for arbitrary limits of integration.

[21] Interested readers might refer to Noll, W., (1974) The foundations of mechanics and thermodynamics: Springer-Verlag, and therein to The foundations of classical mechanics in the light of recent advances in continuum mechanics.

[22] See also Section 2.1.2.

REMARK 2.9.1. To illustrate the essence of Exercise 2.1 and following Remark 2.1.1, note that

$$\int_0^{2\pi} \sin x \, \mathrm{d}x = 0 = \int_0^{2\pi} \cos x \, \mathrm{d}x$$

does not imply that, in general, $\sin x = \cos x$, since the limits are not arbitrary.

EXERCISE 2.2. [23]Show that the equation of continuity (2.1.8), namely,

$$\frac{\partial \rho(\mathbf{x},t)}{\partial t} + \nabla \cdot (\rho \mathbf{v}) = 0, \tag{2.9.1}$$

can be written as

$$\frac{\mathrm{D}\rho(\mathbf{x},t)}{\mathrm{D}t} + \rho \nabla \cdot \mathbf{v}(\mathbf{x},t) = 0, \tag{2.9.2}$$

where $\mathrm{D}/\mathrm{D}t$ stands for the material time-derivative operator, namely,

$$\frac{\mathrm{D}}{\mathrm{D}t} := \frac{\partial}{\partial t} + \mathbf{v}(\mathbf{x},t) \cdot \nabla \tag{2.9.3}$$

SOLUTION 2.2. Writing explicitly the divergence on the left-hand side of equation (2.9.1), we have

$$\frac{\partial \rho(\mathbf{x},t)}{\partial t} + \nabla \cdot \Big(\rho(\mathbf{x},t)\, v_1(\mathbf{x},t), \rho(\mathbf{x},t)\, v_2(\mathbf{x},t), \rho(\mathbf{x},t)\, v_3(\mathbf{x},t)\Big)$$
$$= \frac{\partial \rho}{\partial t} + \frac{\partial(\rho v_1)}{\partial x_1} + \frac{\partial(\rho v_2)}{\partial x_2} + \frac{\partial(\rho v_3)}{\partial x_3} = 0.$$

Using the product rule and gathering similar terms, we get

$$\frac{\partial \rho}{\partial t} + \rho\left(\frac{\partial v_1}{\partial x_1} + \frac{\partial v_2}{\partial x_2} + \frac{\partial v_3}{\partial x_3}\right) + [v_1, v_2, v_3] \cdot \left[\frac{\partial}{\partial x_1}, \frac{\partial}{\partial x_2}, \frac{\partial}{\partial x_3}\right] \rho = 0.$$

Recognizing the divergence of $\mathbf{v}$ and the gradient operator, we rewrite this expression as

$$\frac{\partial \rho(\mathbf{x},t)}{\partial t} + \rho \nabla \cdot \mathbf{v}(\mathbf{x},t) + \mathbf{v} \cdot \nabla \rho = \left[\frac{\partial}{\partial t} + \mathbf{v} \cdot \nabla\right] \rho(\mathbf{x},t) + \rho \nabla \cdot \mathbf{v}(\mathbf{x},t) = 0.$$

[23]See also Section 2.1.3.

Invoking expression (2.9.3), we obtain

$$\frac{\mathrm{D}\rho\left(\mathbf{x},t\right)}{\mathrm{D}t}+\rho\nabla\cdot\mathbf{v}\left(\mathbf{x},t\right)=0\,,$$

which is equation (2.9.2), as required.

EXERCISE 2.3. Using a single spatial dimension, show that equation (2.2.5) is a case of Leibniz's integral rule; namely,

$$\frac{\mathrm{d}}{\mathrm{d}t}\int\limits_{a(t)}^{b(t)}f\left(x,t\right)\mathrm{d}x=\int\limits_{a(t)}^{b(t)}\frac{\partial f\left(x,t\right)}{\partial t}\mathrm{d}x+f\left(b\left(t\right),t\right)\frac{\mathrm{d}b}{\mathrm{d}t}-f\left(a\left(t\right),t\right)\frac{\mathrm{d}a}{\mathrm{d}t}\,. \quad (2.9.4)$$

SOLUTION 2.3. Let us write equation (2.2.5) in a single spatial dimension; namely,

$$\frac{\mathrm{d}}{\mathrm{d}t}\int\limits_{a(t)}^{b(t)}\rho\left(x,t\right)\mathscr{A}\left(x,t\right)\mathrm{d}x=\int\limits_{a(t)}^{b(t)}\rho\left(x,t\right)\frac{\mathrm{D}\mathscr{A}\left(x,t\right)}{\mathrm{D}t}\mathrm{d}x\,,$$

where $\mathrm{D}/\mathrm{D}t$ stands for the material time-derivative operator. Invoking its definition (1.3.5) for a single spatial dimension, we write explicitly

$$\frac{\mathrm{d}}{\mathrm{d}t}\int\limits_{a(t)}^{b(t)}\rho\left(x,t\right)\mathscr{A}\left(x,t\right)\mathrm{d}x=\int\limits_{a(t)}^{b(t)}\rho\left(x,t\right)\left(\frac{\partial}{\partial t}+v\left(x,t\right)\frac{\partial}{\partial x}\right)\mathscr{A}\left(x,t\right)\mathrm{d}x$$

$$=\int\limits_{a(t)}^{b(t)}\left[\rho\left(x,t\right)\frac{\partial\mathscr{A}\left(x,t\right)}{\partial t}+\rho\left(x,t\right)v\left(x,t\right)\frac{\partial\mathscr{A}\left(x,t\right)}{\partial x}\right]\mathrm{d}x\,.$$

In view of expression (2.9.4) and anticipating the substitution given by $f\left(x,t\right)=\rho\left(x,t\right)\mathscr{A}\left(x,t\right)$, let us rewrite the right-hand side of the above equation as

$$\int\limits_{a(t)}^{b(t)}\left[\rho\left(x,t\right)\frac{\partial\mathscr{A}\left(x,t\right)}{\partial t}+\frac{\partial\rho\left(x,t\right)}{\partial t}\mathscr{A}\left(x,t\right)+\rho\left(x,t\right)v\left(x,t\right)\frac{\partial\mathscr{A}\left(x,t\right)}{\partial x}\right.$$

$$\left.+\frac{\partial\rho\left(x,t\right)v\left(x,t\right)}{\partial x}\mathscr{A}\left(x,t\right)-\frac{\partial\rho\left(x,t\right)}{\partial t}\mathscr{A}\left(x,t\right)-\frac{\partial\rho\left(x,t\right)v\left(x,t\right)}{\partial x}\mathscr{A}\left(x,t\right)\mathrm{d}x\right].$$

Recognizing the product rule of differentiation, we write

$$\int_{a(t)}^{b(t)} \left\{ \frac{\partial}{\partial t} \left[\rho(x,t) \mathscr{A}(x,t) \right] + \frac{\partial}{\partial x} \left[\rho(x,t) v(x,t) \mathscr{A}(x,t) \right] - \left[\frac{\partial \rho(x,t)}{\partial t} + \frac{\partial \rho(x,t) v(x,t)}{\partial x} \right] \mathscr{A}(x,t) \right\} \mathrm{d}x .$$

Recognizing that the third bracketed term is equation of continuity (2.1.8) in a single spatial dimension, we get

$$\int_{a(t)}^{b(t)} \left\{ \frac{\partial}{\partial t} \left[\rho(x,t) \mathscr{A}(x,t) \right] + \frac{\partial}{\partial x} \left[\rho(x,t) v(x,t) \mathscr{A}(x,t) \right] \right\} \mathrm{d}x ,$$

which we can write as the sum of two integrals to obtain

$$\int_{a(t)}^{b(t)} \frac{\partial}{\partial t} \left[\rho(x,t) \mathscr{A}(x,t) \right] \mathrm{d}x + \int_{a(t)}^{b(t)} \frac{\partial}{\partial x} \left[\rho(x,t) v(x,t) \mathscr{A}(x,t) \right] \mathrm{d}x$$
$$= \int_{a(t)}^{b(t)} \frac{\partial}{\partial t} \left[\rho(x,t) \mathscr{A}(x,t) \right] \mathrm{d}x + \left[\rho(x,t) v(x,t) \mathscr{A}(x,t) \right]_{x=a(t)}^{x=b(t)} .$$

Using the fact that $v(x,t) = \mathrm{d}x/\mathrm{d}t$, we write the above result as

$$\int_{a(t)}^{b(t)} \frac{\partial}{\partial t} \left[\rho(x,t) \mathscr{A}(x,t) \right] \mathrm{d}x + \rho(b(t),t) \frac{\mathrm{d}b(t)}{\mathrm{d}t} \mathscr{A}(b(t),t) - \rho(a(t),t) \frac{\mathrm{d}a(t)}{\mathrm{d}t} \mathscr{A}(a(t),t) .$$

Letting $f(x,t) = \rho(x,t) \mathscr{A}(x,t)$, we write

$$\frac{\mathrm{d}}{\mathrm{d}t} \int_{a(t)}^{b(t)} f(x,t) \, \mathrm{d}x = \int_{a(t)}^{b(t)} \frac{\partial f(x,t)}{\partial t} \mathrm{d}x + f(b(t),t) \frac{\mathrm{d}b(t)}{\mathrm{d}t} - f(a(t),t) \frac{\mathrm{d}a(t)}{\mathrm{d}t} ,$$

which is equation (2.9.4), as required.

EXERCISE 2.4. Provide a geometrical interpretation of expression (2.9.4); namely,

$$\frac{\mathrm{d}}{\mathrm{d}t}\int\limits_{a(t)}^{b(t)} f(x,t)\,\mathrm{d}x = \int\limits_{a(t)}^{b(t)} \frac{\partial f(x,t)}{\partial t}\mathrm{d}x + f(b(t),t)\frac{\mathrm{d}b}{\mathrm{d}t} - f(a(t),t)\frac{\mathrm{d}a}{\mathrm{d}t}. \quad (2.9.5)$$

SOLUTION 2.4. Consider the area under the curve corresponding to f between points a and b. The integral on the right-hand side is the temporal rate of change of the area between given points a and b due to the change of shape of f as a function of time between these two points. The other two terms describe the temporal rate of change due to changes in location of these points.

Consider point b. At this point, the value of the function is $f(b(t),t)$. An infinitesimal area of a rectangle due to the time increment, Δt, is

$$f(b(t),t)\,(b(t+\Delta t)-b(t)).$$

which, in view of the definition of derivative, we write as

$$f(b(t),t)\,\frac{b(t+\Delta t)-b(t)}{\Delta t}\Delta t.$$

If Δt tends to zero, we rewrite this expression as

$$f(b(t),t)\,\frac{\mathrm{d}b}{\mathrm{d}t}\Delta t,$$

which is the second term on the right-hand side of expression (2.9.5) multiplied by Δt. Similarly, $f(a(t),t)\,\mathrm{d}a/\mathrm{d}t$ is the rate of change at a. If $\mathrm{d}b/\mathrm{d}t$ is positive the area is increasing. If $\mathrm{d}a/\mathrm{d}t$ is positive the area is decreasing. Hence, we have the negative sign in front of the last term on the right-hand side of expression (2.9.5).

EXERCISE 2.5. [24]Using the stress tensor, prove the particular case of the following theorem.

THEOREM 2.9.2. *If an mth-rank tensor is linearly related to an nth-rank tensor through a quantity that possesses $n+m$ indices, then this quantity is an $(n+m)$th-rank tensor.*

[24] See also Sections 2.5.2 and 5.2.2.

SOLUTION 2.5. Consider the stress tensor that relates two vectors, namely, the traction and the unit normal vector. Thus, two first-rank tensors are linearly related by a second-rank tensor.

NOTATION 2.9.3. The repeated-index summation notation is used in this proof. Any term in which an index appears twice stands for the sum of all such terms as the index assumes values 1, 2 and 3.

PROOF. The relationship between the components of the traction, $\mathbf{T}$, in two coordinate systems can be stated as

$$\hat{T}_i = a_{ik} T_k \,, \qquad i \in \{1,2,3\} \,,$$

where a_{ij} are the entries of the transformation matrix. Also, the components of the traction, $\mathbf{T}$, are related to the components of the normal vector, $\mathbf{n}$, by the quantity σ, as

$$T_k = \sigma_{kj} n_j \,, \qquad k \in \{1,2,3\} \,.$$

Combining both expressions, we can write

$$\hat{T}_i = a_{ik} \sigma_{kj} n_j \,, \qquad i \in \{1,2,3\} \,.$$

Since $\mathbf{n}$ is a vector, it obeys the inverse transformation laws, namely,

$$n_j = a_{mj} \hat{n}_m \,, \qquad j \in \{1,2,3\} \,.$$

Thus, we can write

$$\hat{T}_i = a_{ik} \sigma_{kj} a_{mj} \hat{n}_m \,, \qquad i \in \{1,2,3\} \,.$$

Since the relationship between the components of the traction, $\mathbf{T}$, and the components of the normal vector, $\mathbf{n}$, are valid for all coordinate systems, we can formally write

$$\hat{T}_i = \hat{\sigma}_{im} \hat{n}_m \,, \qquad i \in \{1,2,3\} \,.$$

Subtracting the latter equation from the former one, we obtain

$$\left(a_{ik} \sigma_{kj} a_{mj} - \hat{\sigma}_{im} \right) \hat{n}_m = 0 \,, \qquad i \in \{1,2,3\} \,.$$

Since the result must hold for any orientation of vector $\mathbf{n}$, as required by the physical argument discussed in this chapter, we get

$$a_{ik}\sigma_{kj}a_{mj} - \hat{\sigma}_{im} = 0, \qquad i,m \in \{1,2,3\},$$

and we can restate it as

$$\hat{\sigma}_{im} = a_{ik}a_{mj}\sigma_{kj}, \qquad i,m \in \{1,2,3\}. \tag{2.9.6}$$

The last expression shows that σ obeys standard transformation rules for a second-rank tensor. Consequently, σ, which linearly relates two vectors, is a second-rank tensor. □

REMARK 2.9.4. The quotient rule, stated in this theorem, is also exemplified by the stress-strain equations (3.2.1), where two second-rank tensors are linearly related by a fourth-rank elasticity tensor, namely,

$$\sigma_{ij} = c_{ijk\ell}\varepsilon_{k\ell}, \qquad i,j \in \{1,2,3\},$$

with $\varepsilon_{k\ell}$ denoting the strain tensor, and where the repeated index assumes values 1, 2 and 3.

EXERCISE 2.6. Using expressions (2.5.1) and (2.5.13), obtain the components of the traction vector acting on the plane whose normal is parallel to the x_1-axis. Compare the results to expression (2.5.14).

SOLUTION 2.6. Following expression (2.5.1), we can immediately write the components of the traction vectors acting on the plane whose normal is parallel to the x_1-axis as $\left[T_1^{(\mathbf{e}_1)}, T_2^{(\mathbf{e}_1)}, T_3^{(\mathbf{e}_1)}\right]$. In view of definition (2.5.2), we can rewrite these components as $[\sigma_{11}, \sigma_{12}, \sigma_{13}]$. Using equation (2.5.13), the traction vector acting on the plane whose unit normal is parallel to the x_1-axis is given by

$$\begin{bmatrix} T_1^{(\mathbf{e}_1)} & T_1^{(\mathbf{e}_2)} & T_1^{(\mathbf{e}_3)} \\ T_2^{(\mathbf{e}_1)} & T_2^{(\mathbf{e}_2)} & T_2^{(\mathbf{e}_3)} \\ T_3^{(\mathbf{e}_1)} & T_3^{(\mathbf{e}_2)} & T_3^{(\mathbf{e}_3)} \end{bmatrix} \begin{bmatrix} 1 \\ 0 \\ 0 \end{bmatrix} = \begin{bmatrix} T_1^{(\mathbf{e}_1)} \\ T_2^{(\mathbf{e}_1)} \\ T_3^{(\mathbf{e}_1)} \end{bmatrix}.$$

In view of definition (2.5.2), we can rewrite these components as

$$\begin{bmatrix} \sigma_{11} \\ \sigma_{12} \\ \sigma_{13} \end{bmatrix},$$

as expected from the property stated in expression (2.5.14).

EXERCISE 2.7. [25]Find the physical SI units of equations of motion (2.6.3). Show that these units are consistent for all terms involved.

SOLUTION 2.7. Consider equations (2.6.3); namely,

$$\sum_{j=1}^{3} \frac{\partial \sigma_{ij}}{\partial x_j} + f_i = \rho \frac{\partial^2 u_i}{\partial t^2}, \qquad i \in \{1,2,3\}.$$

Following the definition of stress as force per unit area, the units of stress tensor are $\left[N/m^2\right]$. Consequently, the units of the first term of the left-hand side are $\left[N/m^3\right]$. In view of f_i being the components of force per unit volume, the units are also $\left[N/m^3\right]$. On the right-hand side, the units of mass density are $\left[kg/m^3\right]$, while the units of acceleration are $\left[m/s^2\right]$, resulting in $\left[kg/\left(m^2 s^2\right)\right]$. Since $[N] = \left[kg\,m/s^2\right]$, the units of the right-hand side are also $\left[N/m^3\right]$, as expected. Thus, the physical units of equations (2.6.3) are $\left[N/m^3\right]$.

EXERCISE 2.8. [26]Using Figure 2.9.1, prove the following theorem.

THEOREM 2.9.5. *If the element within a continuum is not allowed to rotate, then it follows that the stress tensor is symmetric; namely,* $\sigma_{ij} = \sigma_{ji}$, *where* $i, j \in \{1,2,3\}$.

SOLUTION 2.8. Let us proceed with the following proof.

PROOF. [27]Consider a rectangular box that is an element within a continuum. Let the volume of this box, whose edges are parallel to the coordinate axes, be

$$\Delta V = \Delta x_1 \Delta x_2 \Delta x_3 .$$

[25] See also Section 2.6.1.

[26] See also Section 2.7.3.

[27] Readers interested in more details of the argument used herein as well as the consequences of this symmetry might refer to Malvern, L.E., (1969) Introduction to the mechanics of a continuous medium: Prentice-Hall, pp. 77–79.

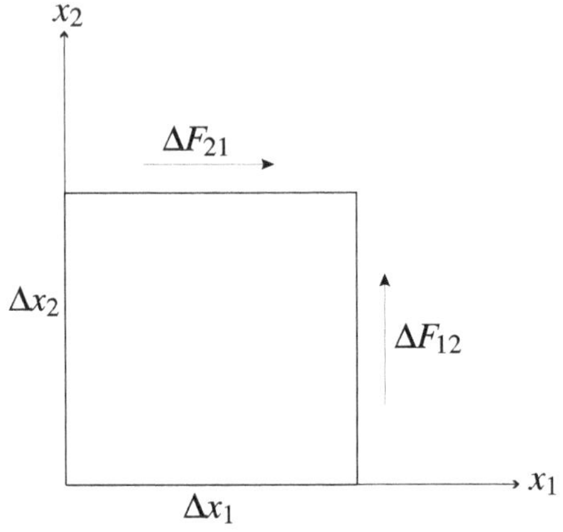

Fig. 2.9.1 Angular momentum: An x_1x_2-cross-section of a rectangular box, $\Delta x_1 \Delta x_2 \Delta x_3$, with the moment-producing forces, ΔF_{12} and ΔF_{21}. The directions of the two forces are perpendicular to one another and their magnitudes are equal. In the x_1x_2-plane, the box remains in equilibrium.

We require that this element of volume does not rotate within the continuum; in other words, we assume this element to be in equilibrium. This requirement implies that the sum of moments acting on this box must be zero. The sum of moments about the x_3-axis is zero if

$$\Delta F_{12} \Delta x_1 = \Delta F_{21} \Delta x_2 . \tag{2.9.7}$$

Using formulations of traction and the stress-tensor components, we can write $\sigma_{1j} = \Delta F_{1j} / \Delta S_1$. Thus, we have

$$\Delta F_{12} = \sigma_{12} \Delta S_1 = \sigma_{12} \Delta x_2 \Delta x_3 , \tag{2.9.8}$$

and

$$\Delta F_{21} = \sigma_{21} \Delta S_2 = \sigma_{21} \Delta x_1 \Delta x_3 . \tag{2.9.9}$$

Inserting expressions (2.9.8) and (2.9.9) into equation (2.9.7), we obtain

$$\sigma_{12} \Delta x_1 \Delta x_2 \Delta x_3 = \sigma_{21} \Delta x_1 \Delta x_2 \Delta x_3 ,$$

which implies

$$\sigma_{12} = \sigma_{21} .$$

Hence, together with the equality of the sum of moments about the x_1-axis and the x_2-axis, we can write

$$\sigma_{ij} = \sigma_{ji}, \qquad i, j \in \{1, 2, 3\},$$

as required. □

CHAPTER 3

Stress-Strain Equations

> ... there is a conjecture that two sets of small motions may be superimposed without interfering with each other in a nonlinear fashion. Another conjecture is that the seismic motions set up by some physical source should be uniquely determined by the combined properties of that source and of the medium of wave propagation. These conjectures, and many others that are generally assumed by seismologists to be true, are properties of infinitesimal motion in classical continuum mechanics for an elastic medium with a linear stress-strain relation;
>
> *Keiiti Aki and Paul G. Richards (1980)*
> *Quantitative seismology: Theory and methods*

Preliminary Remarks

The equations resulting from the fundamental principles discussed in Chapter 2 are valid for any continuum irrespective of its constitution. In other words, they do not explicitly account for distinct properties of a particular material. Also, these equations constitute a system of differential equations that contains more unknowns than equations.

To consider the properties of a particular material and to formulate a determined system of equations that describes the propagation of deformations within that material, we turn our attention to empirical relations that are expressed as constitutive equations, and are based on observations of physical materials. An elastic continuum is defined by the constitutive equations that—within a satisfactory agreement with experiments—state that, for a certain type of material, forces are linearly related to small deformations. Such materials are commonly referred to as elastic.

We begin this chapter with a brief discussion of the rudiments of constitutive equations. Subsequently, we formulate the linear stress-strain equations, which underlie the theory of elasticity used in this book. We express these equations in both tensor and matrix forms. To gain a perspective on the concept of elasticity, we conclude this chapter with a section on anelasticity.

3.1. Rudiments of Constitutive Equations

In continuum mechanics, a constitutive equation is a relation between the stress tensor and another quantity that is an important analogy in quantifying the behaviour of a given material; this quantity could be, for instance, deformation or the rate of deformation.

Constitutive equations also appear beyond continuum mechanics. In electromagnetism, for instance, they relate the displacement field to the electric field and magnetizing field to the magnetic field.

Even though, in general, a constitutive equation is an empirical relation, which does not stem from fundamental principles, it cannot contradict such principles, which include the conservation of mass and balance of linear momentum, in continuum mechanics, and conservation of electric charge, in electromagnetism.

In continuum mechanics, the key purposes of a constitutive equation are the determination of stresses within the continuum and, hence, a representation of a variety of materials by examining different responses to loads. Furthermore, we need these equations to verify the theory of continuum mechanics. In other words, for an empirical validation of the theory, we have to enrich the general theory with constitutive equations that allow us to predict experimental results. In the context of this book, the two quantities related by the constitutive equation are the stress and strain tensors.

[1]There are three principles that constitutive equations must obey. First, the principle of determinism requires that stresses within a body be determined uniquely by the history of its deformation. Secondly, the principle

[1] Readers interested in more detail and elegant exposition might refer to Truesdell, C., (1966) Six lectures on modern natural philosophy: Springer-Verlag, pp. 3–6.

of local action requires that the deformation outside an arbitrarily small neighbourhood of a point may be disregarded in determining stresses at that point. This means that there is no need to invoke action-at-a-distance between stress and strain or to consider a delay due to spatial separation, which would require relativistic continuum mechanics.

These two principles are satisfied trivially in the theory of linear elasticity discussed in this book, where a stress at a given instant does not depend on the history of deformation, and only tractional forces are involved.

Thirdly, the principle of objectivity, also known as the principle of material frame indifference, requires that any two observers find the same stress within a given body. In other words, constitutive equations must be invariant under changes of a reference frame that preserves the essential structure of the manifold of physical experience discussed on page 12; the principle of objectivity limits our study to nonrelativistic continuum mechanics. [2]For further insight, let us quote an example stated by Truesdell in his "Six lectures on modern natural philosophy" that is particularly applicable to the theory of elasticity.

> Take a spring, and on one end hang a weight of one pound. The spring lengthens, say by one inch. Now, lay a spring on a horizontal table, fastening one end to the center, and leaving the weight attached to the other end. Spin the table, and adjust the angular speed until the spring again stretches exactly one inch. [...] For an observer standing on the floor as well as for an observer seated upon that table [...] one inch of extension corresponds to one poundal of force.[3]

In contrast to the general principles, the constitutive equations can be, and often are, contradictory among each other. Their limited validity results in specialized theories, as exemplified by elasticity and by fluid mechanics.

---

[2] Readers interested in a rigorous formulation of the principle of material frame indifference might refer to Marsden, J.E., and Hughes, T.J.R., (1983/1994) Mathematical foundations of elasticity: Dover, p. 8 and pp. 189–199.

[3] Readers interested in the equality of inertial mass and gravitational mass, which is invoked in the above example, might refer to Bunge, M., (1967) Foundations of physics: Springer-Verlag, pp. 207–210.

Also, let us cite the comment of Truesdell that completes the above quote.

> What has been assumed, tacitly, is that the elastic law or constitutive equation of the spring is invariant under rotation. [...] For a person lacking this belief, the experiment would measure nothing at all. (As is usual in "fundamental" experiments, the main point at issue must be conceded before the experiment is started.)

3.2. Formulation of Stress-Strain Equations: Hookean Solid

3.2.1. Introductory comments

Perhaps the best-known constitutive equation is based on Hooke's law of elasticity discovered by Robert Hooke in the middle of the seventeenth century.[4] This law furnishes us with the physical justification for the mathematical theory of linear elasticity.

Ut tensio sic vis—"as the extension, so the force" is a famous statement from Hooke's work of 1676. He described it in more detail by writing that

> the power of any spring is in the same proportion with the tension thereof: that is, if one power stretch or bend it in one space, two will bend in two, three will bend in three, and so forward. And this is the rule or law of Nature, upon which all manner of restituent or springing motion doth proceed.

In an earlier paper, "De potentia restitutiva", Hooke published the results of his experiments with elastic materials and stated that

> it is very evident that the rule or law of Nature in every springing body is, that the force or power thereof to restore itself to its natural position is always proportional to the distance or space it is removed therefrom ...

In the modern terminology of continuum mechanics and in view of Chapters 1 and 2, the linearity of Hooke's law can be stated in the following manner.

---

[4] Readers interested in a historical comments about Hooke's law might refer to Love, A.E.H., (1892/1944) A treatise on the mathematical theory of elasticity: Dover, p. 2.

> At any point of a continuum, each component of the stress tensor is a linear function of all the components of the strain tensor.

This statement is used to formulate stress-strain equations, which are introduced in this chapter. The restoring force is discussed in Chapter 4.

The mathematical justification of Hooke's law and the resulting linear theory stems from the assumption of small deformations since every function is linear on the infinitesimal level. We also assume that the stress tensor is zero for the undeformed state, which corresponds to the strain tensor being zero; there are no constant terms in the stress-strain equations.

3.2.2. Tensor form

Stress-strain equations

At a given point of a three-dimensional continuum, Hooke's law—which relates linearly each stress-tensor component, σ_{ij}, to all strain-tensor components, $\varepsilon_{k\ell}$—can be written as

$$\sigma_{ij} = \sum_{k=1}^{3} \sum_{\ell=1}^{3} c_{ijk\ell}\varepsilon_{k\ell}, \qquad i, j \in \{1, 2, 3\}, \tag{3.2.1}$$

where $c_{ijk\ell}$ are the components of a tensor, known as the elasticity tensor.[5] Equations (3.2.1) are the constitutive equations of a linearly elastic continuum. Since the units of the stress tensor are N/m^2, and the strain tensor is dimensionless, the units of the elasticity tensor are N/m^2.

The essence of the above formulation is the assumption of a relationship given by $\sigma_{ij} = \sigma_{ij}(\varepsilon_{ij})$; in other words, stress is a function of strain only, and vice versa: strain is a function of stress only. If we develop this function in a power series, set the constant term and all the nonlinear terms to zero, we obtain expression (3.2.1). As stated above, the justification for setting the constant term to zero is the assumption of no initial stress, and the justification of setting the nonlinear terms to zero is the assumption of

[5] Elasticity tensor is also commonly referred to as the stiffness tensor. Our nomenclature is consistent with Marsden, J.E., and Hughes, T.J.R., (1983/1994) Mathematical foundations of elasticity: Dover, pp. 9 – 10, and with Marsden, J.E., and Ratiu, T.S., (1999) Introduction to mechanics and symmetry: A basic exposition of classical mechanical systems (2*nd* edition): Springer-Verlag, p. 113.

small gradients of displacement that appear in the definition of strain given by expression (1.4.6). Such a formulation of stress-strain equations (3.2.1) is referred to as Cauchy's approach. We could also formulate these equations in a manner stated on page 116, and referred to as Green's approach.

Tensor $c_{ijk\ell}$ relates two second-rank tensors. Hence, in view of tensor algebra, the elasticity tensor must be a fourth-rank tensor. Consequently, in a three-dimensional continuum, it has $3^4 = 81$ components. However, in view of the symmetries of the stress and strain tensors, the number of independent components is thirty-six, as shown below. The physical meaning of the rank of a tensor can be viewed as the number of directions necessary to measure the corresponding property. Tensor $c_{ijk\ell}$ relates four directions necessary to measure elasticity: two directions of the stress tensor and two directions of the strain tensor, which are the direction of the force together with the normal to the face on which the force is acting and the direction of the displacement vector together with the orientation of the measurement axis, respectively.

The elasticity tensor describes properties of the continuum; hence, it is a property tensor in the study of elasticity in the same manner as a dielectric tensor is a property tensor in the study of electricity. The stress and strain tensors are not property tensors since we can apply various forces to, or cause different deformations of, a given material to study its elasticity. In the same manner, we can apply various electric fields and displacements to study a material's ability to store electric charge.[6]

Consequences of stress-tensor and strain-tensor symmetries

We assume that—at every point of a continuum—the stress tensor is symmetric, namely, $\sigma_{ij} = \sigma_{ji}$, as discussed in Section 2.7, in the context of the balance of angular momentum; consequently, it has six independent components. The strain tensor, by its definition stated in expression (1.4.6), is also symmetric, namely, $\varepsilon_{k\ell} = \varepsilon_{\ell k}$; thus, it has six independent components. To investigate the number of independent components of $c_{ijk\ell}$, which relates the stress and strain tensors, let us consider stress-strain equations (3.2.1).

[6] Readers interested in analogies among mechanical, electrical and thermal variables might refer to Heckmann's diagrams, e.g., Newnham, R.E., (2005) Properties of materials: Anisotropy, symmetry, structure: Oxford University Press, pp. 1–3.

Let us examine consequences of the symmetry of the stress tensor. In view of this symmetry, we can write stress-strain equations (3.2.1) as

$$\sum_{k=1}^{3}\sum_{\ell=1}^{3} c_{ijk\ell}\varepsilon_{k\ell} = \sigma_{ij} = \sigma_{ji} = \sum_{k=1}^{3}\sum_{\ell=1}^{3} c_{jik\ell}\varepsilon_{k\ell}, \qquad i,j \in \{1,2,3\}. \quad (3.2.2)$$

In other words, each double-summation term gives the same value of the stress-tensor, which means that the double-summation terms are equal to one another; their difference is zero. Subtracting the first double-summation term from the second one, we can write

$$\sum_{k=1}^{3}\sum_{\ell=1}^{3} c_{jik\ell}\varepsilon_{k\ell} - \sum_{k=1}^{3}\sum_{\ell=1}^{3} c_{ijk\ell}\varepsilon_{k\ell} = \sum_{k=1}^{3}\sum_{\ell=1}^{3} \left(c_{ijk\ell} - c_{jik\ell}\right)\varepsilon_{k\ell} = 0,$$

where $i,j \in \{1,2,3\}$. For this equation to be satisfied for all strain-tensor components, we require

$$c_{ijk\ell} = c_{jik\ell}, \qquad i,j,k,\ell \in \{1,2,3\}. \quad (3.2.3)$$

Hence, due to the symmetry of the stress tensor, the elasticity tensor is invariant under permutations in the first pair of subscripts.

Let us examine consequences of the symmetry of the strain tensor. The order of k and ℓ has no effect on stress-strain equations (3.2.1) since they are the summation indices. Hence, we can write

$$\sum_{k=1}^{3}\sum_{\ell=1}^{3} c_{ijk\ell}\varepsilon_{k\ell} = \sum_{k=1}^{3}\sum_{\ell=1}^{3} c_{ij\ell k}\varepsilon_{\ell k}, \qquad i,j \in \{1,2,3\}.$$

In view of the symmetry of the strain tensor, we can rewrite it as

$$\sum_{k=1}^{3}\sum_{\ell=1}^{3} c_{ijk\ell}\varepsilon_{k\ell} = \sum_{k=1}^{3}\sum_{\ell=1}^{3} c_{ij\ell k}\varepsilon_{k\ell}, \qquad i,j \in \{1,2,3\},$$

which, we can also state as

$$\sum_{k=1}^{3}\sum_{\ell=1}^{3} c_{ijk\ell}\varepsilon_{k\ell} - \sum_{k=1}^{3}\sum_{\ell=1}^{3} c_{ij\ell k}\varepsilon_{k\ell} = \sum_{k=1}^{3}\sum_{\ell=1}^{3} \left(c_{ijk\ell} - c_{ij\ell k}\right)\varepsilon_{k\ell} = 0,$$

where $i,j \in \{1,2,3\}$. For this equation to be satisfied for all strain-tensor components, we require

$$c_{ijk\ell} = c_{ij\ell k}, \qquad i,j,k,\ell \in \{1,2,3\}. \quad (3.2.4)$$

Hence, due to the symmetry of the strain tensor, the elasticity tensor is invariant under permutations in the second pair of subscripts.

In view of equalities (3.2.3) and (3.2.4), the number of independent components of the elasticity tensor is thirty-six.

Anisotropy and inhomogeneity

In this book, we study continua that are described by stress-strain equations (3.2.1).[7] To understand the description of a continuum that is provided by $c_{ijk\ell}$, consider these stress-strain equations. In view of Chapters 1 and 2, tensors σ_{ij} and $\varepsilon_{k\ell}$ are direction-dependent. Hence, the values of $c_{ijk\ell}$ are intrinsically direction-dependent. Consequently, at a given point of a continuum, these values determine the anisotropic properties of the continuum at this point. Furthermore, if the values of $c_{ijk\ell}$ depend on position, $\mathbf{x}$, the continuum is inhomogeneous. This is explicitly used in stress-strain equations (7.1.2).

Note the following distinction between the continuum model and physical materials. Studying anisotropy and inhomogeneity in physical materials, we observe that anisotropy is rooted in the inhomogeneity of the material. Anisotropy may result from the inhomogeneity exhibited by an atomic structure or crystal lattice. In a seismological context, anisotropy results from the arrangement of grains, layers or fractures in the materials through which seismic waves propagate. Hence, physically, at some scale, anisotropy is linked to inhomogeneity. In the mathematical context of continuum mechanics, however, anisotropy and inhomogeneity are two distinct properties.

3.2.3. Matrix form

Introductory comments

In view of symmetries of the stress and strain tensors, constitutive equations (3.2.1) can be conveniently written in a matrix form containing six

[7] Readers interested in detailed formulations of elasticity, hyperelasticity, linear elasticity, etc., might refer to Marsden, J.E., and Hughes, T.J.R., (1983/1994) Mathematical foundations of elasticity: Dover.

independent equations. This form, which allows us to express elasticity tensor (3.2.1) as an elasticity matrix, is often used in this book.

Note that, although, in some particular cases, the components of a tensor can be written as the entries of a matrix, the matrices and the tensors are distinct mathematical entities.

Elasticity matrix

The thirty-six components of the elasticity tensor that are independent of each other in spite of the symmetries of stress and strain can be written as entries C_{mn} of a 6×6 elasticity matrix, which relates each independent stress-tensor component to the six independent strain-tensor components. To construct this matrix, in view of symmetries (3.2.3) and (3.2.4), it is enough to consider the pairs of (i,j) and (k,ℓ) for $i \leqslant j$ and $k \leqslant \ell$, respectively.

Consider such pairs (i,j), where $i,j \in \{1,2,3\}$. Let us arrange them in the order given by

$$(1,1),(2,2),(3,3),(2,3),(1,3),(1,2).$$

Now, we can replace each pair by a single number m that gives the position of the pair in this list; thus, $m \in \{1,\ldots,6\}$. In other words, we make the following replacement $(i,j) \to m$:

$$(1,1) \to 1, (2,2) \to 2, (3,3) \to 3,$$

$$(2,3) \to 4, (1,3) \to 5, (1,2) \to 6.$$

We can concisely write this replacement as

$$\begin{cases} m = i & \text{if } i = j \\ m = 9-(i+j) & \text{if } i \neq j \end{cases}, \qquad i,j \in \{1,2,3\}. \tag{3.2.5}$$

Considering the analogous pairs (k,ℓ), where $k,\ell \in \{1,2,3\}$, we see that identical replacements can be made. Consequently, we can replace $c_{ijk\ell}$, where $i,j,k,\ell \in \{1,2,3\}$, by C_{mn}, where $m,n \in \{1,\ldots,6\}$, to obtain the elasticity matrix given by

$$\mathbf{C} = \begin{bmatrix} C_{11} & C_{12} & C_{13} & C_{14} & C_{15} & C_{16} \\ C_{21} & C_{22} & C_{23} & C_{24} & C_{25} & C_{26} \\ C_{31} & C_{32} & C_{33} & C_{34} & C_{35} & C_{36} \\ C_{41} & C_{42} & C_{43} & C_{44} & C_{45} & C_{46} \\ C_{51} & C_{52} & C_{53} & C_{54} & C_{55} & C_{56} \\ C_{61} & C_{62} & C_{63} & C_{64} & C_{65} & C_{66} \end{bmatrix}, \tag{3.2.6}$$

which we use below to write the matrix form of the stress-strain equations.

Stress-strain equations

Using matrix (3.2.6) and in view of the symmetries of the stress and strain tensors, equations (3.2.1) can be restated as

$$\begin{bmatrix} \sigma_{11} \\ \sigma_{22} \\ \sigma_{33} \\ \sigma_{23} \\ \sigma_{13} \\ \sigma_{12} \end{bmatrix} = \begin{bmatrix} C_{11} & C_{12} & C_{13} & C_{14} & C_{15} & C_{16} \\ C_{21} & C_{22} & C_{23} & C_{24} & C_{25} & C_{26} \\ C_{31} & C_{32} & C_{33} & C_{34} & C_{35} & C_{36} \\ C_{41} & C_{42} & C_{43} & C_{44} & C_{45} & C_{46} \\ C_{51} & C_{52} & C_{53} & C_{54} & C_{55} & C_{56} \\ C_{61} & C_{62} & C_{63} & C_{64} & C_{65} & C_{66} \end{bmatrix} \begin{bmatrix} \varepsilon_{11} \\ \varepsilon_{22} \\ \varepsilon_{33} \\ 2\varepsilon_{23} \\ 2\varepsilon_{13} \\ 2\varepsilon_{12} \end{bmatrix}. \tag{3.2.7}$$

The factors of 2 result from the symmetry of the strain tensor: for each $k \neq \ell$, the corresponding strain-tensor component appears twice in the summation on the right-hand side of equations (3.2.1)—as $\varepsilon_{k\ell}$ and as $\varepsilon_{\ell k}$—as shown in Exercise 3.1. Also, due to the symmetry of the stress tensor, it is sufficient to consider only six of the nine equations stated in expression (3.2.1). Using formula (3.2.5), we could replace the pairs of subscripts for $\varepsilon_{k\ell}$ and σ_{ij} by single subscripts. However, we keep the original notation of these components to ensure that their physical meaning, discussed in Sections 1.4.3 and 2.5.1, remains clear. In a concise notation, we write stress-strain equations (3.2.7) as

$$\underline{\sigma} = \mathbf{C}\underline{\varepsilon}, \tag{3.2.8}$$

where $\underline{\sigma}$ and $\underline{\varepsilon}$ are six-entry single-column matrices, composed of the stress-tensor and the strain-tensor components, respectively, while $\mathbf{C}$ is a 6×6 matrix.

Writing the stress-strain equations in a matrix notation, as shown in expression (3.2.7), is a convenient way to display the stress-tensor, strain-tensor and elasticity-tensor components. We use this notation extensively in Chapter 5, even though, as discussed in Section 5.2.5, a modification of that expression might be preferable, in particular, for examining the effect of the coordinate-system rotations on the components of the elasticity tensor.

3.3. Determined System

Stress-strain equations furnish us with six additional equations and no new unknowns for the system discussed in Chapter 2. The system is no longer underdetermined.

Note that the strain-tensor components, in accordance with the definition given in expression (1.4.6), may be expressed in terms of the displacement components, u_i, where $i \in \{1,2,3\}$, which are the unknowns used in the equations of motion and the equation of continuity, as illustrated in Exercise 3.2. Thus, in a three-dimensional continuum, we have a system of ten equations for ten unknowns. These equations are the equation of continuity, given by expression (2.8.2), the three equations of motion, given by expressions (2.8.3), (2.8.4) and (2.8.5), and six constitutive equations.

Note that the consistency of this system requires the linearized theory that allows us to ignore the fact that, in principle, equations of motion (2.6.3) refer to the spatial coordinates while definition (1.4.6), which is used in formulating stress-strain equations, refers to the material coordinates. In other words, we ignore the distinction between the spatial and the material coordinates and use the equations of motion and the stress-strain equations in the same system of equations.

3.4. Anelasticity

3.4.1. Introductory comments

For the study presented in this book, we use constitutive equations (3.2.1). In other words, we study physical phenomena in the context of linear elasticity. To see this choice in the scope of other possibilities, we briefly discuss the concept of viscoelasticity.[8]

3.4.2. Viscosity: Stokesian fluid

Constitutive equations (3.2.1) state that deformation is proportional to the applied load; they define a Hookean solid. We could also state that deformation is proportional to the rate of the application of the load. With a certain abuse of notation, let us formally write

$$\sigma(t) = \eta \frac{\mathrm{d}\varepsilon(t)}{\mathrm{d}t}, \qquad (3.4.1)$$

where η is the viscosity parameter. Constitutive equation (3.4.1) defines a Stokesian fluid, named in deference to George Gabriel Stokes, an English physicist. Integrating both sides of this equation, we can rewrite it as

$$\varepsilon(t) = \frac{1}{\eta}\int \sigma(t)\,\mathrm{d}t. \qquad (3.4.2)$$

If we submit an elastic continuum, $\sigma = c\varepsilon$, to a constant load, σ_0, it deforms instantaneously to $\varepsilon = \sigma_0/c$, where c is the elasticity parameter. Thereafter, the elastic continuum remains deformed at σ_0/c as long as the load is applied, and immediately returns to its undeformed state upon the removal of the load; this behaviour is shown in Exercise 3.3.

To study the effect of the constant load on a viscous continuum, let us first write such a load as

$$\sigma(t) = \sigma_0 h(t), \qquad (3.4.3)$$

[8] Readers interested in a more detailed discussion might refer to Borcherdt, R.D., (2009) Viscoelastic waves in layered media: Cambridge University Press, Chapter 1.

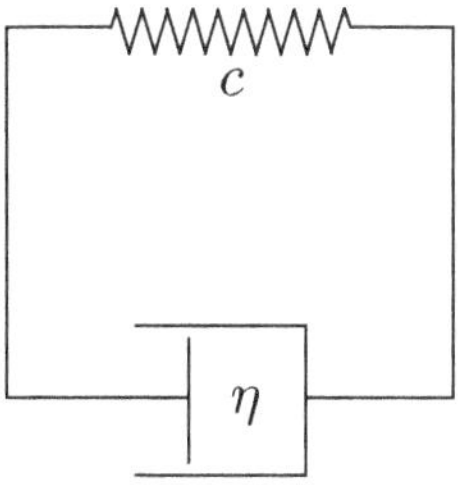

Fig. 3.4.1 A mechanical illustration of the Kelvin-Voigt model; c represents stiffness of the spring and η represents viscosity of a dashpot, whose mathematical analogues are stated in equation (3.4.6).

where h is Heaviside's step, which is defined by

$$h(t) = \begin{cases} 0, & t < 0 \\ 1, & t \geqslant 0 \end{cases} . \tag{3.4.4}$$

To subject a viscous continuum—represented by a Stokesian fluid—to the constant load, we insert expression (3.4.3) into equation (3.4.2) to get

$$\varepsilon(t) = \frac{\sigma_0}{\eta} \int h(t)\,\mathrm{d}t = \frac{\sigma_0}{\eta} \int_0^t \mathrm{d}\tau = \frac{\sigma_0}{\eta} t . \tag{3.4.5}$$

Since $(\sigma_0/\eta)t$ is a straight line whose slope is σ_0/η, this result means that the deformation increases linearly as long as the load is applied. If we remove the load at time t_1, the deformation remains at $(\sigma_0/\eta)t_1$, as shown in Exercise 3.4.

3.4.3. Viscoelasticity: Kelvin-Voigt model

To increase the accuracy of modelling physical materials, it is possible to combine elasticity and viscosity.[9] A common combination, called the Kelvin-Voigt model, consists of combining elastic and viscous elements in parallel. Mechanically, the Kelvin-Voigt model can be represented by a spring and dashpot, respectively, as shown in Figure 3.4.1.

---

[9] Readers interested in a description of several such combinations might refer to Ben-Menahem, A., and Singh, S.J., (1981/2000) Seismic waves and sources: Dover, pp. 850–865.

Before continuing our study, let us gain a perspective on the meaning of arrangements of springs and dashpots by quoting Coleman and Noll from their article on "Foundations of linear viscoelasticity".

> We feel that the physicist's confidence in the usefulness of the theory of infinitesimal viscoelasticity does not stem from the belief that the materials to which the theory is applied are really composed of microscopic networks of springs and dashpots, but comes rather from other considerations. First, there is the observation that the theory works for many real materials. But second, and perhaps more important, is the fact that the theory looks plausible because it seems to be a mathematization of little more than certain intuitive prejudices about smoothness in macroscopic phenomena.

Hence, in the spirit of continuum mechanics, we write the constitutive equation of the Kelvin-Voigt model as

$$\sigma(t) = c\varepsilon(t) + \eta\frac{\mathrm{d}\varepsilon(t)}{\mathrm{d}t}. \tag{3.4.6}$$

As a consequence of a parallel structure, at each instant, the value of strain, ε, is the same for both elements. This is not be the case if we connect these two elements in series, which is known as the Maxwell model, discussed in Section 3.4.4, below. Examining equation (3.4.6), we see that it reduces to $\sigma = c\varepsilon$, if $\eta = 0$, and to $\sigma = \eta\,\mathrm{d}\varepsilon/\mathrm{d}t$, if $c = 0$. In other words, the Hookean solid and the Stokesian fluid are the two extremes of the Kelvin-Voigt model.

Let us examine the behaviour of the Kelvin-Voigt model under a load, σ. Given σ, we can solve differential equation (3.4.6) to get

$$\varepsilon(t) = \frac{\varepsilon(0)}{\exp\left(\frac{c}{\eta}t\right)} + \frac{1}{\eta}\int_0^t \frac{\sigma(\tau)}{\exp\left[\frac{c}{\eta}(t-\tau)\right]}\mathrm{d}\tau, \tag{3.4.7}$$

as shown in Exercise 3.5. We can view the first term on the right-hand side as the strain at a given instant and the second term as the temporal evolution of strain following that instant.

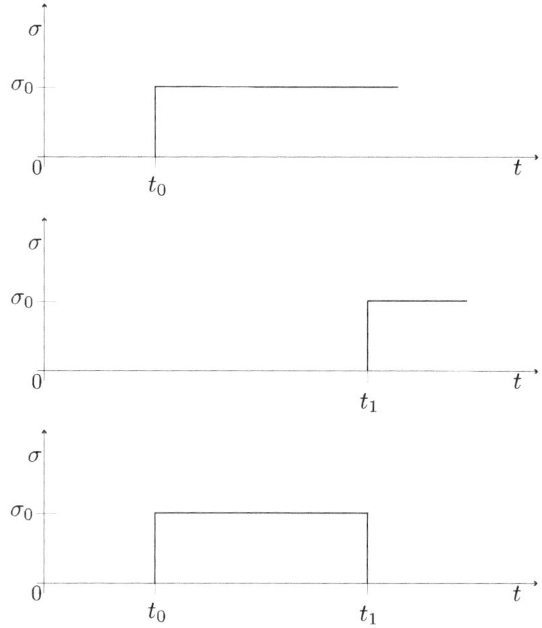

Fig. 3.4.2 Illustration of expression (3.4.8), which is a constant stress expressed as Heaviside's step of value σ_0 , and applied at t_0 and removed at t_1 .

Let us consider a constant load, σ_0 , applied at t_0 and removed at t_1 . We can express such a load as

$$\sigma(t) = \sigma_0 [h(t - t_0) - h(t - t_1)] , \tag{3.4.8}$$

where h is Heaviside's step; expression (3.4.8) is illustrated in Figure 3.4.2, where $\sigma_0 h(t - t_0)$, $\sigma_0 h(t - t_1)$ and their difference correspond to the top, middle and bottom diagrams, respectively. Let us consider the processes of deformation and recovery.

To consider deformation, we let $t_0 = 0$ and set $\varepsilon(0) = 0$ in expression (3.4.7), which means that we assume no deformation prior to $t = 0$. Following expression (3.4.8), we insert the constant value, σ_0 , for $\sigma(\tau)$, in the integrand. Considering the top diagram in Figure 3.4.2, we let the integration limits remain at 0 and t ; thus we write expression (3.4.7) as

$$\varepsilon(t) = \frac{\sigma_0}{\eta} \int_0^t \frac{1}{\exp\left[\frac{c}{\eta}(t-\tau)\right]} \mathrm{d}\tau .$$

Integrating, we obtain

$$\varepsilon(t) = \frac{\sigma_0}{c}\left[1 - \frac{1}{\exp\left(\frac{c}{\eta}t\right)}\right], \qquad t \geqslant 0, \tag{3.4.9}$$

which describes the process of deformation.

To consider the process of recovery, we write expression (3.4.7) as

$$\varepsilon(t) = \varepsilon(t_1) + \frac{\sigma_0}{\eta} \int_{t_1}^t \frac{1}{\exp\left[\frac{c}{\eta}(t-\tau)\right]} \mathrm{d}\tau ,$$

where the first term on the right-hand side is the value of strain at t_1, which—according to expression (3.4.9)—is

$$\varepsilon(t_1) = \frac{\sigma_0}{c}\left[1 - \frac{1}{\exp\left(\frac{c}{\eta}t_1\right)}\right]; \tag{3.4.10}$$

the integration limits correspond to $\sigma = 0$ for $t \geqslant t_1$, as illustrated in the bottom diagram of Figure 3.4.2. Integrating, we obtain

$$\varepsilon(t) = \frac{\sigma_0}{c}\left[\frac{1}{\exp\left(\frac{c}{\eta}(t-t_1)\right)} - \frac{1}{\exp\left(\frac{c}{\eta}t_1\right)}\right], \qquad t \geqslant t_1, \tag{3.4.11}$$

which describes the process of recovery.

Let us discuss these results. If we take the temporal derivative of expression (3.4.9) and evaluate it at $t = 0$, we get $\mathrm{d}\varepsilon/\mathrm{d}t|_{t=0} = \sigma_0/\eta$, which is the rate of deformation of the viscous element, as shown in expression (3.4.5). If we let t tend to infinity in expression (3.4.9), we see that $\varepsilon(t) = \sigma_0/c$, which is tantamount to the result exhibited by the elasticity alone; the presence of viscosity only slows the deformation process. Hence, the initial rate of deformation depends on viscosity, and the final magnitude of deformation on elasticity. However—in the presence of viscosity—the deformation continues for as long as σ_0 is applied, as illustrated in Exercise 3.6 and in agreement with expression (3.4.5).

If we consider expression (3.4.11) at t_1, which is the initial instant of recovery, it becomes expression (3.4.10): the state of strain at that instant. If we take the limit of expression (3.4.11) as t tends to infinity, we get

$$\varepsilon(t) = -\frac{\sigma_0}{c}\frac{1}{\exp\left(\frac{c}{\eta}t_1\right)}, \qquad (3.4.12)$$

which means that the strain present at the initial instant of recovery and associated with viscosity, η, remains. To complete the recovery, $\varepsilon(t) \to 0$, we need $t_1 \to \infty$, to make this term vanish. Examining expressions (3.4.9) and (3.4.10), we see that this requirement implies that—prior to allowing the recovery at t_1—the final deformation must be approached, $\varepsilon(t) = \sigma_0/c$, which means that the effect of viscosity disappears. Otherwise, the presence of viscosity results in a permanent deformation, as indicated by expression (3.4.5).

Be that as it may, if the commencement of deformation is at t_0, its completeness requires $(t - t_0) \to \infty$, and if the commencement of recovery is at t_1, its completeness requires $(t - t_1) \to \infty$. Thus the completeness of deformation followed by recovery, as described by the Kelvin-Voigt model, requires both $t \to \infty$ and $t_1 \to \infty$, with $t \gg t_1$. Such a behaviour is illustrated in Figure 3.4.3, with the infinity approximated by the values of t and t_1 that are large-enough in the context of the parameters of the Kelvin-Voigt model, c and η.

The viscous element slows the processes of deformation and recovery. In that respect, the Kelvin-Voigt model provides a more accurate mathematical analogy than a Hookean solid, whose definition assumes these two processes to be instantaneous. However, Hookean solids, in contrast to the Kelvin-Voigt model, lend themselves conveniently to study of anisotropy, examined in Chapter 5. In general, our model choices, as is common in

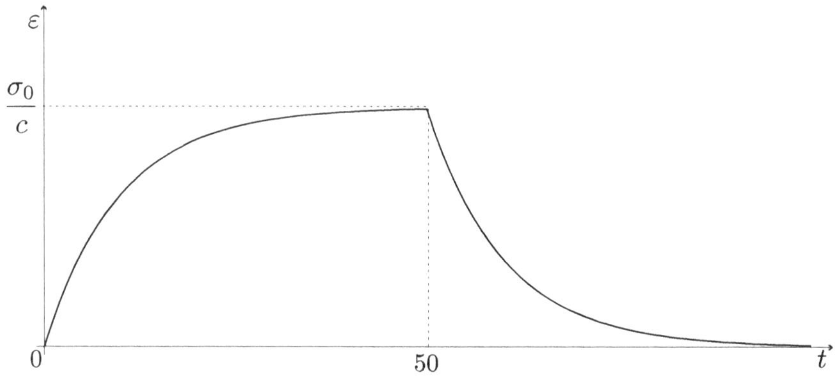

Fig. 3.4.3 The response of the Kelvin-Voigt model with $c = 1$ and $\eta = 10$ to a constant stress, σ_0, applied at $t = 0$ and removed at $t = 50$. Between $t = 0$ and $t = 50$, the response is calculated using expression (3.4.9); for $t \geqslant 50$, it is calculated using expression (3.4.11). The response of the corresponding Hookean solid, which is instantaneous, is shown by the dotted line.

mathematical physics, depend on the context and are also a compromise between the enhanced accuracy and an increased complexity of a model.

3.4.4. Viscoelasticity: Maxwell model

To complete this chapter, let us discuss the Maxwell model, which—contrary to the Kelvin-Voigt model—consists of combining in series the elastic and viscous elements. As a consequence of such an arrangement, at each instant, the value of stress is the same for both elements. We write the constitutive equation of this model as

$$\frac{\mathrm{d}\varepsilon(t)}{\mathrm{d}t} = \frac{1}{\eta}\sigma(t) + \frac{1}{c}\frac{\mathrm{d}\sigma(t)}{\mathrm{d}t}, \tag{3.4.13}$$

where $\varepsilon(t)$ and $\sigma(t)$ are strain and stress as functions of time, and η and c are the viscosity and elasticity parameters, respectively.

To examine deformation as a function of time, we integrate both sides of equation (3.4.13) with respect to t from 0 to t,

$$\varepsilon(t) = \frac{1}{\eta}\int_0^t \sigma(\tau)\,\mathrm{d}\tau + \frac{1}{c}\sigma(t), \qquad t \geqslant 0,$$

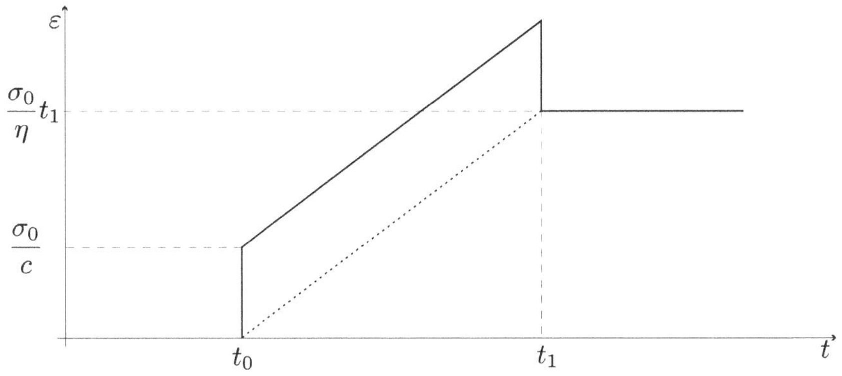

Fig. 3.4.4 The response of the Maxwell model to a constant stress, σ_0, applied at t_0 and removed at t_1. The response of the corresponding Stokesian fluid is shown by the dotted line; it continues, as a solid line, for $t \geqslant t_1$.

where τ is the integration variable. Let us consider a constant stress, $\sigma_0 h(t)$, where $h(t)$ is given in expression (3.4.4); we obtain

$$\varepsilon(t) = \frac{\sigma_0}{\eta}t + \frac{\sigma_0}{c}, \qquad t \geqslant 0.$$

Using translational properties of Heaviside's step, $h(t - t_0)$, where $t_0 \in (0, t)$, we write

$$\varepsilon(t) = \begin{cases} 0, & t < t_0 \\ \dfrac{\sigma_0}{\eta}t + \dfrac{\sigma_0}{c}, & t \geqslant t_0 \end{cases}.$$

which describes the process of deformation initiated by σ_0 shown in the top diagram of Figure 3.4.2. For $t \geqslant t_0$, the first term on the right-hand side corresponds to deformation due to viscosity, which increases linearly with time, and the second to elasticity, which is instantaneous. Upon removal of stress at t_1, the effect of deformation due to elasticity disappears instantaneously, and the effect due to viscosity remains constant at $(\sigma_0/\eta)t_1$.

Deformation and recovery of the Maxwell model are illustrated in Figure 3.4.4. Examining this figure, we see that the response at t_0 and t_1 is due solely to elasticity; the response between these times and after t_1 is due solely to viscosity. Contrary to the Kelvin-Voigt model, the elastic and viscous elements—arranged in series—do not influence one another.

Notably, the response of the Maxwell model for $t \geqslant t_1$ is the same as the response of the Stokesian fluid alone.

Closing Remarks

We use constitutive equations, namely, stress-strain equations (3.2.1) or, equivalently, equations (3.2.7) to obtain a determined system of equations that describes the propagation of deformations in elastic continua. For many seismological studies, the linear equations relating the stress-tensor components and the strain-tensor components agree, within sufficient accuracy, with experimental observations involving small deformations.

Stress-strain equations (3.2.1) and (3.2.7), as well as their equivalent given by equations (5.14.7) in Chapter 5, link the fundamental principles discussed in Chapter 2 with the properties of a particular elastic material. Notably, this link allows us to investigate Cauchy's equations of motion in the context of elastic materials, which leads to the wave equation and the eikonal equation, discussed in Chapters 6 and 7, respectively.

Stress-strain equations (3.2.1) and (3.2.7) describe the continuum whose deformations are linearly related to loads. Also, examining these equations, we see that the vanishing of stress in this continuum is accompanied by the instantaneous disappearance of strain therein, as illustrated in Exercise 3.3. Notably, the instantaneity of deformation and recovery distinguishes Hookean solids from Stokesian fluids and the Kelvin-Voigt and Maxwell models.

For a continuum to represent an elastic material, we require the existence of the restoring force that allows, upon the removal of the load, the return to the undeformed state. In Chapter 4, we investigate the effects of this requirement upon parameters $c_{ijk\ell}$ and C_{mn}.

3.5. Exercises

EXERCISE 3.1. Using stress-strain equations (3.2.1) exemplify the presence of the factor 2 in stress-strain equations (3.2.7).

SOLUTION 3.1. Using stress-strain equations (3.2.1), namely,

$$\sigma_{ij} = \sum_{k=1}^{3} \sum_{\ell=1}^{3} c_{ijk\ell}\varepsilon_{k\ell}, \qquad i,j \in \{1,2,3\},$$

we explicitly write

$$\begin{aligned}\sigma_{23} &= c_{2311}\varepsilon_{11} + c_{2312}\varepsilon_{12} + c_{2313}\varepsilon_{13} + c_{2321}\varepsilon_{21} + c_{2322}\varepsilon_{22} + c_{2323}\varepsilon_{23} \\ &\quad + c_{2331}\varepsilon_{31} + c_{2332}\varepsilon_{32} + c_{2333}\varepsilon_{33}.\end{aligned}$$

In view of the symmetry of the strain tensor, $\varepsilon_{ij} = \varepsilon_{ji}$, which implies that $c_{23ij} = c_{23ji}$, we can write

$$\sigma_{23} = c_{2311}\varepsilon_{11} + c_{2322}\varepsilon_{22} + c_{2333}\varepsilon_{33} + 2c_{2323}\varepsilon_{23} + 2c_{2313}\varepsilon_{13} + 2c_{2312}\varepsilon_{12}.$$

Using formula (3.2.5), we restate this equation in the matrix form, namely,

$$\sigma_{23} = \begin{bmatrix} C_{41} & C_{42} & C_{43} & C_{44} & C_{45} & C_{46} \end{bmatrix} \begin{bmatrix} \varepsilon_{11} \\ \varepsilon_{22} \\ \varepsilon_{33} \\ 2\varepsilon_{23} \\ 2\varepsilon_{13} \\ 2\varepsilon_{12} \end{bmatrix},$$

with the factor 2, as required.

EXERCISE 3.2. [10]Consider a one-dimensional homogeneous continuum. Using stress-strain equations (3.2.7), equation of continuity (2.1.9) and Cauchy's equations of motion (2.6.3) with no body force, write the resulting system of two differential equations.

SOLUTION 3.2. Following equations (3.2.7) and considering a one-dimensional continuum that coincides with the x_1-axis, we can write

$$\sigma_{11} = C_{11}\varepsilon_{11},$$

[10]See also Sections 3.3 and 4.4.2.

which, in view of definition (1.4.6) can be written as

$$\sigma_{11} = C_{11} \frac{\partial u_1(x,t)}{\partial x_1}, \tag{3.5.1}$$

where, due to the homogeneity of the continuum, C_{11} is a constant. The corresponding equation of continuity, whose general form is given by expression (2.1.9), is

$$\frac{\partial \rho(x,t)}{\partial t} + \frac{\partial}{\partial x_1}\left[\rho(x,t)\frac{\partial u_1(x,t)}{\partial t}\right] = 0, \tag{3.5.2}$$

and Cauchy's equation of motion, whose general form is given by expression (2.6.3), is

$$\frac{\partial \sigma_{11}}{\partial x_1} = \rho(x,t)\frac{\partial^2 u_1(x,t)}{\partial t^2}. \tag{3.5.3}$$

Inserting expression (3.5.1) into equation (3.5.3), differentiating and rearranging, we obtain

$$\frac{\partial^2 u_1(x,t)}{\partial x_1^2} = \frac{\rho(x,t)}{C_{11}}\frac{\partial^2 u_1(x,t)}{\partial t^2}. \tag{3.5.4}$$

Equations (3.5.2) and (3.5.4) constitute the required system of two differential equations in two unknowns, namely, $u_1(x,t)$ and $\rho(x,t)$, whose variables are x and t.

REMARK 3.5.1. If the mass-density function, $\rho(x,t)$, is given by a constant, equation (3.5.4) is a one-dimensional wave equation, discussed in Chapter 6.

EXERCISE 3.3. Consider the stress-strain equation for an elastic continuum written formally as

$$\sigma(t) = c\varepsilon(t). \tag{3.5.5}$$

Examine deformation ε that results from the load given by

$$\sigma(t) = \sigma_0[h(t) - h(t-t_1)], \tag{3.5.6}$$

where h is Heaviside's step.

REMARK 3.5.2. Expression (3.5.6) describes a load whose magnitude is equal to σ_0. This load is applied at $t = 0$ and removed at $t = t_1$.

SOLUTION 3.3. Let us write strain as a function of stress; in other words,

$$\varepsilon(t) = \frac{1}{c}\sigma(t).$$

Inserting herein expression (3.5.6), we obtain

$$\varepsilon(t) = \frac{\sigma_0}{c}\left[h(t) - h(t-t_1)\right].$$

Examining this result in view of the definition of Heaviside's step, we see that the continuum is deformed by the value of $\varepsilon = \sigma_0/c$ at $t = 0$, and it returns to its undeformed state—namely, $\varepsilon = 0$—at $t = t_1$. We also see that in a Hookean solid, both the deformation—under application of load—and return to the undeformed state—under its removal—are instantaneous.

REMARK 3.5.3. In contrast to Heaviside's step, the process of loading and unloading can be slow, but—at each moment—the response of a Hookean solid is instantaneous. In other words, in accordance with expression (3.5.5), the time dependence of strain and stress is the same.

EXERCISE 3.4. Show that—for a viscous element—the deformation remains at $(\sigma_0/\eta)t_1$ upon the removal of load at time t_1.

SOLUTION 3.4. Inserting expression (3.5.6) into expression (3.4.2), we write

$$\varepsilon(t) = \frac{\sigma_0}{\eta}\int\left[h(t) - h(t-t_1)\right]\mathrm{d}t.$$

Following the properties of Heaviside's step, we get

$$\varepsilon(t) = \frac{\sigma_0}{\eta}\left(\int_0^t \mathrm{d}\tau - \int_{t_1}^t \mathrm{d}\tau\right) = \frac{\sigma_0}{\eta}\left[t - (t-t_1)\right] = \frac{\sigma_0}{\eta}t_1,$$

as required.

EXERCISE 3.5. Given function σ, solve equation (3.4.6), namely,

$$\sigma(t) = c\varepsilon(t) + \eta\frac{\mathrm{d}\varepsilon(t)}{\mathrm{d}t},$$

where c and η are constants.

SOLUTION 3.5. This is a linear first-order ordinary differential equation. We would like to integrate $\mathrm{d}\varepsilon/\mathrm{d}t$ to obtain $\varepsilon(t)$. To achieve that, we multiply both sides by $\varsigma(t)$ to get

$$\frac{1}{\eta}\sigma(t)\varsigma(t) = \frac{\mathrm{d}\varepsilon(t)}{\mathrm{d}t}\varsigma(t) + \varepsilon(t)\frac{c}{\eta}\varsigma(t) . \tag{3.5.7}$$

Then, we consider

$$\frac{\mathrm{d}}{\mathrm{d}t}[\varepsilon(t)\varsigma(t)] = \frac{\mathrm{d}\varepsilon(t)}{\mathrm{d}t}\varsigma(t) + \varepsilon(t)\frac{\mathrm{d}\varsigma(t)}{\mathrm{d}t} . \tag{3.5.8}$$

We wish to find η such that the right-hand sides of these equations are equal to one another. Thus, we require

$$\frac{\mathrm{d}\varsigma(t)}{\mathrm{d}t} = \frac{c}{\eta}\varsigma(t) ,$$

which we can rewrite as

$$\frac{\frac{\mathrm{d}\varsigma(t)}{\mathrm{d}t}}{\varsigma(t)} = \frac{c}{\eta} .$$

Integrating both sides, we get

$$\ln|\varsigma(t)| = \frac{c}{\eta}t + C ,$$

where C is the integration constant. Solving for ς, we get

$$\varsigma(t) = \exp\left(\frac{c}{\eta}t + C\right) = K\exp\left(\frac{c}{\eta}t\right) , \tag{3.5.9}$$

where $K = \exp C$; for our use of ς, we can let $C = 0$, and hence $K = 1$. Since the right-hand sides of equations (3.5.7) and (3.5.8) are equal to one another if $\varsigma = \exp[(c/\eta)t]$, we can equate also the left-hand sides of these equations by writing

$$\frac{\mathrm{d}}{\mathrm{d}t}\left[\varepsilon(t)\exp\left(\frac{c}{\eta}t\right)\right] = \frac{1}{\eta}\exp\left(\frac{c}{\eta}t\right)\sigma(t) .$$

Integrating both sides with respect to the integration variable, τ, we get

$$\varepsilon(\tau)\exp\left(\frac{c}{\eta}\tau\right)\Big|_0^t = \frac{1}{\eta}\int_0^t \exp\left(\frac{c}{\eta}\tau\right)\sigma(\tau)\,\mathrm{d}\tau ,$$

with the lower limit of integration referring to the onset of load at $t = 0$. Evaluating the left-hand side, we get

$$\varepsilon(t)\exp\left(\frac{c}{\eta}t\right) - \varepsilon(0) = \frac{1}{\eta}\int_0^t \exp\left(\frac{c}{\eta}\tau\right)\sigma(\tau)\,\mathrm{d}\tau .$$

Solving for $\varepsilon(t)$, we obtain

$$\varepsilon(t) = \frac{\varepsilon(0)}{\exp\left(\frac{c}{\eta}t\right)} + \frac{1}{\eta}\int_0^t \frac{\sigma(\tau)}{\exp\left[\frac{c}{\eta}(t-\tau)\right]}\mathrm{d}\tau ,$$

which is the required solution.

EXERCISE 3.6. Reduce expression (3.4.9) for the case of absence of elasticity: pure viscosity. Interpret this result and compare it to the case of absence of viscosity: pure elasticity.

SOLUTION 3.6. In the absence of elasticity, $c = 0$, the term in front of the bracket in expression (3.4.9) tends to infinity and the term in brackets tends to zero. To avoid this indeterminate result, we invoke the initial terms of the Maclaurin series to write

$$\begin{aligned}
\varepsilon(t) &= \frac{\sigma_0}{c}\left[1 - \exp\left(-\frac{c}{\eta}t\right)\right] = \frac{\sigma_0}{c}\left[1 - \sum_{n=0}^{\infty}\frac{\left(-\frac{c}{\eta}t\right)^n}{n!}\right] \\
&= \frac{\sigma_0}{c}\left[1 - \left(1 - \frac{c}{\eta}t + \frac{1}{2}\left(\frac{c}{\eta}t\right)^2 - \frac{1}{6}\left(\frac{c}{\eta}t\right)^3 + \dots\right)\right] \\
&= \sigma_0\left(\frac{t}{\eta} - c\frac{1}{2}\left(\frac{t}{\eta}\right)^2 + c^2\frac{1}{6}\left(\frac{t}{\eta}\right)^3 - \dots\right), \qquad t \geqslant 0 .
\end{aligned}$$

If $c = 0$, we obtain

$$\varepsilon(t) = \frac{\sigma_0}{\eta}t , \qquad t \geqslant 0 .$$

which is expression (3.4.5); it corresponds to the Stokesian fluid. The deformation continues as long as σ_0 is applied, as discussed in Exercise 3.4. By contrast, in the absence of viscosity, expression (3.4.9) is reduced to $\varepsilon = \sigma_0/c$, which is pure elasticity; the deformation is instantaneous at $t = 0$, as discussed in Exercise 3.3.

CHAPTER 4

Strain Energy

> Ce qui fait la beauté d'une œuvre d'art, ce n'est pas la simplicité de ses parties, c'est plutôt une sorte d'harmonie globale qui donne à l'ensemble un aspect d'unité et d'homogénéité malgré la complication parfois très grande des détails. [...] La beauté des théories scientifiques nous paraît essentiellement de la même nature: elle s'impose quand, dominant sans cesse les raisonnements et les calculs, se retrouve partout une même idée centrale qui unifie et vivifie tout le corps de la doctrine.[1]
>
> *Louis de Broglie (1941)*
> *Continue et discontinue en physique moderne*

Preliminary remarks

When a material undergoes a deformation, energy is expended to deform it. In view of balance of energy, the energy expended must be converted into another form of energy. Elasticity of an actual material results from the fact

1 What makes the beauty of a work of art is not the simplicity of its parts, it is rather a kind of global harmony that gives to the whole an aspect of unity and homogeneity in spite of, at times, very large complications of details. [...] The beauty of scientific theories is of the same nature: this beauty is striking when, constantly dominating the reasoning and the calculations, one finds everywhere the same central idea that unifies and inspires the entire body of the formulation.

that a large part of the expended energy associated with the deformation is converted to potential energy stored within the deformed material. For elastic continua, we assume that all the expended energy is stored within the strained continuum. We refer to this energy as strain energy.

It is important to emphasize that the existence of strain energy, which allows the strained continuum to regain its initial state upon the removal of the load, is the defining property of an elastic continuum. The mathematical expression of this physical entity is the strain-energy function.

We begin this chapter with the derivation of the strain-energy function. Subsequently, in view of this function, we obtain another symmetry of the elasticity tensor, beyond the ones shown in Chapter 3. Then we derive the physical constraints on the components of the elasticity tensor; these constraints arise from the strain-energy function. This chapter concludes with the system of equations describing the behaviour of elastic continua.

4.1. Strain-energy Function

For elastic continua, we assume that all the expended energy is stored in the strained continuum as a potential energy. In other words, we are dealing with a conservative system. We wish to formulate the corresponding potential-energy function.

To motivate our formulation, consider a force, $\mathbf{F}$, acting on a conservative system to increase the potential energy, $U(\mathbf{x})$, of this system. We can write the components of such a force as $\partial U/\partial x_i = F_i$, where $i \in \{1,2,3\}$. By analogy, let us postulate

$$\frac{\partial W(\varepsilon)}{\partial \varepsilon_{ij}} = \sigma_{ij}, \qquad i,j \in \{1,2,3\}, \tag{4.1.1}$$

where—in the context of elasticity theory— W is the potential-energy function of a conservative system. In other words, we postulate that the stress tensor is derived from this scalar function.

To emphasize the fundamental importance of expression (4.1.1), let us comment on stress-strain equations (3.2.1). To obtain these equations in Chapter 3, we use Cauchy's approach, which assumes the existence of a function given by $\sigma_{ij} = \sigma_{ij}(\varepsilon_{ij})$. Also, we could view expression (4.1.1) as another point of departure; such a method is called Green's approach. The essence of such a formulation is the existence of function W, which underlies the concept of elasticity. For Green's approach, we would expand W in a power series of ε_{ij}, and retain only the quadratic terms.[2]

Herein, to obtain the explicit expression for W, we use stress-strain equations (3.2.1) to write expression (4.1.1) as

$$\frac{\partial W(\varepsilon)}{\partial \varepsilon_{ij}} = \sum_{k=1}^{3} \sum_{\ell=1}^{3} c_{ijk\ell} \varepsilon_{k\ell}, \qquad i, j \in \{1,2,3\}. \tag{4.1.2}$$

Integrating both sides of equations (4.1.2) with respect to ε_{ij}, we obtain

$$W(\varepsilon) = \frac{1}{2} \sum_{i=1}^{3} \sum_{j=1}^{3} \sum_{k=1}^{3} \sum_{\ell=1}^{3} c_{ijk\ell} \varepsilon_{k\ell} \varepsilon_{ij}, \tag{4.1.3}$$

where we set the integration constant to zero. The vanishing of this constant results from the convention that, for unstrained continua, $W = 0$.

W in expression (4.1.3) is the strain-energy function. It is the desired potential-energy function that corresponds to elastic continua subjected to infinitesimal strains.[3] W has the units of energy per volume.

Examining expression (4.1.3), we recognize that strain energy is given by a homogeneous function of degree 2 in the strain-tensor components.[4] The fact that the strain-energy function is homogeneous of degree 2 in the ε_{ij}, follows from definition A.1.1, which is discussed in Appendix A. This property of the strain-energy function is illustrated in Exercise 4.2. As shown in that exercise, expression (4.1.3) can be viewed as a second-degree polynomial in the strain-tensor components where both the constant term and the linear term vanish. A mathematical application of the homogeneity of W is illustrated in Exercise 4.3.

---

[2] Readers interested in similarities and distinctions between Cauchy's and Green's approaches to formulate stress-strain equations (3.2.1) might refer to Graff, K.F., (1975/1991) Wave motion in elastic solids: Dover, pp. 590–591.

[3] Readers interested in a general formulation for finite strains might refer to Malvern, L.E., (1969) Introduction to the mechanics of a continuous medium: Prentice-Hall., pp. 282–285.

[4] Both terms "degree" and "order" are commonly used to describe the homogeneity of a function. In this book, we use the former term since it refers to the value of the exponent and, hence, is consistent with other uses of this term, such as "degree of a polynomial".

4.2. Strain-energy Function and Elasticity-tensor Symmetry

4.2.1. Fundamental considerations

The existence of the strain-energy function, which defines an elastic continuum, implies the invariance of the elasticity tensor, $c_{ijk\ell}$, under permutations of pairs of subscripts ij and $k\ell$. This can be derived in the following manner.

Let us return to equations (4.1.2). Differentiating both sides of these equations with respect to $\varepsilon_{k\ell}$, we obtain

$$\frac{\partial^2 W(\varepsilon)}{\partial \varepsilon_{k\ell}\partial \varepsilon_{ij}} = c_{ijk\ell}, \qquad i,j,k,\ell \in \{1,2,3\}. \tag{4.2.1}$$

Now, let us invoke the equality of mixed partial derivatives, which states that, if W is a well-behaved function, the order of differentiation is interchangeable.[5] In view of expression (4.2.1), this implies that

$$c_{ijk\ell} = c_{k\ell ij}, \qquad i,j,k,\ell \in \{1,2,3\}. \tag{4.2.2}$$

Hence, we conclude that the elasticity tensor is invariant under permutations of pairs of subscripts ij and $k\ell$.

We can also justify symmetry $c_{ijk\ell} = c_{k\ell ij}$ in the following manner.[6] Recalling that σ_{ij} and ε_{ij} are associated with force and displacement, respectively, we can write the element of work as

$$\sum_{i=1}^{3}\sum_{j=1}^{3}\sigma_{ij}\,\mathrm{d}\varepsilon_{ij}.$$

---

[5] The equality of mixed partial derivatives is often used in this book. We can state it by the following theorem.

THEOREM 4.2.1. *Let $f = f(x,y)$. Assume that the partial derivatives $\partial f/\partial x$, $\partial f/\partial y$, $\partial^2 f/\partial x\partial y$ and $\partial^2 f/\partial y\partial x$ exist and are continuous. Then*

$$\frac{\partial^2 f}{\partial x\partial y} = \frac{\partial^2 f}{\partial y\partial x}.$$

Readers interested in the proof of this theorem might refer to Lang, S., (1973) Calculus of several variables: Addison-Wesley Publishing Co., pp. 110–111, or to Stewart, J., (1995), Multivariable calculus (3*rd* edition): Brooks/Cole Publishing Co., p. A2.

[6] Readers interested in this formulation might also refer to Ting, T.C.T., (1996) Anisotropic elasticity: Theory and applications: Oxford University Press, p. 33.

In view of the balance of energy, this element of work equals the total differential of W; namely,

$$\mathrm{d}W=\sum_{i=1}^{3}\sum_{j=1}^{3}\sigma_{ij}\,\mathrm{d}\varepsilon_{ij}\,. \tag{4.2.3}$$

Note that the requirement for the element of work to be a total differential results from the fact that the value of work must be independent of the integration path. The physical justification for this is that the work cannot depend on the path of deformations, but only on the difference between the initial and final states. Otherwise—since we can deform the material following one path and let it return to its initial state along a different path—if the amount of energy were not the same for all paths, we could create or destroy energy.

Invoking stress-strain equations (3.2.1), we write expression (4.2.3) as

$$\mathrm{d}W=\sum_{i=1}^{3}\sum_{j=1}^{3}\sum_{k=1}^{3}\sum_{\ell=1}^{3}c_{ijk\ell}\varepsilon_{k\ell}\,\mathrm{d}\varepsilon_{ij}\,. \tag{4.2.4}$$

Since this has to be a total differential, we require that[7]

$$\frac{\partial}{\partial\varepsilon_{st}}\sum_{k=1}^{3}\sum_{\ell=1}^{3}c_{ijk\ell}\varepsilon_{k\ell}=\frac{\partial}{\partial\varepsilon_{ij}}\sum_{k=1}^{3}\sum_{\ell=1}^{3}c_{stk\ell}\varepsilon_{k\ell}\,,\qquad i,j,s,t\in\{1,2,3\}\,,$$

which gives

$$c_{ijst}=c_{stij}\,,\qquad i,j,s,t\in\{1,2,3\}\,.$$

Note that, for the partial derivatives, indices i,j,s,t specify the variable of differentiations between $\partial/\partial\varepsilon_{11}$ and $\partial/\partial\varepsilon_{33}$. Upon renaming the indices, we can write

$$c_{ijk\ell}=c_{k\ell ij}\,,\qquad i,j,k,\ell\in\{1,2,3\}\,, \tag{4.2.5}$$

which are equations (4.2.2). Hence, we have another approach to obtain equations (4.2.2). Also, in view of conditions (4.2.5), we can integrate

---

[7] Readers interested in this requirement might refer to Courant, R., and John, F., (1974/1989) Introduction to calculus and analysis: Springer-Verlag, Vol. II, p. 84, or to Zill, D.G., and Cullen, M.R., (1997) Differential equations with boundary-value problems (4*th* edition): Brooks/Cole Publishing Company, p. 39.

both sides of equation (4.2.4) to obtain

$$W = \frac{1}{2} \sum_{i=1}^{3} \sum_{j=1}^{3} \sum_{k=1}^{3} \sum_{\ell=1}^{3} c_{ijk\ell} \varepsilon_{k\ell} \varepsilon_{ij} , \tag{4.2.6}$$

which is expression (4.1.3), as expected.

4.2.2. Elasticity parameters

In view of the symmetries of the stress and strain tensors, the eighty-one components of the elasticity tensor, $c_{ijk\ell}$, consist of only thirty-six independent components, as shown in Section 3.2.2. Furthermore, since every elastic continuum must obey equations (4.1.1), conditions (4.2.2) provide additional constraints on this tensor. Specifically, in view of the strain-energy function and the resulting symmetry stated in expression (4.2.2), the number of independent components is twenty-one.

Since there are no more reductions in the number of independent components of the elasticity tensor that stem from the fundamental equations, we refer to these remaining twenty-one independent components as the elasticity parameters. These parameters—together with mass density—fully describe a linearly elastic continuum.

4.2.3. Matrix form of stress-strain equations

As shown in Sections 3.2.2 and 3.2.3, we can express stress-strain equations in both the tensor and matrix forms. Recalling expressions (3.2.5) and (4.2.2), we see that $c_{ijk\ell} = c_{k\ell ij}$ implies $C_{mn} = C_{nm}$ since switching ij with $k\ell$ is tantamount to switching m with n. In other words, the elasticity matrix is symmetric; namely,

$$\mathbf{C} = \begin{bmatrix} C_{11} & C_{12} & C_{13} & C_{14} & C_{15} & C_{16} \\ C_{12} & C_{22} & C_{23} & C_{24} & C_{25} & C_{26} \\ C_{13} & C_{23} & C_{33} & C_{34} & C_{35} & C_{36} \\ C_{14} & C_{24} & C_{34} & C_{44} & C_{45} & C_{46} \\ C_{15} & C_{25} & C_{35} & C_{45} & C_{55} & C_{56} \\ C_{16} & C_{26} & C_{36} & C_{46} & C_{56} & C_{66} \end{bmatrix} . \tag{4.2.7}$$

Thus, the number of independent entries of **C** is reduced from thirty-six, used in equations (3.2.7), to twenty-one, stated in matrix (4.2.7).

In view of conditions (4.2.2) and resulting matrix (4.2.7), stress-strain equations (3.2.7) become

$$\begin{bmatrix} \sigma_{11} \\ \sigma_{22} \\ \sigma_{33} \\ \sigma_{23} \\ \sigma_{13} \\ \sigma_{12} \end{bmatrix} = \begin{bmatrix} C_{11} & C_{12} & C_{13} & C_{14} & C_{15} & C_{16} \\ C_{12} & C_{22} & C_{23} & C_{24} & C_{25} & C_{26} \\ C_{13} & C_{23} & C_{33} & C_{34} & C_{35} & C_{36} \\ C_{14} & C_{24} & C_{34} & C_{44} & C_{45} & C_{46} \\ C_{15} & C_{25} & C_{35} & C_{45} & C_{55} & C_{56} \\ C_{16} & C_{26} & C_{36} & C_{46} & C_{56} & C_{66} \end{bmatrix} \begin{bmatrix} \varepsilon_{11} \\ \varepsilon_{22} \\ \varepsilon_{33} \\ 2\varepsilon_{23} \\ 2\varepsilon_{13} \\ 2\varepsilon_{12} \end{bmatrix}. \qquad (4.2.8)$$

Equations (4.2.8) are the matrix form of the stress-strain equations for a general elastic continuum that obeys Hooke's law.

4.2.4. Coordinate transformations

The strain-energy function allows us an insight into properties of $c_{ijk\ell}$, and, in general, into the fundamental properties of tensors. Also, W gives us an insight into continua that exhibit symmetries in the context of anisotropy.

W is a scalar; hence, its value is invariant under coordinate transformations. To achieve this invariance, the values of the elasticity parameters, $c_{ijk\ell}$ and C_{mn}, change depending on the orientation of the coordinate system. In other words, the values of the components ensure that W is invariant under coordinate transformations. Hence, in general, expression (4.2.6) that contains particular values of $c_{ijk\ell}$ holds only for the corresponding orientation of the system.[8] An analogous property is exhibited by the components of a vector: their values must change to satisfy the requirement that its length, which is a scalar, is invariant under coordinate transformations.

If, for a particular continuum, expression (4.2.6) with a given set of elasticity parameters holds for several orientations of the coordinate system, this continuum possesses rotation symmetries, which lead to

---

[8] Interested readers might refer to Malvern, L.E., (1969) Introduction to the mechanics of a continuous medium: Prentice-Hall, p. 285.

further simplifications of matrix (4.2.7). Such symmetries are discussed in Chapter 5, and the invariance of W under coordinate transformations is explicitly used in Section 5.10.2.

4.3. Stability Conditions

4.3.1. Physical justification

Strain-energy function, W, given in expression (4.1.3), is formulated in terms of parameters $c_{ijk\ell}$, where $i,j,k,\ell \in \{1,2,3\}$. This function provides the sole fundamental constraints on these parameters. These constraints are called stability conditions since they constitute a mathematical statement of the fact that it is necessary to expend energy to deform a material. In other words, if energy is not expended, the material remains stable in its undeformed state.

In general, energy is a positive quantity. By convention, the strain energy of an undeformed continuum is zero. Thus, the strain-energy function must be a positive quantity that vanishes only in the undeformed state.[9]

4.3.2. Mathematical formulation

Mathematically, the stability conditions are equivalent to the positive-definiteness of the elasticity matrix. This can be derived in the following manner.

In view of expression (4.1.3) and by equivalence of stress-strain equations (3.2.1) and (3.2.7), we can write the strain-energy function as

$$W = \frac{1}{2}(\mathbf{C}\underline{\varepsilon}) \cdot \underline{\varepsilon}, \tag{4.3.1}$$

where $\mathbf{C}$ is matrix (4.2.7), and $\underline{\varepsilon}$ is the strain matrix, shown explicitly in equation (3.2.7). In view of Section 4.3.1, we require that

$$(\mathbf{C}\underline{\varepsilon}) \cdot \underline{\varepsilon} \geqslant 0, \tag{4.3.2}$$

where the equality sign corresponds to the case where $\underline{\varepsilon} = \mathbf{0}$, which is the undeformed state. Expression (4.3.2) states the positive-definiteness

[9] Readers interested in a more detailed description might refer to Musgrave, M.J.P., (1990) On the constraints of positive-definite strain energy in anisotropic media: Q. J. Mech. Appl. Math., **43**, Part 4, 605 – 621.

of matrix $\mathbf{C}$. In other words, matrix $\mathbf{C}$ is positive-definite if and only if $(\mathbf{C}\underline{\varepsilon}) \cdot \underline{\varepsilon} > 0$, for all $\underline{\varepsilon}$, such that $\underline{\varepsilon} \neq \mathbf{0}$.

4.3.3. Constraints on elasticity parameters

To formulate the conditions of positive-definiteness of the elasticity matrix, we can use either of the following theorems of linear algebra[10]; namely,

THEOREM 4.3.1. *A real symmetric matrix is positive-definite if and only if the determinants of all its leading principal minors, including the determinant of the matrix itself, are positive.*

or

THEOREM 4.3.2. *A real symmetric matrix is positive-definite if and only if all its eigenvalues are positive.*

Since matrix (4.2.7) is symmetric, the stability conditions can be conveniently formulated based on Theorems 4.3.1 and 4.3.2, as shown in Exercises 5.7 and 5.18, respectively. Among these conditions, we find that

$$C_{mm} > 0, \qquad m \in \{1, \ldots, 6\}, \tag{4.3.3}$$

which implies that all the main-diagonal entries of the elasticity matrix must be positive, as shown in Exercise 4.5.[11]

Stability conditions cannot be violated. However, as shown in Exercise 5.18, interesting and, perhaps, nonintuitive results are allowable within the stability conditions.

---

[10]Readers interested in proofs of Theorem 4.3.1 and Theorem 4.3.2 might refer to Ayres, F., (1962) Matrices: Schaum's Outlines, McGraw-Hill, Inc., p. 142, and to Morse P.M., and Feshbach H., (1953) Methods of theoretical physics: McGraw-Hill, Inc., pp. 771–774, respectively.

Note that—in view of the fact that every symmetric matrix can be diagonalized—Theorem 4.3.2 follows from Theorem 4.3.1.

[11]Readers interested in the expressions for the stability conditions for particular continua may derive them from the corresponding stiffness matrices, shown in Chapter 5, or refer to Fedorov, F.I., (1968) Theory of elastic waves in crystals: Plenum Press, New York, p. 16 and p. 33.

4.4. System of Equations for Elastic Continua

4.4.1. Elastic continua

To state a complete system of equations describing behaviour of our continua, we note that linearly elastic continua are defined by stress-strain equations (3.2.1), namely,

$$\sigma_{ij} = \sum_{k=1}^{3} \sum_{\ell=1}^{3} c_{ijk\ell} \varepsilon_{k\ell}, \qquad i, j \in \{1,2,3\}, \tag{4.4.1}$$

where, in view of equations (3.2.3), (3.2.4) and (4.2.2), we require

$$c_{ijk\ell} = c_{jik\ell} = c_{ij\ell k} = c_{k\ell ij}, \qquad i, j, k, \ell \in \{1,2,3\}. \tag{4.4.2}$$

We recall that symmetries (4.4.2) result from definition (1.4.6), which implies $\varepsilon_{k\ell} = \varepsilon_{\ell k}$, as well as from the balance of angular momentum, stated in expression (2.7.2), namely,

$$\frac{\mathrm{d}}{\mathrm{d}t} \iiint\limits_{V(t)} \left(\mathbf{x} \times \rho \frac{\mathrm{d}\mathbf{u}}{\mathrm{d}t} \right) \mathrm{d}V = \iint\limits_{S(t)} (\mathbf{x} \times \mathbf{T})\, \mathrm{d}S + \iiint\limits_{V(t)} (\mathbf{x} \times \mathbf{f})\, \mathrm{d}V,$$

which implies $\sigma_{ij} = \sigma_{ji}$. Symmetries (4.4.2) result also from the existence of the strain-energy function that is given by expression (4.1.3), namely,

$$W(\varepsilon) = \frac{1}{2} \sum_{i=1}^{3} \sum_{j=1}^{3} \sum_{k=1}^{3} \sum_{\ell=1}^{3} c_{ijk\ell} \varepsilon_{k\ell} \varepsilon_{ij}, \tag{4.4.3}$$

and must satisfy equation (4.1.1); namely,

$$\frac{\partial W(\varepsilon)}{\partial \varepsilon_{ij}} = \sigma_{ij}, \qquad i, j \in \{1,2,3\}. \tag{4.4.4}$$

As expected, this formulation is consistent with the following statement from the classic book of Augustus Edward Hugh Love (1892) "A treatise on the mathematical theory of elasticity".

> When a body is slightly strained by gradual application of a load, and the temperature remains constant, the stress components are linear functions of the strain components [equations (4.4.1)], and they are also partial differential

coefficients of a function W of the strain components [equations (4.4.4)]. The strain-energy function, W, is therefore a homogeneous quadratic function of the strain components [equation (4.4.3)].

4.4.2. Governing equations

We can now show that the behaviour of the continuum discussed in this book is governed by a system of ten equations and ten unknowns. The unknowns of this system are ρ, u_1, u_2, u_3, σ_{11}, σ_{12}, σ_{13}, σ_{22}, σ_{23}, σ_{33}, while a given continuum is defined in terms of twenty-one elasticity parameters, $C_{mn} = C_{nm}$, where $m, n \in \{1, \ldots, 6\}$.

Note that four among the ten equations result from the fundamental principles, which are given by the conservation of mass, stated in equation (2.1.7), namely,

$$\iiint\limits_V \left[\frac{\partial \rho}{\partial t} + \nabla \cdot (\rho \mathbf{v})\right] \mathrm{d}V = 0,$$

and the balance of linear momentum, stated in equation (2.4.5), namely,

$$\iiint\limits_{V(t)} \rho \frac{\mathrm{d}^2 \mathbf{u}}{\mathrm{d}t^2} \mathrm{d}V = \iint\limits_{S(t)} \mathbf{T} \mathrm{d}S + \iiint\limits_{V(t)} \mathbf{f} \mathrm{d}V.$$

The remaining six equations are constitutive equations, which provide a phenomenological description of actual materials.

Explicitly, we can write this system as

$$\left\{\begin{array}{l}
\dfrac{\partial}{\partial t}\rho=-\left[\dfrac{\partial}{\partial x_1}\left(\rho\dfrac{\partial}{\partial t}u_1\right)+\dfrac{\partial}{\partial x_2}\left(\rho\dfrac{\partial}{\partial t}u_2\right)+\dfrac{\partial}{\partial x_3}\left(\rho\dfrac{\partial}{\partial t}u_3\right)\right]\\
\rho\dfrac{\partial^2}{\partial t^2}u_1=\dfrac{\partial}{\partial x_1}\sigma_{11}+\dfrac{\partial}{\partial x_2}\sigma_{12}+\dfrac{\partial}{\partial x_3}\sigma_{13}\\
\rho\dfrac{\partial^2}{\partial t^2}u_2=\dfrac{\partial}{\partial x_1}\sigma_{12}+\dfrac{\partial}{\partial x_2}\sigma_{22}+\dfrac{\partial}{\partial x_3}\sigma_{23}\\
\rho\dfrac{\partial^2}{\partial t^2}u_3=\dfrac{\partial}{\partial x_1}\sigma_{13}+\dfrac{\partial}{\partial x_2}\sigma_{23}+\dfrac{\partial}{\partial x_3}\sigma_{33}\\
\sigma_{11}=C_{11}\varepsilon_{11}+C_{12}\varepsilon_{22}+C_{13}\varepsilon_{33}+2C_{14}\varepsilon_{23}+2C_{15}\varepsilon_{13}+2C_{16}\varepsilon_{12}\\
\sigma_{22}=C_{12}\varepsilon_{11}+C_{22}\varepsilon_{22}+C_{23}\varepsilon_{33}+2C_{24}\varepsilon_{23}+2C_{25}\varepsilon_{13}+2C_{26}\varepsilon_{12}\\
\sigma_{33}=C_{13}\varepsilon_{11}+C_{23}\varepsilon_{22}+C_{33}\varepsilon_{33}+2C_{34}\varepsilon_{23}+2C_{35}\varepsilon_{13}+2C_{36}\varepsilon_{12}\\
\sigma_{23}=C_{14}\varepsilon_{11}+C_{24}\varepsilon_{22}+C_{34}\varepsilon_{33}+2C_{44}\varepsilon_{23}+2C_{45}\varepsilon_{13}+2C_{46}\varepsilon_{12}\\
\sigma_{13}=C_{15}\varepsilon_{11}+C_{25}\varepsilon_{22}+C_{35}\varepsilon_{33}+2C_{45}\varepsilon_{23}+2C_{55}\varepsilon_{13}+2C_{56}\varepsilon_{12}\\
\sigma_{12}=C_{16}\varepsilon_{11}+C_{26}\varepsilon_{22}+C_{36}\varepsilon_{33}+2C_{46}\varepsilon_{23}+2C_{56}\varepsilon_{13}+2C_{66}\varepsilon_{12}
\end{array}\right., \tag{4.4.5}$$

where, by definition (1.4.6), $\varepsilon_{ij}=\left(\partial u_i/\partial x_j+\partial u_j/\partial x_i\right)/2$. The first equation is equation of continuity (2.8.2), which results from the conservation of mass. The following three equations are Cauchy's equations of motion (2.8.3), (2.8.4) and (2.8.5), which result from the balance of linear momentum. The last six equations are stress-strain equations (4.2.8), which contain the elasticity parameters that describe a given continuum.

Note that, invoking system (4.4.5) to study physical materials, we can consider directly only C_{mn} and ρ as properties of a continuum that represents a given material. Indirectly, the values of C_{mn} and ρ can indicate other properties, such as layering and fractures.

In a properly chosen coordinate system, discussed in Section 5.6.3, different materials exhibit different values of the elasticity parameters. These values are determined experimentally and characterize a given material. Often, the goal of seismological studies is to determine the values of the elasticity parameters and mass density of the subsurface based on experimental data mediated by a theoretical formulation; in other words, these data are examined in the context of a given theory.

Note that the last six equations of system (4.4.5) can be substituted into the second, the third and the fourth equations to obtain a system of four partial differential equations for four unknowns, namely, $\rho(\mathbf{x},t)$, $u_1(\mathbf{x},t)$, $u_2(\mathbf{x},t)$, $u_3(\mathbf{x},t)$ in the position variables, $\mathbf{x}=[x_1,x_2,x_3]$, and the time variable, t. For a one-dimensional case, a system of partial differential equations is exemplified in Exercise 3.2. Also, this substitution of stress-strain equations into Cauchy's equations of motion is used extensively in Chapters 6 and 7 to formulate equations of motions specifically for elastic continua.

Closing Remarks

For linearly elastic continua, stress is a linear function of strain, which depends on twenty-one elasticity parameters, as shown in expressions (4.2.8). Furthermore, these elastic continua possess strain energy, which is expressed as a quadratic function of strain, shown in expression (4.1.3). Thus, for instance, doubling the strain within the continuum doubles the stress within it, while it quadruples the energy stored within.

In our studies, we assume that the elasticity parameters have no temperature dependence, which is tantamount to our assuming in formulating system (4.4.5) that the process of deformation is isothermal. In other words, this process occurs at a constant temperature. Due to low thermal conductivity of most subsurface materials, we could argue that seismic wave propagation is better approximated by an adiabatic process, where no heat enters or leaves the element of volume. However, we choose the simplicity of the isothermal approach since the difference in experimental determination of elasticity parameters between the isothermal and adiabatic approaches is only of the order of one percent.[12]

Our formulation of elasticity parameters is rooted in a mathematical concept of an elastic continuum, not in fundamental physical theories of matter. The continuum formulation of these parameters is consistent with that of condensed-matter physics where, according to common physical

[12]Readers interested in temperature dependence might refer to Brekhovskikh, L.M., and Goncharov, V., (1982/1994) Mechanics of continua and wave dynamics: Springer-Verlag, pp. 45–47, to Fung, Y.C., (1977) A first course in continuum mechanics: Prentice-Hall, Inc., pp. 169–170, to Grant, F.S., and West, G.F., (1965) Interpretation theory in applied geophysics: McGraw-Hill Book Co., pp. 30 – 31, to Timoshenko, S.P., and Goodier, J.N., (1934/1987) Theory of elasticity: McGraw-Hill Publishing Company, p. 244, and to Udías, A., (1999) Principles of seismology: Cambridge University Press, pp. 17–18.

knowledge, materials are composed of nuclei and electrons. In that context, elasticity parameters represent interactions among the nuclei and electrons within a material. They can be examined by considering the total energy associated with the changes of volume and shape, which result from forces acting on every atom.[13] However, the consistency of various formulations does not imply that one could derive continuum mechanics from microscopic considerations. It is an axiomatic theory that is free of logical contradictions and whose physical applicability is justified *a posteriori*.

A given elastic continuum can possess particular symmetries, which further reduce the number of independent elasticity parameters required to describe it. Such symmetries are discussed in Chapter 5.

4.5. Exercises

EXERCISE 4.1. Using the one-dimensional case illustrated by a spring constant, k, show that for, elastic continua, strain energy is equal to the area below the graph of $F = kx$.

Herein, without affecting the result, we ignore the negative sign on the right-hand side, which states that the force is exerted in the opposite direction of the stretching; in other words, the spring force is pulling back.

SOLUTION 4.1. Following the definition of work and energy in a conservative system, we can write

$$W = \int \mathbf{F} \cdot \mathrm{d}\mathbf{x}, \tag{4.5.1}$$

where $\mathbf{F}$ denotes force and $\mathrm{d}\mathbf{x}$ denotes an element of displacement. The one-dimensional stress-strain equation can be written as $F = kx$, where k

[13] Readers interested in certain relationships between the continuum formulations and the atomic scale associated with the study of condensed-matter physics might refer to Aoki, H., Syono, Y., and Hemley, R.J., (editors), (2000) Physics meets mineralogy: Condensed-matter physics in geosciences: Cambridge University Press.

is an elasticity parameter, commonly known as the spring constant. Thus,

$$W = \int_0^{\Delta x} kx \, \mathrm{d}x = \frac{1}{2} k (\Delta x)^2 ,$$

where $x = 0$ corresponds to the unstrained state while $x = \Delta x$ corresponds to the strained state. This is equal to the triangular area below the straight line, kx, spanned between $x = 0$ and $x = \Delta x$.

NOTATION 4.5.1. In Exercise 4.2, for convenience, we denote the strain-tensor components using single subscripts.

EXERCISE 4.2. [14]Consider the strain-energy function to be a second-degree polynomial given by

$$W = C_0 + \sum_{n=1}^{6} C_n \varepsilon_n + \frac{1}{2} \sum_{n=1}^{6} \sum_{m=1}^{6} C_{nm} \varepsilon_n \varepsilon_m , \tag{4.5.2}$$

where ε_ℓ is an entry of matrix $\underline{\varepsilon}$, given in equation (3.2.7). Show explicitly that the first two terms vanish and, hence, W is homogeneous of degree 2 in the strain-tensor components. Note that since tensor ε_{ij} is symmetric, we only need one index, $i = 1, 2, \ldots, 6$, to represent all components.

SOLUTION 4.2. Expression (4.5.2) can be explicitly written as

$$W = C_0$$
$$+ C_1 \varepsilon_1 + C_2 \varepsilon_2 + C_3 \varepsilon_3 + C_4 \varepsilon_4 + C_5 \varepsilon_5 + C_6 \varepsilon_6$$

[14] See also Section 4.1.

$$+\frac{1}{2}(C_{11}\varepsilon_1^2+C_{21}\varepsilon_2\varepsilon_1+C_{31}\varepsilon_3\varepsilon_1+C_{41}\varepsilon_4\varepsilon_1+C_{51}\varepsilon_5\varepsilon_1+C_{61}\varepsilon_6\varepsilon_1$$
$$+C_{12}\varepsilon_1\varepsilon_2+C_{22}\varepsilon_2^2+C_{32}\varepsilon_3\varepsilon_2+C_{42}\varepsilon_4\varepsilon_2+C_{52}\varepsilon_5\varepsilon_2+C_{62}\varepsilon_6\varepsilon_2$$
$$+C_{13}\varepsilon_1\varepsilon_3+C_{23}\varepsilon_2\varepsilon_3+C_{33}\varepsilon_3^2+C_{43}\varepsilon_4\varepsilon_3+C_{35}\varepsilon_5\varepsilon_3+C_{63}\varepsilon_6\varepsilon_3$$
$$+C_{14}\varepsilon_1\varepsilon_4+C_{24}\varepsilon_2\varepsilon_4+C_{34}\varepsilon_3\varepsilon_4+C_{44}\varepsilon_4^2+C_{54}\varepsilon_5\varepsilon_4+C_{64}\varepsilon_6\varepsilon_4$$
$$+C_{15}\varepsilon_1\varepsilon_5+C_{25}\varepsilon_2\varepsilon_5+C_{35}\varepsilon_3\varepsilon_5+C_{45}\varepsilon_4\varepsilon_5+C_{55}\varepsilon_5^2+C_{65}\varepsilon_6\varepsilon_5$$
$$+C_{16}\varepsilon_1\varepsilon_6+C_{26}\varepsilon_2\varepsilon_6+C_{36}\varepsilon_3\varepsilon_6+C_{46}\varepsilon_4\varepsilon_6+C_{56}\varepsilon_5\varepsilon_6+C_{66}\varepsilon_6^2)\,.$$

We assume that W vanishes for the unstrained state. Thus, $\varepsilon_1=\varepsilon_2=\varepsilon_3=\varepsilon_4=\varepsilon_5=\varepsilon_6=0$, implies $W=0$; consequently, $C_0=0$. Also, following expression (4.1.1) we require that $\sigma_m=\partial W/\partial\varepsilon_m$. Consider, for instance, $m=5$; we can specifically write

$$\sigma_5=\frac{\partial W}{\partial\varepsilon_5}$$
$$=C_5+\frac{1}{2}[(C_{15}+C_{51})\,\varepsilon_1+(C_{25}+C_{52})\,\varepsilon_2+(C_{35}+C_{53})\,\varepsilon_3$$
$$+(C_{45}+C_{54})\,\varepsilon_4+2C_{55}\varepsilon_5+(C_{56}+C_{65})\,\varepsilon_6]\,.$$

No strain implies no stress and, hence, $\varepsilon_1=\varepsilon_2=\varepsilon_3=\varepsilon_4=\varepsilon_5=\varepsilon_6=0\Longrightarrow\sigma_5=0$. Thus, it follows that $C_5=0$. Analogously, considering σ_1, σ_2, σ_3, σ_4 and σ_6, we obtain $C_1=C_2=C_3=C_4=C_6=0$. Thus,

$$W(\varepsilon_m)=\frac{1}{2}(C_{11}\varepsilon_1^2+C_{21}\varepsilon_2\varepsilon_1+C_{31}\varepsilon_3\varepsilon_1+C_{41}\varepsilon_4\varepsilon_1+C_{51}\varepsilon_5\varepsilon_1+C_{61}\varepsilon_6\varepsilon_1$$
$$+C_{12}\varepsilon_1\varepsilon_2+C_{22}\varepsilon_2^2+C_{32}\varepsilon_3\varepsilon_2+C_{42}\varepsilon_4\varepsilon_2+C_{52}\varepsilon_5\varepsilon_2+C_{62}\varepsilon_6\varepsilon_2$$
$$+C_{13}\varepsilon_1\varepsilon_3+C_{23}\varepsilon_2\varepsilon_3+C_{33}\varepsilon_3^2+C_{43}\varepsilon_4\varepsilon_3+C_{35}\varepsilon_5\varepsilon_3+C_{63}\varepsilon_6\varepsilon_3$$
$$+C_{14}\varepsilon_1\varepsilon_4+C_{24}\varepsilon_2\varepsilon_4+C_{34}\varepsilon_3\varepsilon_4+C_{44}\varepsilon_4^2+C_{54}\varepsilon_5\varepsilon_4+C_{64}\varepsilon_6\varepsilon_4$$
$$+C_{15}\varepsilon_1\varepsilon_5+C_{25}\varepsilon_2\varepsilon_5+C_{35}\varepsilon_3\varepsilon_5+C_{45}\varepsilon_4\varepsilon_5+C_{55}\varepsilon_5^2+C_{65}\varepsilon_6\varepsilon_5$$
$$+C_{16}\varepsilon_1\varepsilon_6+C_{26}\varepsilon_2\varepsilon_6+C_{36}\varepsilon_3\varepsilon_6+C_{46}\varepsilon_4\varepsilon_6+C_{56}\varepsilon_5\varepsilon_6+C_{66}\varepsilon_6^2)\,, \quad (4.5.3)$$

which is a homogeneous function of degree 2 in the ε_m, as required.

REMARK 4.5.2. Examining expression (4.5.3), we observe that

$$W(c\varepsilon_m) = c^2 W(\varepsilon_m),$$

where c is a real number. Hence, in view of Definition A.1.1 stated in Appendix A, W is homogeneous of degree 2 in the ε_m.

EXERCISE 4.3. [15]Using the property of the strain-energy function, W, stated in expression (4.1.1), and in view of W being homogeneous of degree 2 in the ε_{ij}, show that

$$W = \frac{1}{2}\sum_{i=1}^{3}\sum_{j=1}^{3}\sigma_{ij}\varepsilon_{ij}. \tag{4.5.4}$$

SOLUTION 4.3. Since W is homogeneous of degree 2 in the ε_{ij}, by Theorem A.2.1 stated in Appendix A, we can write

$$\sum_{i=1}^{3}\sum_{j=1}^{3}\frac{\partial W}{\partial \varepsilon_{ij}}\varepsilon_{ij} = 2W.$$

In view of expression (4.1.1), we can rewrite the above expression as

$$\sum_{i=1}^{3}\sum_{j=1}^{3}\sigma_{ij}\varepsilon_{ij} = 2W,$$

which immediately yields expression (4.5.4), as required.

EXERCISE 4.4. Assuming that $c_{ijk\ell} = c_{k\ell ij}$, derive expression (4.1.1), namely,

$$\frac{\partial W}{\partial \varepsilon_{ij}} = \sigma_{ij}, \qquad i,j \in \{1,2,3\},$$

directly from expression (4.5.4).

SOLUTION 4.4. Differentiating expression (4.5.4) with respect to a particular strain-tensor component $\varepsilon_{k\ell}$, and recalling that stress is a function of

[15]See also Section 4.1.

strain, we obtain

$$\frac{\partial W}{\partial \varepsilon_{k\ell}} = \frac{1}{2}\sum_{i=1}^{3}\sum_{j=1}^{3}\left(\frac{\partial \sigma_{ij}}{\partial \varepsilon_{k\ell}}\varepsilon_{ij} + \sigma_{ij}\frac{\partial \varepsilon_{ij}}{\partial \varepsilon_{k\ell}}\right)$$

$$= \frac{1}{2}\sum_{i=1}^{3}\sum_{j=1}^{3}\left(\frac{\partial \sigma_{ij}}{\partial \varepsilon_{k\ell}}\varepsilon_{ij} + \sigma_{ij}\delta_{ik}\delta_{j\ell}\right), \qquad k,\ell \in \{1,2,3\}.$$

Using stress-strain equations (3.2.1), and recalling that $c_{ijk\ell}$ are independent of strain, we can write

$$\frac{\partial \sigma_{ij}}{\partial \varepsilon_{k\ell}} = c_{ijk\ell}, \qquad i,j,k,\ell \in \{1,2,3\}.$$

Consequently, using the fact that $c_{ijk\ell} = c_{k\ell ij}$, we obtain

$$\frac{\partial W}{\partial \varepsilon_{k\ell}} = \frac{1}{2}\sum_{i=1}^{3}\sum_{j=1}^{3}\left(c_{ijk\ell}\varepsilon_{ij} + \sigma_{ij}\delta_{ik}\delta_{j\ell}\right)$$

$$= \frac{1}{2}\left(\sum_{i=1}^{3}\sum_{j=1}^{3} c_{k\ell ij}\varepsilon_{ij} + \sum_{i=1}^{3}\sum_{j=1}^{3}\sigma_{ij}\delta_{ik}\delta_{j\ell}\right),$$

where $k,\ell \in \{1,2,3\}$. Again, in view of equations (3.2.1) and (4.2.2) as well as using the properties of Kronecker's delta, we obtain

$$\frac{\partial W}{\partial \varepsilon_{k\ell}} = \frac{1}{2}\left(\sigma_{k\ell} + \sigma_{k\ell}\right) = \sigma_{k\ell}, \qquad k,\ell \in \{1,2,3\},$$

where, in view of the arbitrariness of the subscript symbol, we obtain expression (4.1.1), as required.

EXERCISE 4.5. [16]Using expression (4.3.1), justify that the main-diagonal entries of the elasticity matrix must be always positive.

[16] See also Section 4.3.3.

SOLUTION 4.5. Consider expression (4.3.1). In view of equations (4.2.8), we can write

$$W=\frac{1}{2}\left(\begin{bmatrix} C_{11} & C_{12} & C_{13} & C_{14} & C_{15} & C_{16} \\ C_{12} & C_{22} & C_{23} & C_{24} & C_{25} & C_{26} \\ C_{13} & C_{23} & C_{33} & C_{34} & C_{35} & C_{36} \\ C_{14} & C_{24} & C_{34} & C_{44} & C_{45} & C_{46} \\ C_{15} & C_{25} & C_{35} & C_{45} & C_{55} & C_{56} \\ C_{16} & C_{26} & C_{36} & C_{46} & C_{56} & C_{66} \end{bmatrix}\begin{bmatrix} \varepsilon_{11} \\ \varepsilon_{22} \\ \varepsilon_{33} \\ 2\varepsilon_{23} \\ 2\varepsilon_{13} \\ 2\varepsilon_{12} \end{bmatrix}\right)\cdot\begin{bmatrix} \varepsilon_{11} \\ \varepsilon_{22} \\ \varepsilon_{33} \\ 2\varepsilon_{23} \\ 2\varepsilon_{13} \\ 2\varepsilon_{12} \end{bmatrix}.$$

Let the strain matrix, $\underline{\varepsilon}$, have only one nonzero entry. For instance, we can write

$$W=\frac{1}{2}\left(\begin{bmatrix} C_{11} & C_{12} & C_{13} & C_{14} & C_{15} & C_{16} \\ C_{12} & C_{22} & C_{23} & C_{24} & C_{25} & C_{26} \\ C_{13} & C_{23} & C_{33} & C_{34} & C_{35} & C_{36} \\ C_{14} & C_{24} & C_{34} & C_{44} & C_{45} & C_{46} \\ C_{15} & C_{25} & C_{35} & C_{45} & C_{55} & C_{56} \\ C_{16} & C_{26} & C_{36} & C_{46} & C_{56} & C_{66} \end{bmatrix}\begin{bmatrix} 0 \\ 0 \\ 0 \\ 1 \\ 0 \\ 0 \end{bmatrix}\right)\cdot\begin{bmatrix} 0 \\ 0 \\ 0 \\ 1 \\ 0 \\ 0 \end{bmatrix}$$

$$=\frac{1}{2}\left[C_{14},C_{24},C_{34},C_{44},C_{45},C_{46}\right]\cdot\left[0,0,0,1,0,0\right]^{T}=\frac{1}{2}C_{44}.$$

Similarly, for all other single, nonzero entries, $W=C_{ii}/2$. Hence, the positive value of the strain-energy function for all possible nonzero strains requires $C_{ii}>0$.

CHAPTER 5

Material Symmetry

> Symmetry is a vast subject, significant in art and nature. Mathematics lies at its root, and it would be hard to find a better one on which to demonstrate the working of the mathematical intellect.
>
> *Hermann Weyl (1952)*
> *Symmetry*

Preliminary remarks

[1]Materials can possess certain symmetries. In the context of our studies, this means that we can measure a property of a material in several different orientations of the coordinate system and obtain the same result each time. In other words, we are unable to detect the transformations of the reference coordinate system by mechanical experiments. This invariance to the orientation of the coordinate system is called material symmetry. In a properly chosen coordinate system, the form of the elasticity matrix allows us to recognize the symmetry of this continuum. This symmetry is indicative of the properties exhibited by the material represented by this continuum.

We begin this chapter with the formulation of transformations of the coordinate system and the effect of these transformations on the stress-strain equations. Then we formulate the condition that allows us to obtain

[1] This chapter originated in the work of Bos, L.P., Gibson, P.C., Kotchetov, M., Slawinski, M.A., (2004) Classes of anisotropic media: a tutorial: Studia geophysica and geodætica, **48**, pp. 265–287.

the elasticity matrix of a continuum that is invariant under a given transformation of coordinates. Subsequently, we study the eight symmetries of an elastic continuum, which are all the possible symmetries of the elasticity tensor.

5.1. Orthogonal Transformations

5.1.1. Transformation matrix

To study material symmetries, we need to use the transformation of an orthonormal coordinate system in the $x_1x_2x_3$-space. Such a transformation is given by

$$\hat{\mathbf{x}} = \mathbf{A}\mathbf{x}, \tag{5.1.1}$$

where $\mathbf{x} = [x_1, x_2, x_3]^T$ and $\hat{\mathbf{x}} = [\hat{x}_1, \hat{x}_2, \hat{x}_3]^T$ are the original and transformed coordinate systems, respectively, and $\mathbf{A}$ is the transformation matrix. Equation (5.1.1) is the matrix form of equations (1.6.4), shown in Exercise 1.4.

We are interested specifically in distance-preserving transformations, namely, rotations and reflections. In other words, these transformations allow us to change the orientation of the continuum without deforming it. Such transformations are represented by orthogonal matrices, that is, by square matrices given by

$$\mathbf{A} = \begin{bmatrix} A_{11} & A_{12} & A_{13} \\ A_{21} & A_{22} & A_{23} \\ A_{31} & A_{32} & A_{33} \end{bmatrix} \tag{5.1.2}$$

that satisfy the orthogonality condition, namely, $\mathbf{A}^T\mathbf{A} = \mathbf{I}$, which is equivalent to $\mathbf{A}^T = \mathbf{A}^{-1}$, where ^{T} denotes transpose. This orthogonality condition is discussed in Exercise 5.2.

Note that the determinants of these transformation matrices are the Jacobians of the coordinate transformations. This is illustrated in Exercise 5.1.

5.1.2. Symmetry group

Expressing the elasticity matrix of a given continuum in a conveniently chosen orthonormal coordinate system allows us to recognize the material symmetries of that continuum, as discussed in Sections 5.5–5.12. In other words, we can recognize the transformations that belong to the symmetry group of that continuum, which can be stated by the following definition.

DEFINITION 5.1.1. The set of all orthogonal transformations given by matrices $\mathbf{A}$ to which the elastic properties of a given continuum are invariant is called the symmetry group of that continuum.

According to the group theory, if the elastic properties are invariant under orthogonal transformations given by matrices $\mathbf{A}_1$ and $\mathbf{A}_2$, they are also invariant to product $\mathbf{A}_1\mathbf{A}_2$. Furthermore, if these properties are invariant to $\mathbf{A}$, they are also invariant to $\mathbf{A}^{-1}$. These invariances are the reason for our invoking the notion of a group in Definition 5.1.1.[2]

5.2. Transformation of Coordinates

5.2.1. Introductory comments

Recall that the properties of our continuum are formulated in terms of the stress-strain equations. To investigate the material symmetries of a given continuum, we study the stress-strain equations in the context of the orthogonal transformations of the orthonormal coordinate system.

5.2.2. Transformation of stress-tensor components

The components of the stress tensor expressed as a 3×3 matrix transform according to

$$\hat{\sigma} = \mathbf{A}\sigma\mathbf{A}^T, \tag{5.2.1}$$

where $\mathbf{A}$ stands for matrix (5.1.2) and the accent symbolizes the transformed entity. This is the matrix form of transformation (2.9.6), which is proved in Exercise 2.5.

[2] Readers interested in physical aspects of the group theory, which is the study of invariants and symmetries, might refer to Arfken, G.B, and Weber, H.J., (2001) Mathematical methods for physicists (5*th* edition): Harcourt/Academic Press, pp. 237–301.

Following matrix (2.5.2), the stress-tensor components can be written as a square matrix given by

$$\sigma = \begin{bmatrix} \sigma_{11} & \sigma_{12} & \sigma_{13} \\ \sigma_{12} & \sigma_{22} & \sigma_{23} \\ \sigma_{13} & \sigma_{23} & \sigma_{33} \end{bmatrix}, \qquad (5.2.2)$$

which, herein—in view of Theorem 2.7.1—is symmetric; $\hat{\sigma}$ looks analogous to σ.

Note that since a second-rank tensor has two indices, it is convenient to write it as a matrix, even though tensors and matrices are distinct mathematical entities. Herein, we use the fact that under the orthogonal transformations, the entries of a matrix behave like the components of a second-rank tensor.

In this chapter, we would like to use the elasticity matrix, which is given in expression (4.2.7). The corresponding stress-strain equations are given in expression (4.2.8) and involve stress-tensor components as a single-column matrix, $\underline{\sigma}$, namely,

$$\underline{\sigma} = [\sigma_{11}, \sigma_{22}, \sigma_{33}, \sigma_{23}, \sigma_{13}, \sigma_{12}]^T . \qquad (5.2.3)$$

Thus, we wish to obtain a transformation equation for $\underline{\sigma}$ that is equivalent to equation (5.2.1). Since transformation (5.2.1) is linear, it can be rewritten as a multiplication of $\underline{\sigma}$ by a matrix. In other words, we can write

$$\underline{\hat{\sigma}} = \underline{\mathbf{A}}\,\underline{\sigma}, \qquad (5.2.4)$$

where $\underline{\mathbf{A}}$ is a 6×6 transformation matrix. To find the entries of $\underline{\mathbf{A}}$, we substitute the elements of the standard basis of the space of symmetric 3×3 matrices for σ.[3]

Consider the first element of the basis, namely,

$$\sigma = \mathbf{E}_{11},$$

[3] Readers interested in the underlying aspects of linear algebra might refer to Anton, H., (1973) Elementary linear algebra: John Wiley & Sons, pp. 237–238, and to Ayres, F., (1962) Matrices: Schaum's Outlines, McGraw-Hill, p. 88 and p. 94.

where $\mathbf{E}_{ij}$ denotes the matrix with unity in the position (i, j) and with zeros elsewhere. In other words, we write

$$\sigma = \begin{bmatrix} 1 & 0 & 0 \\ 0 & 0 & 0 \\ 0 & 0 & 0 \end{bmatrix} . \tag{5.2.5}$$

Now, using equation (5.2.1) and matrix (5.1.2), we can compute the corresponding entries of $\hat{\sigma}$ as

$$\begin{aligned} \hat{\sigma} &= \begin{bmatrix} A_{11} & A_{12} & A_{13} \\ A_{21} & A_{22} & A_{23} \\ A_{31} & A_{32} & A_{33} \end{bmatrix} \begin{bmatrix} 1 & 0 & 0 \\ 0 & 0 & 0 \\ 0 & 0 & 0 \end{bmatrix} \begin{bmatrix} A_{11} & A_{12} & A_{13} \\ A_{21} & A_{22} & A_{23} \\ A_{31} & A_{32} & A_{33} \end{bmatrix}^T \\ &= \begin{bmatrix} A_{11}A_{11} & A_{11}A_{21} & A_{11}A_{31} \\ A_{11}A_{21} & A_{21}A_{21} & A_{21}A_{31} \\ A_{11}A_{31} & A_{21}A_{31} & A_{31}A_{31} \end{bmatrix} . \end{aligned} \tag{5.2.6}$$

For conciseness, using Kronecker's delta, we can begin by writing the entries of matrix (5.2.5) as

$$\sigma_{ij} = \delta_{i1}\delta_{j1} , \qquad i, j \in \{1,2,3\} ,$$

and, hence, write relation (5.2.6) as

$$\begin{aligned} \hat{\sigma}_{k\ell} &= \sum_{i=1}^{3}\sum_{j=1}^{3} A_{ki}\sigma_{ij}A^T_{j\ell} = \sum_{i=1}^{3}\sum_{j=1}^{3} A_{ki}\delta_{i1}\delta_{j1}A_{\ell j} \\ &= A_{k1}A_{\ell 1} , \qquad k, \ell \in \{1,2,3\} . \end{aligned}$$

To consider the matrix form of stress-strain equations, in view of expression (5.2.3) and taking $(k,\ell) = (1,1)\,,\ (2,2)\,,\ (3,3)\,,\ (2,3)\,,\ (1,3)\,,$

$(1,2)$, we obtain

$$\underline{\hat{\sigma}} = \begin{bmatrix} A_{11}A_{11} \\ A_{21}A_{21} \\ A_{31}A_{31} \\ A_{21}A_{31} \\ A_{11}A_{31} \\ A_{11}A_{21} \end{bmatrix} . \tag{5.2.7}$$

Column matrix (5.2.7) is the result of unfolding the entries of the rightmost square matrix in expression (5.2.6) in the order given by $(1,1)$, $(2,2)$, $(3,3)$, $(2,3)$, $(1,3)$ and $(1,2)$.

Since, herein, σ is chosen as shown in expression (5.2.5), we write

$$\underline{\sigma} = [1,0,0,0,0,0]^T .$$

In view of equation (5.2.4), $\underline{\hat{\sigma}}$ given in expression (5.2.7) is the first column of $\underline{\mathbf{A}}$. Analogously, we can compute the second and the third columns of $\underline{\mathbf{A}}$ by considering $\sigma = \mathbf{E}_{22}$ and $\sigma = \mathbf{E}_{33}$, respectively.

To find the fourth column of $\underline{\mathbf{A}}$, we use

$$\sigma = \mathbf{E}_{23} + \mathbf{E}_{32} ,$$

which is equivalent to

$$\sigma = \begin{bmatrix} 0 & 0 & 0 \\ 0 & 0 & 1 \\ 0 & 1 & 0 \end{bmatrix} . \tag{5.2.8}$$

Using Kronecker's delta, we can write the entries of matrix (5.2.8), as

$$\sigma_{ij} = \delta_{i2}\delta_{j3} + \delta_{i3}\delta_{j2} , \qquad i,j \in \{1,2,3\} .$$

Then, using equation (5.2.1), the corresponding entries for $\hat{\sigma}$ can be computed as

$$\begin{aligned} \hat{\sigma}_{k\ell} &= \sum_{i=1}^{3}\sum_{j=1}^{3} A_{ki}\sigma_{ij}A^T_{j\ell} = \sum_{i=1}^{3}\sum_{j=1}^{3} A_{ki}\delta_{i2}\delta_{j3}A_{\ell j} + \sum_{i=1}^{3}\sum_{j=1}^{3} A_{ki}\delta_{i3}\delta_{j2}A_{\ell j} \\ &= A_{k2}A_{\ell 3} + A_{k3}A_{\ell 2} , \qquad k,\ell \in \{1,2,3\} . \end{aligned}$$

To consider the matrix form of stress-strain equations, in view of expression (5.2.3) and taking $(k,\ell) = (1,1)$, $(2,2)$, $(3,3)$, $(2,3)$, $(1,3)$, $(1,2)$, we obtain

$$\underline{\hat{\sigma}} = \begin{bmatrix} 2A_{12}A_{13} \\ 2A_{22}A_{23} \\ 2A_{32}A_{33} \\ A_{22}A_{33}+A_{23}A_{32} \\ A_{12}A_{33}+A_{13}A_{32} \\ A_{12}A_{23}+A_{13}A_{22} \end{bmatrix}, \tag{5.2.9}$$

which is the fourth column of $\underline{\mathbf{A}}$. Analogously, we can compute the fifth and the sixth columns of $\underline{\mathbf{A}}$ by considering $\sigma = \mathbf{E}_{13} + \mathbf{E}_{31}$ and $\sigma = \mathbf{E}_{12} + \mathbf{E}_{21}$, respectively.

Now, putting together the six columns of $\underline{\mathbf{A}}$, we obtain

$$\underline{\mathbf{A}} =$$
$$\begin{bmatrix} A_{11}A_{11} & A_{12}A_{12} & A_{13}A_{13} & 2A_{12}A_{13} & 2A_{11}A_{13} & 2A_{11}A_{12} \\ A_{21}A_{21} & A_{22}A_{22} & A_{23}A_{23} & 2A_{22}A_{23} & 2A_{21}A_{23} & 2A_{21}A_{22} \\ A_{31}A_{31} & A_{32}A_{32} & A_{33}A_{33} & 2A_{32}A_{33} & 2A_{31}A_{33} & 2A_{31}A_{32} \\ A_{21}A_{31} & A_{22}A_{32} & A_{23}A_{33} & A_{22}A_{33}+A_{23}A_{32} & A_{21}A_{33}+A_{23}A_{31} & A_{21}A_{32}+A_{22}A_{31} \\ A_{11}A_{31} & A_{12}A_{32} & A_{13}A_{33} & A_{12}A_{33}+A_{13}A_{32} & A_{11}A_{33}+A_{13}A_{31} & A_{11}A_{32}+A_{12}A_{31} \\ A_{11}A_{21} & A_{12}A_{22} & A_{13}A_{23} & A_{12}A_{23}+A_{13}A_{22} & A_{11}A_{23}+A_{13}A_{21} & A_{11}A_{22}+A_{12}A_{21} \end{bmatrix}, \tag{5.2.10}$$

which is the desired transformation matrix for the stress-tensor components given by matrix (5.2.3). Thus, given transformation matrix (5.1.2), whose entries are A_{ij}, we can immediately write the corresponding $\underline{\mathbf{A}}$ using matrix (5.2.10). Matrix (5.2.10) was also formulated by Bond (1943).[4]

[4] Readers interested in a mathematical justification for matrix (5.2.10) and its modifications given in matrix (5.2.15) and—in particular—in matrix (5.2.19), which herein are obtained formally by steps leading to expressions (5.2.7) and (5.2.9), might refer to pages 126–127 of Bóna A., Bucataru, I., Slawinski, M.A., (2008) Space of $SO(3)$-orbits of elasticity tensors. Archives of Mechanics **60**(2), 121–136.

5.2.3. Transformation of strain-tensor components

The components of the strain-tensor expressed as a 3×3 matrix transform according to

$$\hat{\varepsilon} = \mathbf{A}\varepsilon\mathbf{A}^T , \tag{5.2.11}$$

where $\mathbf{A}$ stands for matrix (5.1.2) and the accent symbolizes the transformed entity. Equation (5.2.11) is the matrix form of equation (1.6.6), which is shown in Exercise 1.4. Herein, the strain-tensor components are a square matrix given by

$$\varepsilon = \begin{bmatrix} \varepsilon_{11} & \varepsilon_{12} & \varepsilon_{13} \\ \varepsilon_{12} & \varepsilon_{22} & \varepsilon_{23} \\ \varepsilon_{13} & \varepsilon_{23} & \varepsilon_{33} \end{bmatrix} , \tag{5.2.12}$$

whose symmetry results from definition (1.4.6); $\hat{\varepsilon}$ looks analogous to ε.

To consider the matrix form of stress-strain equations, we wish to rewrite the strain-tensor components as a single-column matrix in a manner similar to that shown in Section 5.2.2. As shown in stress-strain equations (4.2.8), the single column matrix, $\underline{\varepsilon}$, is formulated with factors of 2; namely,

$$\underline{\varepsilon} = \left[\varepsilon_{11}, \varepsilon_{22}, \varepsilon_{33}, 2\varepsilon_{23}, 2\varepsilon_{13}, 2\varepsilon_{12}\right]^T . \tag{5.2.13}$$

Hence, the corresponding transformation matrix differs from expression (5.2.10).[5] To account for the factors of 2, we can write

$$\underline{\hat{\varepsilon}} = \mathbf{F}\underline{\mathbf{A}}\mathbf{F}^{-1}\underline{\varepsilon}, \tag{5.2.14}$$

---

[5] If the stress-strain equations were written as a vector equation in the six-dimensional space, we would not require two distinct transformation matrices. Interested readers might refer to Chapman, C., (2004) Fundamentals of seismic wave propagation: Cambridge University Press, pp. 92–94, and to the footnote on page 145.

where $\underline{\mathbf{A}}$ is matrix (5.2.10) and

$$\mathbf{F} = \begin{bmatrix} 1 & 0 & 0 & 0 & 0 & 0 \\ 0 & 1 & 0 & 0 & 0 & 0 \\ 0 & 0 & 1 & 0 & 0 & 0 \\ 0 & 0 & 0 & 2 & 0 & 0 \\ 0 & 0 & 0 & 0 & 2 & 0 \\ 0 & 0 & 0 & 0 & 0 & 2 \end{bmatrix} .$$

Thus, the transformation matrix for the strain-tensor components, given by matrix (5.2.13), can be explicitly written as

$$\mathbf{M_A} = \mathbf{F}\underline{\mathbf{A}}\mathbf{F}^{-1} =$$

$$\begin{bmatrix} A_{11}A_{11} & A_{12}A_{12} & A_{13}A_{13} & A_{12}A_{13} & A_{11}A_{13} & A_{11}A_{12} \\ A_{21}A_{21} & A_{22}A_{22} & A_{23}A_{23} & A_{22}A_{23} & A_{21}A_{23} & A_{21}A_{22} \\ A_{31}A_{31} & A_{32}A_{32} & A_{33}A_{33} & A_{32}A_{33} & A_{31}A_{33} & A_{31}A_{32} \\ 2A_{21}A_{31} & 2A_{22}A_{32} & 2A_{23}A_{33} & A_{22}A_{33}+A_{23}A_{32} & A_{21}A_{33}+A_{23}A_{31} & A_{21}A_{32}+A_{22}A_{31} \\ 2A_{11}A_{31} & 2A_{12}A_{32} & 2A_{13}A_{33} & A_{12}A_{33}+A_{13}A_{32} & A_{11}A_{33}+A_{13}A_{31} & A_{11}A_{32}+A_{12}A_{31} \\ 2A_{11}A_{21} & 2A_{12}A_{22} & 2A_{13}A_{23} & A_{12}A_{23}+A_{13}A_{22} & A_{11}A_{23}+A_{13}A_{21} & A_{11}A_{22}+A_{12}A_{21} \end{bmatrix} \tag{5.2.15}$$

and expression (5.2.14) can be restated as

$$\underline{\hat{\varepsilon}} = \mathbf{M_A}\,\underline{\varepsilon}\,. \tag{5.2.16}$$

Consequently, given transformation matrix (5.1.2), whose entries are A_{ij}, we can immediately write the corresponding $\mathbf{M_A}$ using matrix (5.2.15).

5.2.4. Stress-strain equations in transformed coordinates

Now, having formulated $\underline{\hat{\sigma}}$ and $\underline{\hat{\varepsilon}}$, which are given by expressions (5.2.4) and (5.2.16), respectively, we can formally write the stress-strain equations in transformed coordinates as

$$\underline{\hat{\sigma}} = \hat{\mathbf{C}}\underline{\hat{\varepsilon}}\,.$$

Explicitly, we can write these equations as

$$\begin{bmatrix} \hat{\sigma}_{11} \\ \hat{\sigma}_{22} \\ \hat{\sigma}_{33} \\ \hat{\sigma}_{23} \\ \hat{\sigma}_{13} \\ \hat{\sigma}_{12} \end{bmatrix} = \begin{bmatrix} \hat{C}_{11} & \hat{C}_{12} & \hat{C}_{13} & \hat{C}_{14} & \hat{C}_{15} & \hat{C}_{16} \\ \hat{C}_{12} & \hat{C}_{22} & \hat{C}_{23} & \hat{C}_{24} & \hat{C}_{25} & \hat{C}_{26} \\ \hat{C}_{13} & \hat{C}_{23} & \hat{C}_{33} & \hat{C}_{34} & \hat{C}_{35} & \hat{C}_{36} \\ \hat{C}_{14} & \hat{C}_{24} & \hat{C}_{34} & \hat{C}_{44} & \hat{C}_{45} & \hat{C}_{46} \\ \hat{C}_{15} & \hat{C}_{25} & \hat{C}_{35} & \hat{C}_{45} & \hat{C}_{55} & \hat{C}_{56} \\ \hat{C}_{16} & \hat{C}_{26} & \hat{C}_{36} & \hat{C}_{46} & \hat{C}_{56} & \hat{C}_{66} \end{bmatrix} \begin{bmatrix} \hat{\varepsilon}_{11} \\ \hat{\varepsilon}_{22} \\ \hat{\varepsilon}_{33} \\ 2\hat{\varepsilon}_{23} \\ 2\hat{\varepsilon}_{13} \\ 2\hat{\varepsilon}_{12} \end{bmatrix}, \tag{5.2.17}$$

where, as discussed in Chapter 4, the elasticity matrix is symmetric due to the strain-energy function.

Recall that a continuum is formulated in terms of the stress-strain equations. Consequently, an examination of equations (4.2.8) and (5.2.17) leads to the following definition.

DEFINITION 5.2.1. The elastic properties of a continuum are invariant under an orthogonal transformation if $\mathbf{C} = \hat{\mathbf{C}}$, in other words, if the transformed elasticity matrix is identical to the original elasticity matrix.

Thus, material symmetry is exhibited by a change of the reference coordinate system that is undetectable by any mechanical experiment.

5.2.5. On matrix forms

Hooke's law stated by expression (4.2.8) results from writing stress-strain equations (3.2.1) explicitly in components, using formula (3.2.5) and symmetries (4.4.2). There are other notations that allow one to express the stress-strain equations in a matrix form; notably, a notation for which both stress and strain have the same form, as shown in Exercise 5.3. In this notation, both σ and ε are expressed as coordinate vectors relative to the same basis:

$$\begin{bmatrix} 1 & 0 & 0 \\ 0 & 0 & 0 \\ 0 & 0 & 0 \end{bmatrix}, \begin{bmatrix} 0 & 0 & 0 \\ 0 & 1 & 0 \\ 0 & 0 & 0 \end{bmatrix}, \begin{bmatrix} 0 & 0 & 0 \\ 0 & 0 & 0 \\ 0 & 0 & 1 \end{bmatrix}, \frac{1}{\sqrt{2}} \begin{bmatrix} 0 & 0 & 0 \\ 0 & 0 & 1 \\ 0 & 1 & 0 \end{bmatrix}, \frac{1}{\sqrt{2}} \begin{bmatrix} 0 & 0 & 1 \\ 0 & 0 & 0 \\ 1 & 0 & 0 \end{bmatrix}, \frac{1}{\sqrt{2}} \begin{bmatrix} 0 & 1 & 0 \\ 1 & 0 & 0 \\ 0 & 0 & 0 \end{bmatrix}.$$

The above basis is orthonormal: the scalar product of each two components—which is the sum of products of the corresponding entries—is zero, and the Frobenius norm of each component—which is the square root of the sum of squares of its entries—is unity. For matrix (5.14.6) of Exercise 5.3, namely,[6]

$$\begin{bmatrix} C_{11} & C_{12} & C_{13} & \sqrt{2}C_{14} & \sqrt{2}C_{15} & \sqrt{2}C_{16} \\ C_{12} & C_{22} & C_{23} & \sqrt{2}C_{24} & \sqrt{2}C_{25} & \sqrt{2}C_{26} \\ C_{13} & C_{23} & C_{33} & \sqrt{2}C_{34} & \sqrt{2}C_{35} & \sqrt{2}C_{36} \\ \sqrt{2}C_{14} & \sqrt{2}C_{24} & \sqrt{2}C_{34} & 2C_{44} & 2C_{45} & 2C_{46} \\ \sqrt{2}C_{15} & \sqrt{2}C_{25} & \sqrt{2}C_{35} & 2C_{45} & 2C_{55} & 2C_{56} \\ \sqrt{2}C_{16} & \sqrt{2}C_{26} & \sqrt{2}C_{36} & 2C_{46} & 2C_{56} & 2C_{66} \end{bmatrix}, \tag{5.2.18}$$

the transformation matrix—which is analogous to transformation (5.2.15) for matrix (4.2.7)—is

$$\begin{bmatrix} A_{11}^2 & A_{12}^2 & A_{13}^2 & \sqrt{2}A_{12}A_{13} & \sqrt{2}A_{11}A_{13} & \sqrt{2}A_{11}A_{12} \\ A_{21}^2 & A_{22}^2 & A_{23}^2 & \sqrt{2}A_{22}A_{23} & \sqrt{2}A_{21}A_{23} & \sqrt{2}A_{21}A_{22} \\ A_{31}^2 & A_{32}^2 & A_{33}^2 & \sqrt{2}A_{32}A_{33} & \sqrt{2}A_{31}A_{33} & \sqrt{2}A_{31}A_{32} \\ \sqrt{2}A_{21}A_{31} & \sqrt{2}A_{22}A_{32} & \sqrt{2}A_{23}A_{33} & A_{22}A_{33}+A_{23}A_{32} & A_{21}A_{33}+A_{23}A_{31} & A_{21}A_{32}+A_{22}A_{31} \\ \sqrt{2}A_{11}A_{31} & \sqrt{2}A_{12}A_{32} & \sqrt{2}A_{13}A_{33} & A_{12}A_{33}+A_{13}A_{32} & A_{11}A_{33}+A_{13}A_{31} & A_{11}A_{32}+A_{12}A_{31} \\ \sqrt{2}A_{11}A_{21} & \sqrt{2}A_{12}A_{22} & \sqrt{2}A_{13}A_{23} & A_{12}A_{23}+A_{13}A_{22} & A_{11}A_{23}+A_{13}A_{21} & A_{11}A_{22}+A_{12}A_{21} \end{bmatrix}. \tag{5.2.19}$$

and, in contrast to transformation (5.2.15), is the same for both stress and strain. Since both stress and strain are transformed by the same matrix, this notation is convenient for an advanced study of symmetry groups of elasticity tensors and distances in the space of these tensors, as exemplified in Exercise 5.9.[7]

Note that, as expected, matrices (5.2.10) and (5.2.15) as well as matrix (5.2.19) have an identical form except for factors of 2 and $\sqrt{2}$.

Using either notation, one must remember that it stems from the equation containing a fourth-rank tensor in three dimensions. For instance, to

[6] Readers interested in such a formulation might refer to Chapman, C., (2004) Fundamentals of seismic wave propagation: Cambridge University Press, pp. 92–94.

[7] Readers interested in using matrix (5.2.18) to study symmetry groups of the elasticity tensor might refer to Bóna, A., Bucataru, I., Slawinski, M.A. (2007) Coordinate-free classification of elasticity tensor: Journal of Elasticity **87**(2-3), 109–132, and readers interested in the concept of distance in the space of tensors might refer to Kochetov, M., Slawinski, M.A. (2009) On obtaining effective transversely isotropic elasticity tensors: Journal of Elasticity **94**(1), 1–13.

state the number of components $c_{ijk\ell}$ that a given entry, C_{mn}, represents, we must invoke symmetries (4.4.2), which are used to construct elasticity matrices. As an example: C_{11} represents only c_{1111} but C_{56} represents c_{1213}, c_{2113}, c_{1231}, c_{1312}, c_{3112}, c_{1321}, c_{3112}, and c_{3121}. Another insight into expressing three-dimensional operations by 6×6 matrices is provided by the discussion on page 148, below Lemma 5.3.2.

5.3. Condition for Material Symmetry

In view of Definition 5.2.1, the invariance to an orthogonal transformation imposes certain conditions on the elasticity matrix. For the transformed and the original matrices to be identical to one another, they must possess a particular form. Herein, we study a method where, given an orthogonal transformation, we can find the elasticity matrix that is invariant under this transformation and, hence, describe the material symmetry exhibited by a particular continuum. This method is stated in the following theorem.

THEOREM 5.3.1. *The elastic properties of a Hookean solid are invariant under an orthogonal transformation given by* $\mathbf{A}$, *if and only if*

$$\mathbf{C} = \mathbf{M}_{\mathbf{A}}^{T}\mathbf{C}\mathbf{M}_{\mathbf{A}}, \tag{5.3.1}$$

where $\mathbf{C}$ *is the elasticity matrix and* $\mathbf{M}_{\mathbf{A}}$ *is matrix (5.2.15).*

PROOF. Consider stress-strain equations (3.2.8); namely,

$$\underline{\sigma} = \mathbf{C}\underline{\varepsilon}, \tag{5.3.2}$$

which are expressed in terms of the original coordinate system. In the transformed coordinate system, these equations are written as

$$\underline{\hat{\sigma}} = \hat{\mathbf{C}}\underline{\hat{\varepsilon}}. \tag{5.3.3}$$

Consider equation (5.3.3). Substituting expressions (5.2.4) and (5.2.16) into equation (5.3.3), we obtain

$$\underline{\mathbf{A}}\underline{\sigma} = \hat{\mathbf{C}}\mathbf{M}_{\mathbf{A}}\underline{\varepsilon}.$$

Multiplying both sides by $\underline{\mathbf{A}}^{-1}$, we get

$$\underline{\sigma} = \underline{\mathbf{A}}^{-1}\hat{\mathbf{C}}\mathbf{M}_{\mathbf{A}}\underline{\varepsilon}.$$

According to Lemma 5.3.2 shown below, $\underline{\mathbf{A}}^{-1} = \mathbf{M}_{\mathbf{A}}^{T}$. Hence, we can write

$$\underline{\sigma} = \mathbf{M}_{\mathbf{A}}^{T}\hat{\mathbf{C}}\mathbf{M}_{\mathbf{A}}\,\underline{\varepsilon}. \tag{5.3.4}$$

Examining equations (5.3.2) and (5.3.4), we conclude that they both hold for any $\underline{\varepsilon}$, if and only if

$$\mathbf{C} = \mathbf{M}_{\mathbf{A}}^{T}\hat{\mathbf{C}}\mathbf{M}_{\mathbf{A}},$$

which is the relation between $\mathbf{C}$ and $\hat{\mathbf{C}}$ under transformation matrix $\mathbf{A}$. In view of Definition 5.2.1, invariance with respect to $\mathbf{A}$ means that

$$\mathbf{C} = \mathbf{M}_{\mathbf{A}}^{T}\mathbf{C}\mathbf{M}_{\mathbf{A}},$$

which is expression (5.3.1), as required. □

LEMMA 5.3.2. *Let $\underline{\mathbf{A}}$ be given by matrix (5.2.10) and $\mathbf{M}_{\mathbf{A}}$ be given by matrix (5.2.15). It follows that $\underline{\mathbf{A}}^{-1} = \mathbf{M}_{\mathbf{A}}^{T}$.*

PROOF. Recall that $\mathbf{A}$ is an orthogonal matrix. Let $\hat{\mathbf{x}} = \mathbf{A}\mathbf{x}$ be the transformed coordinate system. Consider expression $\sigma\mathbf{x}$, which, in view of expression (5.2.1), can be stated in the $\hat{\mathbf{x}}$-coordinates as $(\mathbf{A}\sigma\mathbf{A}^{\mathrm{T}})\hat{\mathbf{x}}$. Thus, in terms of the $\hat{\mathbf{x}}$-coordinates, the stress-tensor components become

$$\hat{\sigma} = \mathbf{A}\sigma\mathbf{A}^{T}. \tag{5.3.5}$$

Let us calculate $\underline{\mathbf{A}}^{-1}$. Since $\mathbf{A}$ is an orthogonal matrix, namely, $\mathbf{A}^{T} = \mathbf{A}^{-1}$, we can rewrite equation (5.3.5) as

$$\sigma = \mathbf{A}^{T}\hat{\sigma}\mathbf{A}. \tag{5.3.6}$$

Thus, in a manner analogous to that used to obtain expression (5.2.4), we can rewrite expression (5.3.6) in the desired notation, as

$$\underline{\sigma} = \underline{\mathbf{A}^{T}}\,\underline{\hat{\sigma}}, \tag{5.3.7}$$

where $\underline{\mathbf{A}^{T}}$ is constructed as matrix (5.2.10), but with the entries $A_{ij}^{T} = A_{ji}$ of $\mathbf{A}^{T}$ used in place of the entries A_{ij} of $\mathbf{A}$. Note that the order of operations matters; namely, $\underline{\mathbf{A}^T} \neq \underline{\mathbf{A}}^T$.[8] Comparing expression (5.2.4) with expression (5.3.7), we see that $\underline{\mathbf{A}^T} = \underline{\mathbf{A}}^{-1}$. Hence, we can write the inverse of matrix $\underline{\mathbf{A}}$ explicitly, as

$$\underline{\mathbf{A}}^{-1} =$$

$$\begin{bmatrix} A_{11}A_{11} & A_{21}A_{21} & A_{31}A_{31} & 2A_{21}A_{31} & 2A_{11}A_{31} & 2A_{11}A_{21} \\ A_{12}A_{12} & A_{22}A_{22} & A_{32}A_{32} & 2A_{22}A_{32} & 2A_{12}A_{32} & 2A_{12}A_{22} \\ A_{13}A_{13} & A_{23}A_{23} & A_{33}A_{33} & 2A_{23}A_{33} & 2A_{13}A_{33} & 2A_{13}A_{23} \\ A_{12}A_{13} & A_{22}A_{23} & A_{32}A_{33} & A_{22}A_{33}+A_{32}A_{23} & A_{12}A_{33}+A_{32}A_{13} & A_{12}A_{23}+A_{22}A_{13} \\ A_{11}A_{13} & A_{21}A_{23} & A_{31}A_{33} & A_{21}A_{33}+A_{31}A_{23} & A_{11}A_{33}+A_{31}A_{13} & A_{11}A_{23}+A_{21}A_{13} \\ A_{11}A_{12} & A_{21}A_{22} & A_{31}A_{32} & A_{21}A_{32}+A_{31}A_{22} & A_{11}A_{32}+A_{31}A_{12} & A_{11}A_{22}+A_{21}A_{12} \end{bmatrix}. \tag{5.3.8}$$

Comparing the entries of matrices (5.3.8) and (5.2.15), we notice that the former one is equal to the transpose of the latter, as required. □

Matrix (5.3.8) is the inverse of matrix (5.2.10) in the sense that $\underline{\mathbf{A}}^{-1}\underline{\mathbf{A}} = \underline{\mathbf{A}}\,\underline{\mathbf{A}}^{-1} = I$, provided that $\mathbf{A}$ is an orthogonal transformation, which imposes conditions on $\underline{\mathbf{A}}$, since $\underline{\mathbf{A}}$ is a 6×6 matrix that represents an orthogonal transformation in three dimensions. To clarify this sentence, let us consider two entries of the matrix resulting from the product of matrices (5.2.10) and (5.3.8): a main-diagonal entry and an off-main-diagonal entry, say, $(3,3)$ and $(4,3)$. The former is $(A_{31}^2 + A_{32}^2 + A_{33}^2)^2$, which is the squared length of the third row of $\mathbf{A}$, which is stated in expression (5.1.2). Since orthogonal transformations require that $\mathbf{A}$ be orthonormal, its columns and rows—viewed as vectors—must be of unit length; hence, $(A_{31}^2 + A_{32}^2 + A_{33}^2)^2 = 1$, as expected for an identity matrix. The latter is $(A_{21}A_{31} + A_{22}A_{32} + A_{23}A_{33})(A_{31}^2 + A_{32}^2 + A_{33}^2)$. The first parenthesis is the scalar product of the second and third row of matrix (5.1.2). The orthonormality of $\mathbf{A}$ implies that its columns and rows, viewed as vectors, are orthogonal to each other; hence, $A_{21}A_{31} + A_{22}A_{32} + A_{23}A_{33} = 0$, and the off-main-diagonal entry is zero, as expected for an identity matrix.

---

[8] The reason for this noncommutativity of operations is that the components of the stress and strain tensors are expressed in different bases. This is not the case of the formulation described in Section 5.2.5, where these operations are commutative and, hence, the transpose of matrix (5.2.19) is tantamount to the transpose of matrix (5.1.2)—a consequence of the notation presented in that section.

In a similar manner, one can verify that all entries of the matrix resulting from the product of matrices (5.2.10) and (5.3.8) satisfy these requirements. Consequently, matrix (5.3.8) is the inverse of matrix (5.2.10) in the sense that these matrices represent orthogonal transformations in three dimensions.

Expression (5.3.1) is a concise statement of conditions that the entries of the elasticity matrix must obey for the continuum described by stress-strain equations (4.2.8) to be invariant under an orthogonal transformation. Given transformation (5.1.2), expression (5.3.1) is convenient to apply since it contains twenty-one linear equations for C_{mn}. Furthermore, considering transformations (5.8.1) and (5.6.1), which are discussed below, matrix $\mathbf{M}_A$ is significantly simplified, since $A_{13} = A_{23} = A_{31} = A_{32} = 0$, while $A_{33} = \pm 1$.

In view of Section 5.2.5, we state a corollary of Theorem 5.3.1.

COROLLARY 5.3.3. *The elastic properties of a Hookean solid are invariant under an orthogonal transformation given by* $\mathbf{A}$, *if and only if*

$$\mathbf{C} = \tilde{\mathbf{A}}^T \mathbf{C} \tilde{\mathbf{A}}, \tag{5.3.9}$$

where $\mathbf{C}$ *is the elasticity tensor whose components are matrix (5.2.18) and* $\tilde{\mathbf{A}}$ *is matrix (5.2.19).*

REMARK 5.3.4. The validity of conditions (5.3.1) and (5.14.26) is not affected if we write $\mathbf{C} = \mathbf{M}_{\mathbf{A}} \mathbf{C} \mathbf{M}_{\mathbf{A}}^T$ and $\mathbf{C} = \tilde{\mathbf{A}} \mathbf{C} \tilde{\mathbf{A}}^T$, respectively; the corresponding expressions vary by the sense of rotation, which does not influence the symmetry condition.

As indicated in Section 5.2.5, expressions (5.2.18) and (5.2.19) are convenient in many applications. Hence—particularly for examinations of subtleties of the rotational invariance—the condition provided by this corollary is more convenient than condition (5.3.1).

5.4. Point Symmetry

Let us illustrate condition (5.3.1) by describing the material symmetry that is valid for all continua described by stress-strain equations (4.2.8). In the following theorem, we show that at every point, an elastic continuum is invariant under the reflection through the origin of the coordinate system located at this point. Such a reflection is described by the transformation matrix given by

$$\mathbf{A}_{-\mathbf{I}} := \begin{bmatrix} -1 & 0 & 0 \\ 0 & -1 & 0 \\ 0 & 0 & -1 \end{bmatrix} = -\mathbf{I}. \tag{5.4.1}$$

THEOREM 5.4.1. *At every point, a continuum given by stress-strain equations (4.2.8) is invariant under the reflection about the origin of a coordinate system that is located at that point.*

PROOF. Consider transformation matrix (5.4.1). Matrix (5.2.15) becomes

$$\mathbf{M}_{\mathbf{A}_{-\mathbf{I}}} = \begin{bmatrix} 1&0&0&0&0&0 \\ 0&1&0&0&0&0 \\ 0&0&1&0&0&0 \\ 0&0&0&1&0&0 \\ 0&0&0&0&1&0 \\ 0&0&0&0&0&1 \end{bmatrix} = \mathbf{I}.$$

Hence, condition (5.3.1) becomes

$$\mathbf{C} = \mathbf{I}^T \mathbf{C} \mathbf{I},$$

which is identically satisfied for any $\mathbf{C}$. □

This means that the symmetry group of every continuum contains $\mathbf{A}_{-\mathbf{I}}$.

5.5. Generally Anisotropic Continuum

A generally anisotropic continuum is the most general continuum describable by stress-strain equations (4.2.8). The elasticity matrix of a generally anisotropic continuum is given by

$$\mathbf{C}_{\mathrm{GEN}} = \begin{bmatrix} C_{11} & C_{12} & C_{13} & C_{14} & C_{15} & C_{16} \\ C_{12} & C_{22} & C_{23} & C_{24} & C_{25} & C_{26} \\ C_{13} & C_{23} & C_{33} & C_{34} & C_{35} & C_{36} \\ C_{14} & C_{24} & C_{34} & C_{44} & C_{45} & C_{46} \\ C_{15} & C_{25} & C_{35} & C_{45} & C_{55} & C_{56} \\ C_{16} & C_{26} & C_{36} & C_{46} & C_{56} & C_{66} \end{bmatrix}. \tag{5.5.1}$$

The only symmetry exhibited by a generally anisotropic continuum is point symmetry. Hence, a generally anisotropic continuum is described by an elasticity matrix that contains twenty-one independent entries.

5.6. Monoclinic Continuum

5.6.1. Elasticity matrix

A continuum whose symmetry group contains a reflection about a plane through the origin is said to be monoclinic. For convenience, let us choose the coordinate system such that this reflection takes place about the x_1x_2-plane, which means, along the x_3-axis.

Consider the orthogonal transformation that is represented by matrix (5.1.2) in the form given by

$$\mathbf{A} = \begin{bmatrix} \cos\Theta & \sin\Theta & 0 \\ -\sin\Theta & \cos\Theta & 0 \\ 0 & 0 & -1 \end{bmatrix}. \tag{5.6.1}$$

Matrix (5.6.1), whose determinant is equal to negative unity, corresponds to the composition of two transformations, namely, rotation by angle Θ about the x_3-axis and reflection about the x_1x_2-plane. To consider the reflection alone, we let $\Theta = 0$ to obtain

$$\mathbf{A}_3 = \begin{bmatrix} 1 & 0 & 0 \\ 0 & 1 & 0 \\ 0 & 0 & -1 \end{bmatrix}. \tag{5.6.2}$$

Following expression (5.2.15), the corresponding matrix $\mathbf{M}_A$ is

$$\mathbf{M}_{A_3} = \begin{bmatrix} 1 & 0 & 0 & 0 & 0 & 0 \\ 0 & 1 & 0 & 0 & 0 & 0 \\ 0 & 0 & 1 & 0 & 0 & 0 \\ 0 & 0 & 0 & -1 & 0 & 0 \\ 0 & 0 & 0 & 0 & -1 & 0 \\ 0 & 0 & 0 & 0 & 0 & 1 \end{bmatrix}.$$

Theorem 5.3.1 requires that the elasticity matrix satisfies condition (5.3.1). This condition requires that

$$\mathbf{C} = \mathbf{M}_{\mathbf{A}_3}^T \mathbf{C} \mathbf{M}_{\mathbf{A}_3},$$

which we can explicitly write as

$$\begin{bmatrix} C_{11} & C_{12} & C_{13} & C_{14} & C_{15} & C_{16} \\ C_{12} & C_{22} & C_{23} & C_{24} & C_{25} & C_{26} \\ C_{13} & C_{23} & C_{33} & C_{34} & C_{35} & C_{36} \\ C_{14} & C_{24} & C_{34} & C_{44} & C_{45} & C_{46} \\ C_{15} & C_{25} & C_{35} & C_{45} & C_{55} & C_{56} \\ C_{16} & C_{26} & C_{36} & C_{46} & C_{56} & C_{66} \end{bmatrix} = \begin{bmatrix} C_{11} & C_{12} & C_{13} & -C_{14} & -C_{15} & C_{16} \\ C_{12} & C_{22} & C_{23} & -C_{24} & -C_{25} & C_{26} \\ C_{13} & C_{23} & C_{33} & -C_{34} & -C_{35} & C_{36} \\ -C_{14} & -C_{24} & -C_{34} & C_{44} & C_{45} & -C_{46} \\ -C_{15} & -C_{25} & -C_{35} & C_{45} & C_{55} & -C_{56} \\ C_{16} & C_{26} & C_{36} & -C_{46} & -C_{56} & C_{66} \end{bmatrix}.$$

The equality of these two matrices implies that

$$C_{14} = C_{15} = C_{24} = C_{25} = C_{34} = C_{35} = C_{46} = C_{56} = 0. \tag{5.6.3}$$

Thus, the elasticity matrix of a continuum that possesses a reflection symmetry along the x_3-axis is

$$\mathbf{C}_{\mathrm{MONO}x_3} = \begin{bmatrix} C_{11} & C_{12} & C_{13} & 0 & 0 & C_{16} \\ C_{12} & C_{22} & C_{23} & 0 & 0 & C_{26} \\ C_{13} & C_{23} & C_{33} & 0 & 0 & C_{36} \\ 0 & 0 & 0 & C_{44} & C_{45} & 0 \\ 0 & 0 & 0 & C_{45} & C_{55} & 0 \\ C_{16} & C_{26} & C_{36} & 0 & 0 & C_{66} \end{bmatrix}. \tag{5.6.4}$$

Hence, a monoclinic continuum is described by an elasticity matrix that contains thirteen independent entries.

5.6.2. Vanishing of tensor components

We can recognize the pattern of the vanishing elasticity parameters in Section 5.6.1 by considering the elasticity tensor rather than the elasticity matrix. Recalling expression (3.2.5), we can write expression (5.6.3) as

$$c_{1123} = c_{1113} = c_{2223} = c_{2213} = c_{3323} = c_{3313} = c_{2312} = c_{1312} = 0. \quad (5.6.5)$$

Examining this result, we see that all the components that disappear as a consequence of the invariance to the reflection along the x_3-axis have an odd number of subscript 3.

We can exemplify this pattern in view of tensor algebra. In general, the elasticity tensor transforms according to the rule given by

$$\hat{c}_{mnpr} = \sum_{i=1}^{3}\sum_{j=1}^{3}\sum_{k=1}^{3}\sum_{\ell=1}^{3} A_{mi}A_{nj}A_{pk}A_{rl}c_{ijk\ell}, \qquad m,n,p,r \in \{1,2,3\},$$

where $\mathbf{A}$ is a transformation matrix and $\hat{c}_{mnpr}$ denotes the components of the elasticity tensor in the transformed coordinates. In our case, due to the invariance resulting from the symmetry of the elasticity tensor, $\hat{c}_{mnpr} = c_{mnpr}$, as stated in Definition 5.2.1, we write

$$c_{mnpr} = \sum_{i=1}^{3}\sum_{j=1}^{3}\sum_{k=1}^{3}\sum_{\ell=1}^{3} A_{mi}A_{nj}A_{pk}A_{rl}c_{ijk\ell}, \qquad m,n,p,r \in \{1,2,3\},$$

which is the tensor form of condition (5.3.1). To illustrate the vanishing of a component, let us consider, for instance, c_{1113}. We can write

$$c_{1113} = \sum_{i=1}^{3}\sum_{j=1}^{3}\sum_{k=1}^{3}\sum_{\ell=1}^{3} A_{1i}A_{1j}A_{1k}A_{3l}c_{ijk\ell}.$$

Since matrix (5.6.2) is a diagonal matrix, let us write only the nonzero entries of $\mathbf{A}$ to get

$$c_{1113} = A_{11}A_{11}A_{11}A_{33}c_{1113}.$$

Specifically, we obtain

$$c_{1113} = (1)(1)(1)(-1)c_{1113} = -c_{1113}.$$

This equality requires that $c_{1113} = 0$, as expected.

5.6.3. Natural coordinate system

In general, in an arbitrary coordinate system, all the entries of an elasticity matrix are nonzero. A natural coordinate system is a particular system within which an elasticity matrix has the fewest possible number of nonzero independent entries.

For any continuum, there exists at least three natural coordinate systems.[9] Hence, in principle, we could also rotate our orthonormal coordinate system so as to find a natural coordinate system for a generally anisotropic continuum, discussed above in Section 5.5. This, however, is not a simple task. Yet, in a natural coordinate system, a generally anisotropic continuum can be described by eighteen independent elasticity parameters and three Euler's angles that specify the orientation of this system.

A monoclinic continuum can be conveniently used to illustrate the concept of a natural coordinate system. The coordinate system that is used to formulate matrix (5.6.4) has the x_3-axis coinciding with the normal to the symmetry plane of the continuum. In other words, the x_1x_2-plane coincides with the symmetry plane. The rotation of the coordinate system about the x_3-axis allows us to further reduce the number of elasticity parameters needed to describe a monoclinic continuum. An appropriate rotation reduces matrix (5.6.4) to a new matrix that contains only twelve parameters. This orientation of the coordinate system is a natural coordinate system for a monoclinic continuum.

Rotation of the coordinate axes about the x_3-axis by angle Θ, where the angle is given by

$$\tan(2\Theta) = \frac{2C_{45}}{C_{44} - C_{55}}, \tag{5.6.6}$$

---

[9] Interested readers might refer to Fedorov, F.I., (1968) Theory of elastic waves in crystals: Plenum Press, New York, p. 25 and pp. 110 – 111, to Helbig, K., (1994) Foundations of anisotropy for exploration seismics: Pergamon, pp. 163–170, to Lanczos, C., (1949/1986) The variational principles of mechanics: Dover, p. 373, to Schouten, J.A. (1951/1989) Tensor analysis for physicists: Dover, p. 162., and to Winterstein, D.F., (1990) Velocity anisotropy terminology for geophysicists: Geophysics, **55**, 1070–1088.

leads to a new set of elasticity parameters, which we denote by $\hat{C}_{mn}$.[10] In the new set, $\hat{C}_{45}$ vanishes and elasticity matrix (5.6.4) is reduced to

$$\hat{\mathbf{C}}_{\mathrm{MONO}} = \begin{bmatrix} \hat{C}_{11} & \hat{C}_{12} & \hat{C}_{13} & 0 & 0 & \hat{C}_{16} \\ \hat{C}_{12} & \hat{C}_{22} & \hat{C}_{23} & 0 & 0 & \hat{C}_{26} \\ \hat{C}_{13} & \hat{C}_{23} & \hat{C}_{33} & 0 & 0 & \hat{C}_{36} \\ 0 & 0 & 0 & \hat{C}_{44} & 0 & 0 \\ 0 & 0 & 0 & 0 & \hat{C}_{55} & 0 \\ \hat{C}_{16} & \hat{C}_{26} & \hat{C}_{36} & 0 & 0 & \hat{C}_{66} \end{bmatrix}. \tag{5.6.7}$$

Hence, in a natural coordinate system, a monoclinic continuum is described by twelve independent elasticity parameters and the angle Θ that describes the orientation of the coordinate system and corresponds to Euler's angle.

We can obtain equation (5.6.6) by using the rotation in the x_1x_2-plane to diagonalize the submatrix of interest in matrix (5.6.4); namely,

$$\begin{bmatrix} \hat{C}_{44} & 0 \\ 0 & \hat{C}_{55} \end{bmatrix} = \begin{bmatrix} \cos\Theta & \sin\Theta \\ -\sin\Theta & \cos\Theta \end{bmatrix} \begin{bmatrix} C_{44} & C_{45} \\ C_{45} & C_{55} \end{bmatrix} \begin{bmatrix} \cos\Theta & -\sin\Theta \\ \sin\Theta & \cos\Theta \end{bmatrix}. \tag{5.6.8}$$

We note that considering matrices (5.6.4) and (5.6.7), $C_{mn} \neq \hat{C}_{mn}$, where $m,n \in \{1,\ldots,6\}$. In other words, the rotation about the x_3-axis by the angle (5.6.6) results in a new elasticity matrix to describe the same continuum.

Let us comment on the natural coordinate systems in view of our subsequent work. The orthotropic continuum discussed in Section 5.7 possesses three orthogonal symmetry planes; hence, in accordance with our use of the orthonormal coordinate system, we consider it, *ab initio*, in its natural coordinate system. While discussing the trigonal continuum, in Section 5.8, we again search for its natural coordinate system. The tetragonal continuum, discussed in Section 5.9.1, is obtained as a special case of the orthotropic one; hence, it is already in its natural coordinate system. For a transversely isotropic continuum, which we discuss in Section 5.10, due to its property of rotation invariance, all orthonormal coordinate systems whose one axis

[10]Readers interested in the formulation of expression (5.6.6), in the context presented herein, might refer to Helbig, K., (1994) Foundations of anisotropy for exploration seismics: Pergamon, pp. 82–83, 94–95 and 110–116.

coincides with the rotation axis are natural coordinate systems. Hence, our initial formulation is already set in a natural coordinate system. The cubic continuum, discussed in Section 5.11, can be viewed as a special case of the tetragonal continuum; hence, it is already in its natural coordinate system. Finally, for an isotropic continuum, discussed in Section 5.12, all orthonormal coordinate systems are natural.

In the context of ray theory, natural coordinate systems are associated with pure-mode directions, as discussed in Section 9.2.2. In particular, in Section 9.2.2, expression (5.6.6) is obtained by considering the displacement directions of the three types of waves that propagate along the x_3-axis in a monoclinic continuum.

5.7. Orthotropic Continuum

An orthotropic continuum is a continuum that possesses three orthogonal symmetry planes. For convenience, let us choose the coordinate system such that the symmetry planes coincide with the coordinate planes. This is a natural coordinate system for an orthotropic continuum. Hence, the transformation matrices are given by

$$\mathbf{A}_1 = \begin{bmatrix} -1 & 0 & 0 \\ 0 & 1 & 0 \\ 0 & 0 & 1 \end{bmatrix}, \qquad (5.7.1)$$

$$\mathbf{A}_2 = \begin{bmatrix} 1 & 0 & 0 \\ 0 & -1 & 0 \\ 0 & 0 & 1 \end{bmatrix}, \qquad (5.7.2)$$

and $\mathbf{A}_3$, given by matrix (5.6.2), which correspond to the reflections along the x_1-axis, the x_2-axis and the x_3-axis, respectively.

In view of the properties of the symmetry group, the elasticity matrix of an orthotropic continuum can be obtained using any two of the three symmetry planes. This is shown by the following theorem.

THEOREM 5.7.1. *If a continuum given by stress-strain equations (4.2.8) is invariant under the reflection about two orthogonal planes, it must also be invariant under the reflection about the third orthogonal plane.*

PROOF. Consider a continuum that is invariant under the reflections along the x_1-axis and along the x_3-axis. The corresponding orthogonal transformations are given by matrices (5.7.1) and (5.6.2), respectively. Also, following Theorem 5.4.1, all continua possess point symmetry. In other words, they are invariant under the transformation given by matrix (5.4.1). Since all these transformations belong to the symmetry group of an orthotropic continuum, their products also belong to this group. Consider

$$(\mathbf{A}_1)(\mathbf{A}_3)(\mathbf{A}_{-\mathbf{I}}) = \begin{bmatrix} 1 & 0 & 0 \\ 0 & -1 & 0 \\ 0 & 0 & 1 \end{bmatrix},$$

which we recognize to be matrix (5.7.2) corresponding to the reflections along the x_2-axis. Thus, in view of point symmetry, invariance to the reflections about two orthogonal planes also implies invariance to the reflection about the third orthogonal plane. □

Therefore, to obtain the elasticity matrix of an orthotropic continuum, let us use matrices (5.6.2) and (5.7.2). Matrix (5.6.2) is also used in Section 5.6 where we obtain the relations given in expression (5.6.3); namely,

$$C_{14} = C_{15} = C_{24} = C_{25} = C_{34} = C_{35} = C_{46} = C_{56} = 0. \tag{5.7.3}$$

Using matrix (5.7.2), we write condition (5.3.1) as

$$\mathbf{C} = \mathbf{M}_{A_2}^T \mathbf{C} \mathbf{M}_{A_2}, \tag{5.7.4}$$

where, using matrix (5.2.15), we explicitly write matrix $\mathbf{M}_{A_2}$ as

$$\mathbf{M}_{A_2} = \begin{bmatrix} 1 & 0 & 0 & 0 & 0 & 0 \\ 0 & 1 & 0 & 0 & 0 & 0 \\ 0 & 0 & 1 & 0 & 0 & 0 \\ 0 & 0 & 0 & -1 & 0 & 0 \\ 0 & 0 & 0 & 0 & 1 & 0 \\ 0 & 0 & 0 & 0 & 0 & -1 \end{bmatrix}. \tag{5.7.5}$$

Let us also use—in this condition—$\mathbf{C}$ given by $\mathbf{C}_{\mathrm{MONO}x_3}$, which is the matrix stated in expression (5.6.4). Thus, solving equation (5.7.4), we get the vanishing entries of $\mathbf{C}$ that distinguish the orthotropic continuum from the monoclinic continuum; they are

$$C_{16} = C_{26} = C_{36} = C_{45} = 0\,. \qquad (5.7.6)$$

Combining relations (5.7.3) and (5.7.6), we can write the elasticity matrix for an orthotropic continuum as

$$\mathbf{C}_{\mathrm{ORTHO}x_1x_2x_3} = \begin{bmatrix} C_{11} & C_{12} & C_{13} & 0 & 0 & 0 \\ C_{12} & C_{22} & C_{23} & 0 & 0 & 0 \\ C_{13} & C_{23} & C_{33} & 0 & 0 & 0 \\ 0 & 0 & 0 & C_{44} & 0 & 0 \\ 0 & 0 & 0 & 0 & C_{55} & 0 \\ 0 & 0 & 0 & 0 & 0 & C_{66} \end{bmatrix}. \qquad (5.7.7)$$

Hence, in a natural coordinate system, an orthotropic continuum is described by nine independent elasticity parameters.

To conclude, in view of Section 5.6.2, we can rewrite expression (5.7.6) as

$$c_{1112} = c_{2221} = c_{3312} = c_{2313} = 0\,. \qquad (5.7.8)$$

Examining this result, we see that all the elasticity-tensor components that disappear as a result of the invariance under the reflection along the x_2-axis have an odd number of subscript 2, as expected. Combining expressions (5.6.5) and (5.7.8), which—in view of Theorem 5.7.1—correspond to the elasticity tensor that is invariant under reflections along the x_1-axis, the x_2-axis and the x_3-axis, we see that each vanishing component has two occurrences of an odd number of subscript 1, 2 or 3. This is consistent with Theorem 5.7.1, which implies that each component vanishes independently twice; in other words, it vanishes for reflections along two, among the three axes. Also, while examining expressions (5.6.5) and (5.7.8), we can write the components that vanish as a result of the invariance under the reflection along the x_1-axis; they are: c_{1113}, c_{2213}, c_{3313}, c_{2312}, c_{1112}, c_{2221}, c_{3312} and c_{2313}.

5.8. Trigonal Continuum

5.8.1. Elasticity matrix

A trigonal continuum is a continuum whose symmetry group contains rotations about an axis by Θ, where $\Theta = 2\pi/3$ and $\Theta = 4\pi/3$. To obtain the elasticity matrix for this continuum, we consider the orthogonal transformation that is represented by matrix (5.1.2) in the form given by

$$\mathbf{A}_{x_3\Theta} = \begin{bmatrix} \cos\Theta & \sin\Theta & 0 \\ -\sin\Theta & \cos\Theta & 0 \\ 0 & 0 & 1 \end{bmatrix}. \tag{5.8.1}$$

Matrix (5.8.1), whose determinant is equal to unity, corresponds to rotation by angle Θ about the x_3-axis. Therefore, considering $\Theta = 2\pi/3$, we obtain

$$\mathbf{A}_{x_3 2\pi/3} = \begin{bmatrix} \cos\frac{2\pi}{3} & \sin\frac{2\pi}{3} & 0 \\ -\sin\frac{2\pi}{3} & \cos\frac{2\pi}{3} & 0 \\ 0 & 0 & 1 \end{bmatrix}.$$

Following expression (5.2.15), we write the corresponding 6×6 transformation matrix, $\mathbf{M}_{\mathbf{A}_{x_3 2\pi/3}}$, as

$$\begin{bmatrix} \cos^2\frac{2\pi}{3} & \sin^2\frac{2\pi}{3} & 0 & 0 & 0 & \cos\frac{2\pi}{3}\sin\frac{2\pi}{3} \\ \sin^2\frac{2\pi}{3} & \cos^2\frac{2\pi}{3} & 0 & 0 & 0 & -\sin\frac{2\pi}{3}\cos\frac{2\pi}{3} \\ 0 & 0 & 1 & 0 & 0 & 0 \\ 0 & 0 & 0 & \cos\frac{2\pi}{3} & -\sin\frac{2\pi}{3} & 0 \\ 0 & 0 & 0 & \sin\frac{2\pi}{3} & \cos\frac{2\pi}{3} & 0 \\ -2\cos\frac{2\pi}{3}\sin\frac{2\pi}{3} & 2\sin\frac{2\pi}{3}\cos\frac{2\pi}{3} & 0 & 0 & 0 & \cos^2\frac{2\pi}{3} - \sin^2\frac{2\pi}{3} \end{bmatrix}. \tag{5.8.2}$$

Using matrix (5.8.2) in condition (5.3.1), we obtain

$$C_{16} = C_{26} = C_{34} = C_{35} = C_{36} = C_{45} = 0 \tag{5.8.3}$$

and

$$\begin{aligned} &C_{11} = C_{22}, \quad C_{13} = C_{23}, \quad C_{14} = -C_{56} = -C_{24}, \\ &C_{66} = \frac{1}{2}(C_{22} - C_{12}), \quad C_{15} = C_{46} = -C_{25}, \quad C_{44} = C_{55}. \end{aligned} \tag{5.8.4}$$

We do not need to repeat this process for $\Theta = 4\pi/3$ since for invariance under a such rotation we would obtain the results identical to the ones shown in expressions (5.8.3) and (5.8.4). This is so, because $\mathbf{A}_{x_3 4\pi/3}$ is a morphism of groups $\mathbf{A}_{x_3 2\pi/3}$ and $\mathbf{A}_{x_3 2\pi/3}$. In other words,

$$\mathbf{M}_{\mathbf{A}_{x_3 4\pi/3}} = \mathbf{M}_{\mathbf{A}_{x_3 2\pi/3}} \mathbf{M}_{\mathbf{A}_{x_3 2\pi/3}},$$

where $\mathbf{M}_{\mathbf{A}_{x_3 2\pi/3}}$ is matrix (5.8.2).

Thus, using expressions (5.8.3) and (5.8.4), we can already write the elasticity matrix for the trigonal continuum as

$$\mathbf{C}_{\mathrm{TRIGONAL}_{x_3}} = \begin{bmatrix} C_{11} & C_{12} & C_{13} & C_{14} & C_{15} & 0 \\ C_{12} & C_{11} & C_{13} & -C_{14} & -C_{15} & 0 \\ C_{13} & C_{13} & C_{33} & 0 & 0 & 0 \\ C_{14} & -C_{14} & 0 & C_{44} & 0 & -C_{15} \\ C_{15} & -C_{15} & 0 & 0 & C_{44} & C_{14} \\ 0 & 0 & 0 & -C_{15} & C_{14} & \frac{C_{11} - C_{12}}{2} \end{bmatrix}. \tag{5.8.5}$$

Examining matrix (5.8.5), we might conclude that the trigonal continuum is described by an elasticity matrix with seven independent entries. However, as shown in the section that follows, this number is reduced to six by a rotation of the coordinate system.

5.8.2. Natural coordinate system

With an appropriate rotation of the coordinate system about the x_3-axis, matrix (5.8.5) can be further reduced to a new matrix with only six independent entries. This rotation is given by

$$\tan(3\Theta) = \frac{C_{14}}{C_{15}}, \tag{5.8.6}$$

which, in a manner similar to expression (5.6.6), is a relation between the values of components and orientation of coordinate systems in which a tensor is expressed. If the tensor is expressed with respect to its natural coordinates, $C_{14} = 0$, which implies—in view of expression (5.8.6)—that $\Theta = 0$, as expected. As a result of this rotation, C_{14} vanishes and matrix (5.8.5) becomes

$$\hat{\mathbf{C}}_{\text{TRIGONAL}} = \begin{bmatrix} \hat{C}_{11} & \hat{C}_{12} & \hat{C}_{13} & 0 & \hat{C}_{15} & 0 \\ \hat{C}_{12} & \hat{C}_{11} & \hat{C}_{13} & 0 & -\hat{C}_{15} & 0 \\ \hat{C}_{13} & \hat{C}_{13} & \hat{C}_{33} & 0 & 0 & 0 \\ 0 & 0 & 0 & \hat{C}_{44} & 0 & -\hat{C}_{15} \\ \hat{C}_{15} & -\hat{C}_{15} & 0 & 0 & \hat{C}_{44} & 0 \\ 0 & 0 & 0 & -\hat{C}_{15} & 0 & \dfrac{\hat{C}_{11}-\hat{C}_{12}}{2} \end{bmatrix}.$$

Therefore, in a natural coordinate system, the trigonal continuum is described by six independent elasticity parameters, as illustrated in Exercise 5.5.

5.9. Tetragonal Continuum

5.9.1. Elasticity matrix

A tetragonal continuum is a continuum whose symmetry group contains a four-fold rotation and a reflection through the plane that contains the axis of rotation. Let us choose the coordinate system such that the x_3-axis coincides with the rotation axis; also, for convenience, we let the reflection be along the x_2-axis. This is a natural coordinate system for a tetragonal continuum.

The corresponding transformation matrices are

$$\mathbf{A}_{x_3\pi/2} = \begin{bmatrix} 0 & 1 & 0 \\ -1 & 0 & 0 \\ 0 & 0 & 1 \end{bmatrix}, \tag{5.9.1}$$

which is matrix (5.8.1) with $\Theta = \pi/2$, and matrix (5.7.2). These matrices correspond to, respectively, the rotation about the x_3-axis and to the reflections through the plane that contains that axis, which we choose to be the reflection along the x_2-axis.

Matrix (5.7.2) belongs also to the symmetry group of an orthotropic continuum discussed in Section 5.7, where we obtain the relations given in expression (5.7.6); namely,

$$C_{16} = C_{26} = C_{36} = C_{45} = 0. \tag{5.9.2}$$

These relations also apply to a tetragonal continuum. The additional relations result from matrix (5.9.1). Using matrix (5.9.1), condition (5.3.1) becomes

$$\mathbf{C} = \mathbf{M}^T_{\mathbf{A}_{x_3\pi/2}} \mathbf{C} \mathbf{M}_{\mathbf{A}_{x_3\pi/2}}$$

and results in relations given by

$$C_{22} = C_{11}, C_{23} = C_{13}, C_{55} = C_{44}. \tag{5.9.3}$$

Combining relations (5.9.2) and (5.9.3), we obtain the elasticity matrix of a tetragonal continuum; namely,

$$\mathbf{C}_{\mathrm{TETRA}} = \begin{bmatrix} C_{11} & C_{12} & C_{13} & 0 & 0 & 0 \\ C_{12} & C_{11} & C_{13} & 0 & 0 & 0 \\ C_{13} & C_{13} & C_{33} & 0 & 0 & 0 \\ 0 & 0 & 0 & C_{44} & 0 & 0 \\ 0 & 0 & 0 & 0 & C_{44} & 0 \\ 0 & 0 & 0 & 0 & 0 & C_{66} \end{bmatrix}. \tag{5.9.4}$$

Thus, in a natural coordinate system, only six independent elasticity parameters are needed to describe a tetragonal continuum.

Matrix (5.9.4) is a special case of matrix (5.7.7); further relations are given by expression (5.9.3). The fact of matrix (5.9.4) being a special case

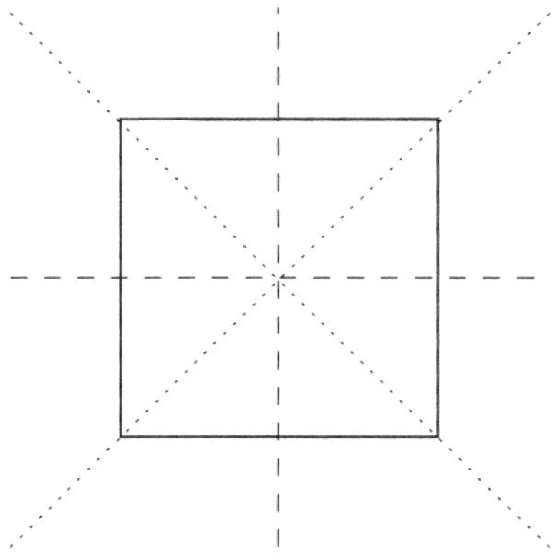

Fig. 5.9.1 Illustration of two tetragonal invariances; one corresponds to $n\pi/2$ rotations of dashed lines and the other to $n\pi/2$ rotations of dotted lines. These cases differ by $n\pi/4$.

of matrix (5.7.7) is to be expected, since the symmetry group of a tetragonal continuum contains the transformations of an orthogonal continuum, together with further constraints provided by matrix (5.9.1).

5.9.2. Natural coordinate system

In a manner similar to the ones discussed for the monoclinic and trigonal continua, we can express a given tetragonal tensor with respect to distinct coordinate systems, whose orientations are associated with the symmetry exhibited by the tensor. In view of the fourfold rotational symmetry, we can consider the square shown in Figure 5.9.1. A convenient choice of the coordinates can be such that the axes coincide with the dashed lines or with the dotted lines. The two orientations differ by $\pi/4$. In either case, the elasticity parameters of a tetragonal tensor are given by matrix (5.9.4).

If the coordinate axes do not coincide with the dashed or dotted lines, but are contained within the rotation plane, there are two more nonzero entries. They are not independent from one another nor from the parameters contained in matrix (5.9.4). They result from a rotation of the coordinate system. To apply such a rotation to matrix (5.9.4), we write it in the form of matrix (5.2.18) as

$$\mathbf{C} = \begin{bmatrix} C_{11} & C_{12} & C_{13} & 0 & 0 & 0 \\ C_{12} & C_{11} & C_{13} & 0 & 0 & 0 \\ C_{13} & C_{13} & C_{33} & 0 & 0 & 0 \\ 0 & 0 & 0 & 2C_{44} & 0 & 0 \\ 0 & 0 & 0 & 0 & 2C_{44} & 0 \\ 0 & 0 & 0 & 0 & 0 & 2C_{66} \end{bmatrix} \tag{5.9.5}$$

to obtain

$$\hat{\mathbf{C}}_{\mathrm{TETRA}} = \underline{\mathbf{A}}\mathbf{C}\underline{\mathbf{A}}^T = \begin{bmatrix} \hat{C}_{11} & \hat{C}_{12} & \hat{C}_{13} & 0 & 0 & \sqrt{2}\hat{C}_{16} \\ \hat{C}_{12} & \hat{C}_{11} & \hat{C}_{13} & 0 & 0 & -\sqrt{2}\hat{C}_{16} \\ \hat{C}_{13} & C_{13} & \hat{C}_{33} & 0 & 0 & 0 \\ 0 & 0 & 0 & 2\hat{C}_{44} & 0 & 0 \\ 0 & 0 & 0 & 0 & 2\hat{C}_{44} & 0 \\ \sqrt{2}\hat{C}_{16} & -\sqrt{2}\hat{C}_{16} & 0 & 0 & 0 & 2\hat{C}_{66} \end{bmatrix}, \tag{5.9.6}$$

which is matrix (5.9.5) expressed—by using matrix (5.2.19)—in a different orientation of the coordinate system. Even though matrix (5.9.6) might appear to have seven independent elasticity parameters, these parameters are expressed in terms of the six parameters stated in matrix (5.9.5). In view of terminology introduced in Section 5.6.3, we refer to components stated in expression (5.9.5) as belonging to a natural system, since there are fewer nonzero components. Natural expressions of a given tetragonal tensor are exemplified in Exercise 5.6.

If the two axes of a coordinate system do not coincide with the rotation plane, the expression of a tetragonal tensor does not exhibit a pattern of zero entries shown in matrices (5.9.5) or (5.9.6).

5.10. Transversely Isotropic Continuum

5.10.1. Elasticity matrix

Now we consider a particularly interesting case. Suppose that a continuum is invariant with respect to a single rotation given by matrix (5.8.1) where Θ is smaller than $\pi/2$. Let us consider, for example, $\Theta = 2\pi/5$, and, hence, assume that the symmetry group contains

$$\mathbf{A}_{x_3 2\pi/5} = \begin{bmatrix} \cos\frac{2\pi}{5} & \sin\frac{2\pi}{5} & 0 \\ -\sin\frac{2\pi}{5} & \cos\frac{2\pi}{5} & 0 \\ 0 & 0 & 1 \end{bmatrix}. \tag{5.10.1}$$

Following condition (5.3.1), the elasticity matrix, $\mathbf{C}$, satisfies the equation given by

$$\mathbf{C} = \mathbf{M}^T_{\mathbf{A}_{x_3 2\pi/5}} \mathbf{C} \mathbf{M}_{\mathbf{A}_{x_3 2\pi/5}}. \tag{5.10.2}$$

The entries of matrix $\mathbf{M}_{\mathbf{A}_{x_3 2\pi/5}}$ are more complicated than the entries of the transformation matrices used in the previous sections, but equation (5.10.2) can still be solved directly to give relations among the entries of $\mathbf{C}$. The solution to condition (5.10.2) is the matrix given by

$$C_{\text{TRANS}} = \begin{bmatrix} C_{11} & C_{12} & C_{13} & 0 & 0 & 0 \\ C_{12} & C_{11} & C_{13} & 0 & 0 & 0 \\ C_{13} & C_{13} & C_{33} & 0 & 0 & 0 \\ 0 & 0 & 0 & C_{44} & 0 & 0 \\ 0 & 0 & 0 & 0 & C_{44} & 0 \\ 0 & 0 & 0 & 0 & 0 & \frac{C_{11}-C_{12}}{2} \end{bmatrix}. \tag{5.10.3}$$

Thus, the requirements that the symmetry group contains $\mathbf{A}_{x_3 2\pi/5}$ results in a continuum that is described by only five independent elasticity parameters.

5.10.2. Rotation invariance

A particularly important property of matrix (5.10.3) is the fact that for any angle Θ, this matrix, without any further simplification, satisfies the equation given by

$$\mathbf{C} = \mathbf{M}^T_{\mathbf{A}_{x_3 \Theta}} \mathbf{C} \mathbf{M}_{\mathbf{A}_{x_3 \Theta}}, \tag{5.10.4}$$

where $\mathbf{A}_{x_3\Theta}$ is given by matrix (5.8.1). This property of matrix (5.10.3) can be directly verified by substituting $\mathbf{M}_{\mathbf{A}_{x_3\Theta}}$, without specifying the value of Θ, and $\mathbf{C}_{\text{TRANS}}$ into the right-hand side of equation (5.10.4). The resulting expression reduces to $\mathbf{C}_{\text{TRANS}}$. Therefore, the invariance of $\mathbf{C}_{\text{TRANS}}$ to the five-fold rotation about a given axis implies invariance to the rotation by any angle about this axis. As stated in Theorem 5.10.1, below, there is nothing special about $2\pi/5$; we could choose any angle smaller than $\pi/2$ to obtain the same elasticity matrix.

To prove Theorem 5.10.1 below, and to see the reason behind it, consider the fact that the material symmetry of a continuum is equivalent to the symmetry of the strain-energy function, $W(\varepsilon)$, as discussed in Section 4.2.4.[11] Since strain energy is a scalar, its value must be the same for all orientations of the coordinate system. In view of expression (4.1.3), namely,

$$W(\varepsilon) = \frac{1}{2}\sum_{i=1}^{3}\sum_{j=1}^{3}\sum_{k=1}^{3}\sum_{\ell=1}^{3} c_{ijk\ell}\varepsilon_{ij}\varepsilon_{k\ell}, \tag{5.10.5}$$

this is, in general, achieved by the values of the components of $c_{ijk\ell}$, which are different for different orientations of the coordinate system in such a way that the value of W remains the same. If a continuum exhibits a given material symmetry, the same values of the components of $c_{ijk\ell}$ give the same value of W for more than one orientation of the coordinate system.

We wish to express the effect of an orthogonal transformation, $\mathbf{A}$, on the strain-energy function. Since the value of strain energy must be the same for the original and the transformed coordinate systems, we can write

$$\hat{W}(\hat{\varepsilon}) = W(\varepsilon), \tag{5.10.6}$$

where the transformed strain-tensor components are given by expression (5.2.11), namely,

$$\hat{\varepsilon} = \mathbf{A}\varepsilon\mathbf{A}^T, \tag{5.10.7}$$

[11] Readers interested in the treatment of symmetries that is based on the strain-energy function might refer to Carcione, J.M., (2001) Wave fields in real media: wave propagation in anisotropic, anelastic and porous media: Pergamon, pp. 2–3, to Epstein, M. and Slawinski, M.A., (1998) On some aspects of the continuum-mechanics context. Revue de l'Institut Français du Pétrole, **53**, No. 5, pp. 673 – 674, to Lanczos, C., (1949/1986) The variational principles of mechanics: Dover, pp. 373–374, and to Macelwane, J.B., and Sohon, F.W., (1936) Introduction to theoretical seismology, Part I: Geodynamics: John Wiley and Sons, Inc., pp. 77–78.

which, for brevity, we denote by $\mathbf{A} \circ \varepsilon$, with $\circ$ standing for the orthogonal-transformation operator.

Note that equation (5.10.6) is always satisfied. Consequently, this equation provides no information about the material symmetry of the continuum. The material symmetry requires that the strain-energy function be invariant under $\mathbf{A}$, namely,

$$W(\hat{\varepsilon}) = W(\varepsilon),$$

which we can write as $W(\varepsilon) = W(\mathbf{A} \circ \varepsilon)$. In other words, material symmetry requires that the strain-energy function be the same for both ε and $\hat{\varepsilon}$.

Herein, we are interested in rotations of the coordinate system; hence, we consider transformation $\mathbf{A}_{x_3\Theta}$, which is given by expression (5.8.1). In view of expressions (5.8.1) and (5.10.7), $\hat{\varepsilon}$ can be regarded as a quadratic trigonometric polynomial in Θ. Hence, $W(\mathbf{A}_{x_3\Theta} \circ \varepsilon)$ is a quartic trigonometric polynomial in Θ.[12]

Now, let the material symmetry be the invariance under rotation by $\Theta = 2\pi/n$, where $n \geqslant 5$. Hence, consider the strain-energy function that is invariant under such rotation. Since the symmetries of a continuum form a group, we conclude that $W(\varepsilon)$ is also invariant under rotations by $2m\pi/n$, where $m \in \{0, 1, \ldots, n-1\}$. In other words, the symmetry group contains rotations given by

$$\mathbf{A}_{x_3 2m\pi/n} = \begin{bmatrix} \cos\frac{2m\pi}{n} & \sin\frac{2m\pi}{n} & 0 \\ -\sin\frac{2m\pi}{n} & \cos\frac{2m\pi}{n} & 0 \\ 0 & 0 & 1 \end{bmatrix}, \qquad \begin{array}{c} n \geqslant 5 \\ m \in \{0, 1, \ldots, n-1\} \end{array}. \tag{5.10.8}$$

So, we can write

$$W(\varepsilon) = W\left(\mathbf{A}_{x_3 2m\pi/n} \circ \varepsilon\right), \qquad \begin{array}{c} n \geqslant 5 \\ m \in \{0, 1, \ldots, n-1\} \end{array}, \tag{5.10.9}$$

[12]Readers interested in trigonometric polynomials might refer to Courant, R., and Hilbert, D., (1924/1989) Methods of mathematical physics: John Wiley & Sons, Vol. I, p. 69–70.

which implies that

$$W(\varepsilon) = \frac{1}{n}\sum_{m=0}^{n-1} W\left(\mathbf{A}_{x_3 2m\pi/n} \circ \varepsilon\right), \qquad n \geqslant 5. \tag{5.10.10}$$

In other words, since equation (5.10.9) holds for any allowable value of m, the sum on the right-hand side of equation (5.10.10) is composed of the identical values of W.

Note that for any $W(\varepsilon)$, not necessarily invariant under transformations (5.10.8), the right-hand side of equation (5.10.10) is called the symmetrization of W with respect to the group of these transformations. Hence, equation (5.10.10) means that if $W(\varepsilon)$ is invariant under rotations (5.10.8), it is equal to its symmetrization with respect to these rotations.

Now we can introduce the key statement that explains why an elasticity matrix invariant to a five-fold rotation about a given axis is necessarily invariant to any rotation about this axis. Since $W(\mathbf{A}_{x_3\Theta} \circ \varepsilon)$ is a quartic trigonometric polynomial, it follows that, for $n \geqslant 5$, we can apply Lemma 5.10.2, below, and rewrite the right-hand side of equation (5.10.10) as

$$\frac{1}{n}\sum_{m=0}^{n-1} W\left(\mathbf{A}_{x_3 2m\pi/n} \circ \varepsilon\right) = \frac{1}{2\pi}\int_0^{2\pi} W(\mathbf{A}_{x_3\Theta} \circ \varepsilon)\,\mathrm{d}\Theta, \qquad n \geqslant 5,$$

which immediately allows us to write

$$W(\varepsilon) = \frac{1}{2\pi}\int_0^{2\pi} W(\mathbf{A}_{x_3\Theta} \circ \varepsilon)\,\mathrm{d}\Theta. \tag{5.10.11}$$

Equation (5.10.11) states that $W(\varepsilon)$ is equal to its symmetrization over all possible rotations. This implies that $W(\varepsilon)$ is invariant under all rotations. Hence, we conclude with the following theorem.

THEOREM 5.10.1. *If $W(\varepsilon)$ is invariant under rotations by angle $2\pi/n$ about a given axis, where $n \geqslant 5$, it is invariant under any rotation about this axis.*

To complete the proof of this theorem, consider the following lemma.

LEMMA 5.10.2. *If f is a trigonometric polynomial of at most degree $n-1$, then*

$$\frac{1}{n}\sum_{m=0}^{n-1} f\left(\frac{2m\pi}{n}\right) = \frac{1}{2\pi}\int_{0}^{2\pi} f(\Theta)\,\mathrm{d}\Theta\,. \tag{5.10.12}$$

PROOF. Consider a basis of the space of trigonometric polynomials of at most degree $n-1$ given by

$$f_r(\Theta) = e^{ir\Theta}\,, \qquad r \in \{-(n-1),\ldots,n-1\}\,.$$

In view of linearity, to prove equation (5.10.12) for f, it suffices to prove it for f_r, where

$$r \in \{-(n-1),\ldots,n-1\}\,.$$

Set

$$z = e^{ir\frac{2\pi}{n}}\,, \qquad r \in \{-(n-1),\ldots,n-1\}\,. \tag{5.10.13}$$

Then, we can write

$$\frac{1}{n}\sum_{m=0}^{n-1} f_r\left(\frac{2m\pi}{n}\right) = \frac{1}{n}\sum_{m=0}^{n-1} z^m\,, \qquad r \in \{-(n-1),\ldots,n-1\}\,. \tag{5.10.14}$$

Examining expression (5.10.13) for the case of $r = 0$, we note that $z = 1$ and, thus, the right-hand side of expression (5.10.14) is equal to 1. Now, for $r \neq 0$, $z \neq 1$ and we can write the right-hand side of expression (5.10.14) as

$$\frac{1}{n}\sum_{m=0}^{n-1} z^m = \frac{1}{n}\frac{z^n - 1}{z - 1}\,. \tag{5.10.15}$$

Examining expression (5.10.13), we also note that $z^n = 1$ and, thus, the right-hand side of expression (5.10.15) is equal to 0. To summarize, we can write the left-hand side of equation (5.10.12) as

$$\frac{1}{n}\sum_{m=0}^{n-1} f_r\left(\frac{2m\pi}{n}\right) = \begin{cases} 0 \text{ if } & r \neq 0 \\ 1 \text{ if } & r = 0 \end{cases}\,. \tag{5.10.16}$$

Performing the integration on the right-hand side of equation (5.10.12), we obtain

$$\frac{1}{2\pi}\int_{0}^{2\pi} e^{ir\Theta}\mathrm{d}\Theta = \begin{cases} 0 \text{ if } & r \neq 0 \\ 1 \text{ if } & r = 0 \end{cases}. \tag{5.10.17}$$

Thus, for $r \in \{-(n-1), \ldots, n-1\}$, expressions (5.10.16) and (5.10.17) are equal to one another and, hence, equation (5.10.12) is valid for polynomials of at most degree $n-1$, as required. □

Note that in the proof of Theorem 5.10.1 we used the fact that $W(\varepsilon)$ is a quadratic polynomial in the strain-tensor components and, hence, $W(\mathbf{A}_{x_3\Theta} \circ \varepsilon)$ is a quartic trigonometric polynomial in Θ. This corresponds to the fact that Theorem 5.10.1 is associated with $c_{ijk\ell}$, which is a fourth-rank tensor. This theorem can be extended to the higher-rank tensors if they are subject to similar transformations. In general, such a rotation invariance was given by Herman in 1945, and is shown in Exercise 5.8.

Since the symmetry of $W(\varepsilon)$ is tantamount to the symmetry of a continuum, we conclude that a continuum described by matrix (5.10.3) is transversely isotropic.

5.11. Cubic Continuum

A cubic continuum is a continuum whose symmetry group contains rotations by $\Theta = \pi/2$ about two axes that are orthogonal to one another. Herein, let us choose the x_3-axis and x_1-axis.

To consider the rotation about the x_3-axis, we set $\Theta = \pi/2$ in matrix (5.8.1) to get

$$\mathbf{A}_{x_3\pi/2} = \begin{bmatrix} 0 & 1 & 0 \\ -1 & 0 & 0 \\ 0 & 0 & 1 \end{bmatrix},$$

which is matrix (5.9.1) used already in Section 5.9.1 to discuss the tetragonal continuum. Following expression (5.2.15), we write the corresponding

6×6 transformation matrix, which is

$$\mathbf{M}_{\mathbf{A}_{x_3\pi/2}} = \begin{bmatrix} 0 & 1 & 0 & 0 & 0 & 0 \\ 1 & 0 & 0 & 0 & 0 & 0 \\ 0 & 0 & 1 & 0 & 0 & 0 \\ 0 & 0 & 0 & 0 & -1 & 0 \\ 0 & 0 & 0 & 1 & 0 & 0 \\ 0 & 0 & 0 & 0 & 0 & -1 \end{bmatrix}. \tag{5.11.1}$$

Using matrix (5.11.1) in condition (5.3.1), we obtain

$$C_{14} = C_{15} = C_{24} = C_{25} = C_{34} = C_{35} = C_{36} = C_{45} = C_{46} = C_{56} = 0 \tag{5.11.2}$$

and

$$C_{11} = C_{22},\, C_{13} = C_{23},\, C_{16} = -C_{26},\, C_{44} = C_{55}\,. \tag{5.11.3}$$

To consider the rotation about the x_1-axis, we use the transformation matrix given by

$$\mathbf{A}_{x_1\Theta} = \begin{bmatrix} 1 & 0 & 0 \\ 0 & \cos\Theta & -\sin\Theta \\ 0 & \sin\Theta & \cos\Theta \end{bmatrix},$$

which—if evaluated for $\Theta = \pi/2$—becomes

$$\mathbf{A}_{x_1\pi/2} = \begin{bmatrix} 1 & 0 & 0 \\ 0 & 0 & -1 \\ 0 & 1 & 0 \end{bmatrix}.$$

The corresponding 6×6 transformation matrix is

$$\mathbf{M}_{\mathbf{A}_{x_1\pi/2}} = \begin{bmatrix} 1 & 0 & 0 & 0 & 0 & 0 \\ 0 & 0 & 1 & 0 & 0 & 0 \\ 0 & 1 & 0 & 0 & 0 & 0 \\ 0 & 0 & 0 & -1 & 0 & 0 \\ 0 & 0 & 0 & 0 & 0 & 1 \\ 0 & 0 & 0 & 0 & -1 & 0 \end{bmatrix}. \tag{5.11.4}$$

Using matrix (5.11.4) in condition (5.3.1), we obtain

$$C_{14} = C_{15} = C_{16} = C_{25} = C_{26} = C_{35} = C_{36} = C_{45} = C_{46} = C_{56} = 0 \tag{5.11.5}$$

and

$$C_{12} = C_{13}, C_{22} = C_{33}, C_{24} = -C_{34}, C_{55} = C_{66}\,. \tag{5.11.6}$$

Combining expressions (5.11.2), (5.11.3), (5.11.5) and (5.11.6), we obtain the elasticity matrix for the cubic continuum. This matrix is

$$\mathbf{C}_{\mathrm{CUBIC}_{x_1x_3}} = \begin{bmatrix} C_{11} & C_{13} & C_{13} & 0 & 0 & 0 \\ C_{13} & C_{11} & C_{13} & 0 & 0 & 0 \\ C_{13} & C_{13} & C_{11} & 0 & 0 & 0 \\ 0 & 0 & 0 & C_{44} & 0 & 0 \\ 0 & 0 & 0 & 0 & C_{44} & 0 \\ 0 & 0 & 0 & 0 & 0 & C_{44} \end{bmatrix},$$

which contains three independent elasticity parameters.

Examining this derivation, we see that cubic symmetry can be viewed as resulting from combining two tetragonal symmetries for the two rotation axes that are orthogonal to one another.

In view of work presented in the sections above, we see that the cubic continuum is a special case of the tetragonal continuum, tetragonal continuum is a special case of the orthotropic continuum, which in turn is a special case of the monoclinic continuum. In other words, the symmetry group of the monoclinic continuum is contained in the symmetry group of the cubic continuum. In Section 5.13, we study the relations among all symmetry classes.

5.12. Isotropic Continuum

5.12.1. Elasticity matrix

A continuum whose symmetry group contains all orthogonal transformations is said to be isotropic. For an isotropic continuum, all coordinate systems are natural coordinate systems and, hence, no particular orientation is required.

Since the symmetry group of an isotropic continuum contains all orthogonal transformations, it must contain all rotations about the x_3-axis. Thus, the elasticity matrix of an isotropic continuum has, at least, the simplicity of the form shown in matrix (5.10.3). Consider also the invariance to the transformation that exchanges the x_1 and x_3 coordinates; namely,

$$\mathbf{A}_{x_1x_3} = \begin{bmatrix} 0 & 0 & 1 \\ 0 & 1 & 0 \\ 1 & 0 & 0 \end{bmatrix} .$$

Following condition (5.3.1), we obtain the equation given by

$$\mathbf{C} = \mathbf{M}_{A_{x_1x_3}}^T \mathbf{C} \mathbf{M}_{A_{x_1x_3}} ,$$

which imposes the additional relations; namely,

$$C_{11} = C_{33}, C_{12} = C_{13}, C_{44} = C_{66} .$$

Incorporating these relations into matrix (5.10.3), we obtain

$$\mathbf{C}_{\mathrm{ISO}} = \begin{bmatrix} C_{11} & C_{11}-2C_{44} & C_{11}-2C_{44} & 0 & 0 & 0 \\ C_{11}-2C_{44} & C_{11} & C_{11}-2C_{44} & 0 & 0 & 0 \\ C_{11}-2C_{44} & C_{11}-2C_{44} & C_{11} & 0 & 0 & 0 \\ 0 & 0 & 0 & C_{44} & 0 & 0 \\ 0 & 0 & 0 & 0 & C_{44} & 0 \\ 0 & 0 & 0 & 0 & 0 & C_{44} \end{bmatrix} . \qquad (5.12.1)$$

Hence, the elasticity matrix of an isotropic continuum contains only two independent elasticity parameters; namely, C_{11} and C_{44}. Furthermore, as shown in Exercise 5.11, the elasticity matrix of an isotropic continuum is symmetric without invoking the existence of the strain-energy function.

REMARK 5.12.1. For consistency—between the elasticity matrix for transverse isotropy, which is stated in expression (5.10.3) and isotropy, discussed herein—we might write matrix (5.12.1) as

$$\left[\begin{array}{cccccc} C_{11} & C_{12} & C_{12} & 0 & 0 & 0 \\ C_{12} & C_{11} & C_{12} & 0 & 0 & 0 \\ C_{12} & C_{12} & C_{11} & 0 & 0 & 0 \\ 0 & 0 & 0 & \dfrac{C_{11}-C_{12}}{2} & 0 & 0 \\ 0 & 0 & 0 & 0 & \dfrac{C_{11}-C_{12}}{2} & 0 \\ 0 & 0 & 0 & 0 & 0 & \dfrac{C_{11}-C_{12}}{2} \end{array}\right].$$

5.12.2. Lamé's parameters

The two independent elasticity parameters that describe an isotropic continuum are often expressed as

$$\left\{\begin{array}{c} \lambda := C_{11} - 2C_{44} \\ \mu := C_{44} \end{array}\right. . \tag{5.12.2}$$

The two parameters, λ and μ, are called Lamé's parameters. Their physical meaning is described in Section 5.12.4. Mathematically, these two parameters are closely related to the eigenvalues of elasticity matrix (5.12.1). If we write matrix (5.12.1) in the form discussed in Exercise 5.3, namely,

$$\left[\begin{array}{cccccc} C_{11} & C_{11}-2C_{44} & C_{11}-2C_{44} & 0 & 0 & 0 \\ C_{11}-2C_{44} & C_{11} & C_{11}-2C_{44} & 0 & 0 & 0 \\ C_{11}-2C_{44} & C_{11}-2C_{44} & C_{11} & 0 & 0 & 0 \\ 0 & 0 & 0 & 2C_{44} & 0 & 0 \\ 0 & 0 & 0 & 0 & 2C_{44} & 0 \\ 0 & 0 & 0 & 0 & 0 & 2C_{44} \end{array}\right], \tag{5.12.3}$$

its eigenvalues are given by $3C_{11} - 4C_{44}$ and by $2C_{44}$, which is repeated five times. In general, for the anisotropic continua, the eigenvalues of an elasticity matrix play an important role in identifying the symmetry class since—in contrast to the elasticity parameters—they are invariant under the orthogonal transformations.[13]

Using the definition of Lamé's parameters (5.12.2), we can rewrite matrix (5.12.1) as

$$\mathbf{C}_{\text{LAMÉ}} = \begin{bmatrix} \lambda+2\mu & \lambda & \lambda & 0 & 0 & 0 \\ \lambda & \lambda+2\mu & \lambda & 0 & 0 & 0 \\ \lambda & \lambda & \lambda+2\mu & 0 & 0 & 0 \\ 0 & 0 & 0 & \mu & 0 & 0 \\ 0 & 0 & 0 & 0 & \mu & 0 \\ 0 & 0 & 0 & 0 & 0 & \mu \end{bmatrix} . \qquad (5.12.4)$$

5.12.3. Tensor formulation

Using matrix (5.12.4) and in view of equations (3.2.1), we can write the stress-strain equations for an isotropic continuum as

$$\sigma_{ij} = \lambda \delta_{ij} \sum_{k=1}^{3} \varepsilon_{kk} + 2\mu\varepsilon_{ij}, \qquad i,j \in \{1,2,3\}, \qquad (5.12.5)$$

where $\sum \varepsilon_{kk}$ is the dilatation defined in expression (1.4.18). This formulation is used to derive the wave equation in Chapter 6.

Since, in equations (3.2.1), the elasticity tensor, $c_{ijk\ell}$, is a fourth-rank tensor, the number of elasticity parameters for an isotropic continuum can also be derived directly from the mathematical properties of a fourth-rank tensor and the concept of an isotropic tensor.

Note that an isotropic tensor is a tensor whose components are the same in all coordinate systems.

The general form of an isotropic fourth-rank tensor is

$$a_{ijk\ell} = \lambda \delta_{ij}\delta_{k\ell} + \xi \delta_{ik}\delta_{j\ell} + \eta \delta_{i\ell}\delta_{jk}, \qquad i,j,k,\ell \in \{1,2,3\}, \qquad (5.12.6)$$

where λ, ξ and η are constants. In other words, an isotropic fourth-rank tensor is stated in terms of three constants that do not depend on the

[13]Readers interested in the use of the eigenvalues of the elasticity matrix for identifying the symmetry class might refer to Bóna, A., Bucataru, I., Slawinski, M.A. (2007) Coordinate-free classification of elasticity tensor: Journal of Elasticity **87**(2-3), 109–132, and to Rychlewski, J., (1985) On Hooke's law: Prikl. Matem. Mekhan., **48** (3), 420–435.

choice of the coordinate system. In elasticity theory, since the strain tensor is symmetric, the most general isotropic elasticity tensor is given by expression (5.12.5), which contains only two constants, λ and μ, where, as shown in Exercise 5.12, $\mu := (\xi + \eta)/2$.

By examining stress-strain equations (5.12.5) in the context of tensor analysis, we can see that they correspond to the isotropic formulation since they retain the same form for all orthogonal transformations. To gain insight into this statement, we rewrite these equations using definition (1.4.6) as

$$\sigma_{ij} = \lambda \delta_{ij} \sum_{k=1}^{3} \frac{\partial u_k}{\partial x_k} + \mu \left(\frac{\partial u_i}{\partial x_j} + \frac{\partial u_j}{\partial x_i} \right), \qquad i, j \in \{1, 2, 3\}. \qquad (5.12.7)$$

Equations (5.12.7) are invariant under the coordinate transformations. We immediately see that the summation term is $\nabla \cdot \mathbf{u}$, which—being a scalar—is invariant under all coordinate transformations. Using transformation rules for the components of a second-rank tensor, we can also show that, upon the coordinate transformation, the term in parentheses retains the same form. Since both the summation term and the term in parentheses are invariant under the coordinate transformations, stress-strain equations (5.12.5) correspond to isotropic continua.

5.12.4. Physical meaning of Lamé's parameters

We can consider the physical meaning of Lamé's parameters, λ and μ, by examining stress-strain equations (5.12.5).

Lamé's parameter μ is a measure of rigidity. We can see that by setting $\lambda = 0$ and considering ε_{ij} with $i \neq j$. Thus, we can write expressions (5.12.5) as

$$\sigma_{ij} = 2\mu\varepsilon_{ij}, \qquad \begin{matrix} i \neq j \\ i, j \in \{1, 2, 3\} \end{matrix},$$

which, using definition (1.4.6), we can rewrite as

$$\sigma_{ij} = \mu \left(\frac{\partial u_i}{\partial x_j} + \frac{\partial u_j}{\partial x_i} \right), \qquad \begin{matrix} i \neq j \\ i, j \in \{1, 2, 3\} \end{matrix}.$$

In view of Section 1.4.3, we see that μ is a coefficient that relates stress to a change in shape. Thus, Lamé's parameter μ describes the rigidity of the continuum.

The physical meaning of Lamé's parameter λ is less immediate. If we let μ vanish and consider $i = j$, equations (5.12.5) become

$$\lim_{\mu \to 0^+} \sigma_{ij} = \lambda \sum_{k=1}^{3} \varepsilon_{kk} = \lambda\left(\varepsilon_{11} + \varepsilon_{22} + \varepsilon_{33}\right) = \lambda \varphi, \qquad \begin{array}{c} i = j \\ i, j \in \{1,2,3\} \end{array}, \tag{5.12.8}$$

where φ is the dilatation defined in expression (1.4.18). Examination of expression (5.12.8), which can be viewed as corresponding to a fluid, shows that Lamé's parameter λ is akin to the compressibility, κ. Note however that, in view of the positive-definiteness of elasticity matrix (5.12.4), as required by the stability conditions for an elastic solid, discussed in Section 4.3, we require $\mu > 0$. Hence, we treat the vanishing of μ as a limit.

REMARK 5.12.2. The case of $\mu = 0$ is a mathematical statement of no work needed for deformations that do not correspond to compression, which is tantamount to no work needed for deformations that do not result in the change of volume. In other words, shearing deformations, which are tantamount to the change of shape, need no work; this property distinguishes a fluid from a solid. If $\mu = 0$, the elasticity matrix—in contrast to the discussion in Section 4.3—is positive semidefinite.

To study a proper solid, we consider a finite value of μ. We still consider ε_{ij} with $i = j$ and we further assume $\varepsilon_{11} = \varepsilon_{22} = \varepsilon_{33}$. For convenience, let $\varepsilon_{ii} \equiv \tilde{\varepsilon}/3$, where $i \in \{1,2,3\}$. Thus, we can write stress-strain equations (5.12.5) as

$$\sigma_{ij} = \lambda \tilde{\varepsilon} + \frac{2}{3}\mu\tilde{\varepsilon} = \left(\lambda + \frac{2}{3}\mu\right)\tilde{\varepsilon}, \qquad \begin{array}{c} i = j \\ i, j \in \{1,2,3\} \end{array}. \tag{5.12.9}$$

In view of expression (1.4.17) and letting $\Delta V := \check{V} - V$, we can write

$$\tilde{\varepsilon} = \frac{\Delta V}{V}.$$

Also, σ_{ii} is equal to $-P$, where P denotes the difference in hydrostatic pressure. In other words, P is a pressure difference between the pressure associated with the deformation and the pressure at the undeformed state. Thus, we can write expression (5.12.9) as

$$-P=\left(\lambda+\frac{2}{3}\mu\right)\frac{\Delta V}{V}. \tag{5.12.10}$$

To gain insight into the physical meaning of λ, we use the concept of compressibility, κ, that is defined as the relative decrease of volume produced by unit pressure; namely,

$$\kappa:=-\frac{1}{P}\frac{\Delta V}{V}. \tag{5.12.11}$$

Using expression (5.12.10), we can rewrite the compressibility as

$$\kappa=\frac{1}{\lambda+\frac{2}{3}\mu}.$$

Solving for λ, we obtain

$$\lambda=\frac{1}{\kappa}-\frac{2}{3}\mu,$$

where $1/\kappa$ is commonly referred to as the bulk modulus, K, which is tantamount to incompressibility, namely, the ratio of the change in hydrostatic pressure and the relative change in volume.

Thus, in the case of vanishing rigidity, $\mu\to 0^+$, Lamé's parameter λ is the reciprocal of the compressibility, while, in general, λ has a more complicated physical significance given in terms of both the rigidity and compressibility.

5.13. Relations Among Symmetry Classes

The eight symmetry classes are related to each other. In this section, we briefly discuss certain aspects of these relations.

The symmetry group of an isotropic continuum contains all orthogonal transformations. Hence it contains, as its subgroups, all other symmetry classes. The symmetry group of the generally anisotropic continuum contains only two orthogonal transformations: the identity and the point symmetry. These two transformations are contained, as a subgroup, in all symmetry classes. However, not all pairs among the classes possess such a subgroup relation; for instance, neither cubic symmetry is a subgroup

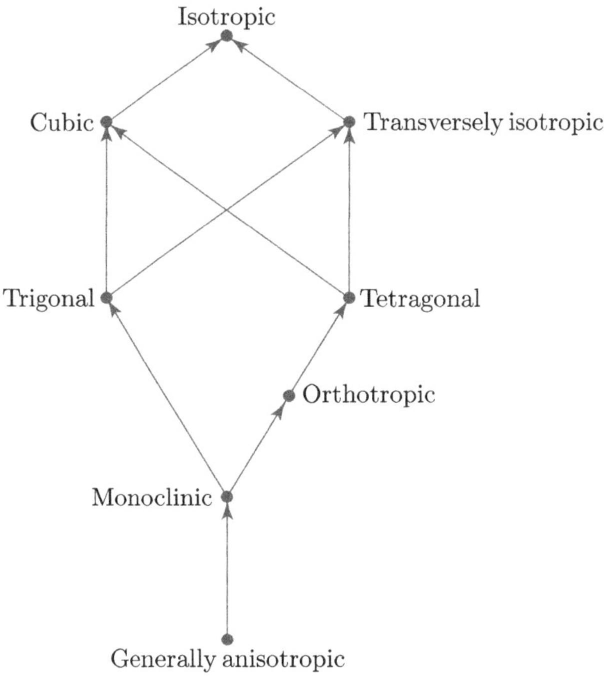

Fig. 5.13.1 Relations among symmetry classes. The arrows state the subgroup relations; for example, Orthotropic is a subgroup of Tetragonal: it has fewer symmetries.

of transversely isotropic symmetry nor vice versa. The set where not all pairs of elements exhibit the property that one is a subgroup of the other is called partial ordering. The set of the symmetry classes of an elastic continuum, which is shown in Figure 5.13.1, is a partially ordered set. The eight symmetry classes discussed above are related to Curie's symmetry groups, which can be represented geometrically as spheres, cylinders and cones.[14]

Another important aspect of a given symmetry class is the number of independent elasticity parameters. However, arranging the classes according to this number has its limitations. We can illustrate these limitations with two examples. A transversely isotropic continuum exhibits five elasticity

[14] Interested readers might refer to Newnham, R.E., (2005) Properties of materials: Anisotropy, symmetry, structure: Oxford University Press, pp. 26–28.

parameters while a cubic one exhibits only three. Yet, as shown above, ordering according to this number cannot be viewed as ordering in terms of the symmetry group. Secondly, both the trigonal and tetragonal continua are described by the same number of independent elasticity parameters, namely six. However, they are distinct continua; their symmetry groups are different from one another and, hence, so are their elasticity matrices.[15]

Closing Remarks

In this chapter, by studying the elasticity matrix, we investigated all eight symmetries of the elasticity tensor. In other words, we investigated the elasticity tensors that correspond to generally anisotropic, monoclinic, orthotropic, trigonal, tetragonal, transversely isotropic, cubic and isotropic continua.[16] Coordinate expressions of the same tensor—but with distinct number of components due to rotation of a coordinate system, discussed in Sections 5.8.2 and 5.9.2—misled certain researchers to count twice the same symmetry class within Hookean solids.[17] In the mathematical language, we should not separate a conjugate symmetry group. This confusion might have resulted also from an insufficient distinction between symmetries of continua, which are tensor symmetries, and crystal symmetries, which are lattice symmetries. Notably, the intuitive and heuristic descriptions associated with lattices might be inappropriate in the context of continua. For instance, Theorem 5.10.1, which we use in Section 5.10.2 and illustrate in Exercise 5.8 has no meaning for crystal symmetries. Also, the use of nomenclature that originates in crystallography to describe symmetries of continua might be misleading. For instance, the use of hexagonal

---

[15] Readers interested in the classification of symmetries for the elasticity tensor might refer to Ting, T.C.T., (1996) Anisotropic elasticity: Theory and applications: Oxford University Press, pp. 40–51.

[16] Readers interested in the existence of the eight symmetry classes might refer to Bóna, A., Bucataru, I., Slawinski, M.A., (2005) Characterization of elasticity-tensor symmetries using SU(2). Journal of Elasticity **75**(3), 267–289, to Bóna, A., Bucataru, I., Slawinski, M.A., (2004) Material symmetries of elasticity tensor. The Quarterly Journal of Mechanics and Applied Mathematics **57**(4), 583–598, and to references therein.

[17] Such a classification is presented by Nye, J.F., (1957/1987) Physical properties of crystals: their representation by tensors and matrices: Oxford University Press, Chapter VIII, Table 9.

symmetry in reference to transverse isotropy has no justification in symmetries of the elasticity tensor, even though it could be argued for within the concept of tessellation.

Studying the symmetries of a continuum provides us with information about the material it represents. For instance, by analyzing seismic data, we can infer information about layering and fractures. Also, knowing the smallest number of independent elasticity parameters required to describe a given continuum provides us with a convenient way to study wave propagation in specific materials. For instance, explicit expressions for wave velocities in a generally anisotropic continuum are complicated. However, if we know that a given material can be adequately described by a continuum that possesses particular symmetries, we reduce the complication of these expressions. Explicit expressions for wave velocities in a transversely isotropic continuum are derived in Chapter 9.

5.14. Exercises

EXERCISE 5.1. [18]Show that the Jacobian that is associated with matrix (5.8.1) is equal to unity.

SOLUTION 5.1. Using matrix (5.8.1), we can write the transformation of coordinates as

$$\begin{bmatrix} \hat{x}_1 \\ \hat{x}_2 \\ \hat{x}_3 \end{bmatrix} = \begin{bmatrix} \cos\Theta & \sin\Theta & 0 \\ -\sin\Theta & \cos\Theta & 0 \\ 0 & 0 & 1 \end{bmatrix} \begin{bmatrix} x_1 \\ x_2 \\ x_3 \end{bmatrix}, \tag{5.14.1}$$

[18] See also Section 5.1.1.

where $\mathbf{x}$ and $\hat{\mathbf{x}}$ are the original and the transformed coordinates, respectively. The Jacobian is given by

$$J := \det \begin{bmatrix} \frac{\partial \hat{x}_1}{\partial x_1} & \frac{\partial \hat{x}_1}{\partial x_2} & \frac{\partial \hat{x}_1}{\partial x_3} \\ \frac{\partial \hat{x}_2}{\partial x_1} & \frac{\partial \hat{x}_2}{\partial x_2} & \frac{\partial \hat{x}_2}{\partial x_3} \\ \frac{\partial \hat{x}_3}{\partial x_1} & \frac{\partial \hat{x}_3}{\partial x_2} & \frac{\partial \hat{x}_3}{\partial x_3} \end{bmatrix} . \tag{5.14.2}$$

Thus, examining equations (5.14.1) and (5.14.2), we see that the determinant of matrix (5.8.1) is the Jacobian. Hence, we immediately obtain

$$J = \det \begin{bmatrix} \cos\Theta & \sin\Theta & 0 \\ -\sin\Theta & \cos\Theta & 0 \\ 0 & 0 & 1 \end{bmatrix} = \cos^2\Theta + \sin^2\Theta = 1 ,$$

as required.

EXERCISE 5.2. [19]Formulate the condition that a transformation matrix must obey in order to preserve the scalar product. Provide a physical interpretation of the result.

SOLUTION 5.2. The scalar product is preserved if, for any two vectors, we can write

$$\hat{\mathbf{v}} \cdot \hat{\mathbf{u}} = \mathbf{v} \cdot \mathbf{u} .$$

In other words, the value of the product is the same in the original and transformed coordinate systems. Using expression (5.1.1), we can rewrite this condition as

$$\mathbf{Av} \cdot \mathbf{Au} = \mathbf{v} \cdot \mathbf{u} , \tag{5.14.3}$$

where $\mathbf{A}$ is a transformation matrix. In terms of components, we write the left-hand side of equation (5.14.3) as

$$\mathbf{Av} \cdot \mathbf{Au} = \sum_{i=1}^{3} \sum_{j=1}^{3} \sum_{k=1}^{3} A_{ij} v_j A_{ik} u_k .$$

[19] See also Section 5.1.1.

We can change the order of this summation to write

$$\mathbf{Av}\cdot\mathbf{Au}=\sum_{i=1}^{3}\sum_{j=1}^{3}\sum_{k=1}^{3}A_{ij}A_{ik}v_ju_k\,.$$

To state this summation in terms of matrix multiplication, we write

$$\mathbf{Av}\cdot\mathbf{Au}=\sum_{i=1}^{3}\sum_{j=1}^{3}\sum_{k=1}^{3}\left(A^T\right)_{ji}A_{ik}v_ju_k=\left(\mathbf{A}^T\mathbf{A}\right)\mathbf{v}\cdot\mathbf{u}\,.$$

Thus, we can write equation (5.14.3) as

$$\mathbf{Av}\cdot\mathbf{Au}=\left(\mathbf{A}^T\mathbf{A}\right)\mathbf{v}\cdot\mathbf{u}=\mathbf{v}\cdot\mathbf{u}\,.$$

Since this equality must be valid for any $\mathbf{u}$, we require that

$$\left(\mathbf{A}^T\mathbf{A}\right)\mathbf{v}=\mathbf{v}\,,$$

which implies that $\mathbf{A}^T\mathbf{A}=\mathbf{I}$. This is the condition that a transformation matrix must obey to preserve the scalar product; we refer to $\mathbf{A}^T\mathbf{A}=\mathbf{I}$ as the orthogonality condition. Physically, since the distance is expressed in terms of the scalar product, the transformations that obey the orthogonality condition are distance-preserving.

EXERCISE 5.3. Rewrite stress-strain equations (4.2.8) in such a manner that both the matrix containing the stress-tensor components and the matrix containing the strain-tensor components have the same form.

SOLUTION 5.3. Both matrices have the same form if we multiply $\underline{\sigma}$ and $\underline{\varepsilon}$ by

$$\mathbf{B}=\begin{bmatrix}1&0&0&0&0&0\\0&1&0&0&0&0\\0&0&1&0&0&0\\0&0&0&\sqrt{2}&0&0\\0&0&0&0&\sqrt{2}&0\\0&0&0&0&0&\sqrt{2}\end{bmatrix}$$

and $\mathbf{B}^{-1}$, respectively. Let us consider equations (4.2.8), which we write as

$$\underline{\sigma}=\mathbf{C}\underline{\varepsilon}.$$

Since we can perform the same operation to both sides of an equation, let us write

$$\mathbf{B}\underline{\sigma} = \mathbf{BC}\underline{\varepsilon}.$$

Since multiplication by the identity matrix does not change an expression, we can write

$$\mathbf{B}\underline{\sigma} = \mathbf{BCI}\underline{\varepsilon}. \tag{5.14.4}$$

Substituting $\mathbf{BB}^{-1}$ for $\mathbf{I}$, we rewrite equation (5.14.4) as

$$\mathbf{B}\underline{\sigma} = (\mathbf{BCB})\left(\mathbf{B}^{-1}\underline{\varepsilon}\right), \tag{5.14.5}$$

where, explicitly,

$$\mathbf{B}\underline{\sigma} = \begin{bmatrix} \sigma_{11} \\ \sigma_{22} \\ \sigma_{33} \\ \sqrt{2}\sigma_{23} \\ \sqrt{2}\sigma_{13} \\ \sqrt{2}\sigma_{12} \end{bmatrix}$$

and

$$\mathbf{B}^{-1}\underline{\varepsilon} = \begin{bmatrix} \varepsilon_{11} \\ \varepsilon_{22} \\ \varepsilon_{33} \\ \sqrt{2}\varepsilon_{23} \\ \sqrt{2}\varepsilon_{13} \\ \sqrt{2}\varepsilon_{12} \end{bmatrix},$$

as required; in other words, both the matrix containing the stress-tensor components and the matrix containing the strain-tensor components have the same form. The corresponding elasticity matrix is

$$\mathbf{BCB} = \begin{bmatrix} C_{11} & C_{12} & C_{13} & \sqrt{2}C_{14} & \sqrt{2}C_{15} & \sqrt{2}C_{16} \\ C_{12} & C_{22} & C_{23} & \sqrt{2}C_{24} & \sqrt{2}C_{25} & \sqrt{2}C_{26} \\ C_{13} & C_{23} & C_{33} & \sqrt{2}C_{34} & \sqrt{2}C_{35} & \sqrt{2}C_{36} \\ \sqrt{2}C_{14} & \sqrt{2}C_{24} & \sqrt{2}C_{34} & 2C_{44} & 2C_{45} & 2C_{46} \\ \sqrt{2}C_{15} & \sqrt{2}C_{25} & \sqrt{2}C_{35} & 2C_{45} & 2C_{55} & 2C_{56} \\ \sqrt{2}C_{16} & \sqrt{2}C_{26} & \sqrt{2}C_{36} & 2C_{46} & 2C_{56} & 2C_{66} \end{bmatrix}. \tag{5.14.6}$$

Thus, stress-strain equations (4.2.8) can be written as

$$\begin{bmatrix} \sigma_{11} \\ \sigma_{22} \\ \sigma_{33} \\ \sqrt{2}\sigma_{23} \\ \sqrt{2}\sigma_{13} \\ \sqrt{2}\sigma_{12} \end{bmatrix} = \begin{bmatrix} C_{11} & C_{12} & C_{13} & \sqrt{2}C_{14} & \sqrt{2}C_{15} & \sqrt{2}C_{16} \\ C_{12} & C_{22} & C_{23} & \sqrt{2}C_{24} & \sqrt{2}C_{25} & \sqrt{2}C_{26} \\ C_{13} & C_{23} & C_{33} & \sqrt{2}C_{34} & \sqrt{2}C_{35} & \sqrt{2}C_{36} \\ \sqrt{2}C_{14} & \sqrt{2}C_{24} & \sqrt{2}C_{34} & 2C_{44} & 2C_{45} & 2C_{46} \\ \sqrt{2}C_{15} & \sqrt{2}C_{25} & \sqrt{2}C_{35} & 2C_{45} & 2C_{55} & 2C_{56} \\ \sqrt{2}C_{16} & \sqrt{2}C_{26} & \sqrt{2}C_{36} & 2C_{46} & 2C_{56} & 2C_{66} \end{bmatrix} \begin{bmatrix} \varepsilon_{11} \\ \varepsilon_{22} \\ \varepsilon_{33} \\ \sqrt{2}\varepsilon_{23} \\ \sqrt{2}\varepsilon_{13} \\ \sqrt{2}\varepsilon_{12} \end{bmatrix}. \tag{5.14.7}$$

EXERCISE 5.4. Consider a continuum whose symmetry group contains the reflection about the x_2x_3-plane. This reflection implies that $\varepsilon_{12} = -\hat{\varepsilon}_{12}$ and $\varepsilon_{13} = -\hat{\varepsilon}_{13}$, as well as $\sigma_{12} = -\hat{\sigma}_{12}$ and $\sigma_{13} = -\hat{\sigma}_{13}$. Using stress-strain equations (4.2.8), show that

$$C_{15} = C_{16} = C_{25} = C_{26} = C_{35} = C_{36} = C_{45} = C_{46} = 0,$$

and state the resulting elasticity matrix $\mathbf{C}_{\mathrm{MONO}x_1}$.

SOLUTION 5.4. Consider the stress-tensor components σ_{12} and $\hat{\sigma}_{12}$. Using stress-strain equations (4.2.8), we can write

$$\sigma_{12} = C_{16}\varepsilon_{11} + C_{26}\varepsilon_{22} + C_{36}\varepsilon_{33} + 2C_{46}\varepsilon_{23} + 2C_{56}\varepsilon_{13} + 2C_{66}\varepsilon_{12}, \tag{5.14.8}$$

and

$$\hat{\sigma}_{12} = C_{16}\hat{\varepsilon}_{11} + C_{26}\hat{\varepsilon}_{22} + C_{36}\hat{\varepsilon}_{33} + 2C_{46}\hat{\varepsilon}_{23} + 2C_{56}\hat{\varepsilon}_{13} + 2C_{66}\hat{\varepsilon}_{12}.$$

The second equation can be expressed in terms of the original strain components as

$$\hat{\sigma}_{12} = C_{16}\varepsilon_{11} + C_{26}\varepsilon_{22} + C_{36}\varepsilon_{33} + 2C_{46}\varepsilon_{23} - 2C_{56}\varepsilon_{13} - 2C_{66}\varepsilon_{12}\,.$$

In view of relations $\sigma_{12} = -\hat{\sigma}_{12}$, and the equality of the stress-strain equations required in view of the assumed symmetry, we obtain

$$\sigma_{12} = -\hat{\sigma}_{12} = -C_{16}\varepsilon_{11} - C_{26}\varepsilon_{22} - C_{36}\varepsilon_{33} - 2C_{46}\varepsilon_{23} + 2C_{56}\varepsilon_{13} + 2C_{66}\varepsilon_{12}\,. \qquad (5.14.9)$$

Equality between (5.14.8) and (5.14.9) requires

$$C_{16} = C_{26} = C_{36} = C_{46} = 0\,.$$

Similarly, for $\sigma_{13} = -\hat{\sigma}_{13}$, we require

$$C_{15} = C_{25} = C_{35} = C_{45} = 0\,.$$

Thus, we obtain

$$\mathbf{C}_{\mathrm{MONO}x_1} = \begin{bmatrix} C_{11} & C_{12} & C_{13} & C_{14} & 0 & 0 \\ C_{12} & C_{22} & C_{23} & C_{24} & 0 & 0 \\ C_{13} & C_{23} & C_{33} & C_{34} & 0 & 0 \\ C_{14} & C_{24} & C_{34} & C_{44} & 0 & 0 \\ 0 & 0 & 0 & 0 & C_{55} & C_{56} \\ 0 & 0 & 0 & 0 & C_{56} & C_{66} \end{bmatrix}, \qquad (5.14.10)$$

as required.

EXERCISE 5.5. [20]Show that both

$$\mathbf{C} = \begin{bmatrix} 8.063584 & 3.369495 & 2.459688 & \sqrt{2}(0.071065) & \sqrt{2}(0.019042) & 0 \\ 3.369495 & 8.063584 & 2.459688 & -\sqrt{2}(0.071065) & -\sqrt{2}(0.019042) & 0 \\ 2.459688 & 2.459688 & 7.084078 & 0 & 0 & 0 \\ \sqrt{2}(0.071065) & -\sqrt{2}(0.071065) & 0 & 2(1.861663) & 0 & -2(0.019042) \\ \sqrt{2}(0.019042) & -\sqrt{2}(0.019042) & 0 & 0 & 2(1.861663) & 2(0.071065) \\ 0 & 0 & 0 & -2(0.019042) & 2(0.071065) & 2(2.347044) \end{bmatrix}$$

[20] See also Section 5.8.2.

and

$$\hat{\mathbf{C}} = \begin{bmatrix} 8.063584 & 3.369495 & 2.459688 & 0 & \sqrt{2}(0.073572) & 0 \\ 3.369495 & 8.063584 & 2.459688 & 0 & \sqrt{2}(-0.073572) & 0 \\ 2.459688 & 2.459688 & 7.084078 & 0 & 0 & 0 \\ 0 & 0 & 0 & 2(1.861663) & 0 & 2(-0.073572) \\ \sqrt{2}(0.073572) & \sqrt{2}(-0.073572) & 0 & 0 & 2(1.861663) & 0 \\ 0 & 0 & 0 & 2(-0.073572) & 0 & 2(2.347044) \end{bmatrix}$$

represent the same trigonal tensor.[21]

SOLUTION 5.5. It suffices to show that the two expressions are related by an orthonormal transformation, which is tantamount to the tensor being expressed in two coordinate systems exhibiting different orientations. In view of Section 5.8.2, from the form of these matrices, we infer that they differ by a rotation about the x_3-axis. To specify the amount of rotation, we invoke formula (5.8.6), namely,

$$\tan(3\Theta) = \frac{C_{14}}{C_{15}} = \frac{0.071065}{0.019042} = 3.732013\,,$$

which implies $\Theta = 25^\circ$. Hence, letting $A_{11} = A_{22} = \cos\Theta$, $A_{12} = -\sin\Theta$, $A_{21} = \sin\Theta$, $A_{13} = A_{23} = A_{31} = A_{32} = 0$ and $A_{33} = 1$, in expression (5.2.19), after a laborious matrix multiplication, made easy by a mathematical software, we show that it is indeed the case, since $\hat{\mathbf{C}} = \underline{\mathbf{A}}\mathbf{C}\underline{\mathbf{A}}^T$. Note that, since $\mathbf{C}$ and $\hat{\mathbf{C}}$ represent the same tensor, both expressions share the same determinant, 20992.35, eigenvalues, $(13.360844, 5.156313, 4.715902, 4.715901, 3.701513, 3.701513)$, and Frobenius norm, 16.64276, which is the square root of the sum of squared entries. Also, note that, among the six eigenvalues, two pairs are the same, as expected for the trigonal symmetry.[22]

---

[21] This tensor is obtained from seismic measurements, as described by Danek, T., Kochetov, M., Slawinski, M.A., (2013) Uncertainty analysis of effective elasticity tensors using quaternion-based global optimization and Monte-Carlo method. The Quarterly Journal of Mechanics and Applied Mathematics **66**(2), 253–272. This is a density-scaled elasticity tensor; its units are $\mathrm{km}^2/\mathrm{s}^2$.

[22] Interested readers might refer to Bóna, A., Bucataru, I., Slawinski, M.A., (2007) Coordinate-free classification of elasticity tensor: Journal of Elasticity **87**(2-3), 109–132.

EXERCISE 5.6. [23]Show that both

$$\mathbf{C}=\begin{bmatrix} 7.972427 & 3.462667 & 2.459467 & 0 & 0 & 0 \\ 3.462667 & 7.972427 & 2.459467 & 0 & 0 & 0 \\ 2.459467 & 2.459467 & 7.080827 & 0 & 0 & 0 \\ 0 & 0 & 0 & 2(1.862166) & 0 & 0 \\ 0 & 0 & 0 & 0 & 2(1.862166) & 0 \\ 0 & 0 & 0 & 0 & 0 & 2(2.438771) \end{bmatrix}$$

and

$$\hat{\mathbf{C}}=\begin{bmatrix} 8.156348 & 3.278799 & 2.459469 & 0 & 0 & 0 \\ 3.278799 & 8.156348 & 2.459469 & 0 & 0 & 0 \\ 2.459469 & 2.459469 & 7.080820 & 0 & 0 & 0 \\ 0 & 0 & 0 & 2(1.862168) & 0 & 0 \\ 0 & 0 & 0 & 0 & 2(1.862168) & 0 \\ 0 & 0 & 0 & 0 & 0 & 2(2.254901) \end{bmatrix}$$

represent the same tetragonal tensor.[24]

SOLUTION 5.6. It suffices to show that the two expressions are related by an orthonormal transformation, which is tantamount to the tensor being expressed in two coordinate systems exhibiting different orientations. In view of discussion in the context of Figure 5.9.1, we recognize that the natural coordinate systems in which this tensor is expressed are rotated by $n\pi/4$, with respect to one another. Hence, letting $A_{11}=A_{22}=\cos(\pi/4)=\sqrt{2}/2$, $A_{12}=-\sin(\pi/4)=-\sqrt{2}/2$, $A_{21}=\sin(\pi/4)=\sqrt{2}/2$, $A_{13}=A_{23}=A_{31}=A_{32}=0$ and $A_{33}=1$, in expression (5.2.19), after a laborious matrix multiplication, made easy by a mathematical software, we show that it is indeed the case, since $\hat{\mathbf{C}}=\underline{\mathbf{A}}\mathbf{C}\underline{\mathbf{A}}^T$. Note that, since $\mathbf{C}$ and $\hat{\mathbf{C}}$ represent the same tensor, both expressions share the same determinant, 21013.31, eigenvalues, $(13.361363, 5.154548, 4.877560, 4.509753, 3.724336, 3.724336)$, and Frobenius norm, 16.64227, which is the square root of the sum of

[23] See also Section 5.9.2.

[24] $\mathbf{C}$ and $\mathbf{C}'$ represent the tensor obtained from seismic measurements and described by Danek, T., Kochetov, M., Slawinski, M.A., (2013) Uncertainty analysis of effective elasticity tensors using quaternion-based global optimization and Monte-Carlo method. The Quarterly Journal of Mechanics and Applied Mathematics **66**(2). This is a density-scaled elasticity tensor; its units are $\mathrm{km}^2/\mathrm{s}^2$.

squared entries. Also note that two among the six eigenvalues are the same, as expected for the tetragonal symmetry.[25]

EXERCISE 5.7. [26]Find the stability conditions for a transversely isotropic continuum described by matrix (5.10.3).

SOLUTION 5.7. In view of Section 4.3, the stability conditions require that matrix (5.10.3) be positive-definite. Recalling equations (4.3.3), we obtain

$$C_{11} > 0, \tag{5.14.11}$$

$$C_{33} > 0, \tag{5.14.12}$$

$$C_{44} > 0, \tag{5.14.13}$$

and

$$C_{11} > C_{12}. \tag{5.14.14}$$

We notice that matrix (5.10.3) is a direct sum of two submatrices given by

$$\mathbf{C}_1 = \begin{bmatrix} C_{11} & C_{12} & C_{13} \\ C_{12} & C_{11} & C_{13} \\ C_{13} & C_{13} & C_{33} \end{bmatrix},$$

and

$$\mathbf{C}_2 = \begin{bmatrix} C_{44} & 0 & 0 \\ 0 & C_{44} & 0 \\ 0 & 0 & \frac{C_{11}-C_{12}}{2} \end{bmatrix}.$$

Conditions (5.14.13) and (5.14.14) ensure that matrix $\mathbf{C}_2$ is positive-definite. In view of condition (5.14.11), the remaining conditions for the positive-definiteness of matrix $\mathbf{C}_1$ are

$$\det \begin{bmatrix} C_{11} & C_{12} \\ C_{12} & C_{11} \end{bmatrix} > 0, \tag{5.14.15}$$

[25]Interested readers might refer to Bóna, A., Bucataru, I., Slawinski, M.A., (2007) Coordinate-free classification of elasticity tensor: Journal of Elasticity **87**(2-3), 109–132.
[26]See also Section 4.3.3.

and

$$\det \begin{bmatrix} C_{11} & C_{12} & C_{13} \\ C_{12} & C_{11} & C_{13} \\ C_{13} & C_{13} & C_{33} \end{bmatrix} > 0 . \tag{5.14.16}$$

The condition resulting from determinant (5.14.15) is

$$C_{11} > |C_{12}| , \tag{5.14.17}$$

while the condition resulting from determinant (5.14.16) is

$$C_{33} (C_{11} - C_{12}) (C_{11} + C_{12}) > 2C_{13}^2 (C_{11} - C_{12}) .$$

In view of expression (5.14.14), we can rewrite the latter condition as

$$C_{33} (C_{11} + C_{12}) > 2C_{13}^2 . \tag{5.14.18}$$

Also, in view of condition (5.14.12), we have $C_{11} + C_{12} > 0$. Consequently, condition (5.14.17) follows from conditions (5.14.12) and (5.14.18). Thus, all the stability conditions for a transversely isotropic continuum are given by expressions (5.14.11), (5.14.12), (5.14.13), (5.14.14) and (5.14.18).

REMARK 5.14.1. Note that if matrix $\mathbf{C}_1$ is positive-definite, we also have

$$\det \begin{bmatrix} C_{11} & C_{13} \\ C_{13} & C_{33} \end{bmatrix} > 0 , \tag{5.14.19}$$

which we can write as

$$C_{11} C_{33} > C_{13}^2 . \tag{5.14.20}$$

To show that condition (5.14.20) is a consequence of condition (5.14.18), let us rewrite condition (5.14.18) as

$$C_{33} (C_{11} + C_{12}) - 2C_{13}^2 > 0 ,$$

which we restate as

$$a + b > 0 ,$$

where

$$a := C_{11} C_{33} - C_{13}^2$$

and

$$b := C_{12} C_{33} - C_{13}^2 .$$

Using this notation, we can write condition (5.14.20) as $a > 0$. To show that condition (5.14.20) is a consequence of condition (5.14.18), we first show that $a > b$, which is equivalent to showing that

$$C_{11}C_{33} > C_{12}C_{33}. \tag{5.14.21}$$

Inequality (5.14.21) is true due to conditions (5.14.12) and (5.14.14). Hence, since $a+b>0$ and $a-b>0$, by summation we get $2a>0$, which immediately implies that $a>0$, as required.

EXERCISE 5.8. [27]Using the formula for the change of coordinates for the components of a tensor as well as Lemma 5.14.2 below, show that if a tensor of rank n, given by $T_{i_1\ldots i_n}$, is invariant under the $(n+1)$-fold rotation about a given axis, it is invariant under any rotation about this axis.

LEMMA 5.14.2. *Let $P(\Theta)$ be a trigonometric polynomial of at most degree n. If $P(\Theta)$ has a period of $2\pi/(n+1)$, then $P(\Theta)\equiv const$.*

PROOF. Consider a basis of the space of trigonometric polynomials of at most degree n, given by

$$f_r(\Theta) = e^{ir\Theta}, \qquad r\in\{-n,\ldots,n\}. \tag{5.14.22}$$

We can uniquely write

$$P(\Theta) = \sum_{r=-n}^{n} \alpha_r f_r(\Theta), \tag{5.14.23}$$

where α_r are complex numbers. In view of expressions (5.14.22) and (5.14.23), we can write

$$P\left(\Theta+\frac{2\pi}{n+1}\right) = \sum_{r=-n}^{n} \alpha_r f_r\left(\Theta+\frac{2\pi}{n+1}\right) = \sum_{r=-n}^{n} \alpha_r e^{ir2\pi/(n+1)} f_r(\Theta). \tag{5.14.24}$$

Since $P(\Theta)$ has a period of $2\pi/(n+1)$, examining equations (5.14.23) and (5.14.24), we obtain

$$\alpha_r = \alpha_r e^{ir2\pi/(n+1)}, \qquad r\in\{-n,\ldots,n\},$$

[27] See also Section 5.10.2.

which implies that the exponential term is equal to unity. Observing that $e^{ir2\pi/(n+1)} \neq 1$, for $r \in \{-n, \ldots, n\}$, except for $r = 0$, we conclude that $\alpha_r = 0$, except, possibly, α_0, which is a constant. Hence, $P(\Theta)$ is constant. □

SOLUTION 5.8. Consider transformation matrix (5.8.1); namely,

$$\mathbf{A} = \begin{bmatrix} \cos\Theta & \sin\Theta & 0 \\ -\sin\Theta & \cos\Theta & 0 \\ 0 & 0 & 1 \end{bmatrix}. \tag{5.14.25}$$

The transformed tensor components are given by

$$\hat{T}_{i_1 \ldots i_n} = \sum_{j_1=1}^{3} \cdots \sum_{j_n=1}^{3} A_{i_1 j_1} \ldots A_{i_n j_n} T_{j_1 \ldots j_n}, \qquad i_1, \ldots, i_n \in \{1,2,3\}.$$

In view of matrix (5.14.25), we see that $\hat{T}_{i_1 \ldots i_n} = \hat{T}_{i_1 \ldots i_n}(\Theta)$ is a trigonometric polynomial in Θ of at most degree n. Since tensor $T_{i_1 \ldots i_n}$ is invariant under the rotation by the angle $2\pi/(n+1)$, polynomial $\hat{T}_{i_1 \ldots i_n}(\Theta)$ has a period of $2\pi/(n+1)$. Since $\hat{T}_{i_1 \ldots i_n}(\Theta)$ is at most of degree n, it follows from Lemma 5.14.2 that this trigonometric polynomial is constant. This means that $T_{i_1 \ldots i_n}$ is invariant under any rotation, as required.

EXERCISE 5.9. [28]Using the formulation presented in Section 5.2.5, use expression (5.7.7) to obtain the expressions for transverse isotropy whose rotation-symmetry axis coincides with the x_1-axis and one whose rotation-symmetry axis coincides with the x_2-axis.

SOLUTION 5.9. In the context of the notation used in Section 5.2.5, the symmetry condition that is equivalent to Theorem 5.3.1 is given by Corollary 5.3.3,

$$\tilde{\mathbf{C}} = \tilde{\mathbf{A}}^T \tilde{\mathbf{C}} \tilde{\mathbf{A}}, \tag{5.14.26}$$

where $\tilde{\mathbf{C}}$ and $\tilde{\mathbf{A}}$ are given in expressions (5.2.18) and (5.2.19), respectively. The symmetry group of a transversely isotropic continuum contains the transformations of an orthogonal continuum presented in Section 5.7, together with further constraints provided by the rotational invariance about

[28] See also Sections 5.2.5 and 5.10.2.

one of the coordinate axes. In the notation of Section 5.2.5, expression (5.7.7) is

$$\tilde{\mathbf{C}}_{\mathrm{ORTHO}x_1x_2x_3} = \begin{bmatrix} C_{11} & C_{12} & C_{13} & 0 & 0 & 0 \\ C_{12} & C_{22} & C_{23} & 0 & 0 & 0 \\ C_{13} & C_{23} & C_{33} & 0 & 0 & 0 \\ 0 & 0 & 0 & 2C_{44} & 0 & 0 \\ 0 & 0 & 0 & 0 & 2C_{55} & 0 \\ 0 & 0 & 0 & 0 & 0 & 2C_{66} \end{bmatrix}.$$

In general, an orthotropic continuum does not exhibit any preference with respect to the coordinate axes.

In view of Theorem 5.10.1, to consider transverse isotropy whose rotation-symmetry axis coincides with the x_1-axis, we consider a rotation matrix analogous to expression (5.10.8), and given by

$$\mathbf{A}_{x_1 2m\pi/n} = \begin{bmatrix} 1 & 0 & 0 \\ 0 & \cos\frac{2m\pi}{n} & \sin\frac{2m\pi}{n} \\ 0 & -\sin\frac{2m\pi}{n} & \cos\frac{2m\pi}{n} \end{bmatrix}, \qquad \begin{matrix} n \geqslant 5 \\ m \in \{0, 1, \ldots, n-1\} \end{matrix}. \tag{5.14.27}$$

Using entries of matrix (5.14.27) in matrix (5.2.19), we obtain $\tilde{\mathbf{A}}_{x_1 2m\pi/n}$. Herein, according to condition (5.14.26), we write

$$\tilde{\mathbf{C}}_{\mathrm{ORTHO}x_1x_2x_3} = \tilde{\mathbf{A}}^{T}_{x_1 2m\pi/n} \tilde{\mathbf{C}}_{\mathrm{ORTHO}x_1x_2x_3} \tilde{\mathbf{A}}_{x_1 2m\pi/n},$$

which results in

$$C_{33} = C_{22}, \; C_{22} = C_{23} + 2C_{44}, \; C_{13} = C_{12}, \; C_{66} = C_{55};$$

thus, we obtain

$$\tilde{\mathbf{C}}_{\mathrm{TRANS}x_1} = \begin{bmatrix} C_{11} & C_{12} & C_{12} & 0 & 0 & 0 \\ C_{12} & C_{23}+2C_{44} & C_{23} & 0 & 0 & 0 \\ C_{12} & C_{23} & C_{23}+2C_{44} & 0 & 0 & 0 \\ 0 & 0 & 0 & 2C_{44} & 0 & 0 \\ 0 & 0 & 0 & 0 & 2C_{55} & 0 \\ 0 & 0 & 0 & 0 & 0 & 2C_{55} \end{bmatrix}, \qquad (5.14.28)$$

which is analogous to expression (5.10.3), except for the use of notation discussed in Section 5.2.5 and for the rotation-symmetry axis coinciding with the x_1-axis, not with the x_3-axis. Following the same procedure with

$$\mathbf{A}_{x_2 2m\pi/n} = \begin{bmatrix} \cos\frac{2m\pi}{n} & 0 & \sin\frac{2m\pi}{n} \\ 0 & 1 & 0 \\ -\sin\frac{2m\pi}{n} & 0 & \cos\frac{2m\pi}{n} \end{bmatrix}, \qquad \begin{matrix} n \geqslant 5 \\ m \in \{0,1,\ldots,n-1\} \end{matrix},$$

we obtain

$$\tilde{\mathbf{C}}_{\mathrm{TRANS}x_2} = \begin{bmatrix} C_{13}+2C_{55} & C_{12} & C_{13} & 0 & 0 & 0 \\ C_{12} & C_{22} & C_{12} & 0 & 0 & 0 \\ C_{13} & C_{12} & C_{13}+2C_{55} & 0 & 0 & 0 \\ 0 & 0 & 0 & 2C_{44} & 0 & 0 \\ 0 & 0 & 0 & 0 & 2C_{55} & 0 \\ 0 & 0 & 0 & 0 & 0 & 2C_{44} \end{bmatrix}, \qquad (5.14.29)$$

which is analogous to expressions (5.10.3) and (5.14.28), except that the rotation-symmetry axis coincides with the x_2-axis.

EXERCISE 5.10. Show that a continuum whose symmetry contains the rotational invariance along two axes that are perpendicular to one another is isotropic.

SOLUTION 5.10. Let us consider expressions (5.14.28) and (5.14.29), where the rotation-symmetry axes coincide with the x_1-axis and the x_2-axis, respectively. Letting $\tilde{\mathbf{C}}_{\mathrm{TRANS}x_1} = \tilde{\mathbf{C}}_{\mathrm{TRANS}x_2}$, which is equivalent to equating the corresponding entries of matrices (5.14.28) and (5.14.29), we obtain relations between $\tilde{\mathbf{C}}_{\mathrm{TRANS}x_1}$ and $\tilde{\mathbf{C}}_{\mathrm{TRANS}x_2}$ that reduce them to expression (5.12.3), which is isotropic, as required.

A heuristic explanation can be provided by illustrating mechanical properties of a Hookean solid with geometrical shapes. We can view expressions (5.14.28) and (5.14.29) as cylinders along the x_1-axis and the x_2-axis, respectively. A single object that exhibits rotational properties of both such cylinders is a sphere at the origin, which is a geometrical illustration of properties of expression (5.12.3).

Also, in view of Exercise 5.9, we can write $\tilde{\mathbf{C}}_{\mathrm{TRANS}x_1} = \tilde{\mathbf{C}}_{\mathrm{TRANS}x_2}$ as

$$\tilde{\mathbf{A}}^T_{x_1 2m\pi/n} \tilde{\mathbf{C}}_{\mathrm{ORTHO}x_1x_2x_3} \tilde{\mathbf{A}}_{x_1 2m\pi/n} = \tilde{\mathbf{A}}^T_{x_2 2m\pi/n} \tilde{\mathbf{C}}_{\mathrm{ORTHO}x_1x_2x_3} \tilde{\mathbf{A}}_{x_2 2m\pi/n} .$$

Using the fact that — for orthogonal transformations — $\mathbf{A}^{-1} \equiv \mathbf{A}^T$, which means that $\mathbf{A}^T\mathbf{A} = \mathbf{A}\mathbf{A}^T = \mathbf{I}$, where $\mathbf{I}$ is the identity, we multiply both sides by $\tilde{\mathbf{A}}_{x_2 2m\pi/n}$, on the left, and by $\tilde{\mathbf{A}}^T_{x_2 2m\pi/n}$, on the right, to obtain

$$\tilde{\mathbf{C}}_{\mathrm{ORTHO}x_1x_2x_3} = \tilde{\mathbf{A}}_{x_1 2m\pi/n} \tilde{\mathbf{A}}^T_{x_2 2m\pi/n} \tilde{\mathbf{C}}_{\mathrm{ORTHO}x_1x_2x_3} \tilde{\mathbf{A}}_{x_2 2m\pi/n} \tilde{\mathbf{A}}^T_{x_1 2m\pi/n} .$$

The conditions imposed by this equality on $\tilde{\mathbf{C}}_{\mathrm{ORTHO}x_1x_2x_3}$ reduce it to expression (5.12.3), which is isotropic. Notably, the same result is obtained if we apply this equality to a generally anisotropic tensor, $\tilde{\mathbf{C}}$, given in expression (5.2.18). In other words, this equality is a general condition for isotropy.

EXERCISE 5.11. [29]Show that the elasticity matrix of an isotropic continuum is symmetric even without invoking the strain-energy function.

NOTATION 5.14.3. The repeated-index summation notation is used in this solution. Any term in which an index appears twice stands for the sum of all such terms as the index assumes values 1, 2 and 3.

SOLUTION 5.11. In view of Section 4.2, to show the symmetry of the elasticity matrix, $C_{mn} = C_{nm}$, where, $m, n \in \{1, \ldots, 6\}$, it suffices to show

[29]See also Section 5.12.1.

that

$$c_{ijk\ell} = c_{k\ell ij}\,, \qquad i,j,k,\ell \in \{1,2,3\}\,.$$

Recall stress-strain equations (3.2.1); namely,

$$\sigma_{ij} = c_{ijk\ell}\varepsilon_{k\ell}\,, \qquad i,j \in \{1,2,3\}\,, \tag{5.14.30}$$

as well as a particular case of these equations that corresponds to isotropic continua and is given by equations (5.12.5), namely,

$$\sigma_{ij} = \lambda\,\delta_{ij}\varepsilon_{kk} + 2\mu\,\varepsilon_{ij}\,, \qquad i,j \in \{1,2,3\}\,, \tag{5.14.31}$$

where λ and μ are Lamé's parameters. Equating the right-hand sides of equations (5.14.30) and (5.14.31), we can write

$$c_{ijk\ell}\varepsilon_{k\ell} - (\lambda\,\delta_{ij}\varepsilon_{kk} + 2\mu\,\varepsilon_{ij}) = 0\,, \qquad i,j \in \{1,2,3\}\,. \tag{5.14.32}$$

Let us factor out the strain-tensor components, $\varepsilon_{k\ell}$. Using the properties of Kronecker's delta, we can write $\varepsilon_{kk} = \delta_{k\ell}\varepsilon_{k\ell}$ and $\varepsilon_{ij} = \delta_{ik}\delta_{j\ell}\varepsilon_{k\ell}$, where $i,j \in \{1,2,3\}$. Thus, we can rewrite equations (5.14.32) as

$$\left[c_{ijk\ell} - \left(\lambda\,\delta_{ij}\delta_{k\ell} + 2\mu\,\delta_{ik}\delta_{j\ell}\right)\right]\varepsilon_{k\ell} = 0\,, \qquad i,j \in \{1,2,3\}\,. \tag{5.14.33}$$

In general, $\varepsilon_{k\ell}$ is not zero. Thus, for equations (5.14.33) to be always true, we require the expression in brackets to be zero. In other words,

$$c_{ijk\ell} = \lambda\,\delta_{ij}\delta_{k\ell} + 2\mu\,\delta_{ik}\delta_{j\ell}\,, \qquad i,j,k,\ell \in \{1,2,3\}\,.$$

By the commutativity of Kronecker's delta, $\delta_{ij}\delta_{k\ell} = \delta_{k\ell}\delta_{ij}$, while by its symmetry, $\delta_{ik}\delta_{j\ell} = \delta_{ki}\delta_{\ell j}$. Consequently, we can write

$$\begin{aligned} c_{ijk\ell} &= \lambda\,\delta_{ij}\delta_{k\ell} + 2\mu\,\delta_{ik}\delta_{j\ell} \\ &= \lambda\,\delta_{k\ell}\delta_{ij} + 2\mu\,\delta_{ki}\delta_{\ell j} = c_{k\ell ij}\,, \qquad i,j,k,\ell \in \{1,2,3\}\,, \end{aligned}$$

as required.

EXERCISE 5.12. [30]Using Lemma 5.14.5, prove Theorem 5.14.6, stated below.

[30] See also Section 5.12.3.

NOTATION 5.14.4. Repeated-index summation is used in this exercise. Any term in which an index appears twice stands for the sum of all such terms as the index assumes values 1 , 2 and 3 .

LEMMA 5.14.5. [31]*The general isotropic fourth-rank tensor is*

$$a_{ijk\ell} = \lambda\,\delta_{ij}\delta_{k\ell} + \xi\,\delta_{ik}\delta_{j\ell} + \eta\,\delta_{i\ell}\delta_{jk}\,, \qquad i,j,k,\ell \in \{1,2,3\}\,. \quad (5.14.34)$$

THEOREM 5.14.6. *Given the symmetry of the strain tensor, defined in expression (1.4.6), the stress-strain equations for a three-dimensional isotropic continuum are given by expression (5.12.5), namely,*

$$\sigma_{ij} = \lambda\,\delta_{ij}\varepsilon_{kk} + 2\mu\varepsilon_{ij}\,, \qquad i,j \in \{1,2,3\}\,,$$

where $2\mu = \xi + \eta$.

SOLUTION 5.12. Let us proceed with the following proof.

PROOF. Consider stress-strain equations (3.2.1), namely,

$$\sigma_{ij} = c_{ijk\ell}\varepsilon_{k\ell}\,, \qquad i,j \in \{1,2,3\}\,.$$

Inserting expression (5.14.34) for $c_{ijk\ell}$, and using the properties of Kronecker's delta, in view of Lemma 5.14.5, we can write

$$\begin{aligned}\sigma_{ij} &= \left(\lambda\,\delta_{ij}\delta_{k\ell} + \xi\,\delta_{ik}\delta_{j\ell} + \eta\,\delta_{i\ell}\delta_{jk}\right)\varepsilon_{k\ell} \\ &= \lambda\,\delta_{ij}\varepsilon_{kk} + \xi\varepsilon_{ij} + \eta\,\varepsilon_{ji}\,, \qquad i,j \in \{1,2,3\}\,.\end{aligned}$$

Since, by its definition, the strain tensor, $\varepsilon_{k\ell}$, is symmetric, we can write

$$\sigma_{ij} = \lambda\,\delta_{ij}\varepsilon_{kk} + (\xi + \eta)\,\varepsilon_{ij}\,, \qquad i,j \in \{1,2,3\}\,,$$

and, hence, there are only two independent constants in the stress-strain equations for an isotropic continuum. Since the constants are arbitrary, we can set $2\mu = \xi + \eta$, and write

$$\sigma_{ij} = \lambda\,\delta_{ij}\varepsilon_{kk} + 2\mu\varepsilon_{ij}\,, \qquad i,j \in \{1,2,3\}\,,$$

as required. □

[31] Readers interested in a proof of Lemma 5.14.5 might refer to Matthews, P.C., (1998) Vector calculus: Springer, pp. 124–125, and to Semay, C., and Silvestre-Brac, B., (2007) Introduction au calcul tensoriel: Applications à la physique: Dunod, pp. 205, 221, 243.

REMARK 5.14.7. While studying isotropic materials it is common to express the two elasticity parameters, shown in matrices (5.12.1) and (5.12.4), in terms of quantities that possess an immediate physical meaning. In Exercises 5.13–5.18, we use Poisson's ratio, which is defined as

$$\nu := -\frac{\varepsilon_{xx}}{\varepsilon_{zz}} = -\frac{\varepsilon_{yy}}{\varepsilon_{zz}}, \tag{5.14.35}$$

and Young's modulus, which is defined as

$$E := \frac{\sigma_{xx}}{\varepsilon_{xx}} = \frac{\sigma_{yy}}{\varepsilon_{yy}} = \frac{\sigma_{zz}}{\varepsilon_{zz}}.$$

The relations among Poisson's ratio, Young's modulus and Lamé's parameters are given by

$$\lambda = \frac{\mathrm{E}\nu}{(1+\nu)(1-2\nu)}, \tag{5.14.36}$$

and

$$\mu = \frac{\mathrm{E}}{2(1+\nu)}. \tag{5.14.37}$$

EXERCISE 5.13. Consider an isotropic continuum. Subjecting this continuum to a uniaxial stress along the z-axis so that $\sigma_{xx} = \sigma_{yy} = \sigma_{xy} = \sigma_{yz} = \sigma_{zx} = 0$, show that Poisson's ratio is given by

$$\nu = \frac{\lambda}{2(\lambda+\mu)}, \tag{5.14.38}$$

where λ and μ are Lamé's parameters.

SOLUTION 5.13. Following stress-strain equations (5.12.5), which describe isotropic continua, we can write

$$\sigma_{xx} = \lambda\left(\varepsilon_{xx} + \varepsilon_{yy} + \varepsilon_{zz}\right) + 2\mu\varepsilon_{xx} = (\lambda+2\mu)\,\varepsilon_{xx} + \lambda\varepsilon_{yy} + \lambda\varepsilon_{zz} = 0.$$

Dividing both sides by ε_{zz}, we obtain

$$(\lambda+2\mu)\frac{\varepsilon_{xx}}{\varepsilon_{zz}} + \lambda\frac{\varepsilon_{yy}}{\varepsilon_{zz}} + \lambda = 0.$$

Invoking the definition of Poisson's ratio given in expression (5.14.35), we can rewrite the above expression as

$$-(\lambda+2\mu)\,\nu - \lambda\nu + \lambda = -2(\lambda+\mu)\,\nu + \lambda = 0.$$

Hence, solving for ν, we get

$$\nu = \frac{\lambda}{2(\lambda + \mu)},$$

which is expression (5.14.38), as required.

EXERCISE 5.14. [32]Consider an isotropic continuum under a uniaxial stress that leads to small deformations. Using expression (5.14.38), show that no change in volume implies no resistance to change in shape, as stated by $\mu = 0$.

SOLUTION 5.14. Consider a rectangular box with initial dimensions x_1, x_2, and x_3. Its volume is $V = x_1 x_2 x_3$. Let the dimensions after deformation be $x_1 + \Delta x_1$, $x_2 + \Delta x_2$, and $x_3 + \Delta x_3$, where, after the deformation, the original rectangular box remains rectangular. Thus, the volume after deformation is

$$\begin{aligned} \breve{V} &= (x_1 + \Delta x_1)(x_2 + \Delta x_2)(x_3 + \Delta x_3) \\ &\approx x_1 x_2 x_3 + x_2 x_3 \Delta x_1 + x_1 x_3 \Delta x_2 + x_1 x_2 \Delta x_3, \end{aligned} \tag{5.14.39}$$

where the approximation stems from the assumption of small deformations and, consequently, from neglecting the second-order and the third-order terms involving Δx_i, where $i \in \{1, 2, 3\}$. No change in volume implies

$$\breve{V} - V = 0.$$

Using expression (5.14.39), we can write

$$\breve{V} - V = x_2 x_3 \Delta x_1 + x_1 x_3 \Delta x_2 + x_1 x_2 \Delta x_3 = 0.$$

Dividing both sides by $V = x_1 x_2 x_3$, we get

$$\frac{\breve{V} - V}{V} = \frac{\Delta x_1}{x_1} + \frac{\Delta x_2}{x_2} + \frac{\Delta x_3}{x_3} = 0.$$

In view of expression (1.4.17) and denoting $\varepsilon_{11} := \Delta x_1 / x_1$, $\varepsilon_{22} := \Delta x_2 / x_2$, $\varepsilon_{33} := \Delta x_3 / x_3$, we obtain

$$\frac{\breve{V} - V}{V} = \varepsilon_{11} + \varepsilon_{22} + \varepsilon_{33} = 0.$$

[32] See also Section 1.4.3.

Dividing both sides by ε_{33} and invoking the definition of Poisson's ratio, given in expression (5.14.35), we can write

$$\frac{\varepsilon_{11}}{\varepsilon_{33}}+\frac{\varepsilon_{22}}{\varepsilon_{33}}+1=-\nu-\nu+1=0,$$

which implies that the corresponding Poisson's ratio is $\nu=1/2$. Using expression (5.14.38), we obtain

$$\mu=\frac{1-2\nu}{2\nu}\lambda=0,$$

as required.

EXERCISE 5.15. Using equations (5.12.5), show that in an isotropic continuum, the strain-tensor components, ε_{ij}, can be expressed in terms of the stress-tensor components, σ_{ij}, as

$$\varepsilon_{ij}=\frac{1+\nu}{\mathrm{E}}\sigma_{ij}-\frac{\nu}{\mathrm{E}}\delta_{ij}\sum_{k=1}^{3}\sigma_{kk}, \qquad i,j\in\{1,2,3\}, \tag{5.14.40}$$

where ν is Poisson's ratio and E is Young's modulus.

SOLUTION 5.15. Using expressions (5.14.36) and (5.14.37), we can write stress-strain equations (5.12.5) as

$$\sigma_{ij}=\frac{\mathrm{E}\nu}{(1+\nu)(1-2\nu)}\delta_{ij}\sum_{k=1}^{3}\varepsilon_{kk}+\frac{\mathrm{E}}{1+\nu}\varepsilon_{ij}, \qquad i,j\in\{1,2,3\}.$$

Solving for ε_{ij}, we obtain

$$\varepsilon_{ij}=\frac{1+\nu}{\mathrm{E}}\sigma_{ij}-\frac{\nu}{1-2\nu}\delta_{ij}\sum_{k=1}^{3}\varepsilon_{kk}, \qquad i,j\in\{1,2,3\}. \tag{5.14.41}$$

Now, we seek to express strains $\sum_{k=1}^{3}\varepsilon_{kk}$ in terms of stresses. In view of Kronecker's delta and stress-strain equations (5.12.5), we can write all stress-tensor components for which $\delta_{ij}\sum_{k=1}^{3}\varepsilon_{kk}$ does not vanish. They are

$$\sigma_{ii}=\lambda\sum_{k=1}^{3}\varepsilon_{kk}+2\mu\varepsilon_{ii}, \qquad i\in\{1,2,3\}.$$

Writing these three equations explicitly, we get

$$\begin{cases} \sigma_{11} = \lambda \sum\limits_{k=1}^{3} \varepsilon_{kk} + 2\mu\varepsilon_{11} \\ \sigma_{22} = \lambda \sum\limits_{k=1}^{3} \varepsilon_{kk} + 2\mu\varepsilon_{22} \\ \sigma_{33} = \lambda \sum\limits_{k=1}^{3} \varepsilon_{kk} + 2\mu\varepsilon_{33} \end{cases}.$$

Summing these three equations, we obtain

$$\begin{aligned} \sigma_{11} + \sigma_{22} + \sigma_{33} &= 3\lambda \sum_{k=1}^{3} \varepsilon_{kk} + 2\mu\left(\varepsilon_{11} + \varepsilon_{22} + \varepsilon_{33}\right) \\ &= 3\lambda \sum_{k=1}^{3} \varepsilon_{kk} + 2\mu \sum_{k=1}^{3} \varepsilon_{kk} = (3\lambda + 2\mu) \sum_{k=1}^{3} \varepsilon_{kk}\,. \end{aligned}$$

Expressing the left-hand side as a summation, we can write the sought expression

$$\sum_{k=1}^{3} \varepsilon_{kk} = \frac{\sum\limits_{k=1}^{3} \sigma_{kk}}{3\lambda + 2\mu}. \tag{5.14.42}$$

Using expression (5.14.42), we can write expression (5.14.41) as

$$\varepsilon_{ij} = \frac{1+\nu}{\mathrm{E}}\sigma_{ij} - \frac{\nu}{(1-2\nu)(3\lambda + 2\mu)}\delta_{ij} \sum_{k=1}^{3} \sigma_{kk}, \qquad i,j \in \{1,2,3\}\,. \tag{5.14.43}$$

Consider the term in parentheses that contains λ and μ. Using expressions (5.14.36) and (5.14.37), we can write this term as

$$3\lambda + 2\mu = 3\frac{\mathrm{E}\nu}{(1+\nu)(1-2\nu)} + \frac{\mathrm{E}}{1+\nu} = \frac{\mathrm{E}}{1-2\nu}.$$

Hence, expression (5.14.43) becomes

$$\varepsilon_{ij} = \frac{1+\nu}{\mathrm{E}}\sigma_{ij} - \frac{\nu}{\mathrm{E}}\delta_{ij} \sum_{k=1}^{3} \sigma_{kk}, \qquad i,j \in \{1,2,3\}\,,$$

which is expression (5.14.40), as required.

EXERCISE 5.16. Using expression (4.5.4), show that for isotropic continua the strain-energy function can be expressed in terms of the strain-tensor components as

$$W = \frac{\lambda}{2} \sum_{i=1}^{3} \sum_{j=1}^{3} \varepsilon_{ii}\varepsilon_{jj} + \mu \sum_{i=1}^{3} \sum_{j=1}^{3} \varepsilon_{ij}\varepsilon_{ij}, \tag{5.14.44}$$

where λ and μ are Lamé's parameters.

SOLUTION 5.16. Recall expression (4.5.4); namely,

$$W = \frac{1}{2} \sum_{i=1}^{3} \sum_{j=1}^{3} \sigma_{ij}\varepsilon_{ij}. \tag{5.14.45}$$

Also, recall that for an isotropic continuum the stress-strain equations are given by expression (5.12.5); namely,

$$\sigma_{ij} = \lambda \delta_{ij} \sum_{k=1}^{3} \varepsilon_{kk} + 2\mu\varepsilon_{ij}, \qquad i, j \in \{1, 2, 3\}. \tag{5.14.46}$$

Inserting expression (5.14.46) into expression (5.14.45), we obtain

$$W = \frac{1}{2} \sum_{i=1}^{3} \sum_{j=1}^{3} \left[\left(\lambda \delta_{ij} \sum_{k=1}^{3} \varepsilon_{kk} + 2\mu\varepsilon_{ij} \right) \varepsilon_{ij} \right].$$

The properties of Kronecker's delta imply that

$$W = \frac{\lambda}{2} \sum_{i=1}^{3} \sum_{j=1}^{3} \varepsilon_{ii}\varepsilon_{jj} + \mu \sum_{i=1}^{3} \sum_{j=1}^{3} \varepsilon_{ij}\varepsilon_{ij},$$

which is expression (5.14.44), as required.

EXERCISE 5.17. Using expression (4.5.4), show that, for isotropic continua, the strain-energy function can be expressed in terms of the stress-tensor components as

$$W = \frac{1}{2\mathrm{E}} \left[(1+\nu) \sum_{i=1}^{3} \sum_{j=1}^{3} \sigma_{ij}\sigma_{ij} - \nu \sum_{i=1}^{3} \sum_{j=1}^{3} \sigma_{ii}\sigma_{jj} \right], \tag{5.14.47}$$

where ν is Poisson's ratio, and E is Young's modulus.

SOLUTION 5.17. Recall expression (4.5.4); namely,

$$W=\frac{1}{2}\sum_{i=1}^{3}\sum_{j=1}^{3}\sigma_{ij}\varepsilon_{ij}\,. \tag{5.14.48}$$

Also, recall expression (5.14.40); namely,

$$\varepsilon_{ij}=\frac{1+\nu}{\mathrm{E}}\sigma_{ij}-\frac{\nu}{\mathrm{E}}\delta_{ij}\sum_{k=1}^{3}\sigma_{kk}\,,\qquad i,j\in\{1,2,3\}\,, \tag{5.14.49}$$

where ν and E are Poisson's ratio and Young's modulus, respectively. Inserting expression (5.14.49) into expression (5.14.48), we obtain

$$W=\frac{1}{2}\sum_{i=1}^{3}\sum_{j=1}^{3}\left[\sigma_{ij}\left(\frac{1+\nu}{\mathrm{E}}\sigma_{ij}-\frac{\nu}{\mathrm{E}}\delta_{ij}\sum_{k=1}^{3}\sigma_{kk}\right)\right].$$

The properties of Kronecker's delta, δ_{ij}, imply

$$W=\frac{1}{2\mathrm{E}}\left[(1+\nu)\sum_{i=1}^{3}\sum_{j=1}^{3}\sigma_{ij}\sigma_{ij}-\nu\sum_{i=1}^{3}\sum_{j=1}^{3}\sigma_{ii}\sigma_{jj}\right],$$

which is expression (5.14.47), as required.

EXERCISE 5.18. [33] Consider elasticity matrix (5.12.4). Find the range of values for Lamé's parameters that is required by the stability conditions. Express this range in terms of Poisson's ratio. Provide a physical interpretation of this result.

SOLUTION 5.18. Stability conditions require the elasticity matrix to be positive-definite. Matrix (5.12.4) is symmetric. As stated in Theorem 4.3.2, for the positive-definiteness we require all eigenvalues to be positive. Consider the two submatrices, namely,

$$\begin{bmatrix}\lambda+2\mu & \lambda & \lambda\\ \lambda & \lambda+2\mu & \lambda\\ \lambda & \lambda & \lambda+2\mu\end{bmatrix}\qquad\text{and}\qquad\begin{bmatrix}\mu & 0 & 0\\ 0 & \mu & 0\\ 0 & 0 & \mu\end{bmatrix}.$$

[33] See also Section 4.3.3.

We obtain the eigenvalues, Λ_i, by solving

$$\det\left(\begin{bmatrix}\lambda+2\mu & \lambda & \lambda\\ \lambda & \lambda+2\mu & \lambda\\ \lambda & \lambda & \lambda+2\mu\end{bmatrix}-\Lambda\begin{bmatrix}1&0&0\\0&1&0\\0&0&1\end{bmatrix}\right)=0,$$

and

$$\det\left(\mu\begin{bmatrix}1&0&0\\0&1&0\\0&0&1\end{bmatrix}-\Lambda\begin{bmatrix}1&0&0\\0&1&0\\0&0&1\end{bmatrix}\right)=0.$$

The eigenvalues are $\Lambda_{1,2}=2\mu$, $\Lambda_3=3\lambda+2\mu$, and $\Lambda_{4,5,6}=\mu$. Positiveness of the eigenvalues means that $\mu>0$ and $\lambda>-2\mu/3$. Recalling expression (5.14.38), we obtain the range of physically acceptable values of Poisson's ratio, namely,

$$\nu\in\left[-1,\frac{1}{2}\right].$$

Physically, for a cylindrical sample and in view of $\nu:=-\varepsilon_{xx}/\varepsilon_{zz}$, the negative value of Poisson's ratio implies that the diminishing of the length of the cylinder along the z-axis is accompanied by the shortening of the radius along the x-axis. For most solids, we would expect a more limited range, namely, $\nu\in[0,1/2]$, where the diminishing of the length is accompanied by the extension of the radius.

Part 2

Waves and rays

Introduction to Part 2

> The solution of the equation of motion for an elastic medium results in the existence of elastic waves in its interior. The wave phenomenon is a way of transporting energy without transport of matter. The propagation of energy is, then, a very important aspect of wave propagation.
>
> *Agustín Udías (1999)*
> *Principles of seismology*

In *Part 1*, we derived Cauchy's equations of motion, the equation of continuity, and formulated the stress-strain equations for elastic continua. These equations form a determined system, which allows us to describe the behaviour of such continua.

In *Part 2*, we combine Cauchy's equations of motion with the stress-strain equations to formulate the equations of motion in elastic continua. In the particular case of isotropic homogeneous continua, these equations are wave equations, which possess analytic solutions. However, in anisotropic inhomogeneous continua, we are unable to formulate equations of motion that possess analytic solutions. Hence, we choose to study these equations in terms of the high-frequency approximation, which results in ray theory. This approach allows us to study rays, wavefronts, traveltimes and amplitudes of signals that propagate within such a continuum. Although the resulting expressions are exact only for the case of an infinitely high frequency of a signal, the experimental results agree well with the theoretical predictions, provided that the properties of the continuum do not change significantly within the wavelength of the signal.

Ray methods form an important theoretical platform for seismological studies. They allow us to formulate problems in the context of such mathematical tools as differential geometry and the calculus of variations While referring to the ray solution in their volumes on "Quantitative seismology: Theory and methods," Aki and Richards state that

> [it] provides the basis for routine interpretation of most seismic body waves, and it always provides a guide to more sophisticated methods, should they be necessary.

However, in view of this being an approximate solution, we must be aware of its limitations. Grant and West, in their book on "Interpretation theory in applied geophysics," state that

> it is often surprising to observe how uncritically their [ray methods] validity in seismological problems is accepted.

Rays, as a scientific entity, can be traced to the work of Willebrord Snell who, at the turn of the sixteenth and seventeenth century, formulated the law of refraction. The mathematical underpinnings of ray theory were established by William Rowan Hamilton in the first half of the nineteenth century.[34] The formulation of rays in terms of asymptotic series, which is the platform for our studies, is associated with the work of Carl Runge, Arnold Sommerfeld and Pieter Debye, at the beginning of twentieth century, as well as Vassily M. Babich and Joseph B. Keller in the middle of the twentieth century. Further work, specifically in the context of seismic rays, was done by Vlastislav Červený.

To conclude the introduction to waves and rays, let us comment on the concept of a wave, which—although nicely exemplified by a vibrating string and easily illustrated by a sinusoid—is a richer subject than

[34] Readers interested in Hamilton's formulation of the ray theory might refer to Hankins, T.L., (1980) Sir William Rowan Hamilton: The Johns Hopkins University Press, pp. 59–95.

commonly understood. A wave is a field-theory analogy to a motion of a particle. Consistently with this analogy, wave propagation must exhibit a finite-speed and causality. Hence, waves are the solutions of the time-dependent field theory that obey these conditions.

CHAPTER 6

Equations of Motion: Isotropic Homogeneous Continua

From the study of nature there arose that class of partial differential equations that is at the present time the most thoroughly investigated and probably the most important in the general structure of human knowledge, namely, the equations of mathematical physics.

Sergei L. Sobolev and Olga A. Ladyzenskaya (1969)
Partial differential equations in Mathematics
(editors: Aleksandrov et al.)

Preliminary Remarks

Having formulated system (4.4.5)—a system of equations to describe the behaviour of an elastic continuum—we wish to write Cauchy's equations of motion explicitly in the context of the stress-strain equations for such a continuum. This way, we commence our study of wave phenomena in an elastic continuum.

We begin by choosing the simplest type of elastic continuum, namely an isotropic homogeneous one, and, hence, we derive the corresponding equations of motion, which lead to the wave equations. In the process of formulating these equations, we learn about the existence of the two types of waves that can propagate in isotropic continua. Furthermore, we obtain the expressions for the speed of these waves as functions of the properties of the continuum.

We begin this chapter by combining Cauchy's equations of motion (2.8.1) with constitutive equations (5.12.5). This formulation results in the derivation of the wave equations.[1] To gain insight into these equations, we study them in the context of plane waves and displacement potentials. We also investigate the solutions of the wave equations, including solutions in various spatial dimensions and nondifferentiable solutions. We conclude this chapter with examples of extensions of the standard form of the wave equation that take into account aspects of anisotropy and of inhomogeneity.

6.1. Wave Equations

6.1.1. Equation of motion

To derive the wave equation, assume that a given three-dimensional continuum is isotropic and homogeneous. Thus, we consider the corresponding stress-strain equations given by expression (5.12.5), namely,

$$\sigma_{ij} = \lambda \delta_{ij} \sum_{k=1}^{3} \varepsilon_{kk} + 2\mu \varepsilon_{ij}, \qquad i,j \in \{1,2,3\}, \tag{6.1.1}$$

where λ and μ are constants. We also consider Cauchy's equations of motion (2.8.1), namely,

$$\sum_{j=1}^{3} \frac{\partial \sigma_{ij}}{\partial x_j} = \rho \frac{\partial^2 u_i}{\partial t^2}, \qquad i \in \{1,2,3\}. \tag{6.1.2}$$

We wish to combine stress-strain equations (6.1.1) with equations of motion (6.1.2) to get the equations of motion in an isotropic homogeneous

[1] Readers interested in the wave equations and their solutions might refer to Slawinski, M.A. (2021) Waves and rays in seismology: Answers to unasked questions (3*rd* edition): World Scientific, Chapter 5.

continuum. In other words, we substitute expression (6.1.1) into equations (6.1.2) to obtain

$$\rho\frac{\partial^2 u_i}{\partial t^2}=\sum_{j=1}^{3}\frac{\partial}{\partial x_j}\left(\lambda\delta_{ij}\sum_{k=1}^{3}\varepsilon_{kk}+2\mu\varepsilon_{ij}\right)$$
$$=\sum_{j=1}^{3}\left(\delta_{ij}\lambda\sum_{k=1}^{3}\frac{\partial\varepsilon_{kk}}{\partial x_j}+2\mu\frac{\partial\varepsilon_{ij}}{\partial x_j}\right),\qquad i\in\{1,2,3\}.\qquad(6.1.3)$$

Now, we wish to express the right-hand side of equations (6.1.3) in terms of the displacement vector, $\mathbf{u}$. Invoking the definition of the strain tensor, given in expression (1.4.6), we can rewrite equations (6.1.3) as

$$\rho\frac{\partial^2 u_i}{\partial t^2}=\sum_{j=1}^{3}\left[\delta_{ij}\lambda\sum_{k=1}^{3}\frac{\partial}{\partial x_j}\left(\frac{\partial u_k}{\partial x_k}\right)+\mu\frac{\partial}{\partial x_j}\left(\frac{\partial u_i}{\partial x_j}+\frac{\partial u_j}{\partial x_i}\right)\right],$$
$$i\in\{1,2,3\}.$$

Using the property of Kronecker's delta, we obtain

$$\rho\frac{\partial^2 u_i}{\partial t^2}=\lambda\sum_{k=1}^{3}\frac{\partial}{\partial x_i}\left(\frac{\partial u_k}{\partial x_k}\right)+\mu\sum_{j=1}^{3}\frac{\partial}{\partial x_j}\left(\frac{\partial u_i}{\partial x_j}+\frac{\partial u_j}{\partial x_i}\right),\qquad i\in\{1,2,3\}.$$

Using the linearity of the differential operators, we can rewrite these equations as

$$\rho\frac{\partial^2 u_i}{\partial t^2}=\lambda\sum_{j=1}^{3}\frac{\partial^2 u_j}{\partial x_i\partial x_j}+\mu\sum_{j=1}^{3}\frac{\partial^2 u_i}{\partial x_j^2}+\mu\sum_{j=1}^{3}\frac{\partial^2 u_j}{\partial x_j\partial x_i},\qquad i\in\{1,2,3\},$$

where, in the first summation, for the summation indices, we let $k=j$. Using the equality of mixed partial derivatives, we obtain

$$\rho\frac{\partial^2 u_i}{\partial t^2}=(\lambda+\mu)\sum_{j=1}^{3}\frac{\partial^2 u_j}{\partial x_i\partial x_j}+\mu\sum_{j=1}^{3}\frac{\partial^2 u_i}{\partial x_j^2}$$
$$=(\lambda+\mu)\frac{\partial}{\partial x_i}\sum_{j=1}^{3}\frac{\partial u_j}{\partial x_j}+\mu\left(\sum_{j=1}^{3}\frac{\partial^2}{\partial x_j^2}\right)u_i,\qquad i\in\{1,2,3\}.$$
$$(6.1.4)$$

We can use vector calculus to concisely state equations (6.1.4). Consider the right-hand side of these equations. The first summation term is the divergence of $\mathbf{u}$, namely, $\nabla\cdot\mathbf{u}$, while the second summation term is the Laplacian, namely, ∇^2. Consequently, we can rewrite equations (6.1.4) as

$$\rho\frac{\partial^2 u_i}{\partial t^2}=(\lambda+\mu)\frac{\partial}{\partial x_i}\nabla\cdot\mathbf{u}+\mu\nabla^2 u_i,\qquad i\in\{1,2,3\}. \tag{6.1.5}$$

We write the three equations stated in expression (6.1.5) as

$$\rho\frac{\partial^2}{\partial t^2}\begin{bmatrix}u_1\\u_2\\u_3\end{bmatrix}=(\lambda+\mu)\begin{bmatrix}\dfrac{\partial(\nabla\cdot\mathbf{u})}{\partial x_1}\\\dfrac{\partial(\nabla\cdot\mathbf{u})}{\partial x_2}\\\dfrac{\partial(\nabla\cdot\mathbf{u})}{\partial x_3}\end{bmatrix}+\mu\nabla^2\begin{bmatrix}u_1\\u_2\\u_3\end{bmatrix}.$$

Noticing that the first matrix on the right-hand side involves the gradient operator, we can concisely state the three equations shown in expression (6.1.5) as

$$\rho\frac{\partial^2\mathbf{u}}{\partial t^2}=(\lambda+\mu)\nabla(\nabla\cdot\mathbf{u})+\mu\nabla^2\mathbf{u}. \tag{6.1.6}$$

This is the equation of motion that applies to isotropic homogeneous continua.

To proceed from expression (6.1.5) to expression (6.1.6), we assume that $\mathbf{u}$ is expressed in Cartesian coordinates, which allow us to use the equality given by $(\nabla^2\mathbf{u})_i=\nabla^2 u_i$. In general, for a vector field, the ith component of $\nabla^2\mathbf{u}$ is not equal to the Laplacian of the ith component of $\mathbf{u}$.[2] However—even though, for convenience, we use Cartesian coordinates in the above derivation—equation (6.1.6), above, identity (6.1.7) and equation (6.1.8), below, are valid in curvilinear coordinates, provided that a proper form of the vector Laplacian is used.

We wish to write equation (6.1.6) in a form that allows us to express it in terms of the dilatation and the rotation vector, in accordance with their definitions stated in Chapter 1. Using the vector identity given by

$$\nabla^2\mathbf{a}=\nabla(\nabla\cdot\mathbf{a})-\nabla\times(\nabla\times\mathbf{a}), \tag{6.1.7}$$

[2] Readers interested in this statement might refer to Feynman, R.P., Leighton, R.B., and Sands, M., (1963/1989) Feynman's lectures on physics: Addison-Wesley Publishing Co., Vol. II, p. 2-12.

and letting $\mathbf{a} = \mathbf{u}$, we can rewrite equation (6.1.6) as

$$\rho \frac{\partial^2 \mathbf{u}}{\partial t^2} = (\lambda + 2\mu)\nabla(\nabla \cdot \mathbf{u}) - \mu\nabla \times (\nabla \times \mathbf{u}) . \tag{6.1.8}$$

Equation (6.1.8) contains information about the deformations expressed in terms of the divergence and the curl operators. Recalling the definitions of the dilatation and the rotation vector, given by expressions (1.4.18) and (1.5.2), respectively, we can immediately write

$$\rho \frac{\partial^2 \mathbf{u}}{\partial t^2} = (\lambda + 2\mu)\nabla\varphi - \mu\nabla \times \Psi . \tag{6.1.9}$$

Equation (6.1.9) describes the propagation of the deformations in terms of both dilatation and the rotation vector in an isotropic homogeneous continuum. It describes the propagation related to both the change in volume and the change in shape. The divergence operator is associated with the change of volume while the curl operator is associated with the change in shape.

6.1.2. Wave equation for *P* waves

To gain insight into the types of waves that propagate in an isotropic homogeneous continuum, we wish to split equation (6.1.9) into its parts, which are associated with the dilatation and with the rotation vector.[3]

To obtain the wave equation for *P* waves, we take the divergence of equation (6.1.9). Since in a homogeneous continuum λ and μ are constants, we can write

$$\nabla \cdot \left[\rho \frac{\partial^2 \mathbf{u}}{\partial t^2}\right] = (\lambda + 2\mu)\nabla \cdot \nabla\varphi - \mu\nabla \cdot \nabla \times \Psi . \tag{6.1.10}$$

The factor of μ disappears since $\nabla \cdot \nabla \times \Psi = 0$ for all Ψ. Considering the factor of $\lambda + 2\mu$ and invoking the definition of the Laplacian, we can write

$$\nabla \cdot \nabla\varphi = \left[\frac{\partial}{\partial x_1}, \frac{\partial}{\partial x_2}, \frac{\partial}{\partial x_3}\right] \cdot \left[\frac{\partial}{\partial x_1}, \frac{\partial}{\partial x_2}, \frac{\partial}{\partial x_3}\right] \varphi = \nabla^2\varphi .$$

[3] Readers interested in an elegant formulation of *P* and *S* waves might refer to Rochester, M.G., (2010) Note on the necessary conditions for *P* and *S* wave propagation in a homogeneous isotropic elastic solid: Journal of Elasticity, **98** (1), 111–114. Also, Section 6.4, below, is based on that publication.

Consequently, equation (6.1.10) becomes

$$\nabla \cdot \left[\rho \frac{\partial^2 \mathbf{u}}{\partial t^2} \right] = (\lambda + 2\mu) \nabla^2 \varphi . \tag{6.1.11}$$

Let us consider the left-hand side of equation (6.1.11). In a homogeneous continuum, the mass density, ρ, is a constant. Hence—in view on the linearity of the differential operators—we can take ρ outside of the divergence. Also—in view of the equality of mixed partial derivatives—we can interchange time and space derivatives. Thus, we get

$$\rho \frac{\partial^2 \varphi}{\partial t^2} = (\lambda + 2\mu) \nabla^2 \varphi ,$$

where, on the left-hand side, we used again definition (1.4.18). Rearranging, we obtain

$$\nabla^2 \varphi = \frac{1}{\frac{\lambda + 2\mu}{\rho}} \frac{\partial^2 \varphi}{\partial t^2} , \tag{6.1.12}$$

which is the wave equation whose wave function is given by dilatation, $\varphi(\mathbf{x}, t) = \nabla \cdot \mathbf{u}(\mathbf{x}, t)$.

Equation (6.1.12) is the wave equation for P waves. As shown in Section 6.5,

$$v := \sqrt{\frac{\lambda + 2\mu}{\rho}} \tag{6.1.13}$$

is the propagation speed. In view of Section 5.12, the presence of both Lamé's parameters in expression (6.1.13) suggests that P waves subject the continuum to both a change in volume and a change in shape.

In view of definition (1.4.18), P waves are sometimes referred to as dilatational waves. Also, since the dilatation, φ, is the relative change in volume, they are sometimes referred to as pressure waves. Furthermore, since the speed of P waves is always greater than the speed of S waves, which are discussed below, in earthquake observations, P waves are sometimes referred to as primary waves.

6.1.3. Wave equation for S waves

To obtain the wave equation for S waves, we take the curl of equation (6.1.9) and write

$$\nabla \times \left[\rho \frac{\partial^2 \mathbf{u}}{\partial t^2} \right] = (\lambda + 2\mu) \nabla \times \nabla \varphi - \mu \nabla \times \nabla \times \Psi . \tag{6.1.14}$$

The factor of $\lambda + 2\mu$ disappears since $\nabla \times \nabla \varphi = \mathbf{0}$ for all φ. Recalling definition (1.5.2) and considering the constancy of the mass density, ρ — in view of the linearity of the differential operators as well as the equality of mixed partial derivatives—we get

$$\rho \frac{\partial^2 \Psi}{\partial t^2} = -\mu \nabla \times [\nabla \times \Psi] . \tag{6.1.15}$$

Invoking vector-calculus identity (6.1.7) and letting $\mathbf{a} = \Psi$, we can write equation (6.1.15) as

$$\rho \frac{\partial^2 \Psi}{\partial t^2} = -\mu \left[\nabla (\nabla \cdot \Psi) - \nabla^2 \Psi \right] .$$

In view of definition (1.5.2) and the vanishing of the divergence of a curl, the first term in brackets disappears. Hence, we obtain

$$\nabla^2 \Psi = \frac{1}{\frac{\mu}{\rho}} \frac{\partial^2 \Psi}{\partial t^2} , \tag{6.1.16}$$

which is the wave equation whose wave function is given by the rotation vector, $\Psi(\mathbf{x}, t) = \nabla \times \mathbf{u}(\mathbf{x}, t)$.

Equation (6.1.16) is the wave equation wave for S waves. As shown in Section 6.5,

$$v := \sqrt{\frac{\mu}{\rho}} \tag{6.1.17}$$

is the propagation speed. In view of Section 5.12, the presence of the single Lamé's parameter, namely, μ, in expression (6.1.17), suggests that S waves subject the continuum to a change in shape. Also, due to the vanishing of rigidity in fluids, we can conclude that the propagation of S waves is limited to solids.

In view of definition (1.5.2), S waves are sometimes referred to as rotational waves. Since the rotation vector is given by $\Psi = \nabla \times \mathbf{u}$, we conclude that $\nabla \cdot \Psi = 0$. If the divergence of a vector field vanishes, this vector field is volume-preserving; hence, S waves are sometimes referred to as the equivoluminal waves. In English, the justification for the letter S is due to the fact that these waves are often referred to as shear waves. Also, due to the fact that the speed of S waves is always smaller than the speed of P waves, in earthquake observations, S waves are sometimes referred to as secondary waves.

6.1.4. Physical interpretation

Examining equations (6.1.12) and (6.1.16) in view of definitions (1.4.18) and (1.5.2), respectively, we see that both of them contain the three components of the displacement vector, $\mathbf{u}$. However, in view of definition (1.4.6), we see that equation (6.1.12) contains only partial derivatives given by $\partial u_i / \partial x_i$. In view of definition (1.5.1), we see that equation (6.1.16) contains only $\partial u_i / \partial x_j$, for which $i \neq j$. Recalling expression (1.5.3), we write

$$\frac{\partial (u_1, u_2, u_3)}{\partial (x_1, x_2, x_3)} := \begin{bmatrix} \frac{\partial u_1}{\partial x_1} & \frac{\partial u_1}{\partial x_2} & \frac{\partial u_1}{\partial x_3} \\ \frac{\partial u_2}{\partial x_1} & \frac{\partial u_2}{\partial x_2} & \frac{\partial u_2}{\partial x_3} \\ \frac{\partial u_3}{\partial x_1} & \frac{\partial u_3}{\partial x_2} & \frac{\partial u_3}{\partial x_3} \end{bmatrix}.$$

This Jacobian matrix can be viewed as a second-rank tensor associated with the deformation gradient. Equation (6.1.12) contains only the main-diagonal entries, equation (6.1.16) only the offdiagonal ones. Let us interpret the role of the terms that do not appear explicitly in a given wave equation; in other words, the role of the main-diagonal entries for equation (6.1.16) and of the offdiagonal for equation (6.1.12).

In equation (6.1.12), $\partial u_i / \partial x_i$, which constitute φ, describe the restoring force that results in wave propagation. Terms $\partial u_i / \partial x_j$ with $i \neq j$, which are not part of this equation, prevent any local rotation during propagation; in other words, they ensure that $\nabla \times \mathbf{u} = \mathbf{0}$.

In equation (6.1.16), the terms that constitute $\Psi = \nabla \times \mathbf{u}$ describe the restoring force, and the other ones prevent any change in volume; in other words, they ensure that $\nabla \cdot \mathbf{u} = 0$.

To justify these statements, let us consider Figure 1.4.1. After deformation, the horizontal edge can be written as vector

$$\left[\left(1+\frac{\partial u_1}{\partial X_1}\right)\Delta X_1,\quad \frac{\partial u_2}{\partial X_1}\Delta X_1,\quad 0\right], \tag{6.1.18}$$

and the vertical edge as

$$\left[\frac{\partial u_1}{\partial X_2}\Delta X_2,\quad \left(1+\frac{\partial u_2}{\partial X_2}\right)\Delta X_2,\quad 0\right]. \tag{6.1.19}$$

Examining expression (6.1.18), we see that the horizontal edge is rotated counterclockwise in the X_1X_2-plane by

$$\arctan\frac{\dfrac{\partial u_2}{\partial X_1}\Delta X_1}{\Delta X_1+\dfrac{\partial u_1}{\partial X_1}\Delta X_1} = \arctan\frac{\dfrac{\partial u_2}{\partial X_1}}{1+\dfrac{\partial u_1}{\partial X_1}} \approx \frac{\partial u_2}{\partial X_1},$$

where we assume $\partial u_1/\partial X_1 \ll 1$. Similarly, by expression (6.1.19), it follows that the vertical edge is rotated clockwise by $\partial u_1/\partial X_2$.

Let us recall Figure 1.4.2. If $\nabla \times \mathbf{u} = [\partial u_2/\partial X_1 - \partial u_1/\partial X_2, 0, 0] = \mathbf{0}$, the two angles are equal to one another, which implies that the diagonal of the deformed rectangle is aligned with the diagonal of the undeformed rectangle. Thus, the deformed rectangle is not rotated. In general, $\nabla \times \mathbf{u} = \mathbf{0}$ means that the deformation does not result in rotation.

Let us illustrate the meaning of $\nabla \cdot \mathbf{u} = 0$. The area of the deformed rectangle is the magnitude of the cross product of vectors (6.1.18) and (6.1.19), which is

$$\left(1+\frac{\partial u_1}{\partial X_1}\right)\left(1+\frac{\partial u_2}{\partial X_2}\right)\Delta X_1\Delta X_2 - \frac{\partial u_2}{\partial X_1}\frac{\partial u_1}{\partial X_2}\Delta X_1\Delta X_2,$$

which is the only nonzero component of the cross product. Considering only the first-order terms in partial derivatives, we rewrite this area as

$$\left(1+\frac{\partial u_1}{\partial X_1}+\frac{\partial u_2}{\partial X_2}\right)\Delta X_1\Delta X_2.$$

If $\partial u_1/\partial X_1 + \partial u_2/\partial X_2 = 0$, the above expression reduces to $\Delta X_1 \Delta X_2$, which implies that there is no change in area. In general, $\nabla \cdot \mathbf{u} = 0$ means that the deformation does not result in change of volume.

6.2. Plane Waves

In general, equations (6.1.4) are complicated partial differential equations. This shows that even in isotropic homogeneous continua, the description of wave phenomena constitutes a serious mathematical problem. We can simplify these equations by introducing certain abstract mathematical entities that allow us to describe particular aspects of wave phenomena. While studying wave propagation in homogeneous media, we can consider plane waves. These are the waves for which the components of the displacement vector are functions of the direction of propagation only.

To gain insight into the concept of plane waves, let us revisit equations (6.1.12) and (6.1.16). Let the plane waves propagate along the x_1-axis. Thus, in view of the properties of plane waves and following expression (2.4.3), we write the displacement vector as

$$\mathbf{u} = [u_1(x_1,t), u_2(x_1,t), u_3(x_1,t)] .$$

Since all the partial derivatives of $\mathbf{u}$ with respect to x_2 and x_3 vanish, equations (6.1.4) become

$$\rho \frac{\partial^2 u_1}{\partial t^2} = (\lambda + 2\mu) \frac{\partial^2 u_1}{\partial x_1^2},$$

$$\rho \frac{\partial^2 u_2}{\partial t^2} = \mu \frac{\partial^2 u_2}{\partial x_1^2},$$

and

$$\rho \frac{\partial^2 u_3}{\partial t^2} = \mu \frac{\partial^2 u_3}{\partial x_1^2}.$$

After algebraic manipulations, we can write

$$\frac{\partial^2 u_1}{\partial x_1^2} = \frac{1}{\frac{\lambda + 2\mu}{\rho}} \frac{\partial^2 u_1}{\partial t^2}, \tag{6.2.1}$$

$$\frac{\partial^2 u_2}{\partial x_1^2} = \frac{1}{\frac{\mu}{\rho}} \frac{\partial^2 u_2}{\partial t^2}, \tag{6.2.2}$$

and

$$\frac{\partial^2 u_3}{\partial x_1^2} = \frac{1}{\frac{\mu}{\rho}} \frac{\partial^2 u_3}{\partial t^2}. \tag{6.2.3}$$

Consider equation (6.2.1). Recall expression (1.4.18), which in this case becomes

$$\varphi = \frac{\partial u_1}{\partial x_1}. \tag{6.2.4}$$

Taking the derivative of equation (6.2.1) with respect to x_1, we obtain

$$\frac{\partial^3 u_1}{\partial x_1^3} = \frac{1}{\frac{\lambda + 2\mu}{\rho}} \frac{\partial^3 u_1}{\partial x_1 \partial t^2}. \tag{6.2.5}$$

Using expression (6.2.4) in equation (6.2.5), we obtain

$$\frac{\partial^2 \varphi}{\partial x_1^2} = \frac{1}{\frac{\lambda + 2\mu}{\rho}} \frac{\partial^2 \varphi}{\partial t^2}, \tag{6.2.6}$$

which is a plane-wave form of equation (6.1.12). Examining equations (6.2.4) and (6.2.6), we recognize that the displacement and the direction of propagation are parallel to one another, which is the key property of *P* waves in isotropic continua. This property is also shown in Exercise 9.5.

Now, consider equations (6.2.2) and (6.2.3). Recall expression (1.5.2), which in this case becomes

$$\Psi = \left[0, -\frac{\partial u_3}{\partial x_1}, \frac{\partial u_2}{\partial x_1}\right]. \tag{6.2.7}$$

Taking the derivative of equations (6.2.3) and (6.2.2) with respect to x_1, and writing them as a vector, we obtain

$$\begin{bmatrix} 0 \\ -\dfrac{\partial^3 u_3}{\partial x_1^3} \\ \dfrac{\partial^3 u_2}{\partial x_1^3} \end{bmatrix} = \frac{1}{\frac{\mu}{\rho}} \begin{bmatrix} 0 \\ -\dfrac{\partial^3 u_3}{\partial x_1\,\partial t^2} \\ \dfrac{\partial^3 u_2}{\partial x_1\,\partial t^2} \end{bmatrix}.$$

Using the equality of mixed partial derivatives and expression (6.2.7), we obtain

$$\frac{\partial^2 \Psi}{\partial x_1^2} = \frac{1}{\frac{\mu}{\rho}} \frac{\partial^2 \Psi}{\partial t^2}, \tag{6.2.8}$$

which is a plane-wave form of equation (6.1.16). Examining equations (6.2.7) and (6.2.8), we recognize that the displacements and the direction of propagation are orthogonal to each other, which is the key property of S waves in isotropic continua. This property is also shown in Exercise 9.7.

Plane waves are an approximation that allows us to study, in homogeneous media, a wavefield that results from a distant source. Notably, in Chapter 10, we use plane waves to study reflection and transmission of waves at an interface separating two anisotropic homogeneous half-spaces. For close sources, we can construct a wavefield as a superposition of plane waves. In such an approach, there is a constructive interference in the regions where the plane waves coincide and a destructive interference outside of these regions.

While studying inhomogeneous media, the behaviour of seismic waves cannot be conveniently described using plane waves and their superposition. For such studies, we introduce in Section 6.11.4 another abstract mathematical entity—a seismic ray, which belongs to the realm of asymptotic methods and provides us with a different perspective to study seismic wavefields.

6.3. Displacement Potentials

6.3.1. Helmholtz's decomposition

In Sections 6.1.2 and 6.1.3, we derived the wave equations for P and S waves, respectively. To derive the wave equation for P waves, which we expressed in terms of scalar function φ, we took the divergence of Cauchy's equations of motion. To derive the wave equation for S waves, which we expressed in terms of vector function Ψ, we took the curl of these equations. Herein, we obtain equations that correspond to P and S waves by using Helmholtz's method of separating a vector function into its scalar and vector potentials. We obtain these equations by inserting the potentials into Cauchy's equations of motion.

According to Helmholtz's theorem[4], a differentiable function $\mathbf{u}(\mathbf{x},t)$ can be decomposed into

$$\mathbf{u}(\mathbf{x},t) = \nabla\mathscr{P}(\mathbf{x},t) + \nabla\times\mathbf{S}(\mathbf{x},t)\,, \tag{6.3.1}$$

where $\mathscr{P}$ and $\mathbf{S} = [S_1, S_2, S_3]$ are called the scalar and vector potentials, respectively. Following the definitions of the gradient and curl operators, we can explicitly write the components of $\mathbf{u}$ as

$$u_1(\mathbf{x},t) = \frac{\partial\mathscr{P}}{\partial x_1} + \frac{\partial S_3}{\partial x_2} - \frac{\partial S_2}{\partial x_3}\,,$$

$$u_2(\mathbf{x},t) = \frac{\partial\mathscr{P}}{\partial x_2} + \frac{\partial S_1}{\partial x_3} - \frac{\partial S_3}{\partial x_1}$$

and

$$u_3(\mathbf{x},t) = \frac{\partial\mathscr{P}}{\partial x_3} + \frac{\partial S_2}{\partial x_1} - \frac{\partial S_1}{\partial x_2}\,,$$

which constitute a system of differential equations. This system does not have a unique solution. It is common to consider also another equation; namely,

$$\nabla\cdot\mathbf{S}(\mathbf{x},t) = 0\,. \tag{6.3.2}$$

[4] Readers interested in Helmholtz's theorem might refer to Arfken, G.B, and Weber, H.J., (2001) Mathematical methods for physicists (5*th* edition): Harcourt/Academic Press, pp. 96–101.

since we can always find $\mathscr{P}$ and $\mathbf{S}$ that satisfy the system composed of equations (6.3.1) and (6.3.2).[5] Introducing equation (6.3.2) does not result in a unique determination of $\mathbf{S}$. It only reduces possible choices of this vector, as shown below.

We use expressions (6.3.1) and (6.3.2) in Sections 6.3.3 and 6.3.4. In the next section, we justify our introducing equation (6.3.2).

6.3.2. Gauge transformation

Let us consider equation (6.3.2) in the context of equation (6.3.1). We are allowed to set $\nabla \cdot \mathbf{S} = 0$ since, in view of properties of the vector operators, $\mathbf{S}$ used in equation (6.3.1) is determined up to a gradient, ∇f, where $f(\mathbf{x})$ is any differentiable function. In mathematical physics, changing $\mathbf{S}$ by adding ∇f to it belongs to the so-called gauge transformations, which are any systematic transformations of the potentials that leaves the fields invariant. Let

$$\tilde{\mathbf{S}} = \mathbf{S} + \nabla f. \tag{6.3.3}$$

Taking the curl of both sides of equation (6.3.3), using the linearity of the differential operator and the vanishing of the curl of a gradient, we obtain

$$\nabla \times \tilde{\mathbf{S}} = \nabla \times (\mathbf{S} + \nabla f) = \nabla \times \mathbf{S}.$$

Examining this result in the context of expression (6.3.1), we see that the same $\mathbf{u}$ is obtained using either $\mathbf{S}$ or $\mathbf{S} + \nabla f$. We can use this freedom of choice to set $\nabla \cdot \tilde{\mathbf{S}} = 0$. This is tantamount to finding f such that $\nabla^2 f = -\nabla \cdot \mathbf{S}$. To reach this conclusion, we took the divergence of both sides of equation (6.3.3) to get

$$\nabla \cdot \tilde{\mathbf{S}} = \nabla \cdot (\mathbf{S} + \nabla f(\mathbf{x})) = \nabla \cdot \mathbf{S} + \nabla^2 f = 0. \tag{6.3.4}$$

Since both $\tilde{\mathbf{S}}$ and $\mathbf{S}$ result in the same $\mathbf{u}$, we have justified our adding equation (6.3.2) to system (6.3.1).[6]

---

[5] Interested readers might refer to Box 4.2 in Aki, K., and Richards, P.G., (2002) Quantitative seismology (*2nd* edition): University Science Books.

[6] Readers interested in a philosophical meaning of the fact that vector potentials, $\tilde{\mathbf{S}}$ and $\mathbf{S}$, which cannot be distinguished physically, result in the same displacement vector, $\mathbf{u}$, which is a measurable entity, might refer to Brown, J.R., (1994) Smoke and mirrors: How science reflects reality: Routledge, pp. 142–159.

In particular, it is interesting to notice by examining equation (6.3.4), that if f is a harmonic function—in other words, if f is a solution of $\nabla^2 f = 0$—then, $\nabla \cdot \tilde{\mathbf{S}} = \nabla \cdot \mathbf{S}$. Consequently, in view of equation (6.3.3), we can always add to $\mathbf{S}$ the gradient of a harmonic function.

6.3.3. Equation of motion

To study the equations of motion in terms of the displacement potentials, we insert expression (6.3.1) into equation (6.1.6) and write

$$\rho \frac{\partial^2 (\nabla \mathscr{P} + \nabla \times \mathbf{S})}{\partial t^2} = (\lambda + \mu) \nabla [\nabla \cdot (\nabla \mathscr{P} + \nabla \times \mathbf{S})] + \mu \nabla^2 (\nabla \mathscr{P} + \nabla \times \mathbf{S}).$$

Using the vanishing of the divergence of a curl and the definition of the Laplacian, we obtain

$$\begin{aligned} \rho \frac{\partial^2 (\nabla \mathscr{P} + \nabla \times \mathbf{S})}{\partial t^2} &= (\lambda + \mu) \nabla (\nabla \cdot \nabla \mathscr{P}) + \mu \nabla^2 (\nabla \mathscr{P} + \nabla \times \mathbf{S}) \\ &= (\lambda + \mu) \nabla \left(\nabla^2 \mathscr{P}\right) + \mu \nabla^2 (\nabla \mathscr{P} + \nabla \times \mathbf{S}) . \end{aligned}$$

Using the linearity of the differential operators and the fact that in a homogeneous continuum ρ, λ and μ are constants, as well as using the equality of mixed partial derivatives, we can rewrite this equation as

$$\begin{aligned} &\nabla \left(\rho \frac{\partial^2 \mathscr{P}}{\partial t^2}\right) + \nabla \times \left(\rho \frac{\partial^2 \mathbf{S}}{\partial t^2}\right) \\ &\quad = \nabla \left[(\lambda + \mu) \nabla^2 \mathscr{P}\right] + \nabla \left(\mu \nabla^2 \mathscr{P}\right) + \nabla \times \left(\mu \nabla^2 \mathbf{S}\right) \\ &\quad = \nabla \left[(\lambda + 2\mu) \nabla^2 \mathscr{P}\right] + \nabla \times \left(\mu \nabla^2 \mathbf{S}\right) . \end{aligned}$$

Rearranging, we obtain

$$\nabla \left[(\lambda + 2\mu) \nabla^2 \mathscr{P} - \rho \frac{\partial^2 \mathscr{P}}{\partial t^2}\right] + \nabla \times \left[\mu \nabla^2 \mathbf{S} - \rho \frac{\partial^2 \mathbf{S}}{\partial t^2}\right] = \mathbf{0}, \qquad (6.3.5)$$

which is the equation of motion for isotropic homogeneous continua in terms of the scalar and vector potentials.

6.3.4. *P* and *S* waves

Looking at equation (6.3.5), we see that it is satisfied if

$$\nabla^2 \mathscr{P} - \frac{1}{\frac{\lambda + 2\mu}{\rho}} \frac{\partial^2 \mathscr{P}}{\partial t^2} = 0 \tag{6.3.6}$$

and

$$\nabla^2 \mathbf{S} - \frac{1}{\frac{\mu}{\rho}} \frac{\partial^2 \mathbf{S}}{\partial t^2} = \mathbf{0}, \tag{6.3.7}$$

which, in view of equations (6.1.12) and (6.1.16), appear to be associated with *P* and *S* waves, respectively. A rigorous analysis of this result is associated with Lamé's theorem.[7] Equations (6.3.6) and (6.3.7) are wave equations whose wave functions are the scalar and vector potentials, respectively. Motivated by this observation, we wish to study the relation of the scalar and vector potentials to the two wave equations whose wave functions are given by the dilatation and the rotation vector; namely equations (6.1.12) and (6.1.16), respectively.

As in Section 6.1.2, let us take the divergence of equation (6.3.5). Using the vanishing of the divergence of a curl and the definition of the Laplacian, we obtain

$$\nabla^2 \left[(\lambda + 2\mu) \nabla^2 \mathscr{P} - \rho \frac{\partial^2 \mathscr{P}}{\partial t^2} \right] = 0. \tag{6.3.8}$$

Using the linearity of the differential operator and the equality of mixed partial derivatives, we can rewrite equation (6.3.8) as

$$(\lambda + 2\mu) \nabla^2 \left(\nabla^2 \mathscr{P} \right) - \rho \frac{\partial^2 \left(\nabla^2 \mathscr{P} \right)}{\partial t^2} = 0. \tag{6.3.9}$$

To relate the scalar potential, $\mathscr{P}$, to the dilatation, φ, let us take the divergence of expression (6.3.1). Using the vanishing of the divergence of a curl and recalling definition (1.4.18) as well as the definition of the Laplacian, we obtain

$$\varphi := \nabla \cdot \mathbf{u} = \nabla \cdot \nabla \mathscr{P} \equiv \nabla^2 \mathscr{P}. \tag{6.3.10}$$

[7] Readers interested in this theorem might refer to Aki, K., and Richards, P.G., (2002) Quantitative seismology (*2nd* edition): University Science Books, pp. 67–69, and to Long, C., (1967) On the completeness of the Lamé potentials: Acta Mechanica, **3**, 371–375.

In other words, the dilatation is equal to the Laplacian of the scalar potential. Using expression (6.3.10), we can rewrite equation (6.3.9) as

$$\nabla^2\varphi = \frac{1}{\frac{\lambda+2\mu}{\rho}}\frac{\partial^2\varphi}{\partial t^2}, \tag{6.3.11}$$

which is equation (6.1.12), as expected. Thus, we conclude that the Laplacian of the scalar potential, $\mathscr{P}$, satisfies the wave equation for P waves.

As in Section 6.1.3, let us take the curl of equation (6.3.5). Using the vanishing of the curl of a gradient, we get

$$\nabla\times\nabla\times\left(\mu\nabla^2\mathbf{S}-\rho\frac{\partial^2\mathbf{S}}{\partial t^2}\right)=\mathbf{0}.$$

Recalling identity (6.1.7) and letting **a** denote the term in parentheses, we can rewrite this equation as

$$\nabla\left[\nabla\cdot\left(\mu\nabla^2\mathbf{S}-\rho\frac{\partial^2\mathbf{S}}{\partial t^2}\right)\right]=\nabla^2\left(\mu\nabla^2\mathbf{S}-\rho\frac{\partial^2\mathbf{S}}{\partial t^2}\right).$$

Using the linearity of the differential operators and the equality of mixed partial derivatives, we can rewrite this equation as

$$\nabla\left[\mu\nabla^2(\nabla\cdot\mathbf{S})-\rho\frac{\partial^2(\nabla\cdot\mathbf{S})}{\partial t^2}\right]=\nabla^2\left(\mu\nabla^2\mathbf{S}-\rho\frac{\partial^2\mathbf{S}}{\partial t^2}\right). \tag{6.3.12}$$

In view of equation (6.3.2), $\nabla\cdot\mathbf{S}=0$, equation (6.3.12) becomes

$$\nabla^2\left(\mu\nabla^2\mathbf{S}-\rho\frac{\partial^2\mathbf{S}}{\partial t^2}\right)=\mathbf{0}. \tag{6.3.13}$$

Again using the linearity of the differential operator and the equality of mixed partial derivatives, we can write equation (6.3.13) as

$$\mu\nabla^2\left(\nabla^2\mathbf{S}\right)-\rho\frac{\partial^2\left(\nabla^2\mathbf{S}\right)}{\partial t^2}=\mathbf{0}. \tag{6.3.14}$$

To relate the vector potential, $\mathbf{S}$, to the rotation vector, Ψ, let us take the curl of expression (6.3.1). Using the vanishing of the curl of a gradient and recalling definition (1.5.2), we obtain

$$\Psi := \nabla\times\mathbf{u}=\nabla\times\nabla\times\mathbf{S}.$$

Following identity (6.1.7) and letting $\mathbf{a} = \mathbf{S}$, we get

$$\Psi = \nabla \times (\nabla \times \mathbf{S}) = \nabla (\nabla \cdot \mathbf{S}) - \nabla^2 \mathbf{S}.$$

In view of equation (6.3.2), $\nabla \cdot \mathbf{S} = 0$, we obtain

$$\Psi = -\nabla^2 \mathbf{S}. \tag{6.3.15}$$

In other words, the rotation vector is equal to the negative Laplacian of the vector potential. Using expression (6.3.15), we can rewrite equation (6.3.14) as

$$\nabla^2 \Psi = \frac{1}{\frac{\mu}{\rho}} \frac{\partial^2 \Psi}{\partial t^2}, \tag{6.3.16}$$

which is equation (6.1.16), as expected. Thus, we conclude that under condition (6.3.2) the Laplacian of the vector potential, $\mathbf{S}$, satisfies the wave equation for S waves.

This derivation of equations (6.3.11) and (6.3.16) is analogous to the method for obtaining Maxwell's equations in the electromagnetic theory using the vector and scalar potentials. There are, however, several distinctions between the origins of potentials for the wave equations in elasticity and electromagnetism.[8]

6.4. *P* and *S* Waves in Terms of Displacements

6.4.1. Introductory comments

[9] Examining expression (6.1.8), namely,

$$\rho \frac{\partial^2 \mathbf{u}}{\partial t^2} = (\lambda + 2\mu) \nabla (\nabla \cdot \mathbf{u}) - \mu \nabla \times (\nabla \times \mathbf{u}), \tag{6.4.1}$$

---

[8] Readers interested in the derivation of Maxwell's equations using the potentials might refer to Feynman, R.P., Leighton, R.B., and Sands, M., (1963/1989) Feynman's lectures on physics: Addison-Wesley Publishing Co., Vol. II, pp. 18-9–18-11. Readers interested in distinct origins of potentials in elasticity and electromagnetism might refer to Bóna, A. and Slawinski, M.A., (2021) Wavefronts and rays as characteristics and asymptotics (3*rd* edition): World Scientific, Appendix C, Closing remarks.

[9] This section is based on Rochester, M.G., (2010) Note on the necessary conditions for P and S wave propagation in a homogeneous isotropic elastic solid: Journal of Elasticity, **98**(1), 111–114.

and the formulations of wave equations in Sections 6.1 and 6.3, which result in equations (6.3.11) and (6.3.16), we see that expression (6.4.1) is an equation of motion stated in terms of the displacement vector, $\mathbf{u}$, and expressions (6.3.11) and (6.3.16) are wave equations stated in terms of derivatives of $\mathbf{u}$. In this section, we formulate equations for the P and S waves in terms of the displacement vectors themselves, and we show that the irrotational and equivoluminal displacements imposed by derivatives of $\mathbf{u}$ are the sufficient and necessary conditions for the existence of the P and S waves, respectively.

6.4.2. Sufficient conditions for P and S waves

P waves

If the displacement is irrotational, which means that

$$\nabla \times \mathbf{u} = \mathbf{0}, \tag{6.4.2}$$

equation (6.4.1) becomes

$$\rho \frac{\partial^2 \mathbf{u}}{\partial t^2} = (\lambda + 2\mu) \nabla (\nabla \cdot \mathbf{u}) . \tag{6.4.3}$$

In such a case, identity (6.1.7)—where we let $\mathbf{a} \equiv \mathbf{u}$—becomes

$$\nabla^2 \mathbf{u} = \nabla (\nabla \cdot \mathbf{u}) . \tag{6.4.4}$$

Hence, equation (6.4.3) becomes

$$\nabla^2 \mathbf{u} = \frac{1}{\frac{\lambda + 2\mu}{\rho}} \frac{\partial^2 \mathbf{u}}{\partial t^2}, \tag{6.4.5}$$

which is analogous to expression (6.3.11): the equation for P waves, but stated in terms of $\mathbf{u}$, not $\nabla \cdot \mathbf{u}$. Hence, under condition (6.4.2), equation (6.4.1) implies equation (6.4.5). In other words, given equation (6.4.1), expression (6.4.2) is a sufficient condition for equation (6.4.5).

S waves

If the displacement is equivoluminal, which means that

$$\nabla \cdot \mathbf{u} = 0, \tag{6.4.6}$$

equation (6.4.1) becomes

$$\rho \frac{\partial^2 \mathbf{u}}{\partial t^2} = -\mu \nabla \times (\nabla \times \mathbf{u}) . \tag{6.4.7}$$

In such a case, identity (6.1.7) becomes

$$\nabla^2 \mathbf{u} = -\nabla \times (\nabla \times \mathbf{u}) . \tag{6.4.8}$$

Hence, equation (6.4.3) becomes

$$\nabla^2 \mathbf{u} = \frac{1}{\frac{\mu}{\rho}} \frac{\partial^2 \mathbf{u}}{\partial t^2}, \tag{6.4.9}$$

which is analogous to expression (6.3.16): the equation for *S* waves, but stated in terms of $\mathbf{u}$, not $\nabla \times \mathbf{u}$. Hence, under condition (6.4.6), equation (6.4.1) implies equation (6.4.9). In other words, given equation (6.4.1), expression (6.4.6) is a sufficient condition for equation (6.4.9).

6.4.3. Necessary conditions for *P* and *S* waves

To assert the necessity of conditions (6.4.2) and (6.4.6), we have to show that equation (6.4.5) implies condition (6.4.2) and equation (6.4.9) implies condition (6.4.6).

Following an approach analogous to the one presented in Section 6.1.2, we take the divergence of equation (6.4.1), which, due to $\nabla \cdot \nabla \times \mathbf{u} = 0$, becomes

$$\rho \frac{\partial^2}{\partial t^2} \nabla \cdot \mathbf{u} = (\lambda + 2\mu) \nabla^2 (\nabla \cdot \mathbf{u}) .$$

Using using identity (6.4.4) and the linearity of differential operators, we rewrite it as

$$\left(\nabla^2 - \frac{1}{\frac{\lambda + 2\mu}{\rho}} \frac{\partial^2}{\partial t^2} \right) \nabla \cdot \mathbf{u} = 0 .$$

Examining this equation, we see that either $\nabla\cdot\mathbf{u}=0$ or the result of the differential operator in parentheses acting on $\nabla\cdot\mathbf{u}$ is zero. Denoting $\nabla\cdot\mathbf{u}$ by φ and denoting the operator in parentheses by $\square_P$, we express these alternatives as

$$\varphi=0 \qquad \text{or} \qquad \square_P\,\varphi=0\,. \tag{6.4.10}$$

To present our argument for the necessary condition, we invoke equation (6.4.9), without considering its derivation. Taking the divergence of this equation and using the linearity of differential operators together with the equality of mixed partial derivatives, we obtain $\square_S\,\varphi=0$, where $\square_S$ is the differential operator in expression (6.4.12), below. However, if φ is a general solution for $\square_S\,\varphi=0$, it cannot be a general solution for $\square_P\,\varphi=0$, since these operators differ by a constant factor in front of the temporal derivative. Notably, in view of expression (6.5.6), below, we see that the general solutions are different from one another; in a single spatial dimension, they are $f(x-\sqrt{\mu/\rho}\,t)+g(x+\sqrt{\mu/\rho}\,t)$ and $f(x-\sqrt{(\lambda+2\mu)/\rho}\,t)+g(x-\sqrt{(\lambda-2\mu)/\rho}\,t)$, respectively, and, as shown in Exercise 5.18, for any isotropic Hookean solid, $\mu>0$ and $\lambda>-2\mu/3$, which means that the two waves have different propagation speeds. Thus, we must choose the first alternative, $\varphi=0$, which does not exhibit any inconsistency with equation (6.4.9). This means that equation (6.4.9) implies condition (6.4.6): $\nabla\cdot\mathbf{u}=0$, and, hence, it is a necessary condition.

Following an approach analogous to the one presented in Section 6.1.3, we take the curl of equation (6.4.1), which, due to $\nabla\times\nabla(\nabla\cdot\mathbf{u})=\mathbf{0}$, becomes

$$\rho\frac{\partial^2}{\partial t^2}\nabla\times\mathbf{u}=-\mu\nabla\times(\nabla\times\nabla\times\mathbf{u})\,. \tag{6.4.11}$$

Using identity (6.4.8), we rewrite equation (6.4.11) as

$$\left(\nabla^2-\frac{1}{\frac{\mu}{\rho}}\frac{\partial^2}{\partial t^2}\right)\nabla\times\mathbf{u}=\mathbf{0}\,. \tag{6.4.12}$$

Examining this equation, we see that either $\nabla\times\mathbf{u}=\mathbf{0}$ or the result of the operator in parentheses acting on $\nabla\times\mathbf{u}$ is zero. We express these alternatives as

$$\Psi=\mathbf{0} \qquad \text{or} \qquad \square_S\Psi=\mathbf{0}\,. \tag{6.4.13}$$

Taking the curl of equation (6.4.5), we obtain $\Box_P \Psi = \mathbf{0}$. This result disagrees with the second of alternatives (6.4.13); thus, we must choose the first alternative, which is consistent with equation (6.4.5). This means that equation (6.4.5) implies condition (6.4.2): it is a necessary condition.

To complete this section, using definitions (5.12.2), let us rewrite equations (6.4.5) and (6.4.9) as

$$\nabla^2 \mathbf{u} = \frac{1}{\frac{C_{11}}{\rho}} \frac{\partial^2 \mathbf{u}}{\partial t^2}, \tag{6.4.14}$$

and

$$\nabla^2 \mathbf{u} = \frac{1}{\frac{C_{44}}{\rho}} \frac{\partial^2 \mathbf{u}}{\partial t^2}, \tag{6.4.15}$$

respectively, which is consistent with the notation in Chapter 7, where we discuss equations of motion in anisotropic inhomogeneous continua; equations (6.4.14) and (6.4.15) are special cases of such equations.

6.5. Solutions of Wave Equation for Single Spatial Dimension

6.5.1. d'Alembert's approach

To gain further insights into the physical meaning of equations (6.1.12) and (6.1.16), we study the solution of their generic form, where we do not specify if the wave function corresponds to P waves or to S waves.[10] Consider the initial-value problem given by

$$\frac{\partial^2 u(x,t)}{\partial x^2} - \frac{1}{v^2} \frac{\partial^2 u(x,t)}{\partial t^2} = 0, \tag{6.5.1}$$

[10] Applying Newton's second law of motion, we can derive equation (6.5.1) for either longitudinal waves or transverse waves, which correspond to P waves or S waves, respectively. Readers interested in such derivations of the one-dimensional wave equation for P and S waves might refer to Hanna, J.R., (1982) Fourier series and integrals of boundary value problems: John Wiley and Sons, pp. 109–111 and pp. 121–122, or to Garrity, T.A., (2001) All the mathematics you missed [but need to know for graduate school]: Cambridge University Press, pp. 274 – 277, and to Benson, D.J., (2007) Music: A mathematical offering: Cambridge University Press, pp. 92–94, where derivations for the S waves are presented.

where $u = u(x,t)$ is the wave function and v is a constant. Let the initial conditions be stated by

$$\begin{cases} u(x,t)\big|_{t=0} = \gamma(x) \\ \left.\dfrac{\partial u(x,t)}{\partial t}\right|_{t=0} = \eta(x) \end{cases}. \tag{6.5.2}$$

The following method of solving the wave equation was introduced in 1746 by d'Alembert[11] and further elaborated upon by Euler, with important contributions from Daniel Bernoulli and Lagrange.[12] It is based on the following two lemmas, which allow us to obtain the solution by integration.

LEMMA 6.5.1. *Equation*

$$\frac{\partial^2 u}{\partial x^2} - \frac{1}{v^2}\frac{\partial^2 u}{\partial t^2} = 0,$$

is equivalent to

$$\frac{\partial^2 u(y,z)}{\partial y \partial z} = 0, \tag{6.5.3}$$

where the new coordinates are

$$\begin{cases} y = x + vt \\ z = x - vt \end{cases}. \tag{6.5.4}$$

Details of the derivation of Lemma 6.5.1 are shown in Exercise 6.1.

The form $\partial^2 u(y,z)/\partial y \partial z = 0$ is a normal form of the hyperbolic differential equation, where y and z are referred to as the natural coordinates; informally speaking, we have factorized the differential to obtain a solution by integration, as shown below. For y and z given by constants, $y = x + vt$ and $z = x - vt$ are straight lines in the xt-plane, which are known as the characteristics of the wave equation. Expressions $y(x,t)$ and $z(x,t)$ satisfy

[11] Interested readers might refer to d'Alembert, J., (1747) Recherches sur la courbe que forme une corde tendue mise en vibration: Hist. Acad. Sci. Berlin **3**, 214 – 219.

[12] Readers interested in the history of deriving the wave equation including disagreements among d'Alembert, Euler, Bernoulli and Lagrange in accommodating the initial conditions might refer to Kline, M., (1972) Mathematical thought from ancient to modern times: Oxford University Press, Vol. II, pp. 503–514.

the characteristic equation of wave equation (6.5.1), as shown in Exercise 6.19.[13]

LEMMA 6.5.2. *For equation*

$$\frac{\partial^2 u(y,z)}{\partial y \partial z} = 0,$$

the only form of the solution is

$$u(y,z) = f(y) + g(z), \tag{6.5.5}$$

where f and g are twice-differentiable arbitrary functions.

Details of the derivation of Lemma 6.5.2 are shown in Exercise 6.2.

Combining Lemma 6.5.1 and Lemma 6.5.2, we can state the following corollary.

COROLLARY 6.5.3. *Following Lemma 6.5.1 and Lemma 6.5.2, and using coordinates (6.5.4), we can write the only form of the solution of equation (6.5.1) as*

$$u(x,t) = f(x + vt) + g(x - vt), \tag{6.5.6}$$

where f and g are arbitrary twice-differentiable functions.

Solution (6.5.6) allows arbitrary twice-differentiable functions f and g. Further constraints must be imposed on functions f and g if we wish to obtain a particular solution.

Herein, we wish to obtain a particular form of solution (6.5.6) that satisfies the constraints provided by initial conditions (6.5.2). Inserting expression (6.5.6) into system of equations (6.5.2), we can write

$$\begin{cases} f(x) + g(x) = \gamma(x) \\ vf'(x) - vg'(x) = \eta(x) \end{cases}, \tag{6.5.7}$$

where we used the chain rule with f' and g' denoting the derivatives with respect to arguments $(x + vt)$ and $(x - vt)$, respectively, and evaluated the

[13] Readers interested in normal forms of the hyperbolic equations and its association with characteristics might refer to Morse P.M., and Feshbach H., (1953) Methods of theoretical physics: McGraw-Hill, Inc., Part I, pp. 682–683.

Readers interested in the characteristics and their significance in wave theory might refer to Musgrave, M.J.P., (1970) Crystal acoustics: Introduction to the study of elastic waves and vibrations in crystals: Holden-Day, pp. 68–76.

results at $t = 0$. This system of equations can be solved explicitly for $f(x)$ and $g(x)$. Integrating both sides of the second equation of this system, we obtain

$$\begin{cases} f(x) + g(x) = \gamma(x) \\ f(x) - g(x) = \dfrac{1}{v} \int_{x_0}^{x} \eta(\zeta)\, \mathrm{d}\zeta \end{cases}, \tag{6.5.8}$$

where ζ is the integration variable. Adding the two equations together, we get

$$f(x) = \frac{1}{2}\left[\gamma(x) + \frac{1}{v} \int_{x_0}^{x} \eta(\zeta)\, \mathrm{d}\zeta\right], \tag{6.5.9}$$

while subtracting the second equation from the first one gives us

$$g(x) = \frac{1}{2}\left[\gamma(x) - \frac{1}{v} \int_{x_0}^{x} \eta(\zeta)\, \mathrm{d}\zeta\right]. \tag{6.5.10}$$

Inserting expressions (6.5.9) and (6.5.10) into solution (6.5.6), we write

$$u(x,t) = \frac{1}{2}\left\{\gamma(x+vt) + \gamma(x-vt) + \frac{1}{v}\left[\int_{x_0}^{x+vt} \eta(\zeta)\, \mathrm{d}\zeta - \int_{x_0}^{x-vt} \eta(\zeta)\, \mathrm{d}\zeta\right]\right\}.$$

Using the fact that reversing the limits of integration changes the sign of the integral, we obtain

$$u(x,t) = \frac{1}{2}\left[\gamma(x+vt) + \gamma(x-vt) + \frac{1}{v} \int_{x-vt}^{x+vt} \eta(\zeta)\, \mathrm{d}\zeta\right], \tag{6.5.11}$$

which is the solution of the initial-value problem given by equations (6.5.1) and (6.5.2).

Let us interpret the physical meaning of solution (6.5.11). If we view x as the position variable and t as the time variable, solution (6.5.11) describes propagation of u in the one-dimensional x-space. Solution $u(x,t)$ is completely determined by the differential equation and the initial conditions, which describe the solution at the initial time, $u(x,0) = \gamma(x)$, and the velocity of displacement of u at that instant, $\eta(x)$; in other words, at point x, γ displaces with velocity η.

Herein, the one-dimensional space is homogeneous and infinite. To examine inhomogeneous, finite and semifinite cases, we need to consider conditions at the boundary, as exemplified in Section 6.8, below.[14]

To examine the concept of propagation, let us consider solution (6.5.6) with $g = 0$; namely,

$$u(x,t) = f(x+vt). \tag{6.5.12}$$

We wish to examine the propagation of a given point that belongs to f. Such a point corresponds to a particular value of f. In view of expression (6.5.12), we see that a particular value of $u(x,t)$ remains the same if the value of $x+vt$ stays the same. According to conditions (6.5.2), at time $t=0$, we have $u(x,0)=\gamma(x)$. Let us consider location x_0, at that time. At this location and at that time, we have $\gamma(x_0)$. Let us follow this value of γ. At time t_1, we have $\gamma(x_1+vt_1)$. To follow the same value, we require the constancy of the argument; in other words, $x_0 = x_1 + vt_1$. Solving for x_1, we get $x_1 = x_0 - vt_1$, which means that the value of γ that was at x_0 moved to $x_0 - vt_1$; it moved left along the x-axis by distance vt_1. We conclude that constant v in equation (6.5.1) is the propagation speed.

Returning to solution (6.5.6) and following an argument analogous to the one presented above but with $f = 0$, we conclude that f and g move in opposite directions. In general, f and g are different from one another; they are explicitly stated in terms of the two initial conditions by expressions (6.5.9) and (6.5.10). Examining these expressions, we note that the two functions propagating in the opposite directions are the same if $\eta(x) = 0$; in other words, if there is no initial velocity of displacement. In such a case, the amplitude of displacement at any location and instant is the arithmetic mean of the two identical waves travelling in the opposite directions.[15]

The initial-value problem given by equations (6.5.1) and (6.5.2) does not contain any source; in other words, there is no action of external force.

---

[14]Readers interested in a finite and semifinite cases might refer to Arnold, V.I., (2004) Lectures on partial differential equations: Springer-Verlag, pp. 29–32.

[15]An insightful formulation and discussion of d'Alembert's solution with initial conditions are presented by Benson, D.J., (2007) Music: A mathematical offering: Cambridge University Press, pp. 94–101.

We could incorporate such a force in the initial-value problem by adding a term to equation (6.5.1) to write it as

$$\frac{\partial^2 u(x,t)}{\partial x^2} - \frac{1}{v^2}\frac{\partial^2 u(x,t)}{\partial t^2} = f(x,t), \qquad (6.5.13)$$

where f is a function representing the source.[16]

If the solution exhibits a spherical symmetry—which means that the initial conditions, $\gamma(\mathbf{x})$ and $\eta(\mathbf{x})$, are spherically symmetric, since the wave equation itself, $v^2\nabla^2 u = \partial^2 u/\partial t^2$, is symmetric— d'Alembert's approach can be extended to higher spatial dimensions, as illustrated in Exercises 6.3 and 6.4. In such a case, the general solution is

$$u(r,t) = \frac{1}{r}f(r+vt) + \frac{1}{r}g(r-vt),$$

where the first term on the right-hand side represents a spherically symmetric wave that propagates towards the origin and the second one away from it; r is the distance from the origin, it is the magnitude of $\mathbf{r}$, and it is tantamount to $\sqrt{x_1^2+x_2^2+x_3^2}$, in Cartesian coordinates.

In this manner, the extension discussed in Exercises 6.3 and 6.4 remains formally a problem of a single spatial dimension. The amplitude of the wave propagating towards the origin increases with distance, as the wavefront approaches the origin; the amplitude of the wave propagating away decreases. The change of amplitude is proportional to r^{-1} for both two and three dimensions, which is a consequence, respectively, of a wavefront in two spatial dimensions represented by a circle, which is a one-dimensional entity, and in three dimensions by a two-dimensional sphere. Note that the form of the Laplacian for a function that is independent of the angle is the same for both polar and spherical coordinates, as illustrated in Exercise 6.3.

[16]Readers interested in the solution of the initial-value problem with equation (6.5.13) might refer to Strauss, W.A., (2008) Partial differential equations: An introduction (*2nd* edition): John Wiley and Sons, pp. 71–78.

6.5.2. Directional derivative

To gain further insight into wave equation (6.5.1) and solution (6.5.6), let us rewrite this equation using directional derivatives. We write

$$\left(v^2\frac{\partial^2}{\partial x^2}-\frac{\partial^2}{\partial t^2}\right)u(x,t)=0,$$

where the term in parentheses is a differential operator, which we can rewrite as a composition of two differential operators; namely,[17]

$$\left(v\frac{\partial}{\partial x}-\frac{\partial}{\partial t}\right)\left(v\frac{\partial}{\partial x}+\frac{\partial}{\partial t}\right)u(x,t)=0. \tag{6.5.14}$$

Using the scalar product, we can write each operator as

$$\left([v,-1]\cdot\left[\frac{\partial}{\partial x},\frac{\partial}{\partial t}\right]\right)\left([v,1]\cdot\left[\frac{\partial}{\partial x},\frac{\partial}{\partial t}\right]\right)u(x,t)=0, \tag{6.5.15}$$

where the terms in parentheses have the form of directional derivatives with directions $[v,-1]$ and $[v,1]$. Equation (6.5.15) means that u is constant in these directions. In other words, u is constant along the lines whose slopes are

$$\frac{\mathrm{d}x}{\mathrm{d}t}=\pm v.$$

Solving this ordinary differential equation, we get

$$x=\pm vt+C,$$

where C is the integration constant. Solving for C, we write

$$C=x\mp vt.$$

Since a function of a constant is a constant, the solution of equation (6.5.15) can be written as

$$u(x,t)=f(x+vt)+g(x-vt),$$

which is solution (6.5.6); it states that f and g are constant along $x=\pm vt+C$.

[17] Readers interested in using expression (6.5.14) to obtain d'Alembert's solution via a first-order partial differential equation might refer to Evans, L.C., (1998) Partial differential equations: AMS, pp. 67–68.

In view of equation (6.5.15), we can formulate the solution in a different way. Function u remains unchanged along directions $[v,-1]$ and $[v,1]$, which correspond to lines that are described by their perpendiculars; namely, $[1,v]$ and $[1,-v]$, respectively. Hence, the arguments of f and g are

$$[x,t]\cdot[1,v]=x+vt$$

and

$$[x,t]\cdot[1,-v]=x-vt,$$

respectively.

6.5.3. Well-posed problem

We are interested in knowing whether or not the solution of the initial-value problem given by equations (6.5.1) and (6.5.2) is unique. Also, since this solution results from the initial conditions, we wish to know whether or not it depends smoothly on the initial data. In other words, we wish to verify that the dependence is such that a small change in input affects the solution by a small amount only. We refer to such a solution as a stable solution. A problem consisting of equations that result in a unique and stable solution is called a well-posed problem. Wave equation (6.5.1) together with conditions (6.5.2) constitute a well-posed problem as we see below.

It is important to note that many mathematical physics questions do not constitute well-posed problems; they are ill-posed problems. However, this classical nomenclature does not imply that a well-posed problem is physically more realistic than an ill-posed problem. For instance, inverse problems, which are of great interest in seismology, often do not possess unique solutions.

Let us demonstrate the uniqueness of solution (6.5.11). We demonstrate it explicitly by following the construction of our solution derived in Section 6.5.1.

Consider equation (6.5.1), namely,

$$\frac{\partial^2 u}{\partial x^2}-\frac{1}{v^2}\frac{\partial^2 u}{\partial t^2}=0. \tag{6.5.16}$$

Following Lemma 6.5.1 and Lemma 6.5.2, we obtained Corollary 6.5.3, according to which

$$u(x,t) = f(x+vt) + g(x-vt) \qquad (6.5.17)$$

is the only form of the solution of equation (6.5.16). Using initial conditions (6.5.2), we obtain system (6.5.8), which allows us to uniquely solve for f and g in terms of γ and η. Consequently, we can write solution (6.5.17) in terms of γ and η to obtain expression (6.5.11). Thus, in view of the construction of solution (6.5.11), we conclude that the solution is unique.

Similarly, examining the construction that led to solution (6.5.11), as described in Section 6.5.1, we see that γ and η are linear combinations of the continuous and differentiable functions f and g and of their first derivatives. Hence, functions γ and η, which are the initial conditions, are also continuous. Since the solution depends linearly on γ and on the integral of η, we conclude that a small change in γ or η results in a small change in u. Hence the solution, $u(x,t)$, is stable.

We can also show the uniqueness of the solution of the wave equation by studying the energy of wave function $u(\mathbf{x},t)$ at a given instant in time. Herein, by analogy to the above section, we discuss the energy of a wave function in a single spatial dimension, where $u = u(x,t)$. The presented method, however, can be easily extended to higher dimensions, thereby showing that, in general, the wave equation with appropriate initial conditions is a well-posed problem. Furthermore, we could also define the wave-function energy for equations in dissipative and dispersive media; this aspect of wave phenomena, however, is beyond the scope of this book.[18]

Considering equation (6.5.1), let us define the wave-function energy to be

$$\mathscr{E}(t) := \frac{1}{2}\int_{-\infty}^{\infty}\left[\left(\frac{\partial u}{\partial t}\right)^2 + v^2\left(\frac{\partial u}{\partial x}\right)^2\right]\mathrm{d}x. \qquad (6.5.18)$$

[18]Readers interested in various definitions of energy in the study of partial differential equations might refer to McOwen, R.C., (1996) Partial differential equations: Methods and applications: Prentice-Hall, Inc., pp. 91–97.

In the study of partial differential equations, it is common to define energy as an integral whose integrand is composed of squares of first derivatives of a function, as we did in expression (6.5.18) for function u.

To interpret the physical meaning of this definition, we first note that, as written, the physical units of $\mathscr{E}$ are the product of velocity squared and distance, which are not units of energy. However, if we multiply $\mathscr{E}$ by unit mass per unit length, $[kg/m]$, the units of $\mathscr{E}$ are the product of velocity squared and mass, which are the units of energy. Also, we can write $\mathscr{E}$ as the sum of two integrals, namely, $\int\left[(\partial u/\partial t)^2/2\right]\mathrm{d}x$ and $\int\left[v^2(\partial u/\partial x)^2/2\right]\mathrm{d}x$. We can view the former integral as corresponding to the kinetic energy of displacement. Since $v^2/2$ is a constant, we can view the latter integral as corresponding to the potential energy; in particular, it corresponds to the strain energy that is associated with deformation u.

Differentiating expression (6.5.18) with respect to time and using the fact that limits of integration are fixed, we get

$$\frac{\mathrm{d}\mathscr{E}}{\mathrm{d}t}=\frac{1}{2}\int\limits_{-\infty}^{\infty}\frac{\partial}{\partial t}\left[\left(\frac{\partial u}{\partial t}\right)^2+v^2\left(\frac{\partial u}{\partial x}\right)^2\right]\mathrm{d}x=\int\limits_{-\infty}^{\infty}\left(\frac{\partial u}{\partial t}\frac{\partial^2 u}{\partial t^2}+v^2\frac{\partial u}{\partial x}\frac{\partial^2 u}{\partial t\partial x}\right)\mathrm{d}x.$$

To study this integral, we rewrite it as two integrals, namely,

$$\frac{\mathrm{d}\mathscr{E}}{\mathrm{d}t}=\int\limits_{-\infty}^{\infty}\frac{\partial u}{\partial t}\frac{\partial^2 u}{\partial t^2}\mathrm{d}x+v^2\int\limits_{-\infty}^{\infty}\frac{\partial u}{\partial x}\frac{\partial^2 u}{\partial t\partial x}\mathrm{d}x. \tag{6.5.19}$$

Solving the second integral by parts, we get

$$\int\limits_{-\infty}^{\infty}\frac{\partial u}{\partial x}\frac{\partial^2 u}{\partial t\partial x}\mathrm{d}x=\left.\frac{\partial u}{\partial t}\frac{\partial u}{\partial x}\right|_{x=-\infty}^{x=\infty}-\int\limits_{-\infty}^{\infty}\frac{\partial u}{\partial t}\frac{\partial^2 u}{\partial x^2}\mathrm{d}x.$$

We assume that η, which appears in condition (6.5.2), has compact support. This means that $\eta(x)$ is nonzero on a closed and bounded interval, but is zero everywhere else. Thus, $\partial u/\partial t$ vanishes at both positive and negative infinity of the space variable, x. In other words, there is no displacement velocity infinitely far from the neighbourhood of $x=0$. For the integrated term, $(\partial u/\partial t)(\partial u/\partial x)|_{-\infty}^{\infty}$, to vanish, we must also assume

$\partial u/\partial x$ to be finite at both positive and negative infinity of x. If we assume that function γ has compact support, $\partial u/\partial x = 0$ at $x = \pm\infty$ since the propagation speed, v, is finite. This implies that, at an infinite distance, $u(x,t) = 0$, for all $t < \infty$. Using these assumptions, we can write

$$\int_{-\infty}^{\infty} \frac{\partial u}{\partial x}\frac{\partial^2 u}{\partial t \partial x}\,\mathrm{d}x = -\int_{-\infty}^{\infty} \frac{\partial u}{\partial t}\frac{\partial^2 u}{\partial x^2}\,\mathrm{d}x.$$

Inserting this expression into expression (6.5.19), we get

$$\frac{\mathrm{d}\mathscr{E}}{\mathrm{d}t} = \int_{-\infty}^{\infty} \frac{\partial u}{\partial t}\frac{\partial^2 u}{\partial t^2}\,\mathrm{d}x - v^2 \int_{-\infty}^{\infty} \frac{\partial u}{\partial t}\frac{\partial^2 u}{\partial x^2}\,\mathrm{d}x = \int_{-\infty}^{\infty} \frac{\partial u}{\partial t}\left[\frac{\partial^2 u}{\partial t^2} - v^2\frac{\partial^2 u}{\partial x^2}\right]\mathrm{d}x.$$

Examining the integrand in view of wave equation (6.5.1), we see that the term in brackets vanishes. Hence, we obtain

$$\frac{\mathrm{d}\mathscr{E}}{\mathrm{d}t} = 0, \tag{6.5.20}$$

which implies that $\mathscr{E}$ is constant.

Equation (6.5.20) states that for wave equation (6.5.1), the wave-function energy defined by expression (6.5.18) is conserved.[19] We can state it as the following theorem.

THEOREM 6.5.4. *If $u(x,t)$ is a solution of wave equation (6.5.1) together with initial conditions (6.5.2) that are given by functions with compact support, then energy defined by expression (6.5.18) is conserved. In other words, $\mathscr{E}(t) = \mathscr{E}(0)$ for all t.*

In Exercise 6.5, we prove an analogous theorem for a boundary-value, rather than an initial-value, problem.

We can use Theorem 6.5.4 to show that the solution of the problem given by equations (6.5.1) and (6.5.2) is unique.[20] Since equation (6.5.1)

[19] Readers interested in the energy of function $u(\mathbf{x},t)$ for a three-dimensional wave equation, for a frequency-dispersive wave equation, and for the dissipative wave equation might refer to McOwen, R.C., (1996) Partial differential equations: Methods and applications: Prentice-Hall, Inc., pp. 91–92, 95–96 and 96–97, respectively.

[20] Readers interested in the uniqueness of solution of the wave equation and finiteness of the propagation speed in the context of conservation of energy might refer to Taylor, M.E., (1996) Partial differential equations; Basic theory: Springer-Verlag, pp. 140–148.

is linear, a solution can be composed of a difference of two arbitrary solutions. Each arbitrary solution must obey the initial conditions. Examining the first of conditions (6.5.2), we see that—at $t = 0$—each solution is $\gamma(x)$. This means that—at $t = 0$—the solution composed of a difference of two solutions is zero. If the difference of two arbitrary solutions is zero, these two solutions are equal to one another; in other words, the solution is unique at $t = 0$. Now, using Theorem 6.5.4, we wish to verify this uniqueness for $t > 0$.

Following definition (6.5.18) and using the fact that, at $t = 0$, the solution composed of a difference of two solutions is $u(x,t) = 0$, we see that the corresponding $\mathscr{E}(0) = 0$. Then, Theorem 6.5.4 states that $\mathscr{E}(t) = \mathscr{E}(0) = 0$, for all t. Invoking definition (6.5.18), we can explicitly write

$$\mathscr{E}(t_1) = \frac{1}{2}\int_{-\infty}^{\infty}\left[\left(\left.\frac{\partial u}{\partial t}\right|_{t=t_1}\right)^2 + v^2\left(\left.\frac{\partial u}{\partial x}\right|_{t=t_1}\right)^2\right]\mathrm{d}x = 0,$$

where t_1 is an arbitrary time, $t_1 \in (0,\infty)$. Since $(\partial u/\partial t)^2$, v^2 and $(\partial u/\partial x)^2$ are positive for all t and x, for the integral to vanish we require that

$$\frac{\partial u(x,t)}{\partial t} = 0$$

and

$$\frac{\partial u(x,t)}{\partial x} = 0,$$

for any x and for arbitrary t. This implies that

$$\frac{\mathrm{d}u(x,t)}{\mathrm{d}t} = \frac{\partial u}{\partial x}\frac{\mathrm{d}x}{\mathrm{d}t} + \frac{\partial u}{\partial t} = \frac{\partial u}{\partial x}v + \frac{\partial u}{\partial t} = 0,$$

where we used the fact that v is a finite velocity of propagation. Hence, we can say that solution u is constant. Since u is unique at $t = 0$ while being constant for all $t \in (0,\infty)$, it must be unique for all t.

Thus we can conclude with the following corollary of Theorem 6.5.4.

COROLLARY 6.5.5. *Solution $u(x,t)$ of wave equation (6.5.1) together with conditions (6.5.2) is unique.*

The uniqueness of the solution of a hyperbolic partial differential equation, such as the wave equation, is linked intimately to its initial conditions. This subject is studied within the realm of Cauchy's problems. [21]

6.5.4. Causality, finite propagation speed and sharpness of signals

Let us continue to examine solution (6.5.11), namely,

$$u(x,t) = \frac{1}{2}\left[\gamma(x+vt)+\gamma(x-vt)+\frac{1}{v}\int\limits_{x-vt}^{x+vt}\eta(\zeta)\,\mathrm{d}\zeta\right]. \qquad (6.5.21)$$

For given (x,t), the value of solution u depends on the values of γ at points $x \mp vt$ and on the values of η on interval $[x-vt, x+vt]$. Hence, this interval is the domain of dependence of $u(x,t)$, as illustrated in Figure 6.5.1. To examine the effect of a point source, at x_0, on a location, say x_1, let us consider this figure. Point x_0 belongs to the domains of dependence of points (x,t) that lie in a triangular region whose apex is at x_0 and whose sides have the slopes of $\mp 1/v$. This region is the range of influence of x_0 on solutions $u(x,t)$. The signal generated at x_0 at $t=0$ does not reach x_1 until time t_1. Prior to that instant, the source has no effect at x_1. Thus, we conclude that the process is causal and the propagation speed of the signal is finite; its magnitude is $(x_1 - x_0)/t_1$. There is another important consequence of the range of influence. At x_1, the effect of the signal is observed not only at t_1 but also afterwards. This means that, in general, a sharp signal generated at x_0 does not propagate as a sharp signal—its effect persists after t_1. It is interesting to note that for the acoustic case, where u represents pressure, the constant value of u is not audible since our hearing relies on the change in pressure. Thus, we would hear only the initial arrival of the signal—a sharp-signal effect. For the elastic case, where u represents displacement, the constant value is observable.

The domain of dependence and the range of influence can be viewed as the time reverses of one another. To consider the domain of dependence is to ask about the origin of the phenomena observed at a given instant and

[21] Readers interested in Cauchy's problems and their relation to the wave propagation and rays might refer to Bóna, A. and Slawinski, M.A., (2021) Wavefronts and rays as characteristics and asymptotics (3*rd* edition): World Scientific, Chapter 2.

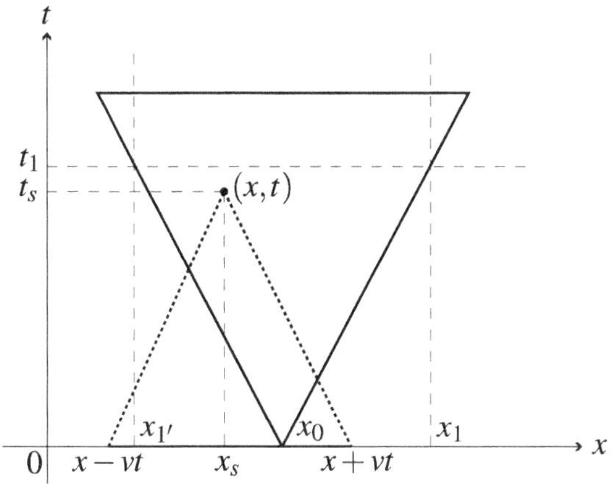

Fig. 6.5.1 The domain of dependence and the range of influence for the wave equation in a single spatial dimension. Consider the initial-value problem given in expressions (6.5.1) and (6.5.2), and its solution (6.5.21) at point (x,t). The point of $\gamma(x)$ located initially at $x+vt$ reaches x_s, from the right, at time t_s; the point located at $x-vt$ reaches the same location from the left. The value of the solution, u, that corresponds to x_s and t_s is the sum of the values of γ at the points initially at $x+vt$ and $x-vt$, and of the integral of η whose limits are $x-vt$ and $x+vt$. Also, at time t_1, the source at x_0 influences the values of the solution between $x_{1'}$ and x_1. The point of $\gamma(x)$ located initially at x_0 reaches $x+vt$ and $x-vt$ at time t_1. The limits of the integral that contributes to the solution are $x-vt$ and $x+vt$. Furthermore, we can conclude that the speed of propagation of a signal is finite, since before time t_1 elapses, point x_1 is not affected by the source. The slopes of the sides of the triangle are $\mp 1/v$.

location. To consider the range of influence is to ask about the regions influenced by an event.[22]

As stated above, in general, a sharp signal generated at a point does not propagate as a sharp signal. It does so in particular cases, however. By setting the initial displacement velocity, η, to zero, we rewrite solution (6.5.21) as

$$u(x,t) = \frac{1}{2}\left[\gamma(x+vt) + \gamma(x-vt)\right];$$

[22]Readers interested in the relation between the domain of dependence and the range of influence in terms of support of $u(x,t)$ might refer to Trèves, F., (1975/2006) Basic linear partial differential equations: Dover, pp. 111–117.

a common example of $\eta = 0$ is the case of a taut string that is pulled, and then released. Herein, the value of the solution depends on points $x \mp vt$ only. Hence, if there is no initial displacement velocity, the effect of the signal does not persist after the signal passes; consequently, sharp signals can propagate.

In Section 6.6, we see that the solution of the wave equation changes with the spatial dimension of the problem being considered; in contrast to in the cases of one and two dimensions, three-dimensional media allow for the propagation of sharp signals. Such a change of behaviour is a particular property of the wave equation; solutions of the heat equation and the steady-state equation, commonly known as Laplace's equation, do not exhibit such changes in physical interpretation due to dimensions.[23]

6.6. Solution of Wave Equation for Two and Three Spatial Dimensions

6.6.1. Introductory comments

Having obtained the solution of the wave equation in one spatial dimension in Section 6.5, we wish to investigate the solutions of the wave equation in two and three spatial dimensions. Let us consider

$$\nabla^2 u(\mathbf{x},t) - \frac{1}{v^2}\frac{\partial^2 u(\mathbf{x},t)}{\partial t^2} = 0, \tag{6.6.1}$$

where ∇^2 is the Laplacian, and the corresponding initial conditions; namely,

$$u(\mathbf{x},t)|_{t=0} = \gamma(\mathbf{x}) \tag{6.6.2}$$

and

$$\left.\frac{\partial u(\mathbf{x},t)}{\partial t}\right|_{t=0} = \eta(\mathbf{x}). \tag{6.6.3}$$

This is a general form of the initial-value problem stated on page 232.

[23]Readers interested in the concept of the range of influence, including an insightful physical consequences of these ranges for the wave equation, the diffusion equation, and Laplace's equation, might refer to Abbott, M.B., (1966) An introduction to the method of characteristics: Elsevier, pp. 16–18 as well as p. 66 and p. 70.

6.6.2. Three spatial dimensions

To consider the case of three spatial dimensions, we write equation (6.6.1) explicitly as

$$\sum_{i=1}^{3} \frac{\partial^2 u(x_1,x_2,x_3,t)}{\partial x_i^2} - \frac{1}{v^2}\frac{\partial^2 u(x_1,x_2,x_3,t)}{\partial t^2} = 0. \tag{6.6.4}$$

To solve this equation, we take its Fourier's transform and the transform of conditions (6.6.2) and (6.6.3) with $\mathbf{x}$ and $\mathbf{k}$ being the transformation variables. As shown in Exercise 6.6, we obtain

$$\frac{\partial^2 \tilde{u}(\mathbf{k},t)}{\partial t^2} + v^2 |\mathbf{k}|^2 \tilde{u}(\mathbf{k},t) = 0. \tag{6.6.5}$$

The corresponding initial conditions are

$$\tilde{u}(\mathbf{k},t)|_{t=0} = \tilde{\gamma}(\mathbf{k}) \tag{6.6.6}$$

and

$$\left.\frac{\partial \tilde{u}(\mathbf{k},t)}{\partial t}\right|_{t=0} = \tilde{\eta}(\mathbf{k}). \tag{6.6.7}$$

As shown in Exercise 6.7, the general solution of equation (6.6.5) is

$$\tilde{u}(\kappa,t) = F(\mathbf{k})\exp\{iv|\mathbf{k}|t\} + G(\mathbf{k})\exp\{-iv|\mathbf{k}|t\}. \tag{6.6.8}$$

To get expressions for F and G, we use conditions (6.6.6) and (6.6.7). Also, as shown in Exercise 6.7, we get

$$F(\mathbf{k}) = \frac{1}{2}\left(\tilde{\gamma}(\mathbf{k}) + \frac{1}{iv|\mathbf{k}|}\tilde{\eta}(\mathbf{k})\right) \tag{6.6.9}$$

and

$$G(\mathbf{k}) = \frac{1}{2}\left(\tilde{\gamma}(\mathbf{k}) - \frac{1}{iv|\mathbf{k}|}\tilde{\eta}(\mathbf{k})\right). \tag{6.6.10}$$

Inserting these expressions into solution (6.6.8), we write

$$\tilde{u}(\mathbf{k},t) = \frac{1}{2}\Bigg(\left(\tilde{\gamma}(\mathbf{k}) + \frac{1}{iv|\mathbf{k}|}\tilde{\eta}(\mathbf{k})\right)\exp\{iv|\mathbf{k}|t\} + \left(\tilde{\gamma}(\mathbf{k}) - \frac{1}{iv|\mathbf{k}|}\tilde{\eta}(\mathbf{k})\right)\exp\left\{-iv|\mathbf{k}|t\right\}\Bigg), \tag{6.6.11}$$

which is the solution of the initial-value problem given by equations (6.6.5), (6.6.6) and (6.6.7). To obtain the solution of the initial-value problem given by equations (6.6.4), (6.6.2) and (6.6.3), we need to find the inverse Fourier's transform of solution (6.6.11). To do so, let us rewrite the solution by factoring out the two initial conditions; namely,

$$\tilde{u}(\mathbf{k},t) = \tilde{\gamma}(\mathbf{k}) \frac{\exp\{iv|\mathbf{k}|t\} + \exp\{-iv|\mathbf{k}|t\}}{2} + \tilde{\eta}(\mathbf{k}) \frac{\exp\{iv|\mathbf{k}|t\} - \exp\{-iv|\mathbf{k}|t\}}{2iv|\mathbf{k}|}.$$

Examining the two fractions, we notice that they are related as follows:

$$\frac{\mathrm{d}}{\mathrm{d}t} \frac{\exp\{iv|\mathbf{k}|t\} - \exp\{-iv|\mathbf{k}|t\}}{2iv|\mathbf{k}|} = \frac{\exp\{iv|\mathbf{k}|t\} + \exp\{-iv|\mathbf{k}|t\}}{2}.$$

Using this relation, we write

$$\tilde{u}(\mathbf{k},t) = \tilde{\gamma}(\mathbf{k}) \frac{\mathrm{d}}{\mathrm{d}t} \frac{\exp\{iv|\mathbf{k}|t\} - \exp\{-iv|\mathbf{k}|t\}}{2iv|\mathbf{k}|} + \tilde{\eta}(\mathbf{k}) \frac{\exp\{iv|\mathbf{k}|t\} - \exp\{-iv|\mathbf{k}|t\}}{2iv|\mathbf{k}|}. \tag{6.6.12}$$

Since each term is a product of two functions, we invoke the fact that a product in the **k**-domain is a convolution in the **x**-domain. As shown in Exercise 6.8, the fractions in solution (6.6.12) are the transforms of the distribution given by

$$\frac{(2\pi)^3}{4\pi v^2 t} \iint\limits_{S(\mathbf{0},vt)} \tau \,\mathrm{d}\zeta = \frac{2\pi^2}{v^2 t} \iint\limits_{S(\mathbf{0},vt)} \tau \,\mathrm{d}\zeta, \tag{6.6.13}$$

where τ is a test function and $\mathrm{d}\zeta$ is the surface element on the sphere, $S(\mathbf{0}, vt)$, that is centred at $\mathbf{x} = \mathbf{0}$ and whose radius is vt. In other words, expression (6.6.13) is the inverse transform of the fraction in solution (6.6.12). As formulated in Exercise 6.9, using expression (6.6.13) and the properties of convolution, we obtain the solution of the initial-value problem given by equations (6.6.4), (6.6.2) and (6.6.3); namely,

$$u(\mathbf{x},t) = \frac{\mathrm{d}}{\mathrm{d}t} \frac{2\pi^2}{v^2 t} \iint\limits_{S(\mathbf{x},vt)} \gamma(\mathbf{y}) \,\mathrm{d}\zeta(\mathbf{y}) + \frac{2\pi^2}{v^2 t} \iint\limits_{S(\mathbf{x},vt)} \eta(\mathbf{y}) \,\mathrm{d}\zeta(\mathbf{y}), \tag{6.6.14}$$

where $\mathbf{y}$ are the variables of integration on the sphere centred at $\mathbf{x}$.

Examining expression (6.6.14), we see that for any point in the $x_1x_2x_3t$-space, solution $u(\mathbf{x},t)$ depends on the values of γ and η whose domain is the surface of a sphere in the $x_1x_2x_3$-space that is centred at the $\mathbf{x}$ coordinates of that point, and whose radius is vt. In a manner analogous to the one discussed in Section 6.5.4, we conclude that a point source in the $x_1x_2x_3$-space influences the points in the $x_1x_2x_3t$-space that are on the surface of the three-dimensional 'cone' embedded in four dimensions whose apex is the source and whose slope is $1/v$. Thus, a signal emitted at $t=0$ arrives at a point located at a distance d from the source at time $t=d/v$. Prior to that instant, there is no effect of the source, which means that the process is causal and the speed of signal propagation is finite. At that instant, the value of u is finite since both integrals in expression (6.6.14) are bounded. Afterwards, again there is no signal. The signal is confined to the spherical shell of radius vt, which is the propagating wavefront. Thus, sharp signals can propagate in three dimensions, which is in agreement with our experience.

6.6.3. Two spatial dimensions

Mathematically, the solution of the wave equation in two dimensions can be viewed as a particular case of the three-dimensional case. Physically, however, the two results have important distinctions.[24]

To investigate the case of two spatial dimensions, we return to equation (6.6.1) and write it explicitly as

$$\sum_{i=1}^{2}\frac{\partial^2 u(x_1,x_2,t)}{\partial x_i^2}-\frac{1}{v^2}\frac{\partial^2 u(x_1,x_2,t)}{\partial t^2}=0, \tag{6.6.15}$$

where the corresponding initial conditions are given by expressions (6.6.2) and (6.6.3) with $\mathbf{s}=[s_1,s_2,0]$, where $\mathbf{s}$ stands for coordinates of the three-dimensional space. Using solution (6.6.14), we can write the solution of

[24]Readers interested in an insightful formulation of the wave equation in two spatial dimensions and its solution might refer to Benson, D.J., (2007) Music: A mathematical offering: Cambridge University Press, pp. 112–115, where the author discusses this solution in terms of Bessel's functions and normal modes for a drum, and to Trèves, F., (1975/2006) Basic linear partial differential equations: Dover, pp. 111–115.

equation (6.6.15) as

$$u(\mathbf{x},t)=\frac{2\pi^2}{v^2t}\iint\limits_{S(\mathbf{x},vt)}\eta(s_1,s_2)\,\mathrm{d}\zeta(\mathbf{s})+\frac{\mathrm{d}}{\mathrm{d}t}\frac{2\pi^2}{v^2t}\iint\limits_{S(\mathbf{x},vt)}\gamma(s_1,s_2)\,\mathrm{d}\zeta(\mathbf{s}).$$

As shown in Exercise 6.10, we can rewrite this solution as

$$u(x_1,x_2,t)=\frac{\mathrm{d}}{\mathrm{d}t}\left(\frac{4\pi^2}{v}\iint\limits_{D(\mathbf{x},vt)}\frac{\gamma(s_1,s_2)}{\sqrt{(vt)^2-\left[(s_1-x_1)^2+(s_2-x_2)^2\right]}}\,\mathrm{d}s_1\,\mathrm{d}s_2\right)$$
$$+\frac{4\pi^2}{v}\iint\limits_{D(\mathbf{x},vt)}\frac{\eta(s_1,s_2)}{\sqrt{(vt)^2-\left[(s_1-x_1)^2+(s_2-x_2)^2\right]}}\,\mathrm{d}s_1\,\mathrm{d}s_2,\tag{6.6.16}$$

where $D(\mathbf{x},vt)$ is the disc that is centred at $\mathbf{x}$ and whose radius is vt.

Examining expression (6.6.16), we see that for any point in the x_1x_2t-space, solution $u(\mathbf{x},t)$ depends on the values of γ and η whose domain is a disc on the x_1x_2-plane that is centred at the $\mathbf{x}$ coordinates of that point, and whose radius is vt. In a manner analogous to the one discussed in Section 6.5.4, we conclude that a point source on the x_1x_2-plane influences the points in the x_1x_2t-space that are both on the surface of and contained within the right circular cone whose apex is the point source and whose slope is $1/v$. The projection onto the x_1x_2-plane of a circular section of this cone is the wavefront at a given time. Thus, a signal emitted at $t=0$ arrives at a point located at a distance d from the source at time $t=d/v$. Prior to that instant, there is no effect of the source, which means that the process is causal and the speed of signal propagation is finite. At that instant, the value of u is infinite since the bracketed terms of the radicands of equation (6.6.16) are equal to $(vt)^2$, and hence the two integrals in expression (6.6.16) are unbounded. Afterwards, the value of u diminishes but remains nonzero forever.[25] Hence, sharp signals do not propagate in two

[25] Readers interested in a quantitative description of fading of the value of u as recorded by the observer in the two-dimensional case might refer to Barton, G., (1989) Elements of Green's functions and propagation: Potentials, diffusion and waves: Oxford Science Publications, pp. 278–282.

dimensions, which we can visualize by picturing a pebble dropped into a pond: the surface is affected by this disturbance after the wavefront has passed.[26]

Thus for all the three cases discussed in Sections 6.5 and 6.6, the process of wave propagation described by the solution of the initial-value problem given by expressions (6.6.1), (6.6.2) and (6.6.3) is causal and the speed of propagation is finite. Also, as shown in Section 6.5.3, the solution of this initial-value problem is unique and stable. However, sharp signals propagate only in three dimensions.

6.7. On Evolution Equation

To gain further insight into the properties of solutions of the wave equation and its initial conditions, let us consider a first-order time-evolution

---

[26] Readers interested in the range of influence for wave propagation in two-dimensional and three-dimensional media, including the consequences for the corresponding wave equations on sharpness of the signal as well as on Huygens' principle, might refer to Kline, M., (1972) Mathematical thought from ancient to modern times: Oxford University Press, Vol 2, pp. 690–691, to McOwen, R.C., (1996) Partial differential equations: Methods and applications: Prentice-Hall, Inc., pp. 83–90 and pp. 130–131, and to Renardy, M., and Rogers, R.C., (1993) An introduction to partial differential equations: Springer-Verlag, pp. 158–159. Also, an insightful description of the wave equation in one spatial dimension is presented in Spivak, M., (1970/1999) A comprehensive introduction to differential geometry: Publish or Perish, Inc., Vol. V., pp. 68 – 71. In Trèves, F., (1975/2006) Basic linear partial differential equations: Dover, p. 115, we read

> Take, for instance, $n = 2$: In concrete terms, this corresponds to a two-dimensional propagation medium, like the surface of a lake or a sea. In the centre of the lake, at time zero, there takes place a very brief but very intense stormlike event ("very" means of course "infinitely"); from there on, a wave, something like a tidal wave, propagates. For a fixed observer at the lake [...] nothing happens until he is reached by the wave [...]. At that moment, he is likely to find himself in a very violent situation: all the energy of the initial "explosion" or storm seems to be concentrated at the *wave front*. If our observer survives, once the wave front passes him by, he will find himself in a smooth "swell", whose amplitude, as measured from the level of the lake at rest, decreases to zero as time goes to $+\infty$.

equation in a single spatial dimension, namely,

$$\frac{\partial^2 u(x,t)}{\partial x^2} - \frac{\partial u(x,t)}{\partial t} = 0, \tag{6.7.1}$$

and the initial condition given by $u(x,t)|_{t=0} = \gamma(x)$. As we see, the solution of this initial-value problem is causal but its speed of propagation is infinite. Furthermore, the mapping $\gamma \to u$ does not have a unique inverse in the limit as $t \to \infty$.

To solve this initial-value problem, we investigate its Fourier's transform in a manner analogous to the one discussed in Section 6.6.2. We get

$$\frac{\partial \tilde{u}(k,t)}{\partial t} + k^2 \tilde{u}(k,t) = 0,$$

which we can solve as if it were a first-order ordinary differential equation in t, with initial condition $\tilde{u}(k,t)|_{t=0} = \tilde{\gamma}(k)$. Hence, in the transformed domain, the solution is

$$\tilde{u}(k,t) = \tilde{\gamma}(k)\exp\{-k^2 t\}.$$

To obtain the solution of the initial-value problem, we proceed to find the inverse transform. We write

$$u(x,t) = \int_{-\infty}^{\infty} \tilde{\gamma}(k)\exp\{-k^2 t\}\exp\{ixk\}\,\mathrm{d}k.$$

Invoking the definition of Fourier's transform, we write

$$\tilde{\gamma}(k) = \frac{1}{2\pi}\int_{-\infty}^{\infty} \gamma(\zeta)\exp\{-i\zeta k\}\,\mathrm{d}\zeta,$$

where ζ is the integration variable. Inserting this result into the above expression for u, we get

$$u(x,t) = \frac{1}{2\pi}\int_{-\infty}^{\infty}\int_{-\infty}^{\infty} \gamma(\zeta)\exp\{-i\zeta k\}\exp\{-k^2 t\}\exp\{ixk\}\,\mathrm{d}\zeta\,\mathrm{d}k.$$

Combining the exponential terms, we write

$$u(x,t) = \frac{1}{2\pi}\int_{-\infty}^{\infty}\int_{-\infty}^{\infty}\gamma(\zeta)\exp\left\{i(x-\zeta)k - k^2t\right\}\mathrm{d}\zeta\,\mathrm{d}k.$$

Exchanging the order of integration and since γ is independent of k, we write

$$u(x,t) = \frac{1}{2\pi}\int_{-\infty}^{\infty}\left(\int_{-\infty}^{\infty}\exp\left\{i(x-\zeta)k - k^2t\right\}\mathrm{d}k\right)\gamma(\zeta)\,\mathrm{d}\zeta.$$

As shown in Exercise 6.11, the inner integral can be written as

$$\int_{-\infty}^{\infty}\exp\left\{i(x-\zeta)k - k^2t\right\}\mathrm{d}k = \frac{\exp\left\{-\frac{(x-\zeta)^2}{4t}\right\}}{\sqrt{t}}\int_{-\infty}^{\infty}e^{-s^2}\,\mathrm{d}s,$$

where s is the variable of integration. Evaluating the definite integral, as shown in Exercise 6.12, we write the inner integral as

$$\int_{-\infty}^{\infty}\exp\left\{i(x-\zeta)k - k^2t\right\}\mathrm{d}k = \sqrt{\frac{\pi}{t}}\exp\left\{-\frac{(x-\zeta)^2}{4t}\right\}.$$

Returning to the above expression for u, we write

$$u(x,t) = \frac{1}{2\sqrt{\pi t}}\int_{-\infty}^{\infty}\exp\left\{-\frac{(x-\zeta)^2}{4t}\right\}\gamma(\zeta)\,\mathrm{d}\zeta, \qquad (6.7.2)$$

which is the solution of the initial-value problem, as stated by the following theorem.

THEOREM 6.7.1. *If $\gamma(x)$ is bounded and continuous on $x \in \mathbb{R}$, then $u(x,t)$ given in expression (6.7.2) is an infinitely differentiable function satisfying—for all $x \in \mathbb{R}$ and $t > 0$—the initial-value problem of equation (6.7.1), and extends continuously to $t = 0$, where $u(x,t) = \gamma(x)$.*

Let us consider the domain of dependence of solution (6.7.2), and the range of influence of a given point on the x-axis. Examining expression (6.7.2) in view of Theorem 6.7.1, we see that for any (x,t) on the xt-plane where

$t > 0$, solution $u(x,t)$ depends on the value of function γ along the entire x-axis. In other words, the domain of dependence is the entire x-axis. By reciprocity, any point on the x-axis influences all the locations on the xt-plane where $t > 0$. In other words, this entire halfplane is the range of influence of such a point. This means that the effect of the initial condition, $\gamma(x)$, upon the solution, $u(x,t)$, is instantaneous; in other words, the solution has an infinite propagation speed. Hence, even if $\gamma(x)$ is confined to a finite domain along the x-axis, its effect for $t > 0$ covers the entire axis. Also, this means that γ affects all points in that halfplane from $t = 0$ to $t = \infty$.

Since, in view of Theorem 6.7.1, solution (6.7.2) is not valid for $t < 0$, the process is causal. Also the forward solution is unique. However, any bounded γ with compact support results in $u \to 0$ as $t \to \infty$; hence, we cannot determine the initial state, γ, from u at $t \to \infty$, which means that the inverse to the mapping $\gamma \to u$ is not unique at infinity—the inverse problem is ill-posed. We note that the wave equation and its initial conditions constitute well-posed forward and inverse problems.

6.8. Solutions of Wave Equation for One-Dimensional Scattering

To introduce wave propagation in inhomogeneous media, let us study a one-dimensional continuum where the wave propagates with speed v_1 at $x \leqslant 0$ and v_2 at $x > 0$. In view of solution (6.5.6), for the wave equations in the aforementioned continuum, namely,

$$\frac{\partial^2 u}{\partial x^2} - \frac{1}{v_1^2}\frac{\partial^2 u}{\partial t^2} = 0, \quad x \leqslant 0,$$

$$\frac{\partial^2 u}{\partial x^2} - \frac{1}{v_2^2}\frac{\partial^2 u}{\partial t^2} = 0, \quad x > 0, \tag{6.8.1}$$

the general solution is

$$u(x,t) = \begin{cases} f_1(x + v_1 t) + g_1(x - v_1 t), & x \leqslant 0 \\ f_2(x + v_2 t) + g_2(x - v_2 t), & x > 0 \end{cases}. \tag{6.8.2}$$

To examine this problem, let us consider only a wave propagating from the left towards the origin, $x = 0$. The effects associated with $x = 0$ are called scattering since at this point the incident wave is separated into a reflected wave and a transmitted wave, as we see below. We can write the expression for the incident wave as

$$w = w(x - v_1 t) , \tag{6.8.3}$$

where w is an arbitrary function that we set to zero for $x > 0$; hence, any wave to the right of the origin is the transmitted wave, transmitted across $x = 0$.

For expression (6.8.3) to represent the initial disturbance, we formulate the initial conditions corresponding to equations (6.8.1) as

$$u|_{t=0} = w(x) \tag{6.8.4}$$

and

$$\left.\frac{\partial u}{\partial t}\right|_{t=0} = -v_1 w'(x) , \tag{6.8.5}$$

where w' denotes the derivative of w.

Examining solution (6.8.2), we see that $f_1 + g_1$ describes the waves on the left-hand side of the origin, and $f_2 + g_2$ describes the waves on the right-hand side. To write it specifically in the context of conditions (6.8.4) and (6.8.5), we have to restate these conditions as

$$w(y) = u(y,0) = \begin{cases} f_1(y) + g_1(y) , & y \leqslant 0 \\ f_2(y) + g_2(y) , & y > 0 \end{cases}$$

and

$$-v_1 w'(y) = \frac{\partial u}{\partial t}(y,0) = \begin{cases} v_1 [f_1'(y) - g_1'(y)] , & y \leqslant 0 \\ v_2 [f_2'(y) - g_2'(y)] , & y > 0 \end{cases} ,$$

respectively.

It was necessary to introduce y for the subsequent derivation to distinguish variable x from the rest of the arguments of functions f_i and g_i, which are $x \pm v_i t$, since expression (6.8.3) with conditions (6.8.4) and (6.8.5) require u at $t = 0$ to be zero for $x > 0$; it was not necessary to do so in the initial-value problem given by equations (6.5.1) and conditions (6.5.2), since u at $t = 0$ is not required to have any particular properties for any given domain of x: no requirements stated in terms of x relate to the arguments expressed by $x \pm v_i t$.

Recalling from expression (6.8.3) that at $t = 0$ and to the right of the origin, $w \equiv 0$, we rewrite these conditions as

$$u(y,0) = \begin{cases} f_1(y) + g_1(y) = w(y), & y \leqslant 0 \\ f_2(y) + g_2(y) = 0, & y > 0 \end{cases}$$

and

$$\frac{\partial u}{\partial t}(y,0) = \begin{cases} f_1'(y) - g_1'(y) = -w'(y), & y \leqslant 0 \\ f_2'(y) - g_2'(y) = 0, & y > 0 \end{cases}.$$

To solve these equations, we group them according to the domains of the functions. Thus, we write

$$\begin{cases} f_1(y) + g_1(y) = w(y), \\ f_1'(y) - g_1'(y) = -w'(y), \end{cases} \quad y \leqslant 0$$

and

$$\begin{cases} f_2(y) + g_2(y) = 0, \\ f_2'(y) - g_2'(y) = 0, \end{cases} \quad y > 0.$$

Integrating the second equations of both sets, we obtain

$$\begin{cases} f_1(y) + g_1(y) = w(y), \\ f_1(y) - g_1(y) = -w(y) + A, \end{cases} \quad y \leqslant 0$$

and

$$\begin{cases} f_2(y) + g_2(y) = 0, \\ f_2(y) - g_2(y) = B, \end{cases} \quad y > 0,$$

where A and B are integration constants. We have two systems of equations. Each system consists of two equations in two unknowns. We can proceed to solve these systems in their corresponding domains of y.

Let us consider the nonpositive arguments. We add the two equations of the first system to get $f_1(y) = A/2$. Hence, $g_1(y) = w(y) - A/2$. Since f_1 is a constant, its value can be incorporated into g_1; in other words, we can let $f_1 = A/2 \equiv 0$, without loss of generality. Thus, $g_1(y) = w(y)$.

Let us consider the positive arguments. We add the two equations of the second system to get $f_2(y) = B/2$. Herein, the constancy of f_2 implies the constancy of g_2, and we can let $f_2(y) = g_2(y) \equiv 0$, without loss of generality.

Using these results, the fact that $v_1 > 0$, $v_2 > 0$ and $t \geqslant 0$, and examining solution (6.8.2), we study the arguments of f_1, g_1, f_2 and g_2 to rewrite the general solution. Let us consider $x \leqslant 0$. The argument of f_1 is $y = x + v_1 t$, which can be either positive or negative. Hence, even though $f_1 \equiv 0$ for negative arguments, we must consider it in the solution. The argument of g_1 is $y = x - v_1 t$, which is negative. Hence, $g_1 = w$. Let us consider $x > 0$. The argument of f_2 is $y = x + v_2 t$, which is positive. Hence, we set $f_2 \equiv 0$. The argument of g_2 is $y = x - v_2 t$, which can be either positive or negative. Hence, even though $g_2 \equiv 0$ for positive arguments, we must consider it in the solution. Thus, we write solution (6.8.2) as

$$u(x,t) = \begin{cases} w(x - v_1 t) + f_1(x + v_1 t), & x \leqslant 0 \\ g_2(x - v_2 t), & x > 0 \end{cases}. \tag{6.8.6}$$

Physically, we can interpret expression (6.8.6) in the following way: w is the incident wave, f_1 propagates in the opposite direction along the same domain of x—it is the reflected wave, g_2 propagates in the same direction as w but on the other side of $x = 0$—it is the transmitted wave; $f_2 \equiv 0$ since there is no wave propagating towards the origin from the right.

In view of the physical interpretation, we could have argued the form of expression (6.8.6) directly from expression (6.8.2) upon introducing expression (6.8.3). However, the rigorous approach presented above provides us with an insight into the relation among the differential equation, its initial conditions and solution.

In expression (6.8.6), f_1 and g_2 are arbitrary functions. However, by invoking physical constraints, we are able to relate them to w; in other words, we obtain relations among the incident, reflected and transmitted waves.

In view of physical considerations, which are discussed in detail in Section 10.2, we require that both the displacement, u, and the stress be continuous along the entire x-axis. In particular, they must be continuous across $x = 0$.

Using expression (6.8.6), we can state the former requirement as the equality at $x = 0$ given by

$$w(-v_1 t) + f_1(v_1 t) = g_2(-v_2 t); \tag{6.8.7}$$

in other words, the sum of displacements of the incident and reflected waves is equal to the displacement of the transmitted wave. To state the latter requirement, we recall stress-strain equations (3.2.1), namely,

$$\sigma_{ij} = \sum_{k=1}^{3} \sum_{\ell=1}^{3} c_{ijk\ell} \varepsilon_{k\ell}, \qquad i, j \in \{1, 2, 3\}.$$

Since we are studying a one-dimensional continuum that coincides with the x-axis, these stress-strain equations are reduced to $\sigma_{11} = c_{1111}\varepsilon_{11}$. Recalling formula (3.2.5) and letting $\sigma_{xx} := \sigma_{11}$ and $\varepsilon_{xx} := \varepsilon_{11}$, we write the constitutive equation for this one-dimensional continuum as

$$\sigma_{xx} = C_{11}\varepsilon_{xx}. \tag{6.8.8}$$

Since a one-dimensional elastic continuum must be isotropic, we use matrix (5.12.4) to write

$$\sigma_{xx} = (\lambda + 2\mu)\,\varepsilon_{xx}.$$

In view of definition (1.4.6), we can restate the above equation as

$$\sigma_{xx} = (\lambda + 2\mu)\frac{\partial u}{\partial x}.$$

Furthermore, since a one-dimensional elastic continuum that is not embedded in a space of higher dimensions can accommodate only longitudinal displacements, we invoke expression (6.1.13) to write

$$\sigma_{xx} = \rho v^2 \frac{\partial u}{\partial x}, \tag{6.8.9}$$

where both ρ and v^2 are positive real numbers since $\lambda + 2\mu$ must be positive, as we can conclude from results of Exercise 5.18. Using expression (6.8.9), we are ready to state the continuity of stress across $x = 0$ as the equality given by

$$\rho_1 v_1^2 \left.\frac{\partial u_1}{\partial x}\right|_{x=0} = \rho_2 v_2^2 \left.\frac{\partial u_2}{\partial x}\right|_{x=0},$$

where u_1 and u_2 are the adjacent displacements on either side of $x = 0$. To write this condition explicitly in terms of expression (6.8.6), we differentiate the appropriate terms to get

$$\rho_1 v_1^2 \left[w'(-v_1 t) + f_1'(v_1 t)\right] = \rho_2 v_2^2 g_2'(-v_2 t), \tag{6.8.10}$$

where symbol $'$ denotes the derivative of a given function with respect to its argument.

We wish to solve equations (6.8.7) and (6.8.10) for f_1 and g_2 in terms of w. In other words, we wish to express the reflected and transmitted waves in terms of the incident wave. To do so, let us impose the equality of time derivatives of displacement at $x = 0$; in other words, the speed of displacement is continuous across $x = 0$—just like the displacement itself, as stated in equation (6.8.7). Thus, differentiating equations in system (6.8.6) with respect to t and equating the results at $x = 0$, we obtain

$$-v_1 w'(-v_1 t) + v_1 f_1'(v_1 t) = -v_2 g_2'(-v_2 t). \tag{6.8.11}$$

Solving for f_1', we get

$$f_1'(v_1 t) = w'(-v_1 t) - \frac{v_2}{v_1} g_2'(-v_2 t).$$

Substituting into equation (6.8.10) and simplifying, we get

$$g_2'(-v_2 t) = \frac{2\rho_1 v_1^2}{v_2(\rho_1 v_1 + \rho_2 v_2)} w'(-v_1 t).$$

Integrating both sides with respect to t and simplifying, we get

$$g_2(-v_2 t) = \frac{2\rho_1 v_1}{\rho_1 v_1 + \rho_2 v_2} w(-v_1 t),$$

where, in view of the initial conditions, the integration constant is zero. Herein, we set $y = -v_2 t$; hence, $t = -y/v_2$ and the argument of w becomes $(v_1/v_2)y$. Thus, we can write

$$g_2(y) = \frac{2\rho_1 v_1}{\rho_1 v_1 + \rho_2 v_2} w\left(\frac{v_1}{v_2} y\right),$$

where product $\rho_i v_i$ is called the acoustic impedance. Similarly, as shown in Exercise 6.13, we obtain

$$f_1(y) = \frac{\rho_1 v_1 - \rho_2 v_2}{\rho_1 v_1 + \rho_2 v_2} w(-y),$$

where $y = v_1 t$.

Thus, we can write expression (6.8.6) as

$$u(x,t) = \begin{cases} w(x - v_1 t) + \dfrac{\rho_1 v_1 - \rho_2 v_2}{\rho_1 v_1 + \rho_2 v_2} w(-(x + v_1 t)), & x \leqslant 0 \\ \\ \dfrac{2\rho_1 v_1}{\rho_1 v_1 + \rho_2 v_2} w\left(\dfrac{v_1}{v_2}(x - v_2 t)\right), & x > 0 \end{cases}, \tag{6.8.12}$$

which is the solution of our scattering problem stated in terms of the incident wave, w, and the properties of the discontinuous medium given by ρ_1, v_1 and ρ_2, v_2.

Having obtained the appropriate form of the solution, let us interpret its physical meaning. For $x \leqslant 0$, the two terms correspond to the incident and the reflected waves, respectively. For $x > 0$, the term corresponds to the transmitted wave.

The factors in front of w are the amplitudes of waves. The incident wave has amplitude set to unity. Examining expressions

$$\frac{\rho_1 v_1 - \rho_2 v_2}{\rho_1 v_1 + \rho_2 v_2} =: A_r \tag{6.8.13}$$

and

$$\frac{2\rho_1 v_1}{\rho_1 v_1 + \rho_2 v_2} =: A_t, \tag{6.8.14}$$

which are the amplitudes of the reflected and transmitted waves, respectively, we see that, if $\rho_1 v_1 \neq \rho_2 v_2$, the amplitude of the reflected wave is nonzero but smaller than the amplitude of the incident wave, while—in agreement with conservation of energy—the amplitude of the transmitted wave can be either smaller or greater than the amplitude of the incident wave; this is further discussed in Section 10.2.2 and Exercise 10.6. Also, expressions (6.8.13) and (6.8.14) are derived in a different way in Exercise 10.4 with a comparison of the two derivations discussed in Exercise 10.5.

The factor v_1/v_2 in the transmitted wave corresponds to the wavelength: The wavelength of the transmitted wave is v_2/v_1 times the wavelength of the incident wave.

Examining solution (6.8.12), we see that upon transmission from a lower to a higher acoustic impedance, the amplitude of the wave decreases. We also see that explicitly mass density has no effect on the wavelength; implicitly, it is contained in the expressions for the speeds of propagation, v_1 and v_2. Upon transmission from a lower to a higher speed, the wavelength decreases.

To complete this section, let us consider two particular cases: no change in acoustic impedance, and no displacement for $x \geqslant 0$.

Symbolically, no change in acoustic impedance is stated as $\rho_1 v_1 = \rho_2 v_2 = \rho v$. In such a case, solution (6.8.12) reduces to

$$u(x,t) = w(x - vt),$$

for all x, as expected; in other words, the wave travels without scattering. Examining expressions (6.8.13) and (6.8.14), we see that $A_r = 0$ and $A_t = 1$; in other words, there is no reflected wave and the transmitted wave propagates without change.

For the second case, let us consider the expression for $x \leqslant 0$ in solution (6.8.6), namely,

$$u(x,t) = w(x - v_1 t) + f_1(x + v_1 t). \tag{6.8.15}$$

Since there is no displacement at $x = 0$, solution (6.8.15) becomes

$$0 = w(-v_1 t) + f_1(v_1 t),$$

which means that

$$f_1(v_1 t) = -w(-v_1 t),$$

for all t. Since, this equation is valid for any $vt \in \mathbb{R}$, we can write $f_1(\zeta) = -w(-\zeta)$, where argument ζ is any real number. This implies that function f_1 is the negative of function w if the corresponding arguments are negatives of one another; in other words, the incident and reflected waves are reverses of one another. To restate it in the form of expression (6.8.15), we let $\zeta = x + v_1 t$ to write

$$f_1(x + v_1 t) = -w(-(x + v_1 t));$$

inserting it into expression (6.8.15), we obtain

$$u(x,t) = w(x - v_1 t) - w(-(x + v_1 t)), \qquad x \leqslant 0.$$

Comparing this result to the first expression in solution (6.8.12), we see that the waves travelling to the right and to the left have the same form and size but are reverses of one another. Herein, the amplitude of the reflected wave given explicitly by expression (6.8.13) is $A_r = -1$, which means that ratio $(\rho_1 v_1)/(\rho_2 v_2)$ tends to zero; in such a case, expression (6.8.14) is $A_t = 0$, as expected. Examining the steps leading from expression (6.8.8) to (6.8.9), with C_{11} being the only elasticity parameter in this continuum, we interpret the result as a perfect rigidity for $x > 0$; the corresponding C_{11} tends to infinity, which justifies the fact that there is no displacement beyond $x = 0$. In Section 10.2.2, we are able to conclude that $A_r = -1$ directly from expression (10.2.10), which stems from the continuity of displacement and is given by $1 + A_r = A_t$, since in the present case it reduces to $1 + A_r = 0$.

6.9. On Weak Solutions of Wave Equation

6.9.1. Introductory comments

In a classical approach to differential equations, we expect the differentiability of solutions to, at least, match the order of the equation. Yet, examining solution (6.5.6), without considering equation (6.5.1), we see that the solution itself does not require f and g to be differentiable; f and g could be even discontinuous. Requirements for differentiability of the solution of the wave equation were the subject of long discussions between d'Alembert and Euler in the second half of the eighteenth century. In these discussions, which lasted for almost thirty years, Euler deemed it necessary from physical considerations of wave propagation to admit nondifferentiable functions as solutions, while d'Alembert strictly required differentiability. In the context of the single spatial dimension, continuity is required to consider displacements on a string that is not broken. A rigorous formulation that allows us to consider nondifferentiable solutions was not available until the middle of the twentieth century. In this section, we briefly study this formulation.

A conventional method to verify that solution (6.5.6) satisfies equation (6.5.1) consists of inserting this solution into the equation. To do so, however, we require that f and g be twice-differentiable. Sergei Sobolev, during his presentation in 1934, stated that

> [t]he class of functions that we can consider as solutions to the wave equation from the classical point of view consists of twice-differentiable functions. But in various practical applications it seems convenient to consider functions with well-defined singularities.

In other words, we might wish to investigate waves that cannot be described by twice-differentiable functions. Consequently, we would like to extend the solutions of wave equation (6.5.1) to incorporate functions that are not differentiable. To do so, we consider the theory of generalized functions, commonly referred to as the theory of distributions, which was formulated by Sergei Sobolev in the first half of the twentieth century and thoroughly developed by Laurent Schwartz in the middle of the twentieth century. The key philosophical point of this theory is that a generalized function is not described by itself alone but by its effect on other functions.

We define the effect of $h(x)$ on $\tau(x)$ by the value of $\int_{-\infty}^{\infty} h(x)\,\tau(x)\,\mathrm{d}x$, where τ is assumed to be infinitely differentiable and compactly supported. In view of this definition, τ is often called the test function. In this formulation, we do not require h to be differentiable. Furthermore, h need not be a function in the classical sense of the term; in general, h is a generalized function, also called a distribution; Dirac's delta is a famous example of such an entity.

Expression $\int_{-\infty}^{\infty} h(x)\,\tau(x)\,\mathrm{d}x$ generalizes our study of functions. If h is a function in the classical sense, we can describe it on $\mathbb{R}^1$ by its values, $h(x)$, for appropriate points $x \in \mathbb{R}^1$, as well as by the values of $\int_{-\infty}^{\infty} h\tau\,\mathrm{d}x$ for appropriate functions τ. If h is not a function, we cannot describe it by its values, but we can still describe it by the values of $\int_{-\infty}^{\infty} h\tau\,\mathrm{d}x$, as exemplified by Dirac's delta, where $\delta(x)$ by itself alone does not make sense, while $\int_{-\infty}^{\infty} \delta(x)\,\tau(x)\,\mathrm{d}x = \tau(0)$ is a well-defined quantity.

6.9.2. Weak derivatives

To study the solutions of the wave equation, which is a partial differential equation, we wish to investigate derivatives of nondifferentiable functions; so called weak derivatives. To do so, let us consider the effect of these derivatives on test functions. Consider $\int_{-\infty}^{\infty} h'(x)\,\tau(x)\,\mathrm{d}x$. Integrating by parts, we obtain

$$\int_{-\infty}^{\infty} h'(x)\,\tau(x)\,\mathrm{d}x = h(x)\,\tau(x)\big|_{-\infty}^{\infty} - \int_{-\infty}^{\infty} h(x)\,\tau'(x)\,\mathrm{d}x.$$

Since τ has compact support, $\lim_{x\to\pm\infty}\tau = 0$; hence, the first term on the right-hand side vanishes, if we consider h to be finite at $\pm\infty$. Thus, we can state the derivative of h as

$$\int_{-\infty}^{\infty} h'(x)\,\tau(x)\,\mathrm{d}x = -\int_{-\infty}^{\infty} h(x)\,\tau'(x)\,\mathrm{d}x. \tag{6.9.1}$$

This is the formula for the weak derivative of order one for the case of a single variable. The use of this formula is exemplified in Exercise 6.14.

To obtain higher-order derivatives for the case of n variables, we can repeat the process of integration by parts to arrive at the general equation for an mth partial derivative, which is given by

$$\int_{\mathbb{R}^n} \frac{\partial^m h(\mathbf{x})}{\partial x_i^m}\,\tau(\mathbf{x})\,\mathrm{d}\mathbf{x} = (-1)^m \int_{\mathbb{R}^n} h(\mathbf{x})\,\frac{\partial^m \tau(\mathbf{x})}{\partial x_i^m}\,\mathrm{d}\mathbf{x}, \tag{6.9.2}$$

where $\int_{\mathbb{R}^n}\mathrm{d}\mathbf{x}$ stands for n integrals over n variables from $-\infty$ to $+\infty$.

Examining equation (6.9.2), we notice the crux of this formulation. To study the effect of the derivatives of h, shown on the left-hand side, we study the expression on the right-hand side where no derivatives of h appear. Hence, we do not need h to be differentiable.

6.9.3. Weak solution of wave equation

We wish to extend the solutions of wave equation (6.5.1) to allow non-differentiable functions. To distinguish the solutions that can be verified directly from the solutions discussed below, we refer to the latter ones as weak solutions.

In view of Section 6.9.2, let us consider the effect of the second derivatives. We write

$$\int_{-\infty}^{\infty}\int_{-\infty}^{\infty}\left\{\left(\frac{\partial^2}{\partial x^2}-\frac{1}{v^2}\frac{\partial^2}{\partial t^2}\right)[f(x+vt)+g(x-vt)]\right\}\tau(x,t)\,\mathrm{d}x\mathrm{d}t\,, \tag{6.9.3}$$

where the term in parentheses is the differential operator of wave equation (6.5.1) and the term in brackets is solution (6.5.6), where we wish to allow f and g to be nondifferentiable.

To verify that $f+g$ is a weak solution of equation (6.5.1), we must show that expression (6.9.3) vanishes for all test functions, τ. Following equation (6.9.2), we write

$$\begin{aligned}&\int_{-\infty}^{\infty}\int_{-\infty}^{\infty}\left\{\left(\frac{\partial^2}{\partial x^2}-\frac{1}{v^2}\frac{\partial^2}{\partial t^2}\right)[f(x+vt)+g(x-vt)]\right\}\tau(x,t)\,\mathrm{d}x\mathrm{d}t\\&\quad=\int_{-\infty}^{\infty}\int_{-\infty}^{\infty}[f(x+vt)+g(x-vt)]\left[\left(\frac{\partial^2}{\partial x^2}-\frac{1}{v^2}\frac{\partial^2}{\partial t^2}\right)\tau(x,t)\right]\mathrm{d}x\mathrm{d}t.\end{aligned} \tag{6.9.4}$$

Recalling that τ is infinitely differentiable, we let $y=x+vt$ and $z=x-vt$ and use Lemma 6.5.1 to rewrite the right-hand side of the above equation as

$$\begin{aligned}&\int_{-\infty}^{\infty}\int_{-\infty}^{\infty}[f(y)+g(z)]\left[\frac{\partial^2\tau(y,z)}{\partial y\partial z}\right]\mathrm{d}y\mathrm{d}z\\&\quad=\int_{-\infty}^{\infty}\int_{-\infty}^{\infty}f(y)\frac{\partial^2\tau(y,z)}{\partial y\partial z}\mathrm{d}y\mathrm{d}z+\int_{-\infty}^{\infty}\int_{-\infty}^{\infty}g(z)\frac{\partial^2\tau(y,z)}{\partial y\partial z}\mathrm{d}y\mathrm{d}z\,.\end{aligned}$$

Changing the order of integration, we get

$$\int_{-\infty}^{\infty}f(y)\int_{-\infty}^{\infty}\frac{\partial^2\tau(y,z)}{\partial y\partial z}\mathrm{d}z\mathrm{d}y+\int_{-\infty}^{\infty}g(z)\int_{-\infty}^{\infty}\frac{\partial^2\tau(y,z)}{\partial y\partial z}\mathrm{d}y\mathrm{d}z\,.$$

To show that wave equation (6.5.1) is satisfied—in the weak sense—by $f+g$, we must show that the above expression vanishes. It suffices to show that

$$\int_{-\infty}^{\infty} \frac{\partial^2 \tau(y,z)}{\partial y \partial z} \mathrm{d}z = 0$$

and

$$\int_{-\infty}^{\infty} \frac{\partial^2 \tau(y,z)}{\partial y \partial z} \mathrm{d}y = 0.$$

Integrating, we obtain

$$\left. \frac{\partial \tau(y,z)}{\partial y} \right|_{z=-\infty}^{z=\infty} = 0$$

and

$$\left. \frac{\partial \tau(y,z)}{\partial z} \right|_{y=-\infty}^{y=\infty} = 0,$$

respectively. Since τ has compact support, all its derivatives—in particular, $\partial\tau/\partial y$ and $\partial\tau/\partial z$—also have compact support. Thus, $\lim_{x\to\pm\infty} \partial\tau/\partial y = \lim_{x\to\pm\infty} \partial\tau/\partial z = 0$, which means that both above equations are satisfied for all test functions, τ, as required. We have verified solution (6.5.6) of wave equation (6.5.1) without requiring differentiability of f or g. In view of expression (6.9.4), we can define the weak solution of the one-dimensional wave equation as $u(x,t)$ that satisfies

$$\int_{-\infty}^{\infty}\int_{-\infty}^{\infty} u(x,t) \left[\frac{\partial^2 \tau(x,t)}{\partial x^2} - \frac{1}{v^2} \frac{\partial^2 \tau(x,t)}{\partial t^2} \right] \mathrm{d}x \mathrm{d}t = 0, \tag{6.9.5}$$

where τ is the test function.

Any differentiable solution of the wave equation satisfies also equation (6.9.5). In other words, any strong solution is also a weak solution. However, since the opposite is not true, we need equation (6.9.5) to study solutions that are not differentiable.

6.10. Reduced Wave Equation

6.10.1. Harmonic-wave trial solution

To motivate the form and the name of the reduced wave equation, let us consider a particular trial solution for the wave equation. Since the wave equation is a partial differential equation, to solve it we often assume a trial solution. For instance, while studying three-dimensional continua, it is common to assume a plane-wave solution. However, we might wish to study a more complicated position dependence of the solution; hence, we would require another trial solution.

Let us consider equation (6.5.1), namely,

$$\frac{\partial^2 u(x,t)}{\partial x^2} - \frac{1}{v^2}\frac{\partial^2 u(x,t)}{\partial t^2} = 0. \tag{6.10.1}$$

If we study an oscillatory motion, we can write our trial solution as

$$u(x,t) = \acute{u}(x)\exp(-i\omega t), \tag{6.10.2}$$

where ω stands for the angular frequency and $\exp(-i\omega t)$ is called the phase factor.[27] The right-hand side of expression (6.10.2) is a standard form of a complex number whose magnitude is $\acute{u}(x)$ and whose phase is $-\omega t$. We can write solution (6.10.2) as

$$\acute{u}(x)\exp(-i\omega t) = \acute{u}(x)\left[\cos(\omega t) - i\sin(\omega t)\right]. \tag{6.10.3}$$

By using this trial solution, we assume that the time dependence of the displacement function, $u(x,t)$, is satisfied by the term in brackets in expression (6.10.3). In other words, we assume that the solution is sinusoidal in time; such waves are called harmonic waves.

Inserting solution (6.10.2) into equation (6.10.1), we obtain

$$\frac{\mathrm{d}^2\acute{u}(x)}{\mathrm{d}x^2} + \left(\frac{\omega}{v}\right)^2 \acute{u}(x) = 0, \tag{6.10.4}$$

as shown in Exercise 6.15. This is the reduced wave equation. Since wave equation (6.10.1) is in a single spatial dimension and since time is not a variable in equation (6.10.4), we obtained an ordinary differential equation. However, to a certain extent, it is a matter of notation: $\acute{u}$ is the solution of equation (6.10.4) for a particular value of ω; hence, $\acute{u}$ is implicitly a function of ω. Using this notation, we emphasize that equation (6.10.4) is equation (6.10.1) with the temporal dependence given by $\exp(-i\omega t)$. Such a reduced wave equation allows us to study particular problems,

[27] In this book, $\exp(\cdot)$ and $e^{(\cdot)}$ are used as synonymous notations.

such as the steady-state problems or standing-wave problems. Notably, the reduced wave equation was used by Hermann von Helmholtz in the mid-nineteenth century to study oscillations in the organ pipes, and it belongs now to the class of Helmholtz's equations. Gustav Kirchhoff continued the work of Helmholtz on the reduced wave equation to study the solution of the initial-value problem, which resulted in a mathematical statement of Huygens' principle.[28]

In the next section, we study the wave equation using Fourier's transform. Therein, we obtain a partial differential equation that is almost equivalent to equation (6.10.4).

6.10.2. Fourier's transform of wave equation

Writing equation (6.10.4) as

$$\frac{\partial^2 \acute{u}}{\partial x^2} + \left(\frac{\omega}{v}\right)^2 \acute{u} = 0,$$

we notice that we could view this equation as Fourier's transform of equation (6.10.1) with t and ω being the transformation variables. Let us investigate this property.

Let us take Fourier's transform of both sides of equation (6.10.1) with t and ω being the variables of transformation. Thus, we write

$$\frac{1}{2\pi}\int_{-\infty}^{\infty}\left[\frac{\partial^2 u(x,t)}{\partial x^2} - \frac{1}{v^2}\frac{\partial^2 u(x,t)}{\partial t^2}\right]\exp(-i\omega t)\,\mathrm{d}t = 0.$$

We write the integral of a difference as a difference of integrals and factor out the constant term, $1/v^2$, to get

$$\frac{1}{2\pi}\int_{-\infty}^{\infty}\frac{\partial^2 u(x,t)}{\partial x^2}\exp(-i\omega t)\,\mathrm{d}t - \frac{1}{v^2}\frac{1}{2\pi}\int_{-\infty}^{\infty}\frac{\partial^2 u(x,t)}{\partial t^2}\exp(-i\omega t)\,\mathrm{d}t = 0. \tag{6.10.5}$$

[28] Readers interested in the role of the reduced wave equation for studies of solutions of the steady-state equations, as well as in the development of mathematical physics, might refer to Kline, M., (1972) Mathematical thought from ancient to modern times: Oxford University Press, Vol. 2, pp. 693–696.

Let us consider the first term in equation (6.10.5). Since the exponential term is not a function of x, we write

$$\frac{1}{2\pi}\int_{-\infty}^{\infty}\frac{\partial^2}{\partial x^2}\left[u(x,t)\exp(-i\omega t)\right]\mathrm{d}t\,,$$

and interchange the integration and differentiation to get

$$\frac{\partial^2}{\partial x^2}\left[\frac{1}{2\pi}\int_{-\infty}^{\infty}u(x,t)\exp(-i\omega t)\,\mathrm{d}t\right];$$

to do so, we use the fact that we take the derivative with respect to x while the limits of integration refer to t. Herein, the term in brackets is the definition of Fourier's transform of function $u(x,t)$, which we denote as $\tilde{u}(x,\omega)$. Let us consider the second term in equation (6.10.5). We evaluate the integral using integration by parts twice—and assuming that $\lim_{t\to\pm\infty}u(x,t)=0$—to get

$$(i\omega)^2\left[\frac{1}{2\pi}\int_{-\infty}^{\infty}u(x,t)\exp(-i\omega t)\,\mathrm{d}t\right].$$

We recognize that the term in brackets is Fourier's transform of function u, namely, $\tilde{u}$. Returning to equation (6.10.5), we rewrite that equation as

$$\frac{\partial^2\tilde{u}(x,\omega)}{\partial x^2}+\left(\frac{\omega}{v}\right)^2\tilde{u}(x,\omega)=0\,, \tag{6.10.6}$$

which is Fourier's transform of equation (6.10.1).

We note that familiarity with properties of the transform allows us to write equation (6.10.6) directly from equation (6.10.1), which is

$$\frac{\partial^2 u(x,t)}{\partial x^2}-\frac{1}{v^2}\frac{\partial^2 u(x,t)}{\partial t^2}=0\,,$$

since the transform of u is $\tilde{u}$, and the transform of its derivative with respect to the transformation variable is $i\omega\tilde{u}$.

Equation (6.10.6) exhibits a similarity to equation (6.10.4). Let us compare these two equations. We note that $\acute{u}$ is a function of x alone, while $\tilde{u}$ is a function of both x and ω. As stated in Section 6.10.1, it is a matter of notation: $\acute{u}$ is the solution of equation (6.10.4) for a particular value of ω; hence, it is implicitly a function of ω—the fact that is explicitly stated for $\tilde{u}$.

To complete this section, we can state that if we consider a three-dimensional equivalent of equation (6.10.1), we can take its Fourier's transform to get

$$\nabla^2 \tilde{u}(\mathbf{x}, \omega) + \left(\frac{\omega}{v}\right)^2 \tilde{u}(\mathbf{x}, \omega) = 0. \qquad (6.10.7)$$

We have shown that the reduced wave equation can be viewed as a particular case of the wave equation subjected to Fourier's transform with t and ω being the transformation variables. In our subsequent work, when invoking the reduced wave equation, we use the wave equation subjected to Fourier's transform. The formulation of the reduced wave equation in terms of a trial solution has provided us with an insight into both its physical meaning and its nomenclature.

6.11. Extensions of Wave Equation

6.11.1. Introductory comments

In Chapter 7, we derive equations of motion in anisotropic inhomogeneous continua. This is accomplished by combining Cauchy's equations of motion with stress-strain equations for generally anisotropic continua and allowing the elasticity parameters to be functions of position. The fundamental derivation shown in Chapter 7 lies at the root of ray theory, which is subsequently studied in this book.

There are, however, certain cases where the standard wave equation, which is derived for isotropic homogeneous continua, can be extended to account for anisotropy and for inhomogeneity. An investigation of such cases is undertaken in this section.

6.11.2. Standard wave equation

In multidimensional continua, wave equation (6.5.1), may be written as

$$\nabla^2 u(\mathbf{x}, t) - \frac{1}{v^2} \frac{\partial^2 u(\mathbf{x}, t)}{\partial t^2} = 0, \qquad (6.11.1)$$

which is a partial differential equation with constant coefficients, where, as shown in Section 6.5, constant v is the magnitude of the velocity of the solution. In equation (6.11.1), $\mathbf{x}$ are the position coordinates. Hence, this equation describes wave propagation in continua characterized by constant speed at all positions $\mathbf{x}$ and in all directions determined by the coordinates. Consequently, this wave equation is valid for isotropic homogeneous continua.

We wish to extend equation (6.11.1) to the anisotropic case. In certain cases, by transforming the coordinates, we can formulate a wave equation that in homogeneous continua associates different velocities with different directions. An example of such an extension, which results in a wave equation for elliptical velocity dependence, is illustrated in Section 6.11.3.

We also wish to extend equation (6.11.1) to the inhomogeneous case. By considering the position dependence $v = v(\mathbf{x})$ and assuming that function $v(\mathbf{x})$ varies slowly with $\mathbf{x}$, we can use an approximation that allows us to describe wave propagation in weakly inhomogeneous continua. This extension of equation (6.11.1) to account for weak inhomogeneity is illustrated in Section 6.11.4 and belongs to the high-frequency approximation.

6.11.3. Wave equation and elliptical velocity dependence

Wave equation

To study an extension of the wave equation to anisotropic cases, consider equation (6.11.1). For convenience, let v be equal to unity. Hence, we can write

$$\nabla^2 u(\mathbf{x},t) = \frac{\partial^2 u(\mathbf{x},t)}{\partial t^2}. \tag{6.11.2}$$

Consider a two-dimensional continuum that is contained in the xz-plane. For $\mathbf{x} = [x, z]$, equation (6.11.2) can be explicitly written as

$$\frac{\partial^2 u(x,z,t)}{\partial x^2} + \frac{\partial^2 u(x,z,t)}{\partial z^2} = \frac{\partial^2 u(x,z,t)}{\partial t^2}. \tag{6.11.3}$$

Let the linear transformation of the position coordinates be such that

$$\acute{u}(x,z,t) = u\left(\frac{x}{v_x}, \frac{z}{v_z}, t\right), \tag{6.11.4}$$

where v_x and v_z are constants. Using the chain rule, as shown in Exercise 6.16, we can write equation (6.11.3) as

$$v_x^2 \frac{\partial^2 \acute{u}(x,z,t)}{\partial x^2} + v_z^2 \frac{\partial^2 \acute{u}(x,z,t)}{\partial z^2} = \frac{\partial^2 \acute{u}(x,z,t)}{\partial t^2}. \tag{6.11.5}$$

Thus, function $\acute{u}$ is the solution of equation (6.11.5).

To illustrate the meaning of constants v_x and v_z, consider transformation (6.11.4) and let

$$u\left(\frac{x}{v_x}, \frac{z}{v_z}, t\right) =: u(\xi, \varsigma, t).$$

If point (ξ, ς) is moving in the $\xi\varsigma$-plane at unit speed, namely,

$$\frac{\mathrm{d}}{\mathrm{d}t}\sqrt{\xi^2 + \varsigma^2} = 1,$$

the solutions $u(\xi, \varsigma, t)$, at different times t, are concentric circles. It follows that, in the xz-plane,

$$\frac{\mathrm{d}}{\mathrm{d}t}\sqrt{\frac{x^2}{v_x^2} + \frac{z^2}{v_z^2}} = 1,$$

and, hence, the solutions $\acute{u}(x,z,t)$, at different times t, are ellipses.

Equation (6.11.5) is the wave equation that describes the wavefront propagation in a two-dimensional homogeneous continuum where the wave is subjected to an elliptical velocity dependence with direction. The semi-axes of the elliptical wavefronts coincide with the coordinate axes and the magnitudes of the wavefront velocities along these axes are given by v_x and v_z, respectively.

Phase velocity

Knowing that v_x and v_z are the magnitudes of the wavefront velocities along the x-axis and the z-axis, respectively, we wish to find the expression for the wavefront velocity in an arbitrary direction. Since the wavefronts are loci of constant phase, the wavefront velocity is referred to as phase velocity.

To solve equation (6.11.5), consider the trial solution given by

$$\acute{u}(x,z,t) = \exp[i\omega(p_x x + p_z z - t)]. \tag{6.11.6}$$

If we consider monochromatic waves, where a given value of ω is constant, loci of constant phase are given by the constancy of the term in parentheses. Thus, wavefronts at time t are straight lines $p_x x + p_z z = t$, where p_x and p_z are the components of vector $\mathbf{p}$ that is normal to a given wavefront. Since x and z have units of distance while t is time, it follows that the units of the components of $\mathbf{p}$ are the units of slowness. In other words, $\mathbf{p}$ is the phase-slowness vector, which describes the slowness with which the wavefront propagates. The envelope of all straight lines $p_x x + p_z z = t$ at time t is an elliptical wavefront. Hence, $\mathbf{p}$ describes the slowness with which the line tangent to the elliptical wavefront propagates.

To examine trial solution (6.11.6), we substitute it into wave equation (6.11.5). We obtain

$$v_x^2 \omega^2 p_x^2 \exp\left[i\omega\left(xp_x + zp_z - t\right)\right] + v_z^2 \omega^2 p_z^2 \exp\left[i\omega\left(xp_x + zp_z - t\right)\right]$$
$$= \omega^2 \exp\left[i\omega\left(xp_x + zp_z - t\right)\right].$$

Dividing by ω^2 and by the exponential term, we can write this equation as

$$v_x^2 p_x^2 + v_z^2 p_z^2 = 1, \tag{6.11.7}$$

where v_x and v_z are the magnitude of the phase velocity along the horizontal and vertical axes, respectively, while p_x and p_z are the components of $\mathbf{p}$ at a given point on the wavefront.

NOTATION 6.11.1. To avoid any confusion, let us clarify the meaning of notation v_x, v_z, p_x and p_z that is used in the entire book. As in expression (6.11.7), v_x and v_z denote constants that define the properties of the velocity field by giving the magnitude of the wavefront velocity along the x-axis and along the z-axis, respectively; v_x and v_z are not the components of a vector. Symbols p_x and p_z, on the other hand, stand for the components of the phase-slowness vector, $\mathbf{p}$.

In other words, p_x and p_z specify the orientation of a wavefront in the velocity field defined by v_x and v_z.

We also observe that—for the wavefront propagating along the x-axis—the magnitudes of p_x and p_z are $1/v_x$ and 0, respectively, while—for the wavefront propagating along the z-axis—they are 0 and $1/v_z$. In general, for elliptical velocity dependence, $|p_x| \in [0, 1/v_x]$ and $|p_z| \in [0, 1/v_z]$, as we can confirm by examining expression (6.11.11), below.

Let us return to expression (6.11.7). To state this expression as a function of the orientation of the wavefront, we can express the phase-slowness vector as

$$\mathbf{p} = [p_x, p_z] = [p(\vartheta)\sin\vartheta, p(\vartheta)\cos\vartheta], \tag{6.11.8}$$

where $p(\vartheta)$ stands for the magnitude of the phase-slowness vector in a given direction ϑ, which is measured between the wavefront normal and the z-axis, and is referred to as the phase angle. Thus, using equation (6.11.8), we can also express the phase angle as

$$\frac{p_x}{p_z} = \frac{\sin\vartheta}{\cos\vartheta} = \tan\vartheta,$$

which means that

$$\vartheta = \arctan\frac{p_x}{p_z}. \tag{6.11.9}$$

Using expression (6.11.8), we can rewrite expression (6.11.7) as

$$[p(\vartheta)]^2 \left(v_x^2 \sin^2\vartheta + v_z^2 \cos^2\vartheta\right) = 1. \tag{6.11.10}$$

Since the magnitude of phase slowness is the reciprocal of the magnitude of phase velocity, expression (6.11.10) can be restated as

$$v(\vartheta) = \frac{1}{p(\vartheta)} = \sqrt{v_x^2 \sin^2\vartheta + v_z^2 \cos^2\vartheta}. \tag{6.11.11}$$

Expression (6.11.11) gives the magnitude of phase velocity as a function of phase angle for the case of elliptical velocity dependence. As shown in Exercise 9.9, *SH* waves in transversely isotropic continua are characterized by elliptical velocity dependence.

Thus, by a linear transformation of the coordinate axes, we obtained an exact formulation of a wave equation for the elliptical velocity dependence. A more sophisticated manipulation of coordinates might allow us to consider wave equations to study complicated anisotropic behaviours in homogeneous continua. In this book, however, we do not pursue this approach. Rather, in Chapter 7, we formulate an approximation to the wave equation that is valid for generally anisotropic continua.

Prior to completing this section, we wish to discuss the meaning of the elliptical velocity dependence with direction. According to wave equation (6.11.5)—with v_x and v_z being the magnitudes of the wavefront velocities along the x-axis and the z-axis, respectively—the wavefronts originating from a point source are elliptical, as shown in discussion that follows expression (6.11.6). This ellipticity is exhibited by the dependence of the magnitude of the phase-slowness vector, $\mathbf{p}$, with direction. This can be illustrated by the polar plot of expression (6.11.11) written as

$$p(\vartheta) = \frac{1}{\sqrt{v_x^2 \sin^2 \vartheta + v_z^2 \cos^2 \vartheta}},$$

which is an ellipse. In other words, phase slowness, p, exhibits elliptical dependence on the phase angle, ϑ. We note that phase velocity does not exhibit elliptical dependence on the phase angle. This can be illustrated by the polar plot of expression (6.11.11) for $v(\vartheta)$, which is not an ellipse, and—for large enough difference between v_x and v_z—is not even a convex curve. Geometrically, the ellipticity of the wavefronts and of the phase slowness is a result of polar reciprocity to be discussed in Section 8.4.3. Analytically, this ellipticity results from the fact that $\mathbf{p}$ and $\dot{\mathbf{x}}$, which are associated with the phase slowness and ray velocity, respectively, are the variables of Legendre's transformation, which is also discussed in Section 8.4. As shown in Chapter 8 in the context of Hamilton's ray equations, it is $\mathbf{p}$ and $\dot{\mathbf{x}}$, rather than phase velocity, that are fundamental entities of seismic theory. Thus, we conclude that the term elliptical velocity dependence with direction—which we introduced to refer to elliptical wavefronts generated by a point source and governed by wave equation (6.11.5)—is tantamount to the dependence of phase slowness on phase angle, as shown above, and to the dependence of ray velocity on ray angle, as discussed in Section 8.4.

6.11.4. Wave equation and weak inhomogeneity

Weak inhomogeneity: Formulation of equation

To study an extension of the wave equation to the inhomogeneous case, consider equation (6.11.1), namely,

$$\nabla^2 u(\mathbf{x},t) - \frac{1}{v^2}\frac{\partial^2 u(\mathbf{x},t)}{\partial t^2} = 0, \tag{6.11.12}$$

which is valid for homogeneous continua with v being a constant denoting the speed of propagation. To extend this equation to inhomogeneous continua, we wish to express v as a function of the position coordinates, $\mathbf{x}$. Consequently, we wish to consider the equation given by

$$\nabla^2 u(\mathbf{x},t) - \frac{1}{[v(\mathbf{x})]^2}\frac{\partial^2 u(\mathbf{x},t)}{\partial t^2} = 0. \tag{6.11.13}$$

Since equation (6.11.13) is a differential equation, it corresponds to local properties of the continuum and can be locally solved for a given $\mathbf{x}$. We can also obtain an approximate global solution to equation (6.11.13) if we assume that function $v(\mathbf{x})$ varies slowly, which means that the inhomogeneity of a continuum is weak. In the seismological context, weak inhomogeneity means that the changes of properties within a single wavelength are negligible.

Replacing v with $v(\mathbf{x})$ is an arbitrary replacement: it is not justified by a derivation from the fundamentals. There is no guarantee that in a weakly inhomogeneous medium, u satisfies equation (6.11.13). Strictly speaking, such an equation should be derived in a manner to be discussed in Section 7.1.

In seismology, we are interested often in studying layered media where the properties vary along only one axis. Considering a three-dimensional continuum, where $\mathbf{x} = [x, y, z]$, we assume often that its properties vary slowly along the z-axis, while remaining the same along the other two axes. It can be shown that, if $v(\mathbf{x}) = v(z)$ varies slowly, equation (6.11.13) is approximately satisfied by the displacements associated with the *SH* waves for all directions of propagation. For the case of *P* and *SV* waves, equation (6.11.13) provides a good approximation only for the displacements of waves propagating near the direction of the z-axis.[29] However, the eikonal equation, which we derive from equation (6.11.13), provides—within the conditions of this derivation—a good approximation for signal trajectories in all directions of propagation.

Weak inhomogeneity: Formulation of solution

To formulate a trial solution of equation (6.11.13), consider the fact that we can write a trial solution of equation (6.11.12) as

$$u(\mathbf{x},t) = A\exp\left[i\omega\left(\mathbf{p}\cdot\mathbf{x}-t\right)\right], \tag{6.11.14}$$

where A is the amplitude of the displacement that varies sinusoidally in space and time as described by $\exp\left[i\omega\left(\mathbf{p}\cdot\mathbf{x}-t\right)\right]$. As stated in Section 6.11.3, $\exp[\cdot]$ is the phase factor, which is constant for a wavefront at time t. In three-dimensional continua, trial solution (6.11.14) is called the plane-wave solution since, for a given time t, $\mathbf{p}\cdot\mathbf{x}=t$ is a plane that corresponds to a moving wavefront. Vector $\mathbf{p}$ is normal to this plane and, as shown in Section 6.11.3, $\mathbf{p}$ is the phase-slowness vector.

If the properties of a three-dimensional continuum vary with position, a planar wavefront is distorted during propagation through this continuum. Consequently, a trial solution of equation (6.11.13) must account for these changes of shape of the wavefront, which also cause changes of amplitude along the wavefront. Using a form analogous to expression (6.11.14), we write

$$u(\mathbf{x},t) = A(\mathbf{x})\exp\left\{i\omega\left[\psi(\mathbf{x})-t\right]\right\}, \tag{6.11.15}$$

where $A(\mathbf{x})$ denotes the amplitude of the displacement—which is allowed to vary along the wavefront—and $\psi(\mathbf{x})$, referred to as the eikonal function, which accounts for the distortions in the shape of the wavefront. Herein, both $A(\mathbf{x})$ and $\psi(\mathbf{x})$ are smooth scalar functions of position coordinates. Examining the phase factor of trial solution (6.11.15) in the context of solutions (6.11.6) and (6.11.14), we see that equation $\psi(\mathbf{x})=t$ represents the moving wavefront. In other words, the level sets of function $\psi(\mathbf{x})$ are

[29] Readers interested in wave propagation in slowly varying vertically nonuniform continua might refer to Grant, F.S., and West, G.F., (1965) Interpretation theory in applied geophysics: McGraw-Hill, Inc., pp. 43–47, 49–50, and to Krebes, E.S., (2004) Seismic theory and methods (Lecture notes): The University of Calgary, pp. 5-8–5-11.

the wavefronts. Since $\mathbf{p}$ is normal to the wavefront, using properties of the gradient, we obtain an important expression, namely,

$$\mathbf{p} = \nabla \psi . \tag{6.11.16}$$

In other words, the phase-slowness vector is the gradient of the eikonal function.

We could write the exponential term as the product of $\exp(i\omega\psi)$ and $\exp(-i\omega t)$; the latter term transforms equation (6.11.13) into the reduced wave equation in a manner shown in Section 6.10.1. This property is illustrated in Exercise 6.20. Following the discussion in Section 6.10—in particular referring to equation (6.10.7)—we take Fourier's transform of equation (6.11.13) with t and ω being the transformation variables to get

$$\nabla^2 \tilde{u}(\mathbf{x},\omega) + \left[\frac{\omega}{v(\mathbf{x})}\right]^2 \tilde{u}(\mathbf{x},\omega) = 0 . \tag{6.11.17}$$

Let us use the trial solution analogous to solution (6.11.15) and given by

$$\tilde{u}(\mathbf{x},\omega) = A(\mathbf{x}) \exp[i\omega\psi(\mathbf{x})] . \tag{6.11.18}$$

We insert trial solution (6.11.18) into equation (6.11.17). Firstly, considering the x_i term of the Laplacian, where $i \in \{1,2,3\}$, we obtain

$$\frac{\partial^2}{\partial x_i^2} A(\mathbf{x}) \exp[i\omega\psi(\mathbf{x})]$$
$$= \exp[i\omega\psi(\mathbf{x})] \left[\frac{\partial^2 A}{\partial x_i^2} + i\omega \left(2\frac{\partial A}{\partial x_i}\frac{\partial \psi}{\partial x_i} + A\frac{\partial^2 \psi}{\partial x_i^2}\right) - \omega^2 A \frac{\partial \psi}{\partial x_i}\frac{\partial \psi}{\partial x_i}\right] .$$

Secondly, considering the second term on the left-hand side of equation (6.11.17), we write

$$\left[\frac{\omega}{v(\mathbf{x})}\right]^2 \tilde{u}(\mathbf{x},\omega) = \left[\frac{\omega}{v(\mathbf{x})}\right]^2 A(\mathbf{x}) \exp\{i\omega[\psi(\mathbf{x})]\} .$$

Consequently, since the exponential term is never zero, equation (6.11.17) becomes

$$\sum_{i=1}^{3} \frac{\partial^2 A}{\partial x_i^2} + A\omega^2 \left(\frac{1}{v^2} - \sum_{i=1}^{3} \frac{\partial \psi}{\partial x_i}\frac{\partial \psi}{\partial x_i}\right) + i\omega \sum_{i=1}^{3} \left(2\frac{\partial A}{\partial x_i}\frac{\partial \psi}{\partial x_i} + A\frac{\partial^2 \psi}{\partial x_i^2}\right) = 0 , \tag{6.11.19}$$

which is a complex-valued function of real variables.

The vanishing of expression (6.11.19), where both A and ψ are assumed to be real, implies the vanishing of both real and imaginary parts. Assuming $\omega \neq 0$ and following the definitions of the gradient operator and the Laplacian, we obtain

$$\begin{cases} \nabla^2 A + A\omega^2 \left[\frac{1}{v^2(\mathbf{x})} - (\nabla\psi)^2 \right] = 0 \\ 2\nabla A \cdot \nabla\psi + A\nabla^2\psi = 0 \end{cases}, \tag{6.11.20}$$

where $(\nabla\psi)^2 := (\partial\psi/\partial x_1)^2 + (\partial\psi/\partial x_2)^2 + (\partial\psi/\partial x_3)^2$. System (6.11.20) corresponds to equation (6.11.17), in the context of trial solution (6.11.18).

Initially, system (6.11.20) might appear not simpler than equation (6.11.17). However, further analysis of the first equation of this system leads to an important simplification.

Eikonal equation

Considering the first equation of system (6.11.20) and assuming that both ω and A are nonzero, we can write it as

$$\frac{\nabla^2 A}{A\omega^2} + \left[\frac{1}{v^2(\mathbf{x})} - (\nabla\psi)^2 \right] = 0. \tag{6.11.21}$$

If we assume the inhomogeneity of the continuum to be weak, this assumption is tantamount to viewing the wavelength as being short with respect to the characteristic distance over which the properties of the continuum change significantly; this is analogous to the frequency being high.[30] In

[30] Readers interested in high-frequency approximation might refer to Bleistein, N., Cohen, J.K., and Stockwell, J.W., (2001) Mathematics of multidimensional seismic imaging, migration, and inversion: Springer-Verlag, pp. 5–7. Therein, while discussing physical applications, the authors state that

> "high frequency" does not refer to absolute values of the frequency content of the waves. What must be considered is the relationship between the *wavelengths* [...] and the natural *length scales* of the medium.

the limit, we let ω tend to infinity, and equation (6.11.21) becomes

$$[\nabla \psi (\mathbf{x})]^2 = \frac{1}{v^2 (\mathbf{x})} . \tag{6.11.22}$$

In view of expression (6.11.16), we can write equation (6.11.22) as

$$p^2 = \frac{1}{v^2 (\mathbf{x})} , \tag{6.11.23}$$

where $p^2 = \mathbf{p} \cdot \mathbf{p}$. Equation (6.11.23) is the eikonal equation for isotropic weakly inhomogeneous continua. It can be viewed—perhaps hastily[31]—as an approximation to wave equation (6.11.13): the eikonal equation as the high-frequency approximation with its physical applications limited to weak inhomogeneity.[32] In Chapter 7, we derive the eikonal equation for anisotropic inhomogeneous continua.

Recall that equation (6.11.13) does not explicitly refer to either P or S waves. Consequently, equation (6.11.23) does not explicitly correspond to either wave. Moreover, in view of the comment on page 276 where we state that replacing v with $v(\mathbf{x})$ in equation (6.11.13) is an arbitrary replacement, our formulation herein is heuristic rather than rigorous: a rigorous and more general formulation is presented in Chapter 7. However, in view of expression (6.1.13) and (6.1.17), if $v(\mathbf{x})$ is a smooth function given by

$$v(\mathbf{x}) = \sqrt{\frac{\lambda (\mathbf{x}) + 2\mu (\mathbf{x})}{\rho (\mathbf{x})}} , \tag{6.11.24}$$

[31] Readers interested in a rigorous formulation of the eikonal equation might refer to Bos, L. and Slawinski, M.A., (2010) Elastodynamic equations: Characteristics, wavefronts and rays. The Quarterly Journal of Mechanics and Applied Mathematics **63**(1), 23–37 and to Bóna, A. and Slawinski, M.A., (2021) Wavefronts and rays as characteristics and asymptotics (3*rd* edition): World Scientific, Chapter 3. Note that herein the elastodynamic equations are tantamount to Cauchy's equations of motion within Hookean solids.

[32] Readers interested in an elegant formulation of the eikonal equation, which does not use weak inhomogeneity but stems from a geometrical argument, might refer to McOwen, R.C., (1996) Partial differential equations: Methods and applications: Prentice-Hall, Inc., pp. 34–35.

equation (6.11.23) can be viewed as corresponding to P waves, and if $v(\mathbf{x})$ is a smooth function given by

$$v(\mathbf{x}) = \sqrt{\frac{\mu(\mathbf{x})}{\rho(\mathbf{x})}}, \qquad (6.11.25)$$

equation (6.11.23) can be viewed as corresponding to S waves. In general, for inhomogeneous continua, equations (6.1.4) cannot be split into two wave equations analogous to equations (6.1.12) and (6.1.16). In other words, the dilatational and rotational waves are coupled due to the inhomogeneity of the medium, as illustrated in Exercise 6.21. However, assuming sufficiently high frequency, there are two distinct wavefronts that propagate in an inhomogeneous continuum with speeds given by expressions (6.11.24) and (6.11.25).

The eikonal equation is a nonlinear partial differential equation. Specifically, it is a first-order and second-degree partial differential equation. In other words, the derivatives are of the first order, while the degree of the exponent is equal to 2. In general, the solution of the eikonal equation requires numerical methods. If the velocity function, v, is constant, the solution of the eikonal equation is also the solution of the corresponding wave equation, as shown in Exercises 6.17 and 7.7. Otherwise, in the cases where $v = v(\mathbf{x})$, the solution of the eikonal equation is not, in general, the solution of the wave equation, and equation (6.11.13) is only an approximation of the wave equation.

Transport equation

The second equation of system (6.11.20), namely,

$$2\nabla A \cdot \nabla \psi + A\nabla^2 \psi = 0, \qquad (6.11.26)$$

is the transport equation. For a given eikonal function, ψ, the transport equation describes the amplitude along the wavefront. Concluding this chapter, let us mention that expression (6.11.18) is a zeroth-order term of the asymptotic series in $\exp(i\omega\psi)/(i\omega)^n$ given by[33]

$$u(\mathbf{x},\omega) \sim \exp[i\omega\psi(\mathbf{x})] \sum_{n=0}^{N} \frac{A_n(\mathbf{x})}{(i\omega)^n},$$

where $\sim$ stands for "is asymptotically equivalent to."[34] Hence, the results presented in Section 6.11.4 belong to the realm of asymptotic methods, which play an important role in seismology. In Section 7.2.4, we discuss briefly ray theory in the context of asymptotic methods.

Closing Remarks

In this chapter, to study wave phenomena, we formulated wave equations. These equations are formulated as special cases of Cauchy's equations of motion for isotropic homogeneous continua. From these equations, we identify two distinct types of waves, namely P and S waves, which propagate with two distinct speeds. In Chapter 7, we formulate Cauchy's equations of motion in the context of anisotropic inhomogeneous continua. Therein, we show the existence of three types of waves.

All waves discussed in this book propagate within the body of a continuum. Consequently, they correspond to the so-called body waves, as opposed to the surface and interface waves, which we do not discuss.

The derivation of the wave equation shown in this chapter is rooted in the balance of linear momentum. This derivation formulates wave propagation as a result of a continuum conserving the linear momentum within

[33] Readers interested in the motivation for choosing this form of the trial solution might refer to Babich, V.M., and Buldyrev, V.S., (1991) Short-wavelength diffraction theory: Asymptotic methods: Springer-Verlag, pp. 10–13, to Bleistein, N., Cohen, J.K., and Stockwell, J.W., (2001) Mathematics of multidimensional seismic imaging, migration, and inversion: Springer-Verlag, pp. 436–437, and to Kennett, B.L.N., (2001) The seismic wavefield, Vol. I: Introduction and theoretical development: Cambridge University Press, pp. 153–154 and 166–167.

[34] For a description of the nature of asymptotic expansions, as well as the ways of obtaining them by the method of steepest descent and the method of stationary phase, readers might refer to Jeffreys, H., and Jeffreys, B., (1946/1999) Methods of Mathematical Physics: Cambridge University Press, pp. 498–507.

For a discussion on an application of asymptotic series to ray theory, readers might refer to Kravtsov, Y.A., and Orlov, Y.I., (1990) Geometrical optics of inhomogeneous media: Springer-Verlag, pp. 7 – 9.

itself. The wave equation can also be derived by invoking other physical principles. For instance, in Chapter 13, its derivation is based on Hamilton's principle, which formulates wave propagation as a result of a continuum restoring itself to the state of equilibrium through the process governed by the principle of stationary action.

The study of solutions for the wave equation motivated several recent developments in mathematics. As a result of these developments, the theory of generalized functions—in particular, the theory of distributions—extends the solutions for the wave equation to include nondifferentiable functions. Also, studies of wave propagation in elastic media have played an important role in the theory of integral equations.[35]

6.12. Exercises

EXERCISE 6.1. Show the details of the derivation of Lemma 6.5.1.

SOLUTION 6.1. For the first term of the wave equation, consider

$$\frac{\partial u}{\partial x}=\frac{\partial u}{\partial y}\frac{\partial y}{\partial x}+\frac{\partial u}{\partial z}\frac{\partial z}{\partial x}.$$

Since, following expression (6.5.4), $\partial y/\partial x=\partial z/\partial x=1$, we obtain

$$\frac{\partial u}{\partial x}=\frac{\partial u}{\partial y}+\frac{\partial u}{\partial z}.$$

[35] Interested readers might refer to Aleksandrov, A.D., Kolmogorov, A.N., Lavrentev, M.A., (editors), (1969/1999) Mathematics: Its content, methods and meaning: Dover, Vol. II, pp. 48 – 54 and Vol. III, pp. 245–250, to Bleistein, N., Cohen, J.K., and Stockwell, J.W., (2001) Mathematics of multidimensional seismic imaging, migration, and inversion: Springer-Verlag, pp. 389–408, and to Demidov, A.S., (2001) Generalized functions in mathematical physics: Main ideas and concepts: Nova Science Publishers, Inc., pp. 41–53.

Consequently,

$$\begin{aligned}\frac{\partial^2 u}{\partial x^2} &= \frac{\partial}{\partial x}\left(\frac{\partial u}{\partial y}+\frac{\partial u}{\partial z}\right)\\ &= \frac{\partial}{\partial y}\left(\frac{\partial u}{\partial y}+\frac{\partial u}{\partial z}\right)\frac{\partial y}{\partial x}+\frac{\partial}{\partial z}\left(\frac{\partial u}{\partial y}+\frac{\partial u}{\partial z}\right)\frac{\partial z}{\partial x}\\ &= \frac{\partial^2 u}{\partial y^2}+\frac{\partial^2 u}{\partial y\partial z}+\frac{\partial^2 u}{\partial z\partial y}+\frac{\partial^2 u}{\partial z^2}\\ &= \frac{\partial^2 u}{\partial y^2}+2\frac{\partial^2 u}{\partial z\partial y}+\frac{\partial^2 u}{\partial z^2},\end{aligned} \tag{6.12.1}$$

where, again, we used the equality given by $\partial y/\partial x = \partial z/\partial x = 1$, and the equality of mixed partial derivatives. Similarly, for the second term of the wave equation, consider

$$\begin{aligned}\frac{\partial u}{\partial t} &= \frac{\partial u}{\partial y}\frac{\partial y}{\partial t}+\frac{\partial u}{\partial z}\frac{\partial z}{\partial t}\\ &= v\frac{\partial u}{\partial y}-v\frac{\partial u}{\partial z}.\end{aligned}$$

Consequently,

$$\begin{aligned}\frac{\partial^2 u}{\partial t^2} &= \frac{\partial}{\partial t}\left(v\frac{\partial u}{\partial y}-v\frac{\partial u}{\partial z}\right)\\ &= \frac{\partial}{\partial y}\left(v\frac{\partial u}{\partial y}-v\frac{\partial u}{\partial z}\right)\frac{\partial y}{\partial t}+\frac{\partial}{\partial z}\left(v\frac{\partial u}{\partial y}-v\frac{\partial u}{\partial z}\right)\frac{\partial z}{\partial t}\\ &= \left(v\frac{\partial^2 u}{\partial y^2}-v\frac{\partial^2 u}{\partial y\partial z}\right)v+\left(v\frac{\partial^2 u}{\partial z\partial y}-v\frac{\partial^2 u}{\partial z^2}\right)(-v)\\ &= v^2\left(\frac{\partial^2 u}{\partial y^2}-\frac{\partial^2 u}{\partial y\partial z}-\frac{\partial^2 u}{\partial z\partial y}+\frac{\partial^2 u}{\partial z^2}\right)\\ &= v^2\left(\frac{\partial^2 u}{\partial y^2}-2\frac{\partial^2 u}{\partial z\partial y}+\frac{\partial^2 u}{\partial z^2}\right).\end{aligned} \tag{6.12.2}$$

where the equality of mixed partial derivatives was used. Inserting expressions (6.12.1) and (6.12.2) into equation (6.5.1), we obtain

$$\begin{aligned}\frac{\partial^2 u}{\partial x^2}-\frac{1}{v^2}\frac{\partial^2 u}{\partial t^2}&=\frac{\partial^2 u}{\partial y^2}+2\frac{\partial^2 u}{\partial z\partial y}+\frac{\partial^2 u}{\partial z^2}-\frac{1}{v^2}v^2\left(\frac{\partial^2 u}{\partial y^2}-2\frac{\partial^2 u}{\partial z\partial y}+\frac{\partial^2 u}{\partial z^2}\right)\\&=\frac{\partial^2 u}{\partial y^2}+2\frac{\partial^2 u}{\partial z\partial y}+\frac{\partial^2 u}{\partial z^2}-\frac{\partial^2 u}{\partial y^2}+2\frac{\partial^2 u}{\partial z\partial y}-\frac{\partial^2 u}{\partial z^2}\\&=4\frac{\partial^2 u}{\partial y\partial z}\\&=0\,,\end{aligned}$$

where the equality of mixed partial derivatives was used. Hence, we conclude that

$$\frac{\partial^2 u}{\partial y\partial z}=0\,,$$

as required.

EXERCISE 6.2. Show the details of the derivation of Lemma 6.5.2.

SOLUTION 6.2. Considering the equality of mixed partial derivatives, we can write

$$\frac{\partial}{\partial y}\left[\frac{\partial u(y,z)}{\partial z}\right]=\frac{\partial}{\partial z}\left[\frac{\partial u(y,z)}{\partial y}\right]=0\,.$$

Consequently, for the second partial derivative to vanish, we require that

$$\left[\frac{\partial u(y,z)}{\partial z}\right]=G(z)\,,$$

on the left-hand side, and

$$\left[\frac{\partial u(y,z)}{\partial y}\right]=F(y)\,,$$

on the right-hand side. In other words, we require that G be a function of z only, while F be a function of y only. Hence, integrating, we obtain

$$\begin{aligned}u(y,z)&=\int F(y)\,\mathrm{d}y\\&=f(y)+a(z)\,,\end{aligned}$$

where $a(z)$ is the integration constant with respect to $\mathrm{d}y$, and

$$u(y,z) = \int G(z)\,\mathrm{d}z$$
$$= g(z) + b(y) ,$$

where $b(y)$ is the integration constant with respect to $\mathrm{d}z$. In view of the arbitrariness of the integration constants, we can denote $a(z) \equiv g(z)$ and $b(y) \equiv f(y)$. Thus, we obtain

$$u(y,z) = f(y) + g(z) ,$$

as required.

EXERCISE 6.3. Formulate the Laplacian in spherical coordinates for a spherically symmetric function.

SOLUTION 6.3. Consider the Laplacian in the Cartesian coordinates,

$$\nabla^2 := \frac{\partial^2}{\partial x^2} + \frac{\partial^2}{\partial y^2} + \frac{\partial^2}{\partial z^2} ,$$

and expressions

$$x = r \sin\alpha \cos\beta , \qquad (6.12.3)$$
$$y = r \sin\alpha \sin\beta ,$$
$$z = r \cos\alpha ;$$

and equivalently,

$$r = \sqrt{x^2 + y^2 + z^2} , \qquad (6.12.4)$$
$$\alpha = \arccos\left(\frac{z}{r}\right) ,$$
$$\beta = \arctan\left(\frac{y}{x}\right) .$$

To find the first term of the operator, we write

$$\frac{\partial}{\partial x} = \frac{\partial r}{\partial x}\frac{\partial}{\partial r} + \frac{\partial \alpha}{\partial x}\frac{\partial}{\partial \alpha} + \frac{\partial \beta}{\partial x}\frac{\partial}{\partial \beta} ,$$

where

$$\frac{\partial r}{\partial x} = \frac{\partial \sqrt{x^2 + y^2 + z^2}}{\partial x} = \frac{x}{\sqrt{x^2 + y^2 + z^2}} ;$$

using expressions (6.12.3) and (6.12.4), we get

$$\frac{\partial r}{\partial x} = \sin\alpha\cos\beta\,.$$

Similarly, $\partial\alpha/\partial x = \cos\alpha\cos\beta/r$ and $\partial\beta/\partial x = -\sin\beta/(r\sin\alpha)$; hence,

$$\frac{\partial}{\partial x} = \sin\alpha\cos\beta\frac{\partial}{\partial r} + \frac{1}{r}\cos\alpha\cos\beta\frac{\partial}{\partial\alpha} - \frac{1}{r}\frac{\sin\beta}{\sin\alpha}\frac{\partial}{\partial\beta}\,.$$

Proceeding in an analogous manner—after laborious but easy computations[36]—we get $\partial^2/\partial x^2$; repeating the process for y and z, we obtain

$$\nabla^2 := \frac{1}{r^2}\frac{\partial}{\partial r}\left(r^2\frac{\partial}{\partial r}\right) + \frac{1}{r^2\sin\alpha}\frac{\partial}{\partial\alpha}\left(\sin\alpha\frac{\partial}{\partial\alpha}\right) + \frac{1}{r^2\sin^2\alpha}\frac{\partial^2}{\partial\beta^2}\,.$$

If a given function is spherically symmetric, $F(r,\alpha,\beta) = F(r)$, the operations of the second term and the third term result in zeros; thus–for such functions—we write the Laplacian as

$$\nabla^2 F = \frac{1}{r^2}\frac{\partial}{\partial r}\left(r^2\frac{\partial F}{\partial r}\right), \qquad (6.12.5)$$

which is the required expression.

EXERCISE 6.4. Using d'Alembert's approach, formulate the wave equation for a spherical wave and find its general solution.

SOLUTION 6.4. Using the product rule, we write expression (6.12.5) as

$$\nabla^2 F = \frac{\partial^2 F}{\partial r^2} + \frac{2}{r}\frac{\partial F}{\partial r},$$

which we rewrite as

$$\nabla^2 F = \frac{1}{r}\frac{\partial}{\partial r}\left(F + r\frac{\partial F}{\partial r}\right) = \frac{1}{r}\frac{\partial^2(rF)}{\partial r^2}\,.$$

The corresponding wave equation can be written as

$$\nabla^2 F - \frac{1}{v^2}\frac{\partial^2 F}{\partial t^2} = \frac{1}{r}\frac{\partial^2(rF)}{\partial r^2} - \frac{1}{v^2}\frac{\partial^2 F}{\partial t^2} = 0\,.$$

[36] Readers interested in more details, expressed in the context of polar coordinates, might refer to Bóna, A. and Slawinski, M.A., (2021) Wavefronts and rays as characteristics and asymptotics (3*rd* edition): World Scientific, Exercise 4.13.

Since r is independent of t, we write

$$\frac{\partial^2(rF)}{\partial r^2}-\frac{1}{v^2}\frac{\partial^2(rF)}{\partial t^2}=0,$$

which is analogous to equation (6.5.1) with $rF\equiv u$; hence, we invoke solution (6.5.6) to write

$$rF=f(r+vt)+g(r-vt),$$

which results in

$$F=\frac{1}{r}f(r+vt)+\frac{1}{r}g(r-vt).$$

The first term on the right-hand side represents a spherical wave that propagates towards the origin and the second one away from it. The amplitude of the former increases with distance from the origin, given by radius r; the amplitude of the latter decreases.

EXERCISE 6.5. Prove the following theorem.

THEOREM 6.12.1. *Solutions of the wave equation given by*

$$\frac{\partial^2 u}{\partial x^2}-\frac{1}{v^2}\frac{\partial^2 u}{\partial t^2}=0, \tag{6.12.6}$$

with the boundary conditions given by

$$u(0,t)=0 \tag{6.12.7}$$

and

$$u(1,t)=0 \tag{6.12.8}$$

that correspond to a string of unit length whose ends are fixed, satisfy the conservation of the wave-function energy that is defined by

$$\mathscr{E}(t):=\int_0^1\left[\left(\frac{\partial u}{\partial t}\right)^2+v^2\left(\frac{\partial u}{\partial x}\right)^2\right]\mathrm{d}x.$$

SOLUTION 6.5. Let us proceed with the following proof.

PROOF. Let us consider energy $\mathscr{E}$ at two arbitrary instances t_1 and t_2. The conservation of energy implies

$$\int_0^1 \left[\left(\left.\frac{\partial u}{\partial t}\right|_{t_1} \right)^2 + v^2 \left(\left.\frac{\partial u}{\partial x}\right|_{t_1} \right)^2 \right] \mathrm{d}x = \int_0^1 \left[\left(\left.\frac{\partial u}{\partial t}\right|_{t_2} \right)^2 + v^2 \left(\left.\frac{\partial u}{\partial x}\right|_{t_2} \right)^2 \right] \mathrm{d}x .$$

To prove this equation is tantamount to showing that

$$\int_0^1 \left[\left(\left.\frac{\partial u}{\partial t}\right|_{t_1} \right)^2 + v^2 \left(\left.\frac{\partial u}{\partial x}\right|_{t_1} \right)^2 \right] \mathrm{d}x - \int_0^1 \left[\left(\left.\frac{\partial u}{\partial t}\right|_{t_2} \right)^2 + v^2 \left(\left.\frac{\partial u}{\partial x}\right|_{t_2} \right)^2 \right] \mathrm{d}x \tag{6.12.9}$$

vanishes for all t. Let us write expression (6.12.9) as

$$\int_0^1 \left\{ \left[\left(\left.\frac{\partial u}{\partial t}\right|_{t_1} \right)^2 + v^2 \left(\left.\frac{\partial u}{\partial x}\right|_{t_1} \right)^2 \right] - \left[\left(\left.\frac{\partial u}{\partial t}\right|_{t_2} \right)^2 + v^2 \left(\left.\frac{\partial u}{\partial x}\right|_{t_2} \right)^2 \right] \right\} \mathrm{d}x .$$

Noticing that we can view the term in braces as a result of a definite integral with respect to t, we can write

$$\int_0^1 \int_{t_2}^{t_1} \frac{\mathrm{d}}{\mathrm{d}t} \left[\left(\frac{\partial u}{\partial t} \right)^2 + v^2 \left(\frac{\partial u}{\partial x} \right)^2 \right] \mathrm{d}t\,\mathrm{d}x . \tag{6.12.10}$$

Using identity (6.12.12) stated in Lemma 6.12.2, below, we can write expression (6.12.10) as

$$2v^2 \int_0^1 \int_{t_2}^{t_1} \frac{\mathrm{d}}{\mathrm{d}x} \left(\frac{\partial u}{\partial t} \frac{\partial u}{\partial x} \right) \mathrm{d}t\,\mathrm{d}x .$$

Changing the order of integration, we get

$$2v^2 \int_{t_2}^{t_1} \int_0^1 \frac{\mathrm{d}}{\mathrm{d}x} \left(\frac{\partial u}{\partial t} \frac{\partial u}{\partial x} \right) \mathrm{d}x\,\mathrm{d}t = 2v^2 \int_{t_2}^{t_1} \left[\frac{\partial u}{\partial t} \frac{\partial u}{\partial x} \right]_{x=0}^{x=1} \mathrm{d}t .$$

Applying the limits to the integrated term, we can write

$$2v^2 \left[\int_{t_2}^{t_1} \left. \left(\frac{\partial u}{\partial t} \frac{\partial u}{\partial x} \right) \right|_{x=1} - \int_{t_2}^{t_1} \left. \left(\frac{\partial u}{\partial t} \frac{\partial u}{\partial x} \right) \right|_{x=0} \right] \mathrm{d}t . \tag{6.12.11}$$

Boundary conditions (6.12.7) and (6.12.8) imply that $\partial u/\partial t|_{x=0} = 0$ and $\partial u/\partial t|_{x=1} = 0$, respectively. Physically, no displacement implies no velocity. Hence, expression (6.12.11) vanishes identically and we conclude that expression (6.12.9) vanishes, as required. Thus, the energy of wave function $u(x,t)$ is conserved. □

LEMMA 6.12.2. *If $u(x,t)$ satisfies wave equation (6.12.6), then the following identity holds:*

$$\frac{\mathrm{d}}{\mathrm{d}t}\left[\left(\frac{\partial u}{\partial t}\right)^2 + v^2\left(\frac{\partial u}{\partial x}\right)^2\right] = 2v^2\frac{\mathrm{d}}{\mathrm{d}x}\left(\frac{\partial u}{\partial t}\frac{\partial u}{\partial x}\right). \tag{6.12.12}$$

PROOF. Let us rewrite identity (6.12.12) as

$$\frac{1}{2}\frac{\mathrm{d}}{\mathrm{d}t}\left[\left(\frac{\partial u}{\partial t}\right)^2 + v^2\left(\frac{\partial u}{\partial x}\right)^2\right] - v^2\frac{\mathrm{d}}{\mathrm{d}x}\left(\frac{\partial u}{\partial t}\frac{\partial u}{\partial x}\right) = 0. \tag{6.12.13}$$

To prove this identity, we need to show that the left-hand side vanishes. Differentiating the left-hand side using the chain rule, we get

$$\frac{1}{2}\left(2\frac{\partial u}{\partial t}\frac{\partial^2 u}{\partial t^2} + 2v^2\frac{\partial u}{\partial x}\frac{\partial^2 u}{\partial t\partial x}\right) - v^2\left(\frac{\partial^2 u}{\partial x\partial t}\frac{\partial u}{\partial x} + \frac{\partial u}{\partial t}\frac{\partial^2 u}{\partial x^2}\right).$$

Using the equality of mixed partial derivatives to cancel the terms containing $\partial^2 u/\partial t\partial x$ and $\partial^2 u/\partial x\partial t$, we get

$$\frac{\partial u}{\partial t}\frac{\partial^2 u}{\partial t^2} - v^2\frac{\partial u}{\partial t}\frac{\partial^2 u}{\partial x^2} = \frac{\partial u}{\partial t}\left(\frac{\partial^2 u}{\partial t^2} - v^2\frac{\partial^2 u}{\partial x^2}\right).$$

If $u(x,t)$ satisfies wave equation (6.12.6), then the term in parentheses in the above expression vanishes identically and we obtain

$$\frac{\partial u}{\partial t}\left(\frac{\partial^2 u}{\partial t^2} - v^2\frac{\partial^2 u}{\partial x^2}\right) = 0.$$

Thus, the left-hand side of equation (6.12.13) vanishes and, hence, identity (6.12.12) is proved, as required. □

EXERCISE 6.6. Perform Fourier's transform of equation (6.6.4); namely,

$$\sum_{i=1}^{3}\frac{\partial^2 u(x_1,x_2,x_3,t)}{\partial x_i^2}-\frac{1}{v^2}\frac{\partial^2 u(x_1,x_2,x_3,t)}{\partial t^2}=0, \qquad (6.12.14)$$

with $\mathbf{x}$ and $\mathbf{k}$ being the variables of transformation.

SOLUTION 6.6. Since equation (6.12.14) is a differential equation, let us consider Fourier's transforms of derivatives. To find the transform of $\partial u(\mathbf{x},t)/\partial x_j$, we write

$$\frac{1}{(2\pi)^3}\iiint\limits_{\mathbb{R}^3}\frac{\partial u(\mathbf{x},t)}{\partial x_j}e^{-i\mathbf{k}\cdot\mathbf{x}}\mathrm{d}\mathbf{x}.$$

Since we are differentiating with respect to a single component of $\mathbf{x}$, let us write explicitly

$$\frac{1}{(2\pi)^3}\int_{-\infty}^{+\infty}\int_{-\infty}^{+\infty}\int_{-\infty}^{+\infty}\frac{\partial u(\mathbf{x},t)}{\partial x_j}e^{-i\mathbf{k}\cdot\mathbf{x}}\mathrm{d}x_j\mathrm{d}x_k\mathrm{d}x_\ell,$$

which allows us to use the integration by parts on the innermost integral. Therein, we let $j,k,\ell\in\{1,2,3\}$ to consider the partial derivative with respect to any of the components of $\mathbf{x}$. Integrating by parts, we get

$$\frac{1}{(2\pi)^3}\int_{-\infty}^{+\infty}\int_{-\infty}^{+\infty}\left(u(\mathbf{x},t)\,e^{-i\mathbf{k}\cdot\mathbf{x}}\Big|_{-\infty}^{+\infty}-\int_{-\infty}^{+\infty}u(\mathbf{x},t)\frac{\partial}{\partial x_j}e^{-i\mathbf{k}\cdot\mathbf{x}}\mathrm{d}x_j\right)\mathrm{d}x_k\mathrm{d}x_\ell.$$

We assume that the integrated term vanishes at $\pm\infty$ to get

$$-\frac{1}{(2\pi)^3}\int_{-\infty}^{+\infty}\int_{-\infty}^{+\infty}\int_{-\infty}^{+\infty}u(\mathbf{x},t)\frac{\partial}{\partial x_j}e^{-i\mathbf{k}\cdot\mathbf{x}}\mathrm{d}x_j\mathrm{d}x_k\mathrm{d}x_\ell.$$

Differentiating the exponential with respect to x_j, we get

$$\frac{1}{(2\pi)^3}\int_{-\infty}^{+\infty}\int_{-\infty}^{+\infty}\int_{-\infty}^{+\infty}u(\mathbf{x},t)\,ik_je^{-i\mathbf{k}\cdot\mathbf{x}}\mathrm{d}x_j\mathrm{d}x_k\mathrm{d}x_\ell,$$

which we rewrite as

$$ik_j \frac{1}{(2\pi)^3} \iiint\limits_{\mathbb{R}^3} u(\mathbf{x},t)\, e^{-i\mathbf{k}\cdot\mathbf{x}} \mathrm{d}\mathbf{x}.$$

We recognize that this expression contains the definition of Fourier's transform. Thus we write this expression as $ik_j \tilde{u}(\mathbf{k},t)$, which is Fourier's transform of $\partial u(\mathbf{x},t)/\partial x_j$. Following a procedure analogous to the one shown above, we obtain the transform of $\partial^2 u(\mathbf{x},t)/\partial x_j^2$; namely,

$$\frac{1}{(2\pi)^3} \iiint\limits_{\mathbb{R}^3} \frac{\partial^2 u(\mathbf{x},t)}{\partial x_j^2} e^{-i\mathbf{k}\cdot\mathbf{x}} \mathrm{d}\mathbf{x} = (ik_j)^2 \tilde{u}(\mathbf{k},t) = -k_j^2 \tilde{u}(\mathbf{k},t). \quad (6.12.15)$$

Having derived the formula for Fourier's transform of the second derivative, let us consider equation (6.12.14). We write

$$\frac{1}{(2\pi)^3} \iiint\limits_{\mathbb{R}^3} \left(\left(\sum_{j=1}^{3} \frac{\partial^2 u(x_1,x_2,x_3,t)}{\partial x_j^2} - \frac{1}{v^2} \frac{\partial^2 u(x_1,x_2,x_3,t)}{\partial t^2} \right) e^{-i\mathbf{k}\cdot\mathbf{x}} \right) \mathrm{d}\mathbf{x}$$
$$= \frac{1}{(2\pi)^3} \iiint\limits_{\mathbb{R}^3} 0 e^{-i\mathbf{k}\cdot\mathbf{x}} \mathrm{d}\mathbf{x}.$$

Using the linearity of the integral operator and the fact that Fourier's transform of zero is zero, we write

$$\frac{1}{(2\pi)^3} \iiint\limits_{\mathbb{R}^3} \sum_{j=1}^{3} \frac{\partial^2 u(x_1,x_2,x_3,t)}{\partial x_j^2} e^{-i\mathbf{k}\cdot\mathbf{x}} \mathrm{d}\mathbf{x}$$
$$- \frac{1}{v^2} \frac{1}{(2\pi)^3} \iiint\limits_{\mathbb{R}^3} \frac{\partial^2 u(x_1,x_2,x_3,t)}{\partial t^2} e^{-i\mathbf{k}\cdot\mathbf{x}} \mathrm{d}\mathbf{x} = 0.$$

Again, using the linearity, we write

$$\frac{1}{(2\pi)^3}\iiint\limits_{\mathbb{R}^3}\frac{\partial^2 u}{\partial x_1^2}e^{-i\mathbf{k}\cdot\mathbf{x}}\mathrm{d}\mathbf{x}+\frac{1}{(2\pi)^3}\iiint\limits_{\mathbb{R}^3}\frac{\partial^2 u}{\partial x_2^2}e^{-i\mathbf{k}\cdot\mathbf{x}}\mathrm{d}\mathbf{x}$$

$$+\frac{1}{(2\pi)^3}\iiint\limits_{\mathbb{R}^3}\frac{\partial^2 u}{\partial x_3^2}e^{-i\mathbf{k}\cdot\mathbf{x}}\mathrm{d}\mathbf{x}$$

$$-\frac{1}{v^2}\frac{1}{(2\pi)^3}\iiint\limits_{\mathbb{R}^3}\frac{\partial^2 u}{\partial t^2}e^{-i\mathbf{k}\cdot\mathbf{x}}\mathrm{d}\mathbf{x}=0\,.$$

Let us consider the first three integrals. In view of expression (6.12.15), we write them as

$$-k_1^2\tilde{u}(\mathbf{k},t)-k_2^2\tilde{u}(\mathbf{k},t)-k_3^2\tilde{u}(\mathbf{k},t)=-\left(k_1^2+k_2^2+k_3^2\right)\tilde{u}(\mathbf{k},t)\,.$$

We recognize that the sum in parentheses is the scalar product of vector $\mathbf{k}$ with itself. Thus it is the squared length of $\mathbf{k}$, which we denote by $|\mathbf{k}|^2$. Let us consider the fourth integral. Using the commutativity of the integral and differential operators, we write

$$-\frac{1}{v^2}\frac{\partial^2}{\partial t^2}\frac{1}{(2\pi)^3}\iiint\limits_{\mathbb{R}^3}u(\mathbf{x},t)\,e^{-i\mathbf{k}\cdot\mathbf{x}}\mathrm{d}\mathbf{x}=0\,.$$

The term to be differentiated is the definition of Fourier's transform of u; namely, $\tilde{u}$. Combining these results, we write Fourier's transform of equation (6.12.14) as

$$|\mathbf{k}|^2\,\tilde{u}(\mathbf{k},t)+\frac{1}{v^2}\frac{\partial^2\tilde{u}(\mathbf{k},t)}{\partial t^2}=0\,,$$

which is equivalent to equation (6.6.5).

EXERCISE 6.7. Solve equation (6.6.5); namely,

$$\frac{\partial^2\tilde{u}(\mathbf{k},t)}{\partial t^2}+v^2\,|\mathbf{k}|^2\,\tilde{u}(\mathbf{k},t)=0\,,$$

with initial conditions given by equations (6.6.6) and (6.6.7); namely,

$$\tilde{u}(\mathbf{k},t)|_{t=0}=\tilde{\gamma}(\mathbf{k})$$

and

$$\left.\frac{\partial \tilde{u}(\mathbf{k},t)}{\partial t}\right|_{t=0} = \tilde{\eta}(\mathbf{k}),$$

respectively.

SOLUTION 6.7. Since the derivatives of $\tilde{u}(\mathbf{k},t)$ are taken only with respect to a single variable, we can approach this partial differential equation as if it were an ordinary differential equation. Invoking the theory of second-order ordinary differential equations with constant coefficients, we write the general solution as a linear combination of $\exp(iv|\mathbf{k}|t)$ and $\exp(-iv|\mathbf{k}|t)$; namely,

$$\tilde{u}(\mathbf{k},t) = F(\mathbf{k})\exp\{iv|\mathbf{k}|t\} + G(\mathbf{k})\exp\{-iv|\mathbf{k}|t\},$$

which is expression (6.6.8). Since we are dealing with a partial differential equation with derivatives in terms of t alone, F and G are constants with respect to t but can depend on $\mathbf{k}$.

To obtain explicit expressions for F and G, we invoke the initial conditions. We get

$$\tilde{u}(\mathbf{k},t)|_{t=0} = F(\mathbf{k}) + G(\mathbf{k}) = \tilde{\gamma}(\mathbf{k})$$

and

$$\begin{aligned}\left.\frac{\partial \tilde{u}(\mathbf{k},t)}{\partial t}\right|_{t=0} &= iv|\mathbf{k}|F(\mathbf{k})\exp\{iv|\mathbf{k}|t\}|_{t=0} \\ &\quad - iv|\mathbf{k}|G(\mathbf{k})\exp\{-iv|\mathbf{k}|t\}|_{t=0} \\ &= iv|\mathbf{k}|F(\mathbf{k}) - iv|\mathbf{k}|G(\mathbf{k}) = \tilde{\eta}(\mathbf{k}).\end{aligned}$$

Let us write this system of equations as

$$\begin{bmatrix} 1 & 1 \\ iv|\mathbf{k}| & -iv|\mathbf{k}| \end{bmatrix}\begin{bmatrix} F(\mathbf{k}) \\ G(\mathbf{k}) \end{bmatrix} = \begin{bmatrix} \tilde{\gamma}(\mathbf{k}) \\ \tilde{\eta}(\mathbf{k}) \end{bmatrix}.$$

Using Cramer's rule, we obtain

$$F(\mathbf{k}) = \frac{\det\begin{bmatrix} \tilde{\gamma}(\mathbf{k}) & 1 \\ \tilde{\eta}(\mathbf{k}) & -iv|\mathbf{k}| \end{bmatrix}}{\det\begin{bmatrix} 1 & 1 \\ iv|\mathbf{k}| & -iv|\mathbf{k}| \end{bmatrix}} = \frac{-iv|\mathbf{k}|\tilde{\gamma}(\mathbf{k}) - \tilde{\eta}(\mathbf{k})}{-2iv|\mathbf{k}|}$$
$$= \frac{1}{2}\left(\tilde{\gamma}(\mathbf{k}) + \frac{1}{iv|\mathbf{k}|}\tilde{\eta}(\mathbf{k})\right)$$

and

$$G(\mathbf{k}) = \frac{\det\begin{bmatrix} 1 & \tilde{\gamma}(\mathbf{k}) \\ iv|\mathbf{k}| & \tilde{\eta}(\mathbf{k}) \end{bmatrix}}{\det\begin{bmatrix} 1 & 1 \\ iv|\mathbf{k}| & -iv|\mathbf{k}| \end{bmatrix}} = \frac{\tilde{\eta}(\mathbf{k}) - iv|\mathbf{k}|\tilde{\gamma}(\mathbf{k})}{-2iv|\mathbf{k}|}$$
$$= \frac{1}{2}\left(\tilde{\gamma}(\mathbf{k}) - \frac{1}{iv|\mathbf{k}|}\tilde{\eta}(\mathbf{k})\right),$$

which are expressions (6.6.9) and (6.6.10), respectively.

EXERCISE 6.8. Find Fourier's transform of distribution (6.6.13), namely,

$$\alpha(\tau) = \frac{(2\pi)^3}{4\pi v^2 t}\iint\limits_{S(\mathbf{0},vt)} \tau \mathrm{d}\zeta. \tag{6.12.16}$$

In other words, find $\tilde{\alpha}(\tau)$, where symbol $\tilde{}$ denotes the transformed entity.

SOLUTION 6.8. Let us consider the definition of Fourier's transform given by

$$\tilde{f}(\mathbf{k}) = \frac{1}{(2\pi)^3}\iiint\limits_{\mathbb{R}^3} f(\mathbf{x})\exp\{-i\mathbf{k}\cdot\mathbf{x}\}\,\mathrm{d}\mathbf{x},$$

where f is a function. If f is considered as a distribution, the effect of this distribution on a test function, τ, is defined as

$$f(\tau) = \iiint\limits_{\mathbb{R}^3} f(\mathbf{x})\,\tau(\mathbf{x})\,\mathrm{d}\mathbf{x},$$

and the corresponding Fourier's transform is defined by letting $\tau = \exp\{-i\mathbf{k}\cdot\mathbf{x}\}/(2\pi)^3$. Distribution (6.12.16) has the effect of Dirac's delta with support on sphere S centred at the origin and with radius vt; in other words, we can view this distribution as a three-dimensional Dirac's delta.[37] We write

$$\tilde{\alpha} = \frac{1}{4\pi v^2 t} \iint\limits_{S(\mathbf{0},vt)} \exp\{-i\mathbf{k}\cdot\mathbf{x}\}\, \mathrm{d}\zeta(\mathbf{x}). \tag{6.12.17}$$

To integrate, we use the spherical coordinates. Since the integrand does not depend on the orientation, we rotate the coordinate system to be such that $\mathbf{k}$ points in the x_1-axis direction. Hence, $\mathbf{k} = [|\mathbf{k}|, 0, 0]$, where $|\mathbf{k}|$ stands for the length of $\mathbf{k}$. Thus, we write expression (6.12.17) as

$$\tilde{\alpha} = \frac{1}{4\pi v^2 t} \iint\limits_{S(\mathbf{0},vt)} \exp\{-i|\mathbf{k}|x_1\}\, \mathrm{d}\zeta(\mathbf{x}). \tag{6.12.18}$$

Recalling that the radius is vt, we write the relation between the coordinates as

$$x_1 = vt\cos\alpha,$$

$$x_2 = vt\sin\alpha\cos\beta$$

and

$$x_3 = vt\sin\alpha\sin\beta,$$

where α is the latitude and β is the azimuth. The element of surface for the sphere is

$$\mathrm{d}\zeta(\mathbf{x}) = (vt)^2 \sin\alpha\, \mathrm{d}\alpha\, \mathrm{d}\beta.$$

Returning to integral (6.12.18), we write this integral in spherical coordinates as

$$\tilde{\alpha}(\tau) = \frac{1}{4\pi v^2 t} \int_0^{2\pi}\int_0^{\pi} \exp\{-i|\mathbf{k}|vt\cos\alpha\}(vt)^2 \sin\alpha\, \mathrm{d}\alpha\, \mathrm{d}\beta.$$

[37] Readers interested in Dirac's delta for two and three dimensions might refer to Barton, G., (1989) Elements of Green's functions and propagation: Potentials, diffusion and waves: Oxford Science Publications, pp. 30–32.

Simplifying, we write

$$\tilde{\alpha}(\tau) = \frac{t}{4\pi}\int_0^{2\pi}\int_0^{\pi} \exp\{-i|\mathbf{k}|vt\cos\alpha\}\sin\alpha\,\mathrm{d}\alpha\,\mathrm{d}\beta.$$

Let us consider the inner integral. Setting $m = \exp\{-i|\mathbf{k}|vt\cos\alpha\}$, we get

$$\frac{1}{i|\mathbf{k}|vt}\int \mathrm{d}m = \frac{m}{i|\mathbf{k}|vt},$$

which means that

$$\begin{aligned}\int_0^{\pi} \exp\{-i|\mathbf{k}|vt\cos\alpha\}\sin\alpha\mathrm{d}\alpha &= \left.\frac{\exp\{-i|\mathbf{k}|vt\cos\alpha\}}{i|\mathbf{k}|vt}\right|_0^{\pi}\\ &= \frac{\exp\{i|\mathbf{k}|vt\} - \exp\{-i|\mathbf{k}|vt\}}{i|\mathbf{k}|vt}.\end{aligned}$$

Thus,

$$\tilde{\alpha}(\tau) = \frac{t}{4\pi}\int_0^{2\pi}\frac{\exp\{i|\mathbf{k}|vt\} - \exp\{-i|\mathbf{k}|vt\}}{i|\mathbf{k}|vt}\mathrm{d}\beta.$$

Since the integrand does not depend on the azimuth, we write

$$\begin{aligned}\tilde{\alpha}(\tau) &= \frac{t}{4\pi}\frac{\exp\{i|\mathbf{k}|vt\} - \exp\{-i|\mathbf{k}|vt\}}{i|\mathbf{k}|vt}\int_0^{2\pi}\mathrm{d}\beta.\\ &= \frac{1}{4\pi}\frac{\exp\{i|\mathbf{k}|vt\} - \exp\{-i|\mathbf{k}|vt\}}{i|\mathbf{k}|v}[\beta]_0^{2\pi}\\ &= \frac{1}{4\pi}\frac{\exp\{i|\mathbf{k}|vt\} - \exp\{-i|\mathbf{k}|vt\}}{i|\mathbf{k}|vt}2\pi,\end{aligned}$$

which is Fourier's transform of distribution (6.12.16). Thus, we write

$$\tilde{\alpha}(\tau) = \frac{\exp\{i|\mathbf{k}|vt\} - \exp\{-i|\mathbf{k}|vt\}}{2i|\mathbf{k}|v}, \tag{6.12.19}$$

which is the fraction in solution (6.6.12).

EXERCISE 6.9. Using the result of Exercise 6.8, show that the inverse Fourier transform of expression (6.6.12); namely,

$$\tilde{u}(\mathbf{k},t)=\tilde{\gamma}(\mathbf{k})\frac{\mathrm{d}}{\mathrm{d}t}\frac{\exp\{iv|\mathbf{k}|t\}-\exp\{-iv|\mathbf{k}|t\}}{2iv|\mathbf{k}|}+\tilde{\eta}(\mathbf{k})\frac{\exp\{iv|\mathbf{k}|t\}-\exp\{-iv|\mathbf{k}|t\}}{2iv|\mathbf{k}|}\qquad(6.12.20)$$

is expression (6.6.14); namely,

$$u(\mathbf{x},t)=\frac{2\pi^2}{v^2t}\iint\limits_{S(\mathbf{x},vt)}\eta(\mathbf{y})\,\mathrm{d}\zeta(\mathbf{y})+\frac{\mathrm{d}}{\mathrm{d}t}\frac{2\pi^2}{v^2t}\iint\limits_{S(\mathbf{x},vt)}\gamma(\mathbf{y})\,\mathrm{d}\zeta(\mathbf{y}).$$

SOLUTION 6.9. Since $\tilde{\gamma}$ does not depend on t, we write expression (6.12.20) as

$$\tilde{u}(\mathbf{k},t)=\frac{\mathrm{d}}{\mathrm{d}t}\left(\tilde{\gamma}(\mathbf{k})\frac{\exp\{iv|\mathbf{k}|t\}-\exp\{-iv|\mathbf{k}|t\}}{2iv|\mathbf{k}|}\right)+\tilde{\eta}(\mathbf{k})\frac{\exp\{iv|\mathbf{k}|t\}-\exp\{-iv|\mathbf{k}|t\}}{2iv|\mathbf{k}|}.$$

As derived in Exercise 6.8, expression (6.12.19) is Fourier's transform of distribution (6.12.16). Thus, we write

$$\tilde{u}(\mathbf{k},t)=\frac{\mathrm{d}}{\mathrm{d}t}\left(\tilde{\gamma}(\mathbf{k})\,\tilde{\alpha}(\mathbf{k})\right)+\tilde{\eta}(\mathbf{k})\,\tilde{\alpha}(\mathbf{k}).$$

Taking the inverse Fourier transform by using the fact that the product of two functions in the **k**-domain is their convolution in the **x**-domain, we write

$$u(\mathbf{x},t)=\frac{\mathrm{d}}{\mathrm{d}t}\left(\gamma(\mathbf{x})\star\alpha(\mathbf{x})\right)+\eta(\mathbf{x})\star\alpha(\mathbf{x}).$$

Invoking the definition of convolution, we write

$$u(\mathbf{x},t)=\frac{\mathrm{d}}{\mathrm{d}t}\iiint\limits_{\mathbb{R}^3}\gamma(\mathbf{x}-\mathbf{y})\,\alpha(\mathbf{y})\,\mathrm{d}\mathbf{y}+\iiint\limits_{\mathbb{R}^3}\eta(\mathbf{x}-\mathbf{y})\,\alpha(\mathbf{y})\,\mathrm{d}\mathbf{y}.$$

In view of the commutativity of convolution, we write

$$u(\mathbf{x},t)=\frac{\mathrm{d}}{\mathrm{d}t}\iiint\limits_{\mathbb{R}^3}\gamma(\mathbf{y})\,\alpha(\mathbf{x}-\mathbf{y})\,\mathrm{d}\mathbf{y}+\iiint\limits_{\mathbb{R}^3}\eta(\mathbf{y})\,\alpha(\mathbf{x}-\mathbf{y})\,\mathrm{d}\mathbf{y},$$

Since α is Dirac's delta on a sphere whose centre is the origin, $\alpha(\mathbf{x}-\mathbf{y})$is the corresponding Dirac's delta with support on the translated sphere whose centre is at $\mathbf{x}$. Recalling expression (6.12.16), we write the effect of this Dirac's delta on γ and η as

$$u(\mathbf{x},t)=\frac{2\pi^2}{v^2t}\left(\frac{\mathrm{d}}{\mathrm{d}t}\iint\limits_{S(\mathbf{x},vt)}\gamma(\mathbf{y})\,\mathrm{d}\zeta(\mathbf{y})+\iint\limits_{S(\mathbf{x},vt)}\eta(\mathbf{y})\,\mathrm{d}\zeta(\mathbf{y})\right),$$

which is equivalent to expression (6.6.14), as required.

EXERCISE 6.10. Using the fact that η is a function of s_1 and s_2 only, write integral

$$\iint\limits_{S(\mathbf{x},vt)}\eta(s_1,s_2)\,\mathrm{d}\zeta(\mathbf{s}), \tag{6.12.21}$$

where S is a sphere with radius vt centred at $\mathbf{x}$, as

$$2vt\iint\limits_{D(\mathbf{x},vt)}\frac{\eta(s_1,s_2)}{\sqrt{(vt)^2-(s_1-x_1)^2-(s_2-x_2)^2}}\mathrm{d}s_1\mathrm{d}s_2, \tag{6.12.22}$$

where D is the disc with radius vt centred at $\mathbf{x}$.

SOLUTION 6.10. Let us write the sphere that is centered at $\mathbf{x}$ and whose radius is vt as

$$(s_1-x_1)^2+(s_2-x_2)^2+(s_3-x_3)^2=(vt)^2, \tag{6.12.23}$$

where $\mathbf{s}$ and $\mathbf{x}$ are coordinates of the three-dimensional space. Since η does not depend on s_3, we consider the s_1s_2-plane. The element of surface for the sphere, $\mathrm{d}\zeta(\mathbf{s})$, projected onto this plane is

$$\mathrm{d}\zeta(\mathbf{s})=\sqrt{1+\left(\frac{\partial s_3}{\partial s_1}\right)^2+\left(\frac{\partial s_3}{\partial s_2}\right)^2}\,\mathrm{d}s_1\mathrm{d}s_2.$$

We solve equation (6.12.23) for s_3 to get

$$s_3=\pm\sqrt{(vt)^2-(s_1-x_1)^2-(s_2-x_2)^2}+x_3,$$

where the signs denote s_3 that corresponds to the upper or lower hemisphere. Since η is independent of s_3, let us use only the upper hemisphere and, later on, multiply our result by 2 . Herein, we write

$$\frac{\partial s_3}{\partial s_1} = \frac{-(s_1 - x_1)}{\sqrt{(vt)^2 - (s_1 - x_1)^2 - (s_2 - x_2)^2}}$$

and

$$\frac{\partial s_3}{\partial s_2} = \frac{-(s_2 - x_2)}{\sqrt{(vt)^2 - (s_1 - x_1)^2 - (s_2 - x_2)^2}}.$$

Consequently,

$$\mathrm{d}\zeta(\mathbf{s}) = \sqrt{1 + \frac{(s_1 - x_1)^2}{(vt)^2 - (s_1 - x_1)^2 - (s_2 - x_2)^2} + \frac{(s_2 - x_2)^2}{(vt)^2 - (s_1 - x_1)^2 - (s_2 - x_2)^2}}\,\mathrm{d}s_1\mathrm{d}s_2.$$

Expressing the radicand using the common denominator, we get

$$\mathrm{d}\zeta(\mathbf{s}) = \frac{vt}{\sqrt{(vt)^2 - (s_1 - x_1)^2 - (s_2 - x_2)^2}}\,\mathrm{d}s_1\mathrm{d}s_2.$$

Hence we write integral (6.12.21) as

$$\iint\limits_{S(\mathbf{x},vt)} \eta(s_1, s_2)\,\mathrm{d}\zeta(\mathbf{s})$$

$$= 2vt \iint\limits_{(s_1-x_1)^2+(s_2-x_2)^2 \leqslant (vt)^2} \frac{\eta(s_1, s_2)}{\sqrt{(vt)^2 - (s_1 - x_1)^2 - (s_2 - x_2)^2}}\,\mathrm{d}s_1\mathrm{d}s_2,$$

where the factor of 2 on the right-hand side accounts for both the upper and lower hemispheres. Denoting $(s_1 - x_1)^2 + (s_2 - x_2)^2 \leqslant (vt)^2$ by $D(\mathbf{x}, vt)$, which stands for the disc with radius vt that is centred at $\mathbf{x}$, we obtain

$$2vt \iint\limits_{D(\mathbf{x},vt)} \frac{\eta(s_1, s_2)}{\sqrt{(vt)^2 - (s_1 - x_1)^2 - (s_2 - x_2)^2}}\,\mathrm{d}s_1\mathrm{d}s_2,$$

which is equation (6.12.22), as required.

EXERCISE 6.11. Show that

$$\int_{-\infty}^{\infty} \exp\left\{i(x-\zeta)k - k^2 t\right\} \mathrm{d}k = \frac{\exp\left\{-\dfrac{(x-\zeta)^2}{4t}\right\}}{\sqrt{t}} \int_{-\infty}^{\infty} \exp\left\{-s^2\right\} \mathrm{d}s. \tag{6.12.24}$$

SOLUTION 6.11. Let $k = a + ib$, where $a \in (-\infty, \infty)$ is the variable of integration and $b \in \mathbb{R}$ is fixed and given by $(x-\zeta)/2t$; hence, $\mathrm{d}k = \mathrm{d}a$. Furthermore, let $a = s/\sqrt{t}$, and hence, $\mathrm{d}a = \mathrm{d}s/\sqrt{t}$ and $\mathrm{d}k = \mathrm{d}s/\sqrt{t}$. Also, $k = s/\sqrt{t} + i(x-\zeta)/2t$, which we write on the common denominator as $k = \left(2s\sqrt{t} + i(x-\zeta)\right)/2t$. Thus, we write the exponential of the above integral as

$$\begin{aligned} i(x-\zeta)k - k^2 t &= i(x-\zeta)\frac{2s\sqrt{t}+i(x-\zeta)}{2t} - \left(\frac{2s\sqrt{t}+i(x-\zeta)}{2t}\right)^2 t \\ &= \frac{2i(x-\zeta)s\sqrt{t}-(x-\zeta)^2}{2t} - \frac{\left(2s\sqrt{t}+i(x-\zeta)\right)^2}{4t}, \end{aligned}$$

which we rewrite as

$$\begin{aligned} &\frac{1}{2t}\left(2i(x-\zeta)s\sqrt{t}-(x-\zeta)^2 - \frac{1}{2}\left(2s\sqrt{t}+i(x-\zeta)\right)^2\right) \\ &= \frac{1}{2t}\Bigg(2i(x-\zeta)s\sqrt{t}-(x-\zeta)^2 \\ &\quad - \frac{1}{2}\left(4s^2 t + 4is\sqrt{t}(x-\zeta) - (x-\zeta)^2\right)\Bigg). \end{aligned}$$

Simplifying, we get

$$\frac{1}{2t}\left(2i(x-\zeta)s\sqrt{t}-(x-\zeta)^2 - 2s^2 t - 2is\sqrt{t}(x-\zeta) + \frac{1}{2}(x-\zeta)^2\right).$$

Canceling and gathering the common terms, we obtain

$$\frac{1}{2t}\left(-\frac{1}{2}(x-\zeta)^2 - 2s^2 t\right) = -s^2 - \frac{(x-\zeta)^2}{4t}.$$

Returning to integral (6.12.24), we write it as

$$\int_{-\infty}^{\infty} \exp\left\{-s^2 - \frac{(x-\zeta)^2}{4t}\right\} \frac{1}{\sqrt{t}} \mathrm{d}s = \int_{-\infty}^{\infty} \exp\left\{-s^2\right\} \exp\left\{-\frac{(x-\zeta)^2}{4t}\right\} \frac{1}{\sqrt{t}} \mathrm{d}s.$$

Since x, ζ and t are independent of s, we get

$$\frac{\exp\left\{-\frac{(x-\zeta)^2}{4t}\right\}}{\sqrt{t}} \int_{-\infty}^{\infty} \exp\left\{-s^2\right\} \mathrm{d}s.$$

EXERCISE 6.12. Evaluate

$$\int_{-\infty}^{\infty} e^{-s^2} \mathrm{d}s. \tag{6.12.25}$$

SOLUTION 6.12. This integrand is not a derivative of an elementary function, which is a function that can be formed from the algebraic, exponential, logarithmic or trigonometric functions by a finite number of operations consisting of addition, subtraction, multiplication and composition of functions. To find the antiderivative, let us consider

$$I = \int_{-\infty}^{\infty} \int_{-\infty}^{\infty} e^{-\left(x^2+y^2\right)} \mathrm{d}x\mathrm{d}y, \tag{6.12.26}$$

to be evaluated in polar coordinates, where

$$x(r, \alpha) = r\cos\alpha$$

and

$$y(r, \alpha) = r\sin\alpha.$$

Since the Jacobian of transformation is

$$\det\begin{bmatrix} \frac{\partial x}{\partial r} & \frac{\partial x}{\partial \alpha} \\ \frac{\partial y}{\partial r} & \frac{\partial y}{\partial \alpha} \end{bmatrix} = \det\begin{bmatrix} \cos\alpha & -r\sin\alpha \\ \sin\alpha & r\cos\alpha \end{bmatrix} = r\left(\cos^2\alpha + \sin^2\alpha\right) = r,$$

we write

$$I=\int_0^{2\pi}\int_0^{\infty} re^{-r^2}\,\mathrm{d}r\mathrm{d}\alpha,$$

with the limits of integration covering the entire xy-plane. Let us consider the inner integral. Letting $r^2=m$, which implies that $\mathrm{d}r=\mathrm{d}m/2r$, we write

$$\int_0^{2\pi}\int_0^{\infty} re^{-r^2}\,\mathrm{d}r=\frac{1}{2}\int_0^{\infty} e^{-m}\mathrm{d}m=\frac{1}{2}\left[-e^{-m}\right]_0^{\infty}=\frac{1}{2}.$$

Hence,

$$I=\frac{1}{2}\int_0^{2\pi}\mathrm{d}\alpha=\pi.$$

Let us return to integral (6.12.26), and write it as

$$I=\int_{-\infty}^{\infty}\int_{-\infty}^{\infty} e^{-x^2}e^{-y^2}\,\mathrm{d}x\mathrm{d}y=\int_{-\infty}^{\infty} e^{-x^2}\mathrm{d}x\int_{-\infty}^{\infty} e^{-y^2}\mathrm{d}y.$$

Since x and y are the dummy variables of integration, we denote them by s, to write

$$I=\left(\int_{-\infty}^{\infty} e^{-s^2}\mathrm{d}s\right)^2,$$

where the integral in parentheses is integral (6.12.25). Thus,

$$\int_{-\infty}^{\infty} e^{-s^2}\mathrm{d}s=\sqrt{I}=\sqrt{\pi},$$

which is the required result.

EXERCISE 6.13. Solve equations (6.8.7) and (6.8.10) for f_1 in terms of w.

SOLUTION 6.13. Solving equation (6.8.7), namely,

$$v_1f_1'(v_1t)-v_1w'(-v_1t)=-v_2g_2'(-v_2t),$$

for g_2', we get

$$g_2'(-v_2 t) = \frac{v_1}{v_2}\left[w'(-v_1 t) - f_1'(v_1 t)\right].$$

Substituting into equation (6.8.10) and simplifying, we obtain

$$(\rho_1 v_1 + \rho_2 v_2) f_1'(v_1 t) = (\rho_2 v_2 - \rho_1 v_1) w'(-v_1 t).$$

Integrating both sides with respect to t and simplifying, we get

$$f_1(v_1 t) = \frac{\rho_1 v_1 - \rho_2 v_2}{\rho_1 v_1 + \rho_2 v_2} w(-v_1 t),$$

as required.

EXERCISE 6.14. [38]Find the derivative of Heaviside's step, which is given by

$$h(t) = \begin{cases} 0, & t < 0 \\ 1, & t \geqslant 0 \end{cases}.$$

SOLUTION 6.14. Using expression (6.9.1), we get

$$\int_{-\infty}^{\infty} h'(t)\,\tau(t)\,\mathrm{d}t = -\int_{0}^{\infty} \tau'(t)\,\mathrm{d}t = -\left.\tau(t)\right|_0^\infty.$$

Since τ has compact support, $\tau(\infty) = 0$ and, hence, $-\left.\tau(t)\right|_0^\infty = \left.\tau(t)\right|_{t=0}$. Thus we can write the derivative of Heaviside's step as

$$\int_{-\infty}^{\infty} h'(t)\,\tau(t)\,\mathrm{d}t = \left.\tau(t)\right|_{t=0}, \tag{6.12.27}$$

which is a well-defined quantity.

REMARK 6.12.3. To interpret this result, let us consider the key property defining Dirac's delta, δ, namely,

$$\int_{-\infty}^{\infty} \delta(t)\,\tau(t)\,\mathrm{d}t = \left.\tau(t)\right|_{t=0}. \tag{6.12.28}$$

[38] See also Section 6.9.2.

Examining expressions (6.12.27) and (6.12.28), we can write

$$\int_{-\infty}^{\infty} h'(t)\,\tau(t)\,\mathrm{d}t = \int_{-\infty}^{\infty} \delta(t)\,\tau(t)\,\mathrm{d}t\,,$$

which means that $h'(t)$ behaves like $\delta(t)$. In the distributional sense, Dirac's delta can be regarded as the derivative of Heaviside's step.

The result of Exercise 6.14 is consistent with our intuition developed in standard differentiation. For $t \neq 0$, the derivative vanishes as it does in classical calculus, namely, $h'(t) = 0$. At the jump discontinuity, where $t = 0$, the derivative corresponds to Dirac's delta, which again is quite intuitive if we regard the value of δ at this point as being infinite.

EXERCISE 6.15. Consider wave equation (6.5.1). Using solution (6.10.2), obtain equation (6.10.4).

SOLUTION 6.15. In view of solution (6.10.2), namely, $u(x,t) = \acute{u}(x)\exp(-i\omega t)$, consider the position derivatives, namely,

$$\frac{\partial^2 u(x,t)}{\partial x^2} = \frac{\partial^2 \acute{u}(x)}{\partial x^2}\exp(-i\omega t)\,, \tag{6.12.29}$$

and the time derivatives, namely,

$$\frac{\partial^2 u(x,t)}{\partial t^2} = -\omega^2 \acute{u}(x)\exp(-i\omega t)\,. \tag{6.12.30}$$

Substituting expressions (6.12.29) and (6.12.30) into equation (6.5.1), and dividing by the exponential factor, we obtain a function of a single variable,

$$\frac{\mathrm{d}^2 \acute{u}(x)}{\mathrm{d}x^2} + \left(\frac{\omega}{v}\right)^2 \acute{u}(x) = 0\,,$$

which is equation (6.10.4), as required.

EXERCISE 6.16. Consider equation (6.11.3). In view of transformation (6.11.4), let

$$\acute{u}(x,z,t) \equiv u(\xi,\varsigma,t)\,, \tag{6.12.31}$$

where $\xi := x/v_x$ and $\varsigma := z/v_z$. Using the chain rule, show that equation (6.11.3) is equivalent to equation (6.11.5).

SOLUTION 6.16. Taking the derivative of both sides of equation (6.12.31) with respect to x, we obtain

$$\frac{\partial \acute{u}(x,z,t)}{\partial x} = \frac{\partial u}{\partial \xi}\frac{\partial \xi}{\partial x} = \frac{1}{v_x}\frac{\partial u(\xi,\varsigma,t)}{\partial \xi},$$

and

$$\frac{\partial^2 \acute{u}(x,z,t)}{\partial x^2} = \frac{1}{v_x^2}\frac{\partial^2 u(\xi,\varsigma,t)}{\partial \xi^2}. \tag{6.12.32}$$

Similarly, taking the derivative of both sides of equation (6.12.31) with respect to z, we obtain

$$\frac{\partial^2 \acute{u}(x,z,t)}{\partial z^2} = \frac{1}{v_z^2}\frac{\partial^2 u(\xi,\varsigma,t)}{\partial \varsigma^2}, \tag{6.12.33}$$

while, taking the derivative of both sides of equation (6.12.31) with respect to t, we get

$$\frac{\partial^2 \acute{u}(x,z,t)}{\partial t^2} = \frac{\partial^2 u(\xi,\varsigma,t)}{\partial t^2}. \tag{6.12.34}$$

We can always write equation (6.11.3) as

$$\frac{\partial^2 u(\xi,\varsigma,t)}{\partial \xi^2} + \frac{\partial^2 u(\xi,\varsigma,t)}{\partial \varsigma^2} = \frac{\partial^2 u(\xi,\varsigma,t)}{\partial t^2}, \tag{6.12.35}$$

where ξ and ς are the variables of differentiation. Substituting expressions from equations (6.12.32), (6.12.33) and (6.12.34) into (6.12.35), we obtain equation (6.11.5), as required.

EXERCISE 6.17. [39]Consider a three-dimensional scalar wave equation given by

$$\nabla^2 u \equiv \frac{\partial^2 u}{\partial x_1^2} + \frac{\partial^2 u}{\partial x_2^2} + \frac{\partial^2 u}{\partial x_3^2} = \frac{1}{v^2}\frac{\partial^2 u}{\partial t^2},$$

where v is the velocity of propagation and t is time. Let the plane-wave solution be $u(\mathbf{x},t) = f(\eta)$, where $\eta = n_1x_1 + n_2x_2 + n_3x_3 - vt$, with n_i being the components of the unit vector that is normal to the wavefront. Show that the plane-wave solution of the wave equation is also a solution

[39] See also Section 6.11.4 and Exercise 7.7.

of its characteristic equation, given by

$$(\nabla u)^2 \equiv \left(\frac{\partial u}{\partial x_1}\right)^2 + \left(\frac{\partial u}{\partial x_2}\right)^2 + \left(\frac{\partial u}{\partial x_3}\right)^2 = \frac{1}{v^2}\left(\frac{\partial u}{\partial t}\right)^2 . \quad (6.12.36)$$

SOLUTION 6.17. Considering the plane-wave solution, we obtain

$$\frac{\partial u}{\partial x_i} = \frac{\partial f}{\partial \eta}\frac{\partial \eta}{\partial x_i} = \frac{\partial f}{\partial \eta} n_i ,$$

and

$$\frac{\partial u}{\partial t} = \frac{\partial f}{\partial \eta}\frac{\partial \eta}{\partial t} = -\frac{\partial f}{\partial \eta} v .$$

Substituting $\partial u/\partial x_i$ and $\partial u/\partial t$ into equation (6.12.36), we can write

$$\left(\frac{\partial f}{\partial \eta} n_1\right)^2 + \left(\frac{\partial f}{\partial \eta} n_2\right)^2 + \left(\frac{\partial f}{\partial \eta} n_3\right)^2 = \frac{1}{v^2}\left(-\frac{\partial f}{\partial \eta} v\right)^2 ,$$

which yields

$$\left(\frac{\partial f}{\partial \eta}\right)^2 \left(n_1^2 + n_2^2 + n_3^2\right) = \left(\frac{\partial f}{\partial \eta}\right)^2 .$$

This equality is justified since for the unit vector, $\mathbf{n}$, we have $n_1^2 + n_2^2 + n_3^2 = 1$.

EXERCISE 6.18. Show that the characteristic equation of wave equation (6.5.1) is, in general, satisfied by $u(x,t) = f(\eta)$, where $\eta = x \pm vt$.

SOLUTION 6.18. We could view the solution of this exercise as a special case of Exercise 6.17. However, if we wish to write it out, we get

$$\frac{\partial u}{\partial x} = \frac{\partial f}{\partial \eta}\frac{\partial \eta}{\partial x} = \frac{\partial f}{\partial \eta}$$

and

$$\frac{\partial u}{\partial t} = \frac{\partial f}{\partial \eta}\frac{\partial \eta}{\partial t} = \pm v \frac{\partial f}{\partial \eta} .$$

In view of equation (6.12.36), the characteristic equation of equation (6.5.1) is

$$\left(\frac{\partial u}{\partial x}\right)^2 - \frac{1}{v^2}\left(\frac{\partial u}{\partial t}\right)^2 = 0 . \quad (6.12.37)$$

Inserting the above expressions for $\partial u/\partial x$ and $\partial u/\partial t$ into the left-hand side of equation (6.12.37), we get

$$\left(\frac{\partial u}{\partial x}\right)^2-\frac{1}{v^2}\left(\frac{\partial u}{\partial t}\right)^2=\left(\frac{\partial f}{\partial \eta}\right)^2-\frac{1}{v^2}\left(\pm v\frac{\partial f}{\partial \eta}\right)^2 .$$

Performing algebraic manipulations, we obtain

$$\left(\frac{\partial f}{\partial \eta}\right)^2-\left(\frac{\partial f}{\partial \eta}\right)^2=0,$$

as required. Thus, functions $f(x\pm vt)$ are solutions of equation (6.12.37), which is the characteristic equation of wave equation (6.5.1).

REMARK 6.12.4. We note that although both wave equation (6.5.1) and characteristic equation (6.12.37) have a general solution that possesses the same form, these equations are not equivalent to one another. Mathematically, equation

$$\frac{\partial u}{\partial x}=\pm\frac{1}{v}\frac{\partial u}{\partial t},$$

whose solutions are functions $u(x,t)=f(x\pm vt)$, does not suffice to describe such wave phenomena as reflection. Also, in the context of this chapter, the above equation is not the result of inserting stress-strain equations into Cauchy's equations of motion. Hence, the above equation is not rooted in physical concepts of the balance of linear momentum.

EXERCISE 6.19. [40]Show that expressions (6.5.4) are particular solutions of the characteristic equation of wave equation (6.5.1).

SOLUTION 6.19. We could view the solution of this exercise as a special case of Exercise 6.18. However, if we wish to write it out, inserting into the left-hand side of characteristic equation (6.12.37) expressions (6.5.4), which we can write as $u(x,t)=x\pm vt$, we get

$$\left[\frac{\partial (x\pm vt)}{\partial x}\right]^2-\frac{1}{v^2}\left[\frac{\partial (x\pm vt)}{\partial t}\right]^2=(1)^2-\frac{1}{v^2}(\pm v)^2 .$$

[40]See also Section 6.5.1.

Performing algebraic manipulations, we obtain $1 - 1 = 0$, as required. Thus, expressions (6.5.4) are solutions of equation (6.12.37), which is the characteristic equation of wave equation (6.5.1).

EXERCISE 6.20. In view of Section 6.10, considering the reduced form of equation (6.11.13) in a single spatial dimension and using the trial solution given by $u(\mathbf{x}) = A(\mathbf{x}) \exp[i\omega\psi(\mathbf{x})]$, obtain system (6.11.20).

SOLUTION 6.20. Considering a single spatial dimension, the reduced form of equation (6.11.13) is

$$\frac{\mathrm{d}^2 u(x)}{\mathrm{d}x^2} + \left[\frac{\omega}{v(x)}\right]^2 u(x) = 0.$$

Inserting a one-dimensional form of the given trial solution into this equation, performing the differentiation and dividing both sides of the resulting equation by the exponential term, we obtain

$$\frac{\mathrm{d}^2 A}{\mathrm{d}x^2} + A\omega^2 \left(\frac{1}{v^2} - \frac{\mathrm{d}\psi}{\mathrm{d}x}\frac{\mathrm{d}\psi}{\mathrm{d}x}\right) + i\omega \left[2\frac{\mathrm{d}A}{\mathrm{d}x}\frac{\mathrm{d}\psi}{\mathrm{d}x} + A\frac{\mathrm{d}^2\psi}{\mathrm{d}x^2}\right] = 0,$$

which is analogous to equation (6.11.19) and, hence, leads to system (6.11.20), as required.

EXERCISE 6.21. [41]Using stress-strain equations (6.1.1) and Cauchy's equations of motion (6.1.2), obtain equations of motion for an isotropic inhomogeneous continuum. Discuss these equations in the context of equations (6.1.4).

SOLUTION 6.21. Considering an inhomogeneous continuum, where Lamé's parameters are functions of position, and in view of definition (1.4.6), we can write equations (6.1.1) as

$$\sigma_{ij} = \lambda(\mathbf{x})\,\delta_{ij} \sum_{k=1}^{3} \frac{\partial u_k}{\partial x_k} + \mu(\mathbf{x}) \left(\frac{\partial u_i}{\partial x_j} + \frac{\partial u_j}{\partial x_i}\right), \qquad i, j \in \{1, 2, 3\}, \tag{6.12.38}$$

which are stress-strain equations for an isotropic inhomogeneous continuum. Considering an inhomogeneous continuum, where mass density is a

[41] See also Sections 6.1.1 and 6.11.4.

function of position, we can write equations (6.1.2) as

$$\rho(\mathbf{x})\frac{\partial^2 u_i}{\partial t^2} = \sum_{j=1}^{3}\frac{\partial \sigma_{ij}}{\partial x_j}, \qquad i \in \{1,2,3\}, \tag{6.12.39}$$

which are equations of motion for an isotropic inhomogeneous continuum. Using equations (6.12.38), we can write equations (6.12.39) as

$$\begin{aligned}\rho(\mathbf{x})\frac{\partial^2 u_i}{\partial t^2} &= \sum_{j=1}^{3}\frac{\partial}{\partial x_j}\left[\lambda(\mathbf{x})\delta_{ij}\sum_{k=1}^{3}\frac{\partial u_k}{\partial x_k} + \mu(\mathbf{x})\left(\frac{\partial u_i}{\partial x_j}+\frac{\partial u_j}{\partial x_i}\right)\right] \\ &= \sum_{j=1}^{3}\frac{\partial}{\partial x_j}\left[\lambda(\mathbf{x})\delta_{ij}\sum_{k=1}^{3}\frac{\partial u_k}{\partial x_k}\right] + \sum_{j=1}^{3}\frac{\partial}{\partial x_j}\left[\mu(\mathbf{x})\left(\frac{\partial u_i}{\partial x_j}+\frac{\partial u_j}{\partial x_i}\right)\right],\end{aligned}$$

where $i \in \{1,2,3\}$. Using the property of Kronecker's delta, we obtain

$$\rho(\mathbf{x})\frac{\partial^2 u_i}{\partial t^2} = \frac{\partial}{\partial x_i}\left[\lambda(\mathbf{x})\sum_{k=1}^{3}\frac{\partial u_k}{\partial x_k}\right] + \sum_{j=1}^{3}\frac{\partial}{\partial x_j}\left[\mu(\mathbf{x})\left(\frac{\partial u_i}{\partial x_j}+\frac{\partial u_j}{\partial x_i}\right)\right],$$

where $i \in \{1,2,3\}$. Letting $k = j$ for the summation index, we can write

$$\rho(\mathbf{x})\frac{\partial^2 u_i}{\partial t^2} = \frac{\partial}{\partial x_i}\left[\lambda(\mathbf{x})\sum_{j=1}^{3}\frac{\partial u_j}{\partial x_j}\right] + \sum_{j=1}^{3}\frac{\partial}{\partial x_j}\left[\mu(\mathbf{x})\left(\frac{\partial u_i}{\partial x_j}+\frac{\partial u_j}{\partial x_i}\right)\right],$$

where $i \in \{1,2,3\}$. Using the product rule and the linearity of differential operators, we obtain

$$\begin{aligned}\rho(\mathbf{x})\frac{\partial^2 u_i}{\partial t^2} &= \frac{\partial \lambda}{\partial x_i}\sum_{j=1}^{3}\frac{\partial u_j}{\partial x_j} + \lambda\frac{\partial}{\partial x_i}\sum_{j=1}^{3}\frac{\partial u_j}{\partial x_j} \\ &\quad + \sum_{j=1}^{3}\frac{\partial \mu}{\partial x_j}\left(\frac{\partial u_i}{\partial x_j}+\frac{\partial u_j}{\partial x_i}\right) + \mu\sum_{j=1}^{3}\left(\frac{\partial}{\partial x_j}\frac{\partial u_i}{\partial x_j}+\frac{\partial}{\partial x_j}\frac{\partial u_j}{\partial x_i}\right),\end{aligned}$$

where $i \in \{1,2,3\}$. Differentiating and using the equality of mixed partial derivatives, we obtain

$$\rho(\mathbf{x})\frac{\partial^2 u_i}{\partial t^2} = \frac{\partial \lambda}{\partial x_i}\sum_{j=1}^{3}\frac{\partial u_j}{\partial x_j} + \lambda\sum_{j=1}^{3}\frac{\partial^2 u_j}{\partial x_i \partial x_j} + \sum_{j=1}^{3}\frac{\partial \mu}{\partial x_j}\left(\frac{\partial u_i}{\partial x_j} + \frac{\partial u_j}{\partial x_i}\right)$$
$$+\mu\sum_{j=1}^{3}\left(\frac{\partial^2 u_i}{\partial x_j^2} + \frac{\partial^2 u_j}{\partial x_i \partial x_j}\right),$$

where $i \in \{1,2,3\}$. Simplifying and rearranging, we get

$$\rho(\mathbf{x})\frac{\partial^2 u_i}{\partial t^2} = (\lambda+\mu)\frac{\partial}{\partial x_i}\sum_{j=1}^{3}\frac{\partial u_j}{\partial x_j} + \mu\left(\sum_{j=1}^{3}\frac{\partial^2}{\partial x_j^2}\right)u_i + \frac{\partial \lambda}{\partial x_i}\sum_{j=1}^{3}\frac{\partial u_j}{\partial x_j}$$
$$+\sum_{j=1}^{3}\frac{\partial \mu}{\partial x_j}\left(\frac{\partial u_i}{\partial x_j} + \frac{\partial u_j}{\partial x_i}\right),$$

where $i \in \{1,2,3\}$. These are equations of motion for an isotropic inhomogeneous continuum.[42]

Examining equations (6.12.40), we notice that if ρ, λ and μ are constants, as is the case for homogeneous continua, equations (6.12.40) reduce to equations (6.1.4), as expected. Also we notice that invoking definitions (1.4.18) and (1.5.2) as well as identity (6.1.7) we can express the displacement vector, $\mathbf{u}$, in the first three terms on the right-hand side, using the dilatation, φ, and the rotation vector, Ψ. Investigating the last term on the right-hand side, however, it can be shown that we cannot express the displacement vector on the right-hand side of equations (6.12.39) using only φ and Ψ. Consequently, we cannot split equations (6.12.40) into two parts that are associated with the dilatation alone and with the rotation vector alone, respectively, as we did in Sections 6.1.2 and 6.1.3 in the case of isotropic homogeneous continua. In other words, the dilatational and rotational waves are coupled due to the inhomogeneity of the continuum.

[42]Readers interested in a solution to these equations might refer to Karal, F.C., and Keller, J.B., (1959) Elastic wave propagation in homogeneous and inhomogeneous media: J. Acoust. Soc. Am., **31** (6), 694–705.

CHAPTER 7

Equations of Motion: Anisotropic Inhomogeneous Continua

> ... an exact solution to a problem in wave phenomena is not an end in itself. Rather, it is the asymptotic solution that provides means of interpretation and a basis for understanding. The exact solution, then, only provides a point of departure for obtaining a meaningful solution.
>
> *Norman Bleistein (1984)*
> *Mathematical methods for wave phenomena*

Preliminary Remarks

In Chapter 6, to study wave phenomena in an isotropic homogeneous continuum, we obtained the equations of motion by invoking Cauchy's equations of motion and using stress-strain equations that correspond to such a continuum. In this chapter, we study wave phenomena in an anisotropic inhomogeneous continuum by following a strategy analogous to that used in Chapter 6. In this study, we learn about the existence of three types of waves that can propagate in anisotropic continua.

We begin this chapter with the derivation of the equations of motion in an anisotropic inhomogeneous continuum. We obtain these equations by combining Cauchy's equations of motion with the stress-strain equations for an anisotropic inhomogeneous continuum. To solve the resulting equations, we use a trial solution. Subsequently, we derive the eikonal equation for anisotropic inhomogeneous continua, which is the fundamental equation of ray theory, to be studied in the subsequent chapters of *Part 2* and in *Part 3*.

7.1. Formulation of Equations

In Chapter 6, the wave equation is derived by considering Cauchy's equations of motion (2.8.1) and expressing the stress-tensor components therein in terms of stress-strain equations (5.12.5), which describe an isotropic homogeneous continuum. In the present chapter, we derive the equations of motion for an anisotropic inhomogeneous continuum by considering Cauchy's equations of motion (2.8.1) and expressing the stress-tensor components therein in terms of the stress-strain equations that describe an anisotropic inhomogeneous continuum.

In view of equations of motion (2.8.1), consider equations

$$\rho(\mathbf{x})\frac{\partial^2 u_i}{\partial t^2}=\sum_{j=1}^{3}\frac{\partial\sigma_{ij}}{\partial x_j},\qquad i\in\{1,2,3\}. \tag{7.1.1}$$

In equations (7.1.1), due to the inhomogeneity of the continuum, mass density, $\rho(\mathbf{x})$, is a function of position; in contrast to for a homogeneous continuum, where ρ is constant. The components of the displacement vector, $\mathbf{u}$, as well as the stress-tensor components, σ_{ij}, are also functions of position. Since $\mathbf{u}$ and σ_{ij} are also functions of $\mathbf{x}$ in homogeneous continua, to emphasize the distinction between a homogeneous and an inhomogeneous continuum, we explicitly state the $\mathbf{x}$-dependence only for the mass density, ρ, and for the elasticity tensor, $c_{ijk\ell}$, below. The stress-strain equations that account for the inhomogeneity of the continuum are expressed by rewriting equations (3.2.1) as

$$\sigma_{ij}=\sum_{k=1}^{3}\sum_{\ell=1}^{3}c_{ijk\ell}(\mathbf{x})\,\varepsilon_{k\ell},\qquad i,j\in\{1,2,3\}, \tag{7.1.2}$$

where the elasticity tensor is a function of position. In view of the properties of the elasticity tensor that were discussed in Chapter 5, $c_{ijk\ell}$ also describes the anisotropy of a given continuum. Thus, $\rho(\mathbf{x})$ and $c_{ijk\ell}(\mathbf{x})$

fully describe a given anisotropic inhomogeneous continuum. Using definition (1.4.6), we can rewrite stress-strain equations (7.1.2) as

$$\sigma_{ij} = \frac{1}{2}\sum_{k=1}^{3}\sum_{\ell=1}^{3} c_{ijk\ell}(\mathbf{x})\left(\frac{\partial u_k}{\partial x_\ell} + \frac{\partial u_\ell}{\partial x_k}\right), \qquad i,j \in \{1,2,3\}. \tag{7.1.3}$$

We wish to combine the equations of motion and the stress-strain equations to obtain the equations of motion in an anisotropic inhomogeneous continuum. Inserting stress-strain equations (7.1.3) into equations of motion (7.1.1), we obtain

$$\begin{aligned}
\rho(\mathbf{x})\frac{\partial^2 u_i}{\partial t^2} &= \sum_{j=1}^{3}\frac{\partial}{\partial x_j}\left[\frac{1}{2}\sum_{k=1}^{3}\sum_{\ell=1}^{3} c_{ijk\ell}(\mathbf{x})\left(\frac{\partial u_k}{\partial x_\ell} + \frac{\partial u_\ell}{\partial x_k}\right)\right] \\
&= \frac{1}{2}\sum_{j=1}^{3}\sum_{k=1}^{3}\sum_{\ell=1}^{3}\left[\frac{\partial c_{ijk\ell}(\mathbf{x})}{\partial x_j}\left(\frac{\partial u_k}{\partial x_\ell} + \frac{\partial u_\ell}{\partial x_k}\right)\right. \\
&\quad \left. + c_{ijk\ell}(\mathbf{x})\left(\frac{\partial^2 u_k}{\partial x_j \partial x_\ell} + \frac{\partial^2 u_\ell}{\partial x_j \partial x_k}\right)\right],
\end{aligned} \tag{7.1.4}$$

where $i \in \{1,2,3\}$. Equations (7.1.4) are equations of motion in anisotropic inhomogeneous continua. For isotropic continua, these equations reduce to equations (6.12.40), and for isotropic homogeneous continua, they further reduce to equations (6.1.4). Equations (7.1.4) are complicated differential equations and, in general, we are unable to find their solutions analytically.

7.2. Formulation of Solutions

7.2.1. Introductory comments

Below, we formulate solutions to equations (7.1.4). To elucidate the meaning of our trial solution, we show three related approaches that lead to similar results.

7.2.2. Trial-solution formulation: General wave

To investigate equations (7.1.4), let us consider the trial solution that is a function of position, $\mathbf{x}$, and time, t, given by

$$\mathbf{u}(\mathbf{x},t) = \mathbf{A}(\mathbf{x}) f(\eta), \tag{7.2.1}$$

where $\mathbf{A}$ is a vector function of position, $\mathbf{x}$, and f is a scalar function whose argument is given by

$$\eta = \omega[\psi(\mathbf{x}) - t], \tag{7.2.2}$$

with ω being a constant with units of frequency. Function $\psi : \mathbb{R}^3 \to \mathbb{R}$ is called the eikonal function. Since both $\mathbf{u}$ and $\mathbf{A}$ have units of distance, we require f to be dimensionless. Since commonly $f(\eta)$ is an exponential function, as in expression (6.11.15), we made η dimensionless.

To see the physical meaning of this trial solution, consider function $f(\eta)$ in the context of trial solutions (6.11.6), (6.11.14) and (6.11.15). We see that f corresponds to the phase factor. Since along the level sets of $\psi(\mathbf{x})$, function f is constant, these level sets correspond to wavefronts. In other words, equation

$$\psi(\mathbf{x}) - t = t_i, \tag{7.2.3}$$

where t_i is a constant, describes a moving wavefront. Function f gives the waveform as a function of time, and $\mathbf{A}$ is the spatially variable amplitude of this waveform.

Inserting trial solution (7.2.1) into equations (7.1.4), using the symmetries of the elasticity tensor, $c_{ijk\ell}$, and the equality of mixed partial derivatives, we obtain

$$\omega^2 \rho(\mathbf{x}) A_i(\mathbf{x}) \frac{\mathrm{d}^2 f}{\mathrm{d}\eta^2}$$

$$= \sum_{j=1}^{3}\sum_{k=1}^{3}\sum_{\ell=1}^{3}\left\{ f\left[\frac{\partial c_{ijk\ell}(\mathbf{x})}{\partial x_j}\frac{\partial A_k}{\partial x_\ell} + c_{ijk\ell}(\mathbf{x})\frac{\partial^2 A_k}{\partial x_j \partial x_\ell}\right]\right.$$

$$+\omega\frac{\mathrm{d}f}{\mathrm{d}\eta}\left[\frac{\partial c_{ijk\ell}(\mathbf{x})}{\partial x_j}A_\ell\frac{\partial \psi}{\partial x_k} c_{ijk\ell}(\mathbf{x})\left(\frac{\partial A_\ell}{\partial x_j}\frac{\partial \psi}{\partial x_k} + \frac{\partial A_k}{\partial x_\ell}\frac{\partial \psi}{\partial x_j} + A_\ell\frac{\partial^2 \psi}{\partial x_j \partial x_k}\right)\right]$$

$$\left. +\omega^2\frac{\mathrm{d}^2 f}{\mathrm{d}\eta^2}\left[c_{ijk\ell}(\mathbf{x}) A_k \frac{\partial \psi}{\partial x_j}\frac{\partial \psi}{\partial x_\ell}\right]\right\},$$

where $i \in \{1,2,3\}$. Taking all terms to the right-hand side of the equations and, therein, gathering the terms with the same derivatives of f, we get

$$0 = \sum_{j=1}^{3}\sum_{k=1}^{3}\sum_{\ell=1}^{3}\left\{ f\left[\frac{\partial c_{ijk\ell}(\mathbf{x})}{\partial x_j}\frac{\partial A_k}{\partial x_\ell} + c_{ijk\ell}(\mathbf{x})\frac{\partial^2 A_k}{\partial x_j \partial x_\ell}\right]\right.$$

$$+\omega\frac{\mathrm{d}f}{\mathrm{d}\eta}\left[\frac{\partial}{\partial x_j}\left(c_{ijk\ell}(\mathbf{x}) A_\ell \frac{\partial \psi}{\partial x_k}\right) + c_{ijk\ell}(\mathbf{x})\frac{\partial A_k}{\partial x_\ell}\frac{\partial \psi}{\partial x_j}\right]$$

$$\left. +\ \omega^2\frac{\mathrm{d}^2 f}{\mathrm{d}\eta^2}\left[c_{ijk\ell}(\mathbf{x}) A_k \frac{\partial \psi}{\partial x_j}\frac{\partial \psi}{\partial x_\ell}\right]\right\} - \omega^2 \rho(\mathbf{x}) A_i(\mathbf{x}) \frac{\mathrm{d}^2 f}{\mathrm{d}\eta^2}, \qquad (7.2.4)$$

where $i \in \{1,2,3\}$.

Concisely, equation (7.2.4) can be stated as

$$a(\mathbf{x}) f'' + b(\mathbf{x}) f' + c(\mathbf{x}) f = 0,$$

where, in view of $f = f(\eta)$ being a single-variable function, we write $f'' := \mathrm{d}^2 f/\mathrm{d}\eta^2$ and $f' := \mathrm{d}f/\mathrm{d}\eta$. For equation (7.2.4) to be satisfied by an arbitrary f with its first and second derivatives, each of the coefficients—$a(\mathbf{x})$, $b(\mathbf{x})$ and $c(\mathbf{x})$—must be zero.[1] As we can see from the physical meaning of trial solution (7.2.1), we require the arbitrariness of f in order to allow any function to describe the waveform. However, we need f, f' and f'' to be linearly independent, as illustrated in Exercise 7.1.

Setting each of the three coefficients to zero and assuming that $\omega \neq 0$, we obtain three systems of three equations, namely,

$$\sum_{j=1}^{3}\sum_{k=1}^{3}\sum_{\ell=1}^{3} c_{ijk\ell}(\mathbf{x})\frac{\partial\psi}{\partial x_j}\frac{\partial\psi}{\partial x_\ell}A_k - \rho(\mathbf{x})A_i = 0, \qquad i \in \{1,2,3\}, \quad (7.2.5)$$

$$\sum_{j=1}^{3}\sum_{k=1}^{3}\sum_{\ell=1}^{3}\left[\frac{\partial}{\partial x_j}\left(c_{ijk\ell}(\mathbf{x})A_\ell\frac{\partial\psi}{\partial x_k}\right) + c_{ijk\ell}(\mathbf{x})\frac{\partial A_k}{\partial x_\ell}\frac{\partial\psi}{\partial x_j}\right] = 0, \quad i \in \{1,2,3\} \quad (7.2.6)$$

and

$$\sum_{j=1}^{3}\sum_{k=1}^{3}\sum_{\ell=1}^{3}\left(\frac{\partial c_{ijk\ell}(\mathbf{x})}{\partial x_j}\frac{\partial A_k}{\partial x_\ell} + c_{ijk\ell}(\mathbf{x})\frac{\partial^2 A_k}{\partial x_j \partial x_\ell}\right) = 0, \qquad i \in \{1,2,3\}, \quad (7.2.7)$$

which correspond to $a(\mathbf{x})$, $b(\mathbf{x})$ and $c(\mathbf{x})$, respectively. Equations (7.2.5), (7.2.6) and (7.2.7) constitute an overdetermined system of equations for $\psi(\mathbf{x})$ and $\mathbf{A}(\mathbf{x})$ that results from inserting trial solution (7.2.1) into equations (7.1.4). Given $\psi(\mathbf{x})$, an exact solution of form (7.2.1) requires function $\mathbf{A}(\mathbf{x})$ to satisfy equations (7.2.5), (7.2.6) and (7.2.7) to allow any form of f in solution (7.2.1). In general, this is impossible; particular conditions must be imposed on $\mathbf{A}$, as illustrated in Exercises 7.2 and 7.3. Such restrictions are avoided by using the asymptotic-series as a trial solution, which is introduced in Section 7.2.4.

7.2.3. Trial-solution formulation: Harmonic wave

Herein, we follow the approach analogous to the one used in Section 6.11.4. It allows us to illustrate that equations (7.2.5), (7.2.6) and (7.2.7) result from the high-frequency approximation.

[1] Readers interested in an analogous formulation of the three vanishing terms of equation (7.2.4) might refer to Červený, V., (2001) Seismic ray theory: Cambridge University Press, p. 55, pp. 57–58 and pp. 62–63.

Recall equations (7.1.4); namely,

$$\rho(\mathbf{x})\frac{\partial^2 u_i(\mathbf{x},t)}{\partial t^2} = \frac{1}{2}\sum_{j=1}^{3}\sum_{k=1}^{3}\sum_{\ell=1}^{3}\left[\frac{\partial c_{ijk\ell}(\mathbf{x})}{\partial x_j}\left(\frac{\partial u_k}{\partial x_\ell}+\frac{\partial u_\ell}{\partial x_k}\right)\right.$$
$$\left.+c_{ijk\ell}(\mathbf{x})\left(\frac{\partial^2 u_k}{\partial x_j\partial x_\ell}+\frac{\partial^2 u_\ell}{\partial x_j\partial x_k}\right)\right], \qquad (7.2.8)$$

where $i\in\{1,2,3\}$. We wish equations (7.2.8) to be expressed in terms of $\mathbf{x}$ and ω in order to consider the limit of ω tending to infinity. For this purpose we perform Fourier's transform of equations (7.2.8) with t and ω being the variables of transformation. Invoking Fourier's transform as illustrated in Section 6.10.2, we get

$$(i\omega)^2\rho(\mathbf{x})\tilde{u}_i(\mathbf{x},\omega) = \frac{1}{2}\sum_{j=1}^{3}\sum_{k=1}^{3}\sum_{\ell=1}^{3}\left[\frac{\partial c_{ijk\ell}(\mathbf{x})}{\partial x_j}\left(\frac{\partial \tilde{u}_k}{\partial x_\ell}+\frac{\partial \tilde{u}_\ell}{\partial x_k}\right)\right.$$
$$\left.+c_{ijk\ell}(\mathbf{x})\left(\frac{\partial^2 \tilde{u}_k}{\partial x_j\partial x_\ell}+\frac{\partial^2 \tilde{u}_\ell}{\partial x_j\partial x_k}\right)\right], \qquad (7.2.9)$$

where

$$\tilde{u}_i(\mathbf{x},\omega) := \frac{1}{2\pi}\int_{-\infty}^{\infty} u_i(\mathbf{x},t)\exp(-i\omega t)\,\mathrm{d}t,$$

and $i\in\{1,2,3\}$. In a manner analogous to those used in Sections 6.11.4 and 7.2.2, let us consider the trial solution given by

$$\tilde{\mathbf{u}}(\mathbf{x},\omega) = \mathbf{A}(\mathbf{x})\exp\{i\omega[\psi(\mathbf{x})]\}, \qquad (7.2.10)$$

which is a generalization of expression (6.11.15) and a particular case of expression (7.2.1). However, with trial solution (7.2.10), we cannot use exactly the method presented in Section 7.2.2, as illustrated in Exercise 7.1. This is due to the fact that—in view of expressions (7.2.1) and (7.2.2)—we see that $\eta = i\omega[\psi(\mathbf{x})]$, and hence $\mathrm{d}^2 f(\eta)/\mathrm{d}\eta^2$, $\mathrm{d}f(\eta)/\mathrm{d}\eta$ and $f(\eta)$ are not linearly independent; they are equal to each other. In the method presented below we take derivatives with respect to x_i, not with respect to the whole argument, η.

Inserting expression (7.2.10) into equations (7.2.9), differentiating and factoring out the common term, $\exp(i\omega\psi)$, we get

$$
\begin{aligned}
&-\omega^2 \exp[i\omega\psi(\mathbf{x})]\,\rho A_i \\
&= \frac{1}{2}\exp[i\omega\psi(\mathbf{x})] \sum_{j=1}^{3}\sum_{k=1}^{3}\sum_{\ell=1}^{3} \\
&\left[\frac{\partial c_{ijk\ell}}{\partial x_j}\left(\left(\frac{\partial A_k}{\partial x_\ell}+\frac{\partial A_\ell}{\partial x_k}+i\omega\left(\frac{\partial \psi}{\partial x_\ell}A_k+\frac{\partial \psi}{\partial x_k}A_\ell\right)\right)\right)\right. \\
&+c_{ijk\ell}\left(\left(\frac{\partial^2 A_k}{\partial x_j\partial x_\ell}+\frac{\partial^2 A_\ell}{\partial x_j\partial x_k}\right)+i\omega\left(\frac{\partial^2 \psi}{\partial x_j\partial x_\ell}A_k\right.\right. \\
&\left.+\frac{\partial \psi}{\partial x_j}\frac{\partial A_\ell}{\partial x_k}+\frac{\partial^2 \psi}{\partial x_j\partial x_k}A_\ell+\frac{\partial \psi}{\partial x_\ell}\frac{\partial A_k}{\partial x_j}+\frac{\partial \psi}{\partial x_k}\frac{\partial A_\ell}{\partial x_j}+\frac{\partial \psi}{\partial x_j}\frac{\partial A_\ell}{\partial x_k}\right) \\
&\left.\left.-\omega^2\left(\frac{\partial \psi}{\partial x_\ell}\frac{\partial \psi}{\partial x_j}A_k+\frac{\partial \psi}{\partial x_k}\frac{\partial \psi}{\partial x_j}A_\ell\right)\right)\right],
\end{aligned}
\tag{7.2.11}
$$

with $i \in \{1,2,3\}$. Canceling the exponential term and considering only the real part, we write

$$
\begin{aligned}
-\omega^2\rho A_i = \frac{1}{2}\sum_{j=1}^{3}\sum_{k=1}^{3}\sum_{\ell=1}^{3}&\left\{\frac{\partial c_{ijk\ell}}{\partial x_j}\left(\frac{\partial A_k}{\partial x_\ell}+\frac{\partial A_\ell}{\partial x_k}\right)\right. \\
&+c_{ijk\ell}\left[\left(\frac{\partial^2 A_k}{\partial x_j\partial x_\ell}+\frac{\partial^2 A_\ell}{\partial x_j\partial x_k}\right)\right. \\
&\left.\left.-\ \omega^2\left(\frac{\partial \psi}{\partial x_\ell}\frac{\partial \psi}{\partial x_j}A_k+\frac{\partial \psi}{\partial x_k}\frac{\partial \psi}{\partial x_j}A_\ell\right)\right]\right\},
\end{aligned}
$$

where $i \in \{1,2,3\}$. Dividing both sides by ω^2 and letting ω tend to infinity, we get

$$
\frac{1}{2}\sum_{j=1}^{3}\sum_{k=1}^{3}\sum_{\ell=1}^{3} c_{ijk\ell}(\mathbf{x})\left(\frac{\partial \psi}{\partial x_\ell}\frac{\partial \psi}{\partial x_j}A_k+\frac{\partial \psi}{\partial x_k}\frac{\partial \psi}{\partial x_j}A_\ell\right)-\rho A_i=0, \quad i \in \{1,2,3\}.
$$

Since j, k and ℓ are summation indices and $c_{ijk\ell}$ is symmetric in k and ℓ, we can write these three equations as

$$\sum_{j=1}^{3}\sum_{k=1}^{3}\sum_{\ell=1}^{3} c_{ijk\ell}(\mathbf{x})\frac{\partial\psi}{\partial x_\ell}\frac{\partial\psi}{\partial x_j}A_k - \rho(\mathbf{x})A_i = 0, \qquad i \in \{1,2,3\},$$

which are equations (7.2.5). In view of the limit of ω, the presented method is the high-frequency approximation.

Considering the imaginary part, we obtain

$$\sum_{j=1}^{3}\sum_{k=1}^{3}\sum_{\ell=1}^{3}\left[\frac{\partial c_{ijk\ell}}{\partial x_j}\left(\frac{\partial\psi}{\partial x_\ell}A_k + \frac{\partial\psi}{\partial x_k}A_\ell\right)\right.$$
$$+ c_{ijk\ell}\left(\frac{\partial^2\psi}{\partial x_j\partial x_\ell}A_k + \frac{\partial\psi}{\partial x_j}\frac{\partial A_\ell}{\partial x_k} + \frac{\partial^2\psi}{\partial x_j\partial x_k}A_\ell\right.$$
$$\left.\left. + \frac{\partial\psi}{\partial x_\ell}\frac{\partial A_k}{\partial x_j} + \frac{\partial\psi}{\partial x_k}\frac{\partial A_\ell}{\partial x_j} + \frac{\partial\psi}{\partial x_j}\frac{\partial A_\ell}{\partial x_k}\right)\right] = 0,$$

where $i \in \{1,2,3\}$, which are equations (7.2.6) with partial derivatives $\partial/\partial x_j$ taken.

Also examining equations (7.2.11), we see that—in a manner analogous to the one used in Section 7.2.2—we could also obtain equations (7.2.5) by setting the coefficients of ω^2 to zero.

7.2.4. Asymptotic-series formulation

To gain insight into the fact that trial solution (7.2.10) is an asymptotic series, and, hence, the resulting equations belong to the asymptotic ray theory, let us consider the following formulation. As discussed in Section 6.11.4, the trial solution is the zeroth-order term of an asymptotic series. Expression (7.2.10) is the zeroth-order term of

$$\tilde{\mathbf{u}}(\mathbf{x},\omega) \sim \exp[i\omega\psi(\mathbf{x})]\left\{\mathbf{A}_0(\mathbf{x}) + \frac{\mathbf{A}_1(\mathbf{x})}{i\omega} + \sum_{n=2}^{\infty}\frac{\mathbf{A}_n(\mathbf{x})}{(i\omega)^n}\right\}, \qquad (7.2.12)$$

where $\sim$ stands for "is asymptotically equivalent to." This is a series whose bases are $\exp(i\omega\psi)/(i\omega)^n$, where $n \in \{0,\ldots,\infty\}$. Series (7.2.12) is an asymptotic series of $\tilde{\mathbf{u}}$; herein, the defining property of the asymptotic series is

$$\lim_{\omega\to\infty}\left\{\frac{(i\omega)^N}{\exp\left[i\omega\psi(\mathbf{x})\right]}\left[\tilde{\mathbf{u}}(\mathbf{x},\omega)-\exp\left[i\omega\psi(\mathbf{x})\right]\sum_{n=0}^{N}\frac{\mathbf{A}_n(\mathbf{x})}{(i\omega)^n}\right]\right\}=0, \tag{7.2.13}$$

for all N. This statement implies that the difference between the exact solution and the partial sum of N terms of the series vanishes as ω tends to infinity—hence, the name "high-frequency approximation";[2] this difference tends to zero more rapidly than $(i\omega)^N/\exp(i\omega\psi)$ tends to infinity. For each function that has an asymptotic series, coefficients $\mathbf{A}_n$ are unique, as illustrated in Exercise 7.4.

Prior to inserting the asymptotic series into the equations of motion, let us explain our motivation for using series (7.2.12). In particular, let us discuss the importance of the exponential term. What is the reason for considering an asymptotic series given by the product of the exponential term and the summation,

$$\exp\left[i\omega\psi(\mathbf{x})\right]\sum_{n=0}^{N}\frac{\mathbf{A}_n(\mathbf{x})}{(i\omega)^n}, \tag{7.2.14}$$

rather than the summation itself? An insight into the physical reason is given by the meanings of ψ and $\mathbf{A}$, which are associated with wavefronts and amplitudes, respectively. This is, however, an insight achieved by examining the solutions to the equations of motion. The mathematical reason to begin with such a series to obtain the solution can be explained as follows. The bases of our asymptotic expansion, which we use in differential equations, are $\exp(i\omega\psi)/(i\omega)^n$. Consider the derivatives of expression (7.2.14). Each time we take a derivative of the exponential term with respect to x_i, we get $i\omega(\partial\psi/\partial x_i)\exp(i\omega\psi)$. The derivatives of expression (7.2.14) are linear combinations of $\exp(i\omega\psi)/(i\omega)^n$; we get

[2] Readers interested in the meaning of the high-frequency approximation might refer to Bos, L. and Slawinski, M.A., (2010) Elastodynamic equations: Characteristics, wavefronts and rays. The Quarterly Journal of Mechanics and Applied Mathematics **63**(1), 23–37 and to Bóna, A. and Slawinski, M.A., (2021) Wavefronts and rays as characteristics and asymptotics (3*rd* edition): World Scientific, Chapter 3. Note that herein the elastodynamic equations are tantamount to equations (7.1.4), which are Cauchy's equations of motion within Hookean solids.

such combinations even if we consider only the zeroth-order term of series (7.2.12), as we can see by examining expression (7.2.11).

Note that the units of the nth term in the denominator of summation (7.2.14) are s^{-n}, where s stands for seconds. Hence, the units of the numerator, $\mathbf{A}_n$, must change with n in such a way that all the terms in the series have the same units.

Inserting the first two terms of series (7.2.12) into the left-hand side of equations (7.2.9), we write

$$(i\omega)^2 \rho(\mathbf{x}) \exp[i\omega\psi(\mathbf{x})] \left[A_{0_i}(\mathbf{x}) + (i\omega)^{-1} A_{1_i}(\mathbf{x})\right], \qquad i \in \{1,2,3\}. \tag{7.2.15}$$

Inserting the first two terms of series (7.2.12) into a first-derivative term on the right-hand side of equations (7.2.9), we write

$$\frac{\partial \tilde{u}_k}{\partial x_\ell} = \frac{\partial}{\partial x_\ell}\left[\exp[i\omega\psi(\mathbf{x})]\left(A_{0_k} + \frac{A_{1_k}}{i\omega}\right)\right].$$

Differentiating, factoring out the common term, $\exp(i\omega\psi)$, and gathering the terms with common powers of $i\omega$, we get

$$\frac{\partial \tilde{u}_k}{\partial x_\ell} = \exp[i\omega\psi(\mathbf{x})]\left[(i\omega)^{-1}\left(\frac{\partial A_{1_k}}{\partial x_\ell}\right) + (i\omega)^0\left(\frac{\partial \psi}{\partial x_\ell}A_{1_k} + \frac{\partial A_{0_k}}{\partial x_\ell}\right) + (i\omega)^1\left(\frac{\partial \psi}{\partial x_\ell}A_{0_k}\right)\right]. \tag{7.2.16}$$

Performing the same operations for $\partial \tilde{u}_\ell / \partial x_k$, adding together the two results, and gathering the terms with common powers of $i\omega$, we get

$$\frac{\partial \tilde{u}_k}{\partial x_\ell} + \frac{\partial \tilde{u}_\ell}{\partial x_k} = \exp[i\omega\psi(\mathbf{x})]\left[(i\omega)^{-1}\left(\frac{\partial A_{1_k}}{\partial x_\ell} + \frac{\partial A_{1_\ell}}{\partial x_k}\right) + (i\omega)^0\left(\frac{\partial \psi}{\partial x_\ell}A_{1_k} + \frac{\partial \psi}{\partial x_k}A_{1_\ell} + \frac{\partial A_{0_k}}{\partial x_\ell} + \frac{\partial A_{0_\ell}}{\partial x_k}\right) + (i\omega)^1\left(\frac{\partial \psi}{\partial x_\ell}A_{0_k} + \frac{\partial \psi}{\partial x_k}A_{0_\ell}\right)\right].$$

To insert the first two terms of series (7.2.12) into a second-derivative term, we write

$$\frac{\partial^2 \tilde{u}_k}{\partial x_j \partial x_\ell} = \frac{\partial}{\partial x_j}\frac{\partial \tilde{u}_k}{\partial x_\ell},$$

where $\partial \tilde{u}_k / \partial x_\ell$ is given in expression (7.2.16). Differentiating, factoring out the common term, $\exp(i\omega\psi)$, and gathering the terms with common powers of $i\omega$, we get

$$\frac{\partial^2 \tilde{u}_k}{\partial x_j \partial x_\ell} = \exp\{i\omega[\psi(\mathbf{x})]\}\left[(i\omega)^{-1}\left(\frac{\partial^2 A_{1_k}}{\partial x_j \partial x_\ell}\right)\right.$$
$$+ (i\omega)^0\left(\frac{\partial^2\psi}{\partial x_j\partial x_\ell}A_{1_k} + \frac{\partial\psi}{\partial x_\ell}\frac{\partial A_{1_k}}{\partial x_j} + \frac{\partial\psi}{\partial x_j}\frac{\partial A_{1_k}}{\partial x_\ell} + \frac{\partial^2 A_{0_k}}{\partial x_j\partial x_\ell}\right)$$
$$+ (i\omega)^1\left(\frac{\partial^2\psi}{\partial x_j\partial x_\ell}A_{0_k} + \frac{\partial\psi}{\partial x_\ell}\frac{\partial\psi}{\partial x_j}A_{1_k} + \frac{\partial\psi}{\partial x_\ell}\frac{\partial A_{0_k}}{\partial x_j} + \frac{\partial\psi}{\partial x_j}\frac{\partial A_{0_k}}{\partial x_\ell}\right)$$
$$\left.+ (i\omega)^2\left(\frac{\partial\psi}{\partial x_\ell}\frac{\partial\psi}{\partial x_j}A_{0_k}\right)\right].$$

Performing the same operations for $\partial^2 \tilde{u}_\ell / \partial x_j \partial x_k$, adding together the two results, and gathering the terms with common powers of $i\omega$, we get

$$\frac{\partial^2 \tilde{u}_k}{\partial x_j \partial x_\ell} + \frac{\partial^2 \tilde{u}_\ell}{\partial x_j \partial x_k}$$
$$= \exp[i\omega\psi(\mathbf{x})]\left[(i\omega)^{-1}\left(\frac{\partial^2 A_{1_k}}{\partial x_j\partial x_\ell} + \frac{\partial^2 A_{1_\ell}}{\partial x_j\partial x_k}\right)\right.$$

$$+ (i\omega)^0\left(\frac{\partial^2\psi}{\partial x_j\partial x_\ell}A_{1_k} + \frac{\partial\psi}{\partial x_\ell}\frac{\partial A_{1_k}}{\partial x_j} + \frac{\partial\psi}{\partial x_j}\frac{\partial A_{1_k}}{\partial x_\ell}\right.$$
$$\left.+ \frac{\partial^2 A_{0_k}}{\partial x_j\partial x_\ell} + \frac{\partial^2\psi}{\partial x_j\partial x_k}A_{1_\ell} + \frac{\partial\psi}{\partial x_k}\frac{\partial A_{1_\ell}}{\partial x_j} + \frac{\partial\psi}{\partial x_j}\frac{\partial A_{1_\ell}}{\partial x_k} + \frac{\partial^2 A_{0_\ell}}{\partial x_j\partial x_k}\right)$$
$$+ (i\omega)^1\left(\frac{\partial^2\psi}{\partial x_j\partial x_\ell}A_{0_k} + \frac{\partial\psi}{\partial x_\ell}\frac{\partial\psi}{\partial x_j}A_{1_k} + \frac{\partial\psi}{\partial x_\ell}\frac{\partial A_{0_k}}{\partial x_j} + \frac{\partial\psi}{\partial x_j}\frac{\partial A_{0_\ell}}{\partial x_k}\right.$$
$$\left.+ \frac{\partial^2\psi}{\partial x_j\partial x_k}A_{0_\ell} + \frac{\partial\psi}{\partial x_k}\frac{\partial\psi}{\partial x_j}A_{1_\ell} + \frac{\partial\psi}{\partial x_k}\frac{\partial A_{0_\ell}}{\partial x_j} + \frac{\partial\psi}{\partial x_j}\frac{\partial A_{0_k}}{\partial x_\ell}\right)$$
$$\left.+ (i\omega)^2\left(\frac{\partial\psi}{\partial x_\ell}\frac{\partial\psi}{\partial x_j}A_{0_k} + \partial x_k\frac{\partial\psi}{\partial x_j}A_{0_\ell}\right)\right].$$

Returning to equations (7.2.9), we can write the right-hand side as

$$\frac{1}{2}\exp\left[i\omega\psi(\mathbf{x})\right]\sum_{j=1}^{3}\sum_{k=1}^{3}\sum_{\ell=1}^{3}\left[a(\mathbf{x})(i\omega)^{2}+b(\mathbf{x})(i\omega)^{1}\right.$$
$$\left.+c(\mathbf{x})(i\omega)^{0}+d(\mathbf{x})(i\omega)^{-1}\right], \tag{7.2.17}$$

where

$$a(\mathbf{x})=c_{ijk\ell}(\mathbf{x})\left(\frac{\partial\psi}{\partial x_\ell}\frac{\partial\psi}{\partial x_j}A_{0_k}+\frac{\partial\psi}{\partial x_k}\frac{\partial\psi}{\partial x_j}A_{0_\ell}\right),$$

$$\begin{aligned}b(\mathbf{x})=&\frac{\partial c_{ijk\ell}(\mathbf{x})}{\partial x_j}\left(\frac{\partial\psi}{\partial x_\ell}A_{0_k}+\frac{\partial\psi}{\partial x_k}A_{0_\ell}\right)\\&+c_{ijk\ell}(\mathbf{x})\left(\frac{\partial^2\psi}{\partial x_j\partial x_\ell}A_{0_k}+\frac{\partial\psi}{\partial x_\ell}\frac{\partial\psi}{\partial x_j}A_{1_k}+\frac{\partial\psi}{\partial x_\ell}\frac{\partial A_{0_k}}{\partial x_j}+\frac{\partial\psi}{\partial x_j}\frac{\partial A_{0_\ell}}{\partial x_k}\right.\\&\left.+\frac{\partial^2\psi}{\partial x_j\partial x_k}A_{0_\ell}+\frac{\partial\psi}{\partial x_k}\frac{\partial\psi}{\partial x_j}A_{1_\ell}+\frac{\partial\psi}{\partial x_k}\frac{\partial A_{0_\ell}}{\partial x_j}+\frac{\partial\psi}{\partial x_j}\frac{\partial A_{0_k}}{\partial x_\ell}\right),\end{aligned}$$

$$\begin{aligned}c(\mathbf{x})=&\frac{\partial c_{ijk\ell}(\mathbf{x})}{\partial x_j}\left(\frac{\partial\psi}{\partial x_\ell}A_{1_k}+\frac{\partial\psi}{\partial x_k}A_{1_\ell}+\frac{\partial A_{0_k}}{\partial x_\ell}+\frac{\partial A_{0_\ell}}{\partial x_k}\right)\\&+c_{ijk\ell}(\mathbf{x})\left(\frac{\partial^2\psi}{\partial x_j\partial x_\ell}A_{1_k}+\frac{\partial\psi}{\partial x_\ell}\frac{\partial A_{1_k}}{\partial x_j}+\frac{\partial\psi}{\partial x_j}\frac{\partial A_{1_k}}{\partial x_\ell}+\frac{\partial^2 A_{0_k}}{\partial x_j\partial x_\ell}\right.\\&\left.+\frac{\partial^2\psi}{\partial x_j\partial x_k}A_{1_\ell}+\frac{\partial\psi}{\partial x_k}\frac{\partial A_{1_\ell}}{\partial x_j}+\frac{\partial\psi}{\partial x_j}\frac{\partial A_{1_\ell}}{\partial x_k}+\frac{\partial^2 A_{0_\ell}}{\partial x_j\partial x_k}\right)\end{aligned}$$

and

$$d(\mathbf{x})=\frac{\partial c_{ijk\ell}(\mathbf{x})}{\partial x_j}\left(\frac{\partial A_{1_k}}{\partial x_\ell}+\frac{\partial A_{1_\ell}}{\partial x_k}\right)+c_{ijk\ell}(\mathbf{x})\left(\frac{\partial^2 A_{1_k}}{\partial x_j\partial x_\ell}+\frac{\partial^2 A_{1_\ell}}{\partial x_j\partial x_k}\right),$$

with $i\in\{1,2,3\}$. Considering the left-hand side of equations (7.2.9) that includes the two terms of series (7.2.12), which is given in expression (7.2.15), we write it as

$$\exp\left[i\omega\psi(\mathbf{x})\right]\left[(i\omega)^{2}\rho(\mathbf{x})A_{0_i}(\mathbf{x})+\omega\rho(\mathbf{x})A_{1_i}(\mathbf{x})\right],\qquad i\in\{1,2,3\}. \tag{7.2.18}$$

Comparing the right-hand and left-hand sides, which are given by expressions (7.2.17) and (7.2.18), respectively, we cancel $\exp(i\omega\psi)$. Then, moving all the terms to one side, we write the resulting coefficient of $(i\omega)^2$ as

$$\frac{1}{2}\sum_{j=1}^{3}\sum_{k=1}^{3}\sum_{\ell=1}^{3}c_{ijk\ell}(\mathbf{x})\left(\frac{\partial\psi}{\partial x_\ell}\frac{\partial\psi}{\partial x_j}A_{0_k}+\frac{\partial\psi}{\partial x_k}\frac{\partial\psi}{\partial x_j}A_{0_\ell}\right)-\rho(\mathbf{x})A_{0_i}(\mathbf{x}),$$
$$i\in\{1,2,3\}.$$

Since j, k and ℓ are summation indices and $c_{ijk\ell}$ is symmetric in k and ℓ, we can write this expression as

$$\sum_{j=1}^{3}\sum_{k=1}^{3}\sum_{\ell=1}^{3}c_{ijk\ell}(\mathbf{x})\frac{\partial\psi}{\partial x_\ell}\frac{\partial\psi}{\partial x_j}A_{0_k}(\mathbf{x})-\rho(\mathbf{x})A_{0_i}(\mathbf{x}),\qquad i\in\{1,2,3\}.$$

For the resulting equations of motion to be satisfied for all $i\omega$, we require all the coefficients to be zero. Again considering the $(i\omega)^2$ coefficient, we write

$$\sum_{j=1}^{3}\sum_{k=1}^{3}\sum_{\ell=1}^{3}c_{ijk\ell}(\mathbf{x})\frac{\partial\psi}{\partial x_\ell}\frac{\partial\psi}{\partial x_j}A_{0_k}-\rho(\mathbf{x})A_{0_i}=0,\qquad i\in\{1,2,3\},\quad(7.2.19)$$

which are equations (7.2.5) in terms of $\mathbf{A}_0$. These equations lead us to the eikonal equation whose solution is $\psi(\mathbf{x})$; this equation is associated with the path of the propagating wave.

We could also write the coefficients of $(i\omega)^{-1}$, $(i\omega)^0$ and $(i\omega)^1$ and set them to zero. The solutions of the corresponding equations are $\mathbf{A}_n(\mathbf{x})$. These equations are associated with the amplitude of the propagating wave and are called transport equations.

Let us return to series (7.2.12). Inserting more terms of this series into equations (7.2.9) would result in terms with higher negative powers of $i\omega$ in expression (7.2.17). Coefficients $a(\mathbf{x})$ and $b(\mathbf{x})$ would remain unchanged; in particular, equations (7.2.5) would remain the same—these equations are always expressed in the zeroth-order term of series (7.2.12). Keeping this property in mind, we can rewrite equations (7.2.19) as

$$\sum_{j=1}^{3}\sum_{k=1}^{3}\sum_{\ell=1}^{3}c_{ijk\ell}(\mathbf{x})\frac{\partial\psi}{\partial x_\ell}\frac{\partial\psi}{\partial x_j}A_k-\rho(\mathbf{x})A_i=0,\qquad i\in\{1,2,3\},$$

which are equations (7.2.5). Coefficient $c(\mathbf{x})$ would have terms with $\mathbf{A}_0$, $\mathbf{A}_1$ and $\mathbf{A}_2$; coefficient $d(\mathbf{x})$ would have terms with $\mathbf{A}_1$, $\mathbf{A}_2$ and $\mathbf{A}_3$, but would no longer have any term with $\mathbf{A}_0$. Each coefficient would have three As. Thus, a new coefficient, $e(\mathbf{x})$, would have terms with $\mathbf{A}_2$, $\mathbf{A}_3$ and $\mathbf{A}_4$. The pattern would continue.

7.3. Eikonal Equation

To obtain $\psi(\mathbf{x})$, we turn our attention to equations (7.2.5) and group the components of vector $\mathbf{A}(\mathbf{x})$. Hence, we rewrite equation (7.2.5) as

$$\sum_{k=1}^{3}\left(\sum_{j=1}^{3}\sum_{\ell=1}^{3}c_{ijk\ell}(\mathbf{x})\frac{\partial\psi}{\partial x_j}\frac{\partial\psi}{\partial x_\ell}-\rho(\mathbf{x})\,\delta_{ik}\right)A_k(\mathbf{x})=\mathbf{0},\qquad i\in\{1,2,3\}. \tag{7.3.1}$$

In view of expression (6.11.16), let us denote

$$p_j:=\frac{\partial\psi}{\partial x_j},\qquad j\in\{1,2,3\}, \tag{7.3.2}$$

where $\mathbf{p}$ is the phase-slowness vector, which describes the slowness of the propagation of the wavefront.

Note that the meaning of $\mathbf{p}$ can be seen by examining expression (7.2.2) and considering a three-dimensional continuum. Therein, ψ is a function relating position variables, x_1, x_2 and x_3, to the traveltime, t. Thus, since ψ has units of time, $p_j:=\partial\psi/\partial x_j$ has units of slowness and the level sets of $\psi(\mathbf{x})$ can be viewed as wavefronts at a given time t. Consequently, in view of properties of the gradient operator, $\mathbf{p}=\nabla\psi(\mathbf{x})$ is a vector whose direction corresponds to the wavefront normal and whose magnitude corresponds to the wavefront slowness.

In view of notation (7.3.2), we can write equations (7.3.1) as

$$\sum_{k=1}^{3}\left(\sum_{j=1}^{3}\sum_{\ell=1}^{3}c_{ijk\ell}(\mathbf{x})\,p_jp_\ell-\rho(\mathbf{x})\,\delta_{ik}\right)A_k(\mathbf{x})=\mathbf{0},\qquad i\in\{1,2,3\}. \tag{7.3.3}$$

Equations (7.3.3) are referred to as Christoffel's equations.

In Chapter 9, we discuss equations (7.3.3) in the context of the particular symmetries of continua, which were introduced in Chapter 5. Therein, we also show that the eigenvalues resulting from these equations are associated with the velocity of the wavefront while the corresponding eigenvectors are the displacement directions. Herein, we study the general form of equations (7.3.3).

We know from linear algebra that equations (7.3.3) have nontrivial solutions, namely, $\mathbf{A} \neq \mathbf{0}$, if and only if

$$\det\left[\sum_{j=1}^{3}\sum_{\ell=1}^{3} c_{ijk\ell}(\mathbf{x})\, p_j p_\ell - \rho(\mathbf{x})\,\delta_{ik}\right] = 0, \qquad i,k \in \{1,2,3\}. \quad (7.3.4)$$

Assuming that $p^2 \neq 0$, we can write determinant (7.3.4) as

$$\left(p^2\right)^3 \det\left[\sum_{j=1}^{3}\sum_{\ell=1}^{3} c_{ijk\ell}(\mathbf{x})\,\frac{p_j p_\ell}{p^2} - \frac{\rho(\mathbf{x})}{p^2}\delta_{ik}\right] = 0, \qquad i,k \in \{1,2,3\},$$

and restate this condition as

$$\det\left[\sum_{j=1}^{3}\sum_{\ell=1}^{3} c_{ijk\ell}(\mathbf{x})\,\frac{p_j p_\ell}{p^2} - \frac{\rho(\mathbf{x})}{p^2}\delta_{ik}\right] = 0, \qquad i,k \in \{1,2,3\}. \quad (7.3.5)$$

Note that $p^2 = 0$ would mean that the slowness of the propagation of the wavefront is zero. This would imply the velocity to be infinite, which is a nonphysical situation. Also, in view of determinant (7.3.4), $p^2 = 0$ would result in $\det\left[\rho(\mathbf{x})\,\delta_{ik}\right] = 0$, which would imply $\rho(\mathbf{x}) = 0$.

Expression (7.3.5) is a polynomial of degree 3 in p^{-2} whose coefficients depend on the direction of the phase-slowness vector, $\mathbf{p}$. Any such polynomial can be factored out as

$$\left[p^2 - \frac{1}{v_1^2(\mathbf{x},\mathbf{p})}\right]\left[p^2 - \frac{1}{v_2^2(\mathbf{x},\mathbf{p})}\right]\left[p^2 - \frac{1}{v_3^2(\mathbf{x},\mathbf{p})}\right] = 0, \quad (7.3.6)$$

where $1/v_i^2$ are the roots of polynomial (7.3.5). The existence of three roots implies the existence of three types of waves, which can propagate in anisotropic continua.

The matrix with entries

$$\Gamma_{ik} = \left[\sum_{j=1}^{3}\sum_{\ell=1}^{3} c_{ijk\ell}(\mathbf{x})\,\frac{p_j p_\ell}{p^2}\right], \qquad i,k \in \{1,2,3\}, \quad (7.3.7)$$

which appears in equation (7.3.5) is called Christoffel's matrix. It is symmetric and positive-definite, as shown in Exercises 7.5 and 7.6, respectively. Since the matrix is symmetric, the roots are real; since it is positive-definite, they are positive and, hence, the values of v_i are real. Terms p_j/p and p_ℓ/p, where p stands for the length of vector $\mathbf{p}$, specify its direction.

Now, let us consider a given root of equation (7.3.6). Each root is the eikonal equation for a given type of wave, namely,

$$p^2 = \frac{1}{v_i^2(\mathbf{x},\mathbf{p})}, \qquad i \in \{1,2,3\}. \tag{7.3.8}$$

Let us examine the meaning of this equation.[3]

Since $p^2 = \mathbf{p}\cdot\mathbf{p}$ is the squared magnitude of the slowness vector, which is normal to the wavefront, then—in view of the wavefronts being the loci of constant phase— v_i is the function describing phase velocity. This velocity is a function of position, $\mathbf{x}$, and the direction of $\mathbf{p}$. Hence, equation (7.3.8) applies to anisotropic inhomogeneous continua and can be viewed as an extension of equation (6.11.23), which is valid for isotropic inhomogeneous continua. As shown above, due to the positive definiteness of Γ, function v_i is real-valued—a property of the body-wave velocities in an elastic continuum, which are discussed in Chapter 9.

Considering two adjacent wavefronts, we can view equation (7.3.8) as an infinitesimal formulation of Huygens' principle.[4]

Note that function v is homogeneous of degree 0 in the p_i. In other words, the orientation of a wavefront is described by the direction of $\mathbf{p}$ and is independent of the length of $\mathbf{p}$. Hence, in equation (7.3.8) we could also write $v_i = v_i(\mathbf{x},\mathbf{n})$, where $\mathbf{n}$ is a unit vector in the direction of $\mathbf{p}$. Notably, we use this notation in Chapter 9. We can explicitly see the homogeneity of function v in Section 6.11.3, where we discussed waves in the context of elliptical velocity dependence. Therein, vector $\mathbf{p}$ appears as

---

[3] Readers interested in the mathematical formulation of the conditions under which the eikonal equation provides a good approximation to the wave equation might refer to Officer, C.B., (1974) Introduction to theoretical geophysics: Springer-Verlag, pp. 204–205.

[4] Readers interested in a formulation relating the eikonal equation to Huygens' principle might refer to Arnold, V.I., (1989) Mathematical methods of classical mechanics (2*nd* edition): Springer-Verlag, pp. 248–252, and to Lanczos, C., (1949/1986) The variational principles in mechanics: Dover, pp. 269–270.

a ratio of its components, namely, the directional dependence is given by expression (6.11.9).

Furthermore, as shown explicitly in Chapter 9, the phase-velocity function can be expressed in terms of the properties of the continuum, namely, its mass density and elasticity parameters. Thus, the eikonal equation relates the magnitude of the slowness with which the wavefront propagates to the properties of the continuum through which it propagates.

In the mathematical context, the eikonal equation is a differential equation. Recalling expressions (7.3.2), we can rewrite equation (7.3.8) as

$$[\nabla\psi(\mathbf{x})]^2 = \frac{1}{v^2(\mathbf{x},\mathbf{p})}. \tag{7.3.9}$$

In general, the eikonal equation is a nonlinear first-order partial differential equation in $\mathbf{x}$ to be solved for the eikonal function, $\psi(\mathbf{x})$. It belongs to the Hamilton-Jacobi class of differential equations.[5]

Equation (7.2.6) is the transport equation. This transport equation possesses a vectorial form that is valid for anisotropic inhomogeneous continua. It is analogous to the scalar transport equation (6.11.26), which is valid for isotropic inhomogeneous continua.

Closing Remarks

In this chapter, while seeking to study the propagation of waves in anisotropic inhomogeneous continua, we follow a strategy analogous to that used in Chapter 6. However, having obtained the equations of motion, we find that we are unable to investigate them analytically. Thus, we utilize a trial solution that leads us to the eikonal equation, which relates the slowness of propagation of the wavefront to the properties of the continuum through which it propagates. In Chapter 8, we continue our study of wave propagation in anisotropic inhomogeneous continua by solving the eikonal equation using the method of characteristics, which leads to Hamilton's ray equations. Notably, since the method of characteristics in the context of the eikonal equation is the subject of the book by Bóna, A. and Slawinski, M.A., (2021) Wavefronts and rays as characteristics and

[5] Readers interested in a mathematical study of the eikonal and transport equations might refer to Taylor, M.E., (1996) Partial differential equations; Basic theory: Springer-Verlag, pp. 79–84 and pp. 440 – 447.

asymptotics (3*rd* edition): World Scientific, readers might refer to that book while studying Chapter 8.

7.4. Exercises

EXERCISE 7.1. Show that trial solution (7.2.1) with $f(\eta) = \exp\eta$, which corresponds to solution (7.2.10), does not allow us to obtain equations (7.2.5), (7.2.6) and (7.2.7).

SOLUTION 7.1. Letting $f(\eta) = \exp\eta$ in expression (7.2.1), inserting the resulting expression into equation (7.1.4), and factoring out the common term,

$$\exp\eta = f = \frac{\mathrm{d}f}{\mathrm{d}\eta} = \frac{\mathrm{d}^2 f}{\mathrm{d}\eta^2},$$

we write

$$\begin{aligned}
\exp\eta \left(\sum_{j=1}^{3}\sum_{k=1}^{3}\sum_{\ell=1}^{3} \left\{ \left[\frac{\partial c_{ijk\ell}(\mathbf{x})}{\partial x_j}\frac{\partial A_k}{\partial x_\ell} + c_{ijk\ell}(\mathbf{x})\frac{\partial^2 A_k}{\partial x_j \partial x_\ell} \right] \right.\right. \\
+ \left[\frac{\partial}{\partial x_j}\left(c_{ijk\ell}(\mathbf{x}) A_\ell \frac{\partial \psi}{\partial x_k} \right) + c_{ijk\ell}(\mathbf{x})\frac{\partial A_k}{\partial x_\ell}\frac{\partial \psi}{\partial x_j} \right] \\
\left.\left. + \left[c_{ijk\ell}(\mathbf{x}) A_k \frac{\partial \psi}{\partial x_j}\frac{\partial \psi}{\partial x_\ell} \right] \right\} - \rho(\mathbf{x}) A_i \right) = 0,
\end{aligned}$$

where $i \in \{1,2,3\}$. Since $\exp\eta \neq 0$, we obtain

$$\begin{aligned}
\sum_{j=1}^{3}\sum_{k=1}^{3}\sum_{\ell=1}^{3} \left\{ \frac{\partial c_{ijk\ell}(\mathbf{x})}{\partial x_j}\frac{\partial A_k}{\partial x_\ell} + \frac{\partial}{\partial x_j}\left(c_{ijk\ell}(\mathbf{x}) A_\ell \frac{\partial \psi}{\partial x_k} \right) \right. \\
\left. + c_{ijk\ell}(\mathbf{x}) \left[\frac{\partial^2 A_k}{\partial x_j \partial x_\ell} + \frac{\partial A_k}{\partial x_\ell}\frac{\partial \psi}{\partial x_j} + A_k \frac{\partial \psi}{\partial x_j}\frac{\partial \psi}{\partial x_\ell} \right] \right\} \\
- \rho(\mathbf{x}) A_i = 0, \qquad i \in \{1,2,3\}.
\end{aligned}$$

These are complicated differential equations for ψ and $\mathbf{A}$. Since f, f' and f'' are not linearly independent, we cannot split their corresponding terms into three distinct equations as we did in Section 7.2.2. Thus, we cannot proceed to solve them by methods that lead to the eikonal and transport equations.

EXERCISE 7.2. Consider the wave equation,

$$\frac{\partial^2 u}{\partial x^2}+\frac{\partial^2 u}{\partial y^2}=\frac{1}{v^2}\frac{\partial^2 u}{\partial t^2},$$

and its trial solution,

$$u(x,y,t)=A(x,y)\,f(\psi(x,y)-t), \tag{7.4.1}$$

where $\psi(x,y)$ is a solution of the corresponding eikonal equation. Find the conditions on $A(x,y)$ for arbitrary f.

SOLUTION 7.2. Differentiating expression (7.4.1), we obtain

$$\frac{\partial^2 u}{\partial x^2}=\frac{\partial^2 A}{\partial x^2}f-f'\left(2\frac{\partial A}{\partial x}\frac{\partial \psi}{\partial x}+A\frac{\partial^2 \psi}{\partial x^2}\right)+Af''\left(\frac{\partial \psi}{\partial x}\right)^2,$$

$$\frac{\partial^2 u}{\partial y^2}=\frac{\partial^2 A}{\partial y^2}f-f'\left(2\frac{\partial A}{\partial y}\frac{\partial \psi}{\partial y}+A\frac{\partial^2 \psi}{\partial y^2}\right)+Af''\left(\frac{\partial \psi}{\partial y}\right)^2$$

and

$$\frac{\partial^2 u}{\partial t^2}=Af'',$$

where prime stands for the derivative with respect to the argument. Substituting these derivatives into the wave equation and rearranging, we obtain

$$f\left(\frac{\partial^2 A}{\partial x^2}+\frac{\partial^2 A}{\partial y^2}\right)-f'\left(2\left(\frac{\partial A}{\partial x}\frac{\partial \psi}{\partial x}+\frac{\partial A}{\partial y}\frac{\partial \psi}{\partial y}\right)+A\left(\frac{\partial^2 \psi}{\partial x^2}+\frac{\partial^2 \psi}{\partial y^2}\right)\right)$$
$$+f''A\left(\left(\frac{\partial \psi}{\partial x}\right)^2+\left(\frac{\partial \psi}{\partial y}\right)^2-\frac{1}{v^2}\right)=0.$$

For this equation to be valid for arbitrary f, we require the three terms in parentheses to be zero.

$$\nabla^2 A=0, \tag{7.4.2}$$

$$2\nabla A\cdot\nabla\psi+A\nabla^2\psi=0 \tag{7.4.3}$$

and

$$(\nabla \psi)^2 = \frac{1}{v^2}.$$

Since the last equation is assumed to be satisfied, the conditions on A are given by the first two equations. In view of the first equation, A must be a harmonic function that satisfies equation (7.4.3).

EXERCISE 7.3. Following Exercise 7.2, find function A that satisfies equations (7.4.2) and (7.4.3) for plane waves, $\psi(x,y) = x/v$, where v is a constant velocity.

SOLUTION 7.3. Since $\psi = x/v$, we get

$$\nabla \psi(x,y) = \left[\frac{1}{v}, 0\right]$$

and

$$\nabla^2 \psi = 0.$$

Inserting these expressions into equation (7.4.3), we get

$$2\left[\frac{\partial A}{\partial x}, \frac{\partial A}{\partial y}\right] \cdot \left[\frac{1}{v}, 0\right] = \frac{2}{v}\frac{\partial A}{\partial x} = 0,$$

which implies $A(x,y) = f(y)$. Using equation (7.4.2), we get

$$\frac{\partial^2 g(y)}{\partial y^2} = 0,$$

which implies

$$g(y) = a_0 + a_1 y.$$

Hence, in view of Exercise 7.2,

$$(a_0 + a_1 y) f\left(\frac{x}{v} - t\right)$$

is a solution of the wave equation for arbitrary f. If $a_1 = 0$, then we obtain a constant-amplitude solution valid for any f.

EXERCISE 7.4. Expression (7.2.13), namely,

$$\lim_{\omega\to\infty}\left\{\frac{(i\omega)^N}{\exp[i\omega\psi(\mathbf{x})]}\left[\tilde{\mathbf{u}}(\mathbf{x},\omega) - \exp[i\omega\psi(\mathbf{x})]\sum_{n=0}^{N}\frac{\mathbf{A}_n(\mathbf{x})}{(i\omega)^n}\right]\right\} = 0, \tag{7.4.4}$$

allows us to determine uniquely all $\mathbf{A}_n$. Determine $\mathbf{A}_0$, $\mathbf{A}_1$ and $\mathbf{A}_2$.

SOLUTION 7.4. For $N=0$ also $n=0$, and we write expression (7.4.4) as

$$\lim_{\omega\to\infty}\left\{\frac{1}{\exp\left[i\omega\psi(\mathbf{x})\right]}\left[\tilde{\mathbf{u}}(\mathbf{x},\omega)-\exp\left[i\omega\psi(\mathbf{x})\right]\mathbf{A}_0(\mathbf{x})\right]\right\}=0,$$

which means that

$$\mathbf{A}_0(\mathbf{x})=\lim_{\omega\to\infty}\frac{\tilde{\mathbf{u}}(\mathbf{x},\omega)}{\exp\left[i\omega\psi(\mathbf{x})\right]}. \tag{7.4.5}$$

For $N=1$, $n=0,1$, and we write expression (7.4.4) as

$$\lim_{\omega\to\infty}\left\{\frac{i\omega}{\exp\left[i\omega\psi(\mathbf{x})\right]}\left[\tilde{\mathbf{u}}(\mathbf{x},\omega)-\exp\left[i\omega\psi(\mathbf{x})\right]\left[\mathbf{A}_0(\mathbf{x})+\frac{\mathbf{A}_1(\mathbf{x})}{i\omega}\right]\right]\right\}=0.$$

We can rewrite it as

$$\lim_{\omega\to\infty}\left\{i\omega\left[\frac{\tilde{\mathbf{u}}(\mathbf{x},\omega)}{\exp\left[i\omega\psi(\mathbf{x})\right]}-\left[\mathbf{A}_0(\mathbf{x})+\frac{\mathbf{A}_1(\mathbf{x})}{i\omega}\right]\right]\right\}=0$$

to get

$$\lim_{\omega\to\infty}\left\{i\omega\left[\frac{\tilde{\mathbf{u}}(\mathbf{x},\omega)}{\exp\left[i\omega\psi(\mathbf{x})\right]}-\mathbf{A}_0(\mathbf{x})\right]-\mathbf{A}_1(\mathbf{x})\right\}=0.$$

This means that

$$\mathbf{A}_1(\mathbf{x})=\lim_{\omega\to\infty}\left\{i\omega\left[\frac{\tilde{\mathbf{u}}(\mathbf{x},\omega)}{\exp\left[i\omega\psi(\mathbf{x})\right]}-\mathbf{A}_0(\mathbf{x})\right]\right\}, \tag{7.4.6}$$

where $\mathbf{A}_0$ is known from equation (7.4.5). For $N=2$, $n=0,1,2$, and we write expression (7.4.4) as

$$\lim_{\omega\to\infty}\left\{\frac{(i\omega)^2}{\exp\left[i\omega\psi(\mathbf{x})\right]}\left[\tilde{\mathbf{u}}(\mathbf{x},\omega)\right.\right.$$
$$\left.\left.-\exp\left[i\omega\psi(\mathbf{x})\right]\left[\mathbf{A}_0(\mathbf{x})+\frac{\mathbf{A}_1(\mathbf{x})}{i\omega}+\frac{\mathbf{A}_2(\mathbf{x})}{(i\omega)^2}\right]\right]\right\}=0.$$

We can rewrite it as

$$\lim_{\omega\to\infty}\left\{(i\omega)^2\left[\frac{\tilde{\mathbf{u}}(\mathbf{x},\omega)}{\exp\left[i\omega\psi(\mathbf{x})\right]}-\mathbf{A}_0(\mathbf{x})\right]-i\omega\mathbf{A}_1(\mathbf{x})-\mathbf{A}_2(\mathbf{x})\right\},$$

which means that

$$\mathbf{A}_2(\mathbf{x}) = \lim_{\omega\to\infty}\left\{(i\omega)^2\left[\frac{\tilde{\mathbf{u}}(\mathbf{x},\omega)}{\exp[i\omega\psi(\mathbf{x})]} - \mathbf{A}_0(\mathbf{x})\right] - i\omega\mathbf{A}_1(\mathbf{x})\right\}, \qquad (7.4.7)$$

where $\mathbf{A}_0$ and $\mathbf{A}_1$ are known from equations (7.4.5) and (7.4.6), respectively. Continuing this process we obtain uniquely all $\mathbf{A}_n$.

EXERCISE 7.5. Show that matrix (7.3.7) is symmetric.

SOLUTION 7.5. To show that matrix (7.3.7) is symmetric, we need to show that $\Gamma_{ik} = \Gamma_{ki}$, which is equivalent to showing that

$$\sum_{j=1}^{3}\sum_{\ell=1}^{3} c_{ijk\ell}(\mathbf{x})\frac{p_j p_\ell}{p^2} = \sum_{j=1}^{3}\sum_{\ell=1}^{3} c_{kji\ell}(\mathbf{x})\frac{p_j p_\ell}{p^2}, \qquad i,k\in\{1,2,3\}. \quad (7.4.8)$$

In view of symmetries (4.2.2), we can write the left-hand side of equation (7.4.8) as

$$\sum_{j=1}^{3}\sum_{\ell=1}^{3} c_{ijk\ell}(\mathbf{x})\frac{p_j p_\ell}{p^2} = \sum_{j=1}^{3}\sum_{\ell=1}^{3} c_{k\ell ij}(\mathbf{x})\frac{p_j p_\ell}{p^2}, \qquad i,k\in\{1,2,3\}.$$

Examining the right-hand sides of the above equations, we see that—since both $j = 1,2,3$ and $\ell = 1,2,3$—each term that appears in

$$\sum_{j=1}^{3}\sum_{\ell=1}^{3} c_{kji\ell}(\mathbf{x})\frac{p_j p_\ell}{p^2}$$

also appears in

$$\sum_{j=1}^{3}\sum_{\ell=1}^{3} c_{k\ell ij}(\mathbf{x})\frac{p_j p_\ell}{p^2}.$$

Thus, we can write

$$\sum_{j=1}^{3}\sum_{\ell=1}^{3} c_{ijk\ell}(\mathbf{x})\frac{p_j p_\ell}{p^2} = \sum_{j=1}^{3}\sum_{\ell=1}^{3} c_{kji\ell}(\mathbf{x})\frac{p_j p_\ell}{p^2}, \qquad i,k\in\{1,2,3\},$$

which is equation (7.4.8), as required. We conclude that Christoffel's matrix (7.3.7) is symmetric as a consequence of symmetries (4.2.2) of the elasticity matrix, which result from the existence of the strain-energy function.

EXERCISE 7.6. Show that matrix (7.3.7) is positive-definite.

SOLUTION 7.6. To show that matrix (7.3.7) is positive-definite, we need to show that

$$\sum_{i=1}^{3}\sum_{k=1}^{3}\left[\sum_{j=1}^{3}\sum_{\ell=1}^{3}c_{ijk\ell}(\mathbf{x})\frac{p_j p_\ell}{p^2}\right]w_i w_k > 0, \tag{7.4.9}$$

for an arbitrary nonzero vector, $\mathbf{w}$. To do so, let us recall equation (4.3.2), which we can rewrite as

$$\sum_{m=1}^{6}\sum_{n=1}^{6}C_{mn}\varepsilon_m\varepsilon_n > 0,$$

where ε is an arbitrary nonzero vector. Invoking formula (3.2.5), we write

$$\sum_{i=1}^{3}\sum_{j=1}^{3}\sum_{k=1}^{3}\sum_{\ell=1}^{3}c_{ijk\ell}\varepsilon_{ij}\varepsilon_{k\ell} > 0. \tag{7.4.10}$$

To proceed, we use the fact that although, in view of definition (1.4.6), ε_{ij} is a symmetric tensor, inequality (7.4.10) remains valid for general second-rank tensors. To show this fact, let us return to Section 1.5 and recall that any second-rank tensor can be written as a sum of symmetric and antisymmetric tensors. As we show below, the summation of the antisymmetric parts vanishes. Consider expression

$$\sum_{i=1}^{3}\sum_{j=1}^{3}\sum_{k=1}^{3}\sum_{\ell=1}^{3}c_{ijk\ell}\eta_{ij}a_{k\ell},$$

where η is an antisymmetric tensor, $\eta_{ij} = -\eta_{ji}$, and $a_{k\ell}$ is an arbitrary one. Thus, we can write

$$\sum_{i=1}^{3}\sum_{j=1}^{3}\sum_{k=1}^{3}\sum_{\ell=1}^{3}c_{ijk\ell}\eta_{ij}a_{k\ell} = \sum_{i=1}^{3}\sum_{j=1}^{3}\sum_{k=1}^{3}\sum_{\ell=1}^{3}\left(-c_{ijk\ell}\eta_{ji}a_{k\ell}\right).$$

Using the symmetry of the elasticity tensor given in expression (3.2.3), which is a result of the symmetry of stress tensor, we write

$$\sum_{i=1}^{3}\sum_{j=1}^{3}\sum_{k=1}^{3}\sum_{\ell=1}^{3}c_{ijk\ell}\eta_{ij}a_{k\ell} = \sum_{j=1}^{3}\sum_{i=1}^{3}\sum_{k=1}^{3}\sum_{\ell=1}^{3}\left(-c_{jik\ell}\eta_{ji}a_{k\ell}\right).$$

Since i and j are summation indices we can rename them so that i becomes j and vice versa. Hence, we write

$$\sum_{i=1}^{3}\sum_{j=1}^{3}\sum_{k=1}^{3}\sum_{\ell=1}^{3} c_{ijk\ell}\eta_{ij}a_{k\ell} = \sum_{i=1}^{3}\sum_{j=1}^{3}\sum_{k=1}^{3}\sum_{\ell=1}^{3}\left(-c_{ijk\ell}\eta_{ij}a_{k\ell}\right)$$
$$= -\sum_{i=1}^{3}\sum_{j=1}^{3}\sum_{k=1}^{3}\sum_{\ell=1}^{3} c_{ijk\ell}\eta_{ij}a_{k\ell},$$

which implies

$$\sum_{i=1}^{3}\sum_{j=1}^{3}\sum_{k=1}^{3}\sum_{\ell=1}^{3} c_{ijk\ell}\eta_{ij}a_{k\ell} = 0.$$

In other words, the summation of the antisymmetric parts vanishes and, hence, inequality (7.4.10) is valid whether or not the second-rank tensor is symmetric. In particular, we can write

$$\sum_{i=1}^{3}\sum_{j=1}^{3}\sum_{k=1}^{3}\sum_{\ell=1}^{3} c_{ijk\ell}a_{ij}a_{k\ell} > 0. \tag{7.4.11}$$

Since a_{ij} is arbitrary, we consider

$$a_{ij} = w_i \frac{p_j}{|p|}$$

to write inequality (7.4.11) as

$$\sum_{i=1}^{3}\sum_{j=1}^{3}\sum_{k=1}^{3}\sum_{\ell=1}^{3} c_{ijk\ell} w_i \frac{p_j}{|p|} w_k \frac{p_\ell}{|p|} > 0.$$

Using the algebraic properties of summation and multiplication, we write

$$\sum_{i=1}^{3}\sum_{k=1}^{3}\left[\sum_{j=1}^{3}\sum_{\ell=1}^{3} c_{ijk\ell}(\mathbf{x})\frac{p_j p_\ell}{p^2}\right] w_i w_k > 0,$$

which is inequality (7.4.9), as required. We conclude that Christoffel's matrix (7.3.7) is positive-definite as a consequence of the positive definiteness of the elasticity matrix, which results from the stability conditions, and the symmetry of the stress tensor, which results from the balance of angular momentum.

EXERCISE 7.7. In view of Exercise 6.17, consider a more general form of the solution that is given by $u(\mathbf{x},t) = f(\eta)$, where $\eta = v_0[\psi(\mathbf{x}) - t]$. Show that the necessary condition for characteristic equation (6.12.36) to be satisfied is the eikonal equation given by

$$(\nabla\psi)^2 = \frac{1}{v^2}.$$

SOLUTION 7.7. Considering the argument of f given by $\eta = v_0[\psi(\mathbf{x}) - t]$, we obtain

$$\frac{\partial u}{\partial x_i} = \frac{\partial f}{\partial \eta}\frac{\partial \eta}{\partial x_i} = v_0 \frac{\partial f}{\partial \eta}\frac{\partial \psi}{\partial x_i}, \qquad i \in \{1,2,3\},$$

and

$$\frac{\partial u}{\partial t} = \frac{\partial f}{\partial \eta}\frac{\partial \eta}{\partial t} = -v_0 \frac{\partial f}{\partial \eta}.$$

Substituting $\partial u/\partial x_i$ and $\partial u/\partial t$ into characteristic equation (6.12.36), we can write

$$\left(v_0 \frac{\partial f}{\partial \eta}\frac{\partial \psi}{\partial x_1}\right)^2 + \left(v_0 \frac{\partial f}{\partial \eta}\frac{\partial \psi}{\partial x_2}\right)^2 + \left(v_0 \frac{\partial f}{\partial \eta}\frac{\partial \psi}{\partial x_3}\right)^2 = \frac{1}{v^2}\left(-\frac{\partial f}{\partial \eta}v_0\right)^2,$$

which yields

$$v_0^2\left(\frac{\partial f}{\partial \eta}\right)^2\left[\left(\frac{\partial \psi}{\partial x_1}\right)^2 + \left(\frac{\partial \psi}{\partial x_2}\right)^2 + \left(\frac{\partial \psi}{\partial x_3}\right)^2\right] = \left(\frac{v_0}{v}\right)^2\left(\frac{\partial f}{\partial \eta}\right)^2.$$

Since, in general, $v_0 \neq 0$ and $\partial f/\partial \eta \neq 0$, we can write

$$\left(\frac{\partial \psi}{\partial x_1}\right)^2 + \left(\frac{\partial \psi}{\partial x_2}\right)^2 + \left(\frac{\partial \psi}{\partial x_3}\right)^2 = \frac{1}{v^2},$$

which is the required eikonal equation.

REMARK 7.4.1. If v is constant, Exercise 7.7 is reduced to Exercise 6.17.

EXERCISE 7.8. Derive eikonal equation (6.11.22) in two spatial dimensions as the requirement describing a surface in the space of the independent variables, x, z and t, on which the initial conditions do not specify uniquely the second partial derivatives for equation (6.11.13) in these two spatial dimensions.

SOLUTION 7.8. In two spatial dimensions, we explicitly write equation (6.11.13) as

$$\frac{\partial^2 u(x,z,t)}{\partial x^2}+\frac{\partial^2 u(x,z,t)}{\partial z^2}=\frac{1}{v^2(x,z)}\frac{\partial^2 u(x,z,t)}{\partial t^2}. \tag{7.4.12}$$

Let the required surface be $\psi(x,z)=t$. On this surface, $u=u(x,z,\psi(x,z))$ and the initial conditions are the value of u given by

$$u(x,z,\psi(x,z))=g(x,z) \tag{7.4.13}$$

and the value of the directional derivative of u given by

$$\frac{\partial u(x,z,\psi(x,z))}{\partial \mathbf{n}}=h(x,z), \tag{7.4.14}$$

where $\mathbf{n}$ is a vector normal to the level-set given by $\psi(x,z)-t=0$, namely,

$$\begin{aligned}\mathbf{n} &= \left[\frac{\partial[\psi(x,z)-t]}{\partial x},\frac{\partial[\psi(x,z)-t]}{\partial z},\frac{\partial[\psi(x,z)-t]}{\partial t}\right]\\ &= \left[\frac{\partial\psi}{\partial x},\frac{\partial\psi}{\partial z},-1\right].\end{aligned}$$

First let us consider the first derivatives, namely, $\partial u/\partial x$, $\partial u/\partial z$ and $\partial u/\partial t$. condition (7.4.13) provides two equations. To get them, we differentiate condition (7.4.13) with respect to x, to get

$$\frac{\partial}{\partial x}u(x,z,\psi(x,z))=\frac{\partial u}{\partial x}+\frac{\partial u}{\partial t}\frac{\partial\psi}{\partial x}=\frac{\partial g}{\partial x},$$

where we used the fact that ψ is tantamount to t. Differentiating with respect to z, we get

$$\frac{\partial}{\partial z}u(x,z,\psi(x,z))=\frac{\partial u}{\partial z}+\frac{\partial u}{\partial t}\frac{\partial\psi}{\partial z}=\frac{\partial g}{\partial z}.$$

We can explicitly write condition (7.4.14) as

$$\begin{aligned}\left[\frac{\partial u}{\partial x},\frac{\partial u}{\partial z},\frac{\partial u}{\partial t}\right]\cdot\mathbf{n} &= \left[\frac{\partial u}{\partial x},\frac{\partial u}{\partial z},\frac{\partial u}{\partial t}\right]\cdot\left[\frac{\partial\psi}{\partial x},\frac{\partial\psi}{\partial z},-1\right]\\ &= \frac{\partial u}{\partial x}\frac{\partial\psi}{\partial x}+\frac{\partial u}{\partial z}\frac{\partial\psi}{\partial z}-\frac{\partial u}{\partial t}=h(x,z),\end{aligned}$$

which is the third equation. Now, we can write these three equations as a system of linear algebraic equations given by

$$\begin{bmatrix} 1 & 0 & \frac{\partial\psi}{\partial x} \\ 0 & 1 & \frac{\partial\psi}{\partial z} \\ \frac{\partial\psi}{\partial x} & \frac{\partial\psi}{\partial z} & -1 \end{bmatrix} \begin{bmatrix} \frac{\partial u}{\partial x} \\ \frac{\partial u}{\partial z} \\ \frac{\partial u}{\partial t} \end{bmatrix} = \begin{bmatrix} \frac{\partial g}{\partial x} \\ \frac{\partial g}{\partial z} \\ h \end{bmatrix}.$$

Since, in general, the right-hand side is not zero, to uniquely solve this system for $\partial u/\partial x$, $\partial u/\partial z$ and $\partial u/\partial t$, we require that the determinant of the coefficient matrix be nonzero; we require that

$$\left(\frac{\partial\psi}{\partial x}\right)^2 + \left(\frac{\partial\psi}{\partial z}\right)^2 + 1 \neq 0. \tag{7.4.15}$$

This condition is always satisfied. Hence, we can always uniquely solve for the first partial derivatives of u. Let us denote these solutions as $\partial u/\partial x :=$ $U_x(x,t)$, $\partial u/\partial z := U_z(x,t)$ and $\partial u/\partial t := U_t(x,t)$.

To find the second partial derivatives, we take partial derivatives of these three solutions with respect to x and with respect to z. We obtain

$$\frac{\partial U_x}{\partial x} = \frac{\partial}{\partial x}\left[\frac{\partial u}{\partial x}(x,z,\psi(x,z))\right] = \frac{\partial^2 u}{\partial x^2} + \frac{\partial^2 u}{\partial x\partial t}\frac{\partial\psi}{\partial x}$$

and

$$\frac{\partial U_x}{\partial z} = \frac{\partial}{\partial z}\left[\frac{\partial u}{\partial x}(x,z,\psi(x,z))\right] = \frac{\partial^2 u}{\partial z\partial x} + \frac{\partial^2 u}{\partial x\partial t}\frac{\partial\psi}{\partial z}.$$

In view of the equality of mixed partial derivatives, $\partial U_x/\partial z = \partial U_z/\partial x$, so we proceed directly to

$$\frac{\partial U_z}{\partial z} = \frac{\partial}{\partial z}\left[\frac{\partial u}{\partial z}(x,z,\psi(x,z))\right] = \frac{\partial^2 u}{\partial z^2} + \frac{\partial^2 u}{\partial z\partial t}\frac{\partial\psi}{\partial z}.$$

Recalling that the third argument, ψ, is tantamount to t, we get

$$\frac{\partial U_t}{\partial x} = \frac{\partial}{\partial x}\left[\frac{\partial u}{\partial t}(x,z,\psi(x,z))\right] = \frac{\partial^2 u}{\partial x\partial t} + \frac{\partial^2 u}{\partial t^2}\frac{\partial\psi}{\partial x}$$

and

$$\frac{\partial U_t}{\partial z} = \frac{\partial^2 u}{\partial z \partial t} + \frac{\partial^2 u}{\partial t^2}\frac{\partial \psi}{\partial z}.$$

As the last equation for the second partial derivatives, we take equation (7.4.12). These six equations can be written as a system of linear algebraic equations given by

$$\begin{bmatrix} 1 & 0 & 0 & 0 & 0 & \frac{\partial \psi}{\partial x} \\ 0 & 0 & 0 & 1 & 0 & \frac{\partial \psi}{\partial z} \\ 0 & 1 & 0 & 0 & \frac{\partial \psi}{\partial z} & 0 \\ 0 & 0 & \frac{\partial \psi}{\partial x} & 0 & 0 & 1 \\ 0 & 0 & \frac{\partial \psi}{\partial z} & 0 & 1 & 0 \\ 1 & 1 & -\frac{1}{v^2} & 0 & 0 & 0 \end{bmatrix} \begin{bmatrix} \frac{\partial^2 u}{\partial x^2} \\ \frac{\partial^2 u}{\partial z^2} \\ \frac{\partial^2 u}{\partial t^2} \\ \frac{\partial^2 u}{\partial x \partial z} \\ \frac{\partial^2 u}{\partial z \partial t} \\ \frac{\partial^2 u}{\partial t \partial x} \end{bmatrix} = \begin{bmatrix} \frac{\partial U_x}{\partial x} \\ \frac{\partial U_x}{\partial z} \\ \frac{\partial U_z}{\partial z} \\ \frac{\partial U_t}{\partial x} \\ \frac{\partial U_t}{\partial z} \\ 0 \end{bmatrix},$$

where we used the equality of mixed partial derivatives.

For this system not to have a unique solution, we require that the determinant of the coefficient matrix be zero. Thus, we require that

$$\left(\frac{\partial \psi}{\partial x}\right)^2 + \left(\frac{\partial \psi}{\partial z}\right)^2 = \frac{1}{v^2(x,z)}, \tag{7.4.16}$$

which is eikonal equation (6.11.22) in two dimensions.

This derivation shows that eikonal equation (7.4.16) is the characteristic equation of equation (7.4.12). Since equation (7.4.12) is a second-order partial differential equation, we require that the initial condition uniquely specify the second partial derivatives. $\psi(x,z)$ is a surface on which we cannot specify the initial conditions so as to find uniquely the second partial derivatives; it is called the characteristic surface.[6] Along this surface, the

---

[6] Readers interested in a rigorous formulation and application of the definition of the characteristic surface might refer to Bóna, A. and Slawinski, M.A., (2021) Wavefronts and rays as characteristics and asymptotics (3*rd* edition): World Scientific, Chapter 2.

behaviour of u is predetermined by the original differential equation, which itself is rooted in the physical laws of equations of motion.

CHAPTER 8

Hamilton's Ray Equations

> It is a common physical knowledge that wavefields, rather than rays, are physical reality. None the less, the traditions to endow rays with certain physical properties, traced back to Descartes times, have been deeply rooted in natural sciences. Rays are discussed as if they were real objects.
>
> *Yuri Aleksandrovich Kravtsov and Yuri Ilyich Orlov (1999)*
> *Caustics, catastrophes and wavefields*

Preliminary Remarks

In Chapter 7, we obtained the eikonal equation that gives us the magnitude of phase slowness as a function of the properties of an anisotropic inhomogeneous continuum through which the wavefront propagates. In this chapter, we focus our attention on the solution of the eikonal equation.

We begin this chapter by using the method of characteristics to solve the eikonal equation, which is a first-order nonlinear partial differential equation. This solution leads to a system of first-order linear ordinary differential equations that describe the curves that form the solution surface in the **xp**-space.[1] These are the characteristic equations. Parametrizing the

[1] In classical mechanics, the **xp**-space corresponds to the momentum phase space. In this book, however, to avoid the confusion with the term "phase" that we use in the specific context of wave phenomena, we do not use this nomenclature.

characteristic equations in terms of time, we obtain Hamilton's ray equations, whose solutions give the trajectory of a signal propagating through an anisotropic inhomogeneous continuum, and which are the key equations of ray theory. Subsequently, we relate the orientations and magnitudes of vectors $\mathbf{p}$ and $\dot{\mathbf{x}}$, which result in expressions relating phase and ray angles as well as phase and ray velocities. We conclude the chapter with two examples that illustrate the Hamiltonian approach.

Readers who are not familiar with Euler's homogeneous-function theorem might find it useful to study this chapter together with Appendix A.

8.1. Method of Characteristics

8.1.1. Level-set functions

The eikonal equation is a first-order nonlinear partial differential equation. It is possible to transform this equation into a system of first-order ordinary differential equations by using the method of characteristics. Then, the solutions of the ordinary differential equations are given as the characteristic curves, which compose the solution surface of the original partial differential equation.

Consider eikonal equation (7.3.8), namely,

$$p^2 = \frac{1}{v^2(\mathbf{x},\mathbf{p})}, \tag{8.1.1}$$

where, $p^2 = \mathbf{p}\cdot\mathbf{p}$, and, in view of definition (7.3.2),

$$p_i := \frac{\partial \psi}{\partial x_i}, \qquad i \in \{1,2,3\}. \tag{8.1.2}$$

We wish to solve this equation for $\mathbf{p}(\mathbf{x})$. In other words, at every point $\mathbf{x}$ we are looking for vector $[\partial\psi/\partial x_1, \partial\psi/\partial x_2, \partial\psi/\partial x_3]$. Once, we get this vector, we can integrate it to obtain ψ—the solution of our partial differential equation.

The method that we are about to describe is specifically designed for the first-order partial differential equations that depend on the independent variables, $\mathbf{x}$, and on the first derivatives of function ψ, namely, $\nabla\psi(\mathbf{x})$. The eikonal equation is such a partial differential equation. This method does not apply to differential equations that also depend on function ψ itself.[2]

The solution of the eikonal equation is a surface in the **xp**-space. We have a choice of several implicit descriptions of this surface as level sets of a function that we denote by $F(\mathbf{x},\mathbf{p})$. The two obvious choices are

$$F(\mathbf{x},\mathbf{p}) = p^2 - \frac{1}{v^2(\mathbf{x},\mathbf{p})}, \tag{8.1.3}$$

and

$$F(\mathbf{x},\mathbf{p}) = p^2 v^2(\mathbf{x},\mathbf{p}). \tag{8.1.4}$$

This way, in view of eikonal equation (8.1.1), the surfaces are the level sets of functions (8.1.3) or (8.1.4), given by

$$F(\mathbf{x},\mathbf{p}) = 0, \tag{8.1.5}$$

and

$$F(\mathbf{x},\mathbf{p}) = 1, \tag{8.1.6}$$

respectively. Since each formulation has different advantages, both are used in various sections of this book.

8.1.2. Characteristic equations

We seek to construct the solution given by $\mathbf{p} = \mathbf{p}(\mathbf{x})$, such that equation (8.1.5) or equation (8.1.6) is satisfied. In both cases, since $F(\mathbf{x},\mathbf{p}(\mathbf{x}))$ is constant, it follows that $\mathrm{d}F = 0$, where F is treated as a function of $\mathbf{x}$ only.

Treating F as a function of $\mathbf{x}$ only, which we denote by $F(\mathbf{x},\mathbf{p}(\mathbf{x}))$, means that we constrain our consideration of F to $\mathbf{p}(\mathbf{x})$—the solution we seek—as opposed to studying $F(\mathbf{x},\mathbf{p})$, which refers to function F in the entire **xp**-space.

---

[2] Readers interested in the method of characteristics for solving general first-order partial differential equations might refer to Bleistein, N., (1984) Mathematical methods for wave phenomena: Academic Press, pp. 1–27, to Courant, R., and Hilbert, D., (1989) Methods of mathematical physics: John Wiley & Sons., Vol. II, Chapter II, to McOwen, R.C., (1996) Partial differential equations: Methods and applications: Prentice-Hall, Inc., pp. 29–38, and to Spivak, M., (1970/1999) A comprehensive introduction to differential geometry: Publish or Perish, Inc., pp. 3–28.

Using the chain rule, we can explicitly state the differential of F as

$$\mathrm{d}F\left[\mathbf{x},\mathbf{p}\left(\mathbf{x}\right)\right]=\sum_{i=1}^{3}\frac{\partial F}{\partial x_i}\,\mathrm{d}x_i+\sum_{i=1}^{3}\sum_{j=1}^{3}\frac{\partial F}{\partial p_j}\frac{\partial p_j}{\partial x_i}\,\mathrm{d}x_i=0\,.$$

Using definition (8.1.2), we can express it in terms of the eikonal function, ψ, as

$$\mathrm{d}F\left[\mathbf{x},\mathbf{p}\left(\mathbf{x}\right)\right]=\sum_{i=1}^{3}\frac{\partial F}{\partial x_i}\,\mathrm{d}x_i+\sum_{i=1}^{3}\sum_{j=1}^{3}\frac{\partial F}{\partial p_j}\frac{\partial}{\partial x_i}\frac{\partial \psi}{\partial x_j}\,\mathrm{d}x_i=0\,.$$

Since $\mathrm{d}x_i\neq 0$, using the equality of mixed partial derivatives and considering a given i, we can factor out $\mathrm{d}x_i$ and write

$$\frac{\partial F}{\partial x_i}+\sum_{j=1}^{3}\frac{\partial F}{\partial p_j}\frac{\partial}{\partial x_j}\frac{\partial \psi}{\partial x_i}=0\,,\qquad i\in\{1,2,3\}\,,$$

which are second-order partial differential equations. Again, using definition (8.1.2), we can rewrite these equations as

$$\frac{\partial F}{\partial x_i}+\sum_{j=1}^{3}\frac{\partial F}{\partial p_j}\frac{\partial p_i}{\partial x_j}=0\,,\qquad i\in\{1,2,3\}\,. \tag{8.1.7}$$

For each $i\in\{1,2,3\}$, we wish to find curves $\left[\mathbf{x}\left(s\right),p_i\left(s\right)\right]$ in the solution surface $p_i=p_i\left(\mathbf{x}\right)$. This way we construct the solution surface as a union of these curves, which are commonly referred to as characteristics. To do so, we desire to obtain vectors tangent to the solution surface. To obtain these vectors, we use geometrical properties of vectors in the context of the solution surface.

Let us consider a given $i\in\{1,2,3\}$. Equation (8.1.7) can be written as a scalar product of two vectors given by

$$\left[\frac{\partial F}{\partial p_1},\frac{\partial F}{\partial p_2},\frac{\partial F}{\partial p_3},-\frac{\partial F}{\partial x_i}\right]\cdot\left[\frac{\partial p_i}{\partial x_1},\frac{\partial p_i}{\partial x_2},\frac{\partial p_i}{\partial x_3},-1\right]=0\,. \tag{8.1.8}$$

Following the properties of the scalar product, we conclude that these two vectors are orthogonal to one another in the four-dimensional $x_1x_2x_3p_i$-space.

For a given $i \in \{1, 2, 3\}$, we can write the solution surface, $p_i = p_i(\mathbf{x})$, as a level set of the function given by

$$g_i(\mathbf{x}, p_i) = p_i(\mathbf{x}) - p_i, \tag{8.1.9}$$

where p_i is the value of function $p_i(\mathbf{x})$ at point $\mathbf{x}$. Herein, since the right-hand side of equation (8.1.9) vanishes identically, $g_i(\mathbf{x}, p_i)$ is a zero set. Using properties of the gradient operator, we can obtain the vector normal to $g_i(\mathbf{x}, p_i)$, namely,

$$\mathbf{n}_i = \nabla g_i = \left[\frac{\partial g_i}{\partial x_1}, \frac{\partial g_i}{\partial x_2}, \frac{\partial g_i}{\partial x_3}, \frac{\partial g_i}{\partial p_i}\right]. \tag{8.1.10}$$

In view of equation (8.1.9), we see that a vector normal to $g_i(\mathbf{x}, p_i)$ is also normal to the solution surface, $p_i = p_i(\mathbf{x})$. Thus, inserting $g_i(\mathbf{x}, p_i)$, given in expression (8.1.9), into expression (8.1.10), we can write

$$\mathbf{n}_i = \left[\frac{\partial(p_i(\mathbf{x}) - p_i)}{\partial x_1}, \frac{\partial(p_i(\mathbf{x}) - p_i)}{\partial x_2}, \frac{\partial(p_i(\mathbf{x}) - p_i)}{\partial x_3}, \frac{\partial(p_i(\mathbf{x}) - p_i)}{\partial p_i}\right],$$

to obtain vector

$$\mathbf{n}_i = \left[\frac{\partial p_i}{\partial x_1}, \frac{\partial p_i}{\partial x_2}, \frac{\partial p_i}{\partial x_3}, -1\right], \tag{8.1.11}$$

which is normal to solution surface, $p_i = p_i(\mathbf{x})$ in the four-dimensional $x_1x_2x_3p_i$-space.

Examining equations (8.1.8) and (8.1.11), we realize that

$$\mathbf{n}_i \perp \left[\frac{\partial F}{\partial p_1}, \frac{\partial F}{\partial p_2}, \frac{\partial F}{\partial p_3}, -\frac{\partial F}{\partial x_i}\right].$$

Thus, for a given $i \in \{1,2,3\}$, vector $[\partial F/\partial p_1, \partial F/\partial p_2, \partial F/\partial p_3, -\partial F/\partial x_i]$ is tangent to the solution surface, $p_i = p_i(\mathbf{x})$. We denote this vector by $\mathbf{t}_i$. Hence, for a given $i \in \{1, 2, 3\}$, we have obtained vectors tangent to the solution surface, as desired.

Curves $[x_1(s), x_2(s), x_3(s), p_i(s)]$ that are in the solution surface and whose tangent vector is $\mathbf{t}_i$ is the solution of a system of first-order ordinary differential equations, namely,

$$\begin{cases} \dfrac{\mathrm{d}x_1(s)}{\mathrm{d}s} = \zeta \dfrac{\partial F}{\partial p_1} \\ \dfrac{\mathrm{d}x_2(s)}{\mathrm{d}s} = \zeta \dfrac{\partial F}{\partial p_2} \\ \dfrac{\mathrm{d}x_3(s)}{\mathrm{d}s} = \zeta \dfrac{\partial F}{\partial p_3} \\ \dfrac{\mathrm{d}p_i(s)}{\mathrm{d}s} = -\zeta \dfrac{\partial F}{\partial x_i} \end{cases},$$

which we can concisely write as

$$\begin{cases} \dfrac{\mathrm{d}x_j(s)}{\mathrm{d}s} = \zeta \dfrac{\partial F}{\partial p_j} \\ \\ \dfrac{\mathrm{d}p_i(s)}{\mathrm{d}s} = -\zeta \dfrac{\partial F}{\partial x_i} \end{cases}, \qquad j \in \{1, 2, 3\}, \tag{8.1.12}$$

where ζ is a scaling factor and s is the parameter along the curve. The choice of ζ determines the parametrization, which we use in Section 8.2. System (8.1.12) describes curves that are associated with solution surface $p_i = p_i(\mathbf{x})$ and—for a given $i \in \{1, 2, 3\}$—exist in a four-dimensional space, $x_1x_2x_3p_i$.

We note that the solutions of system (8.1.12) depend on the initial conditions, which we can write as $x_j(0) = x_j^0$ and $p_i(0) = p_i^0$, where $i, j \in \{1, 2, 3\}$. However, these initial conditions are not arbitrary; they must satisfy the differential equation given in expression (8.1.5) or (8.1.6).

Since the above derivation, which was shown for a given i, must hold for each $i \in \{1, 2, 3\}$, we can write equations (8.1.12) as

$$\begin{cases} \dfrac{\mathrm{d}x_j}{\mathrm{d}s} = \zeta \dfrac{\partial F}{\partial p_j} \\ \\ \dfrac{\mathrm{d}p_i}{\mathrm{d}s} = -\zeta \dfrac{\partial F}{\partial x_i} \end{cases}, \qquad i, j \in \{1, 2, 3\},$$

which, in view of i and j being arbitrary indices, we can restate as

$$\left\{\begin{aligned}\frac{\mathrm{d}x_i}{\mathrm{d}s} &= \zeta\frac{\partial F}{\partial p_i}\\ \frac{\mathrm{d}p_i}{\mathrm{d}s} &= -\zeta\frac{\partial F}{\partial x_i}\end{aligned}\right., \qquad i \in \{1,2,3\}. \tag{8.1.13}$$

The solution of system (8.1.13) are curves that compose solution surface $\mathbf{p} = \mathbf{p}(\mathbf{x})$ in the six-dimensional $\mathbf{xp}$-space. Such curves are the characteristics of eikonal equation (8.1.1).

Hence, three second-order partial differential equations (8.1.7) become six first-order ordinary differential equations (8.1.13). These are the characteristic equations of eikonal equation (8.1.1). The solution of characteristic equations (8.1.13) consists of curves that compose the solution surface of eikonal equation (8.1.1).

8.1.3. Consistency of formulation

As stated in Section 8.1.1, there are two obvious forms of function F. Functions (8.1.3) and (8.1.4) differ in certain aspects, such as their homogeneity with respect to the variables p_i. However, as stated by the following lemma, they both result in the same characteristic equations and, hence, the same characteristic curves.

LEMMA 8.1.1. *Both formulations of the function given by expressions (8.1.3) and (8.1.4) result in the same characteristic curves.*

PROOF. Consider characteristic equations (8.1.13), namely,

$$\left\{\begin{aligned}\frac{\mathrm{d}x_i}{\mathrm{d}s} &= \zeta\frac{\partial F}{\partial p_i}\\ \frac{\mathrm{d}p_i}{\mathrm{d}s} &= -\zeta\frac{\partial F}{\partial x_i}\end{aligned}\right., \qquad i \in \{1,2,3\}.$$

Letting $F = p^2 v^2 (\mathbf{x}, \mathbf{p})$ and setting $\zeta = 1$, we note that equations (8.1.13) become

$$\left\{ \begin{aligned} \frac{\mathrm{d}x_i}{\mathrm{d}s} &= 2\left(p_i v^2 + p^2 v \frac{\partial v}{\partial p_i}\right) \\ \frac{\mathrm{d}p_i}{\mathrm{d}s} &= -2p^2 v \frac{\partial v}{\partial x_i} \end{aligned} \right. , \qquad i \in \{1,2,3\} .$$

If we let $F = p^2 - 1/v^2 (\mathbf{x}, \mathbf{p})$, then equations (8.1.13) become

$$\left\{ \begin{aligned} \frac{\mathrm{d}x_i}{\mathrm{d}s} &= 2\zeta\left(p_i + \frac{1}{v^3}\frac{\partial v}{\partial p_i}\right) \\ \frac{\mathrm{d}p_i}{\mathrm{d}s} &= -2\zeta \frac{1}{v^3}\frac{\partial v}{\partial x_i} \end{aligned} \right. , \qquad i \in \{1,2,3\} .$$

Equating the second equations of each set, we can write

$$p^2 v \frac{\partial v}{\partial x_i} = \zeta \frac{1}{v^3}\frac{\partial v}{\partial x_i}, \qquad i \in \{1,2,3\} .$$

Solving for ζ, we obtain

$$\zeta = \frac{p^2 v \dfrac{\partial v}{\partial x_i}}{\dfrac{1}{v^3}\dfrac{\partial v}{\partial x_i}} = p^2 v^4, \qquad i \in \{1,2,3\} .$$

Substituting $\zeta = p^2 v^4$ into the first equation of the second set, we obtain

$$\frac{\mathrm{d}x_i}{\mathrm{d}s} = 2p^2 v^4 \left(p_i + \frac{1}{v^3}\frac{\partial v}{\partial p_i}\right) = 2\left(p_i p^2 v^4 + p^2 v \frac{\partial v}{\partial p_i}\right), \qquad i \in \{1,2,3\} ,$$

which is equivalent to the first equation of the first set along $p^2 v^2 = 1$. □

Thus, following equations (8.1.13), both equations (8.1.5) and (8.1.6), yield the same characteristic curves, given that $\zeta = v^2$ and $\zeta = 1$, respectively.

8.2. Time Parametrization of Characteristic Equations

8.2.1. General formulation

Different choices of ζ result in different parametrization of the solution curves for the characteristic equations. For seismological studies, it is often convenient to parametrize characteristic equations (8.1.13) in terms of time. Recall equation (7.2.3), which we can rewrite as

$$\psi(\mathbf{x}) = t + t_i ,$$

where t denotes time and t_i is a constant. Differentiating with respect to s, we obtain

$$\frac{\mathrm{d}\psi(\mathbf{x})}{\mathrm{d}s} = \frac{\mathrm{d}t}{\mathrm{d}s} .$$

This equation governs the propagation of $\psi(\mathbf{x})$ along the characteristic curves. The physical interpretation of parameter s depends on the choice of scaling factor ζ in system (8.1.13). If the parameter s is to be equivalent to time, t, we require that

$$\frac{\mathrm{d}\psi(\mathbf{x})}{\mathrm{d}s} = 1 .$$

We can restate the above condition as

$$\frac{\mathrm{d}\psi(\mathbf{x})}{\mathrm{d}s} = \sum_{i=1}^{3} \frac{\partial \psi}{\partial x_i} \frac{\mathrm{d}x_i}{\mathrm{d}s} = 1 . \tag{8.2.1}$$

Using definition (8.1.2), we rewrite condition (8.2.1) as

$$\frac{\mathrm{d}\psi(\mathbf{x})}{\mathrm{d}s} = \sum_{i=1}^{3} p_i \frac{\mathrm{d}x_i}{\mathrm{d}s} = 1 , \tag{8.2.2}$$

which is a condition for the time parametrization of characteristic equations (8.1.13).

8.2.2. Equations with variable scaling factor

To obtain the time parametrization of system (8.1.13) in the context of function (8.1.3), we can write $\mathrm{d}x_i/\mathrm{d}s = \zeta \partial F / \partial p_i$, where $F = p^2 - 1/v^2$, as

$$\frac{\mathrm{d}x_i}{\mathrm{d}s} = 2\zeta \left(p_i + \frac{1}{v^3} \frac{\partial v}{\partial p_i} \right) , \qquad i \in \{1, 2, 3\} .$$

In view of condition (8.2.2), we require that

$$\sum_{i=1}^{3} p_i \frac{\mathrm{d}x_i}{\mathrm{d}s} = 2\zeta \sum_{i=1}^{3} p_i \left(p_i + \frac{1}{v^3} \frac{\partial v}{\partial p_i} \right) = 2\zeta \left(p^2 + \frac{1}{v^3} \sum_{i=1}^{3} p_i \frac{\partial v}{\partial p_i} \right) = 1 .$$

Since v is homogeneous of degree 0 in the p_i, the summation on the right-hand side vanishes by Theorem A.2.1. Thus, we obtain

$$\sum_{i=1}^{3} p_i \frac{\mathrm{d}x_i}{\mathrm{d}s} = 2\zeta p^2 = 1 ,$$

and solving for ζ, we immediately get $\zeta = 1/\left(2p^2\right)$.

Consequently, given function (8.1.3), the system of characteristic equations (8.1.13) that is parametrized in terms of time becomes

$$\left\{ \begin{aligned} \dot{x}_i &= \frac{1}{2p^2} \frac{\partial F}{\partial p_i} \\ \dot{p}_i &= -\frac{1}{2p^2} \frac{\partial F}{\partial x_i} \end{aligned} \right. , \qquad i \in \{1,2,3\} , \tag{8.2.3}$$

where $\dot{x}_i := \mathrm{d}x_i/\mathrm{d}t$ and $\dot{p}_i := \mathrm{d}p_i/\mathrm{d}t$. Equations (8.2.3) are characteristic equations (8.1.13) whose scaling factor is a function of the p_i. In view of eikonal equation (8.1.1), we can also state this scaling factor as $v^2(\mathbf{x}, \mathbf{p})/2$.

An implication of this parametrization is shown in Exercise 8.8. An implication of another parametrization of characteristic equations (8.1.13) in the context of function (8.1.3) is shown in Exercise 8.7.

8.2.3. Equations with constant scaling factor

To obtain the time parametrization of system (8.1.13) in the context of function (8.1.4), we can write $\mathrm{d}x_i/\mathrm{d}s = \zeta \partial F/\partial p_i$, where $F = p^2 v^2$, as

$$\frac{\mathrm{d}x_i}{\mathrm{d}s} = 2\zeta \left(p_i v^2 + p^2 v \frac{\partial v}{\partial p_i} \right) , \qquad i \in \{1,2,3\} .$$

In view of condition (8.2.2), we require

$$\sum_{i=1}^{3} p_i \frac{\mathrm{d}x_i}{\mathrm{d}s} = 2\zeta \sum_{i=1}^{3} p_i \left(p_i v^2 + p^2 v \frac{\partial v}{\partial p_i} \right) = 2\zeta \left(p^2 v^2 + p^2 v \sum_{i=1}^{3} p_i \frac{\partial v}{\partial p_i} \right) = 1 .$$

Following the eikonal equation, the first product in parentheses on the right-hand side is equal to unity. Since v is homogeneous of degree 0 in the p_i, the summation on the right-hand side vanishes by Theorem A.2.1. Thus, we obtain

$$\sum_{i=1}^{3} p_i \frac{\mathrm{d}x_i}{\mathrm{d}s} = 2\zeta = 1,$$

and solving for ζ, we immediately get $\zeta = 1/2$.

Consequently, given function (8.1.4), system (8.1.13) is parametrized in terms of time if

$$\begin{cases} \dot{x}_i = \dfrac{1}{2}\dfrac{\partial F}{\partial p_i} \\ \\ \dot{p}_i = -\dfrac{1}{2}\dfrac{\partial F}{\partial x_i} \end{cases}, \qquad i \in \{1,2,3\}, \tag{8.2.4}$$

where $\dot{x}_i := \mathrm{d}x_i/\mathrm{d}t$ and $\dot{p}_i := \mathrm{d}p_i/\mathrm{d}t$. Equations (8.2.4) are characteristic equations (8.1.13) whose scaling factor is the constant equal to $1/2$.

In view of functions (8.1.3) and (8.1.4), the corresponding scaling factors, $\zeta = v^2/2$ and $\zeta = 1/2$, are consistent with one another. This can be seen by examining the proof of Lemma 8.1.1.

8.2.4. Formulation of Hamilton's ray equations

We now examine systems (8.2.3) and (8.2.4), and choose to proceed with the latter one since, therein, ζ is given by a constant. This constant can be brought inside the differential operator and we can write system (8.2.4) as

$$\begin{cases} \dot{x}_i = \dfrac{\partial}{\partial p_i}\left(\dfrac{F}{2}\right) \\ \\ \dot{p}_i = -\dfrac{\partial}{\partial x_i}\left(\dfrac{F}{2}\right) \end{cases}, \qquad i \in \{1,2,3\}. \tag{8.2.5}$$

Let us denote

$$\mathscr{H} := \frac{F}{2}, \tag{8.2.6}$$

where $\mathscr{H}$ is referred to as the ray-theory Hamiltonian[3]. Now, we can write equations (8.2.5) as

$$\begin{cases} \dot{x}_i = \dfrac{\partial \mathscr{H}}{\partial p_i} \\ \\ \dot{p}_i = -\dfrac{\partial \mathscr{H}}{\partial x_i} \end{cases}, \qquad i \in \{1, 2, 3\} . \tag{8.2.7}$$

Equations (8.2.7) constitute a system of first-order ordinary differential equations in t for $\mathbf{x}(t)$ and $\mathbf{p}(t)$. These equations are Hamilton's ray equations. System (8.2.7) governs the signal trajectories in the $\mathbf{xp}$-space spanned by the position vectors, $\mathbf{x}$, and the phase-slowness vectors, $\mathbf{p}$.

The first set of equations of system (8.2.7) corresponds to the components of vectors tangent to curves $\mathbf{x}(t)$. These curves belong to the physical space. They are the trajectories along which signals propagate and, in the context of ray theory, they are rays.

The second set of equations of system (8.2.7) describes the rate of change of the phase slowness. If $\mathscr{H}$ is not explicitly a function of a given x_i, we obtain $\dot{p}_i = 0$, which implies that p_i is constant along the ray. Hence, in such a case, p_i is a conserved quantity, known as the ray parameter, which is discussed in Chapter 14. Physically, this means that $v(\mathbf{x}, \mathbf{p})$ is not explicitly a function of x_i and, hence, the continuum is homogeneous along that component.

Note that, in the context of Legendre's transformation, discussed in Appendix B, the first set of equations can be viewed as a definition of a variable, while the essence of the physical formulation is contained in the second set of equations.

The ray-theory Hamiltonian, $\mathscr{H}$, resulting from function (8.1.4), can be explicitly stated as

$$\mathscr{H} = \frac{1}{2} p^2 v^2 (\mathbf{x}, \mathbf{p}) . \tag{8.2.8}$$

It is a dimensionless quantity, in contrast to the classical-mechanics Hamiltonian, discussed in Chapter 13, which has units of energy. In view of

[3] In this book we use two distinct Hamiltonians denoted by $\mathscr{H}$ and $\mathbb{H}$. Consequently, in the text, we avoid a generic reference to "the Hamiltonian", unless it is clear from the context which one of the two is considered.

eikonal equation (7.3.8), which states that $p^2 v^2 = 1$, and expression (8.2.8), we require that $\mathscr{H}(\mathbf{x}, \mathbf{p}) = 1/2$, along a ray.

8.3. Physical Interpretation of Hamilton's Ray Equations and Solutions

8.3.1. Equations

Ray velocity

The first set of equations of system (8.2.7), namely, $\dot{x}_i = \partial \mathscr{H} / \partial p_i$, states the components of vector $\dot{\mathbf{x}}$, which is tangent to the ray, $\mathbf{x}(t)$. Since the right-hand sides of these equations are expressed in terms of the p_i, which are the components of the phase-slowness vector, this set of equations relates ray and phase velocities.

Ray orientation

The relations among the $\dot{x}_i$, which appear in the first set of equations of system (8.2.7), give us the orientation of the ray, $\mathbf{x}(t)$. Since the right-hand sides are expressed in terms of the p_i, this set of equations relates ray and wavefront orientations. This is illustrated in Section 8.5.5.

Wavefront orientation

The relations among the $\dot{p}_i$, which appear in the second set of equations of system (8.2.7), namely, $\dot{p}_i = -\partial \mathscr{H} / \partial x_i$, give us the orientation of the wavefront, $\psi(\mathbf{x})$. This is illustrated in Section 8.5.5.

8.3.2. Solutions

Ray

The solution of the first set of equations of system (8.2.7) is $\mathbf{x}(t)$, which is the expression for the ray. This is illustrated in Section 8.5.6 on page 373.

Wavefront velocity

The solution of the second set of equations of system (8.2.7) is $\mathbf{p}(t)$, which is the expression for the vector normal to the wavefront, $\psi(\mathbf{x})$. The magnitude of this vector is the wavefront slowness. This is illustrated in Section 8.5.8.

Traveltime

Since both $\mathbf{x}(t)$ and $\mathbf{p}(t)$ are parametrized by time, t, we can obtain the expression for t by solving any $x_i(t)$ or $p_i(t)$ for t. This is illustrated in Section 8.5.7.

8.4. Relation between p and ẋ

8.4.1. General formulation

We wish to study the relation between the orientations and the magnitudes of vectors $\mathbf{p}$ and $\dot{\mathbf{x}}$. Physically, $\mathbf{p}$ is the vector normal to the wavefront and $\dot{\mathbf{x}}$ is the vector tangent to the ray. Mathematically, the components of these vectors are the variables of Legendre's transformation discussed in Appendix B and used in Chapter 11.

Consider a given point $\mathbf{x}$ of the continuum and, therein, the directional dependence of $\mathscr{H}$. The first set of equations of system (8.2.7) is

$$\dot{x}_i = \frac{\partial \mathscr{H}}{\partial p_i}, \qquad i \in \{1,2,3\}. \tag{8.4.1}$$

Inserting expression (8.2.8), namely,

$$\mathscr{H} = \frac{1}{2} p^2 v^2 (\mathbf{x}, \mathbf{p}),$$

into equations (8.4.1) and using the equality resulting from the eikonal equation, namely, $p^2 v^2 = 1$, we obtain

$$\dot{x}_i = p_i v^2 + \frac{1}{v} \frac{\partial v}{\partial p_i}, \qquad i \in \{1,2,3\}, \tag{8.4.2}$$

where the phase-velocity function, v, is a function of the orientation of vector $\mathbf{p}$. Thus, we have obtained the relation between the components of vectors $\mathbf{p}$ and $\dot{\mathbf{x}}$.

8.4.2. Phase and ray velocities

Vector $\dot{\mathbf{x}}$ is tangent to the ray $\mathbf{x}(t)$. Since, at a given point, this vector corresponds to the velocity of the signal along the ray at that point, we refer to it as ray velocity.[4] We wish to find the magnitude of this vector, which can be written as

$$V := |\dot{\mathbf{x}}| = \sqrt{\dot{\mathbf{x}} \cdot \dot{\mathbf{x}}} = \sqrt{\sum_{i=1}^{3} \dot{x}_i^2}. \tag{8.4.3}$$

Using expression (8.4.2), we can write each term of the summation in radicand (8.4.3) as

$$\begin{aligned}(\dot{x}_i)^2 &= \left(p_i v^2 + \frac{1}{v}\frac{\partial v}{\partial p_i}\right)^2 \\ &= (p_i)^2 v^4 + 2p_i v \frac{\partial v}{\partial p_i} + \frac{1}{v^2}\left(\frac{\partial v}{\partial p_i}\right)^2, \qquad i \in \{1,2,3\}.\end{aligned}$$

Performing the summation of the three terms, we obtain

$$\begin{aligned}\sum_{i=1}^{3}\left[(p_i)^2 v^4 + 2p_i v \frac{\partial v}{\partial p_i} + \frac{1}{v^2}\left(\frac{\partial v}{\partial p_i}\right)^2\right] &= v^4 \sum_{i=1}^{3}(p_i)^2 + 2v\sum_{i=1}^{3} p_i \frac{\partial v}{\partial p_i} \\ &\quad + \frac{1}{v^2}\sum_{i=1}^{3}\left(\frac{\partial v}{\partial p_i}\right)^2 \\ &= v^4 \sum_{i=1}^{3}(p_i)^2 + \frac{1}{v^2}\sum_{i=1}^{3}\left(\frac{\partial v}{\partial p_i}\right)^2,\end{aligned}$$

where, since v is homogeneous of degree 0 in the p_i, the summation of $p_i(\partial v/\partial p_i)$ vanished due to Theorem A.2.1.

[4] In seismology, this quantity is often referred to as the group velocity. Our nomenclature is consistent with Synge, J.L., (1937/1962) Geometrical optics: An introduction to Hamilton's methods: Cambridge University Press, p. 12, and with Born, M., and Wolf, E., (1999) Principles of optics (*7th* edition): Cambridge University Press, pp. 792–795. Also, our nomenclature appears in Winterstein, D.F., (1990) Velocity anisotropy terminology for geophysicists: Geophysics, **55**, 1070–1088, and in Helbig, K., (1994) Foundations of anisotropy for exploration seismics: Pergamon, p. 12.

Thus, in view of equality $p^2v^2 = 1$, we can write expression (8.4.3) as

$$V = \sqrt{v^2 + \frac{1}{v^2}\left(\nabla_{\mathbf{p}} v\right)^2},$$

where $\nabla_{\mathbf{p}} v$ denotes the gradient of the phase-velocity function, v, with respect to the components of the phase-slowness vector, $\mathbf{p}$. Using the chain rule and following the properties of logarithms, we obtain

$$V = \sqrt{v^2 + \left[\nabla_{\mathbf{p}}\left(\ln v\right)\right]^2}. \tag{8.4.4}$$

Expression (8.4.4) gives the magnitude of the signal velocity along the ray $\mathbf{x}(t)$. In expression (8.4.4), the magnitude of the ray velocity, V, is given in terms of the magnitude of the phase velocity, v, as a function of the orientation of the wavefront, given by the wavefront-normal vector, $\mathbf{p}$.

Two-dimensional case

To illustrate expression (8.4.4), consider a two-dimensional continuum that is contained in the x_1x_3-plane. At a given point of the continuum, we can express the orientation of the wavefront-normal vector, $\mathbf{p} = [p_1, p_3]$, in terms of a single angle. This is the phase angle, which, in this two-dimensional continuum, is given by expression (8.5.3), namely,

$$\vartheta = \arctan\frac{p_1}{p_3}. \tag{8.4.5}$$

Hence, using expression (8.4.4), the magnitude of the ray velocity can be expressed in terms of the phase velocity and the phase angle.

Herein, using expression (8.4.4), we can write

$$V = \sqrt{v^2 + \left(\frac{\partial \ln v}{\partial p_1}\right)^2 + \left(\frac{\partial \ln v}{\partial p_3}\right)^2}. \tag{8.4.6}$$

We wish to express differential operators $\partial/\partial p_i$ in terms of the phase angle. Using the chain rule, we can write

$$\frac{\partial}{\partial p_1} = \frac{\partial \vartheta}{\partial p_1}\frac{\partial}{\partial \vartheta} = \frac{\partial \arctan\frac{p_1}{p_3}}{\partial p_1}\frac{\partial}{\partial \vartheta} = \frac{p_3}{p_1^2 + p_3^2}\frac{\partial}{\partial \vartheta} = p_3 v^2 \frac{\partial}{\partial \vartheta}, \tag{8.4.7}$$

where $p_1^2 + p_3^2 = p^2 = 1/v^2$. Similarly, we obtain

$$\frac{\partial}{\partial p_3} = -p_1 v^2 \frac{\partial}{\partial \vartheta} . \tag{8.4.8}$$

Thus, expression (8.4.6) can be written as

$$\begin{aligned} V &= \sqrt{v^2 + \left(p_3 v^2 \frac{\partial \ln v}{\partial \vartheta}\right)^2 + \left(-p_1 v^2 \frac{\partial \ln v}{\partial \vartheta}\right)^2} \\ &= \sqrt{v^2 + \left(p_3^2 + p_1^2\right) v^4 \left(\frac{\partial \ln v}{\partial \vartheta}\right)^2} = \sqrt{v^2 + p^2 v^4 \left(\frac{\partial \ln v}{\partial \vartheta}\right)^2} \\ &= \sqrt{[v(\vartheta)]^2 + [v(\vartheta)]^2 \left(\frac{\partial \ln v(\vartheta)}{\partial \vartheta}\right)^2} . \end{aligned}$$

Following the chain rule, we obtain

$$V(\vartheta) = \sqrt{[v(\vartheta)]^2 + \left[\frac{\partial v(\vartheta)}{\partial \vartheta}\right]^2} , \tag{8.4.9}$$

which gives the magnitude of the ray velocity in terms of the phase velocity as a function of the phase angle.

Since, as shown in Chapter 7, phase velocity is a function of the properties of the continuum—namely, its mass density and the elasticity parameters—expression (8.4.9) gives the magnitude of the ray velocity in terms of these properties and as a function of the phase angle.

8.4.3. Phase and ray angles

To discuss the relation between the orientations of vectors **p** and $\dot{\mathbf{x}}$, consider a two-dimensional continuum that is contained in the x_1x_3-plane. Therein, the phase angle is given by expression (8.4.5). Analogously, we can express the orientation of the vector tangent to the ray, namely, $\dot{\mathbf{x}} = [\dot{x}_1, \dot{x}_3]$, in terms of a single angle. This is the ray angle, which, in this two-dimensional continuum, is given by

$$\theta = \arctan \frac{\dot{x}_1}{\dot{x}_3} . \tag{8.4.10}$$

In this two-dimensional case, expression (8.4.2) is

$$\dot{x}_i = p_i v^2 + \frac{1}{v}\frac{\partial v}{\partial p_i}, \qquad i \in \{1,3\}. \tag{8.4.11}$$

Hence, expression (8.4.10) becomes

$$\tan\theta = \frac{p_1 v^2 + \frac{1}{v}\frac{\partial v}{\partial p_1}}{p_3 v^2 + \frac{1}{v}\frac{\partial v}{\partial p_3}}.$$

We wish to express the differential operators $\partial/\partial p_i$ in terms of the phase angle. Recalling expression (8.4.7) and (8.4.8), we obtain

$$\tan\theta = \frac{p_1 v^2 + p_3 v \frac{\partial v}{\partial \vartheta}}{p_3 v^2 - p_1 v \frac{\partial v}{\partial \vartheta}} = \frac{p_1 + p_3 \frac{1}{v}\frac{\partial v}{\partial \vartheta}}{p_3 - p_1 \frac{1}{v}\frac{\partial v}{\partial \vartheta}}.$$

Recalling expression (8.4.5), we divide both the numerator and the denominator by p_3, to obtain

$$\tan\theta = \frac{\frac{p_1}{p_3} + \frac{1}{v}\frac{\partial v}{\partial \vartheta}}{1 - \frac{p_1}{p_3}\frac{1}{v}\frac{\partial v}{\partial \vartheta}} = \frac{\tan\vartheta + \frac{1}{v}\frac{\partial v}{\partial \vartheta}}{1 - \frac{\tan\vartheta}{v}\frac{\partial v}{\partial \vartheta}}. \tag{8.4.12}$$

Expression (8.4.12) relates the phase and the ray angles.

Note that, in view of standard formulations in polar coordinates, expression (8.4.12) gives the angle θ that corresponds to the vector normal to the curve $1/v(\vartheta)$.[5] We refer to this curve as the phase-slowness curve. Furthermore, as shown in Exercise 11.3, ϑ corresponds to the vector normal to the curve given by $V(\theta)$, which is the ray-velocity curve. This angular relation between the phase-slowness and ray-velocity curves, together with expression (8.4.9), which relates the magnitudes of the phase and ray velocities, results in the property that we call polar reciprocity. In other words, the phase-slowness curve is the polar reciprocal of the ray-velocity

[5] Interested readers might refer to Anton, H., (1984) Calculus with analytic geometry: John Wiley & Sons, pp. 730–731.

curve, and vice-versa. In general, the phase-slowness and ray-velocity surfaces are the polar reciprocals of one another.[6]

The possibility of solving expression (8.4.12) explicitly for ϑ depends on function v. To understand this statement, consider the following description. The explicit solution of expression (8.4.12) for ϑ requires that we can solve expression (8.4.11) for the p_i in terms of the $\dot{x}_i$. Since expression (8.4.11) is derived from expression (8.4.1), we require the solvability of the latter expression for the p_i in terms of the $\dot{x}_i$.

Note that equation (8.4.1) is equivalent to equation (B.3.7) in Appendix B, which relates the variables used in Legendre's transformation. Thus, the possibility of expressing ϑ in terms of θ belongs to the study of this transformation.

For the velocity functions formulated in the context of the elasticity theory, we conjecture that expression (8.4.12) can be explicitly solved for ϑ if and only if v^2 is quadratic in the components of a vector that specifies the orientation of the wavefront. As we see in Section 9.2.3, for *SH* waves in transversely isotropic continua, we get

$$v_{SH}^2 = \frac{1}{\rho}\left[C_{66} + (C_{44} - C_{66})\, n_3^2\right],$$

which is a quadratic function in n_3, with n_3 being a component of the unit vector normal to the wavefront; in other words, $\mathbf{n} = \mathbf{p}/|\mathbf{p}|$. In the seismological context, this quadratic dependence is tantamount to an elliptical velocity dependence. Consequently, an explicit ray-velocity expression, $V = V(\theta)$, where θ is the ray angle, appears to be possible only for elliptical velocity dependence. This expression is illustrated in Exercise 8.5.

---

[6] Interested readers might refer to Arnold, V.I., (1989) Mathematical methods of classical mechanics (2nd edition): Springer-Verlag, pp. 248–252, where the relation between the direction normal to the wavefront and the ray direction is formulated in terms of Huygens' principle, as well as to Born, M., and Wolf, E., (1999) Principles of optics (7th edition): Cambridge University Press, pp. 803–805, and to Helbig, K., (1994) Foundations of anisotropy for exploration seismics: Pergamon, pp. 21–29, where the geometrical properties of the physical concepts are formulated.

8.4.4. Geometrical illustration

In general, at a given point, the direction of a wavefront normal and the direction of a ray tangent are different from one another. Also, considering two wavefronts separated by a given time interval, the magnitudes of phase and ray velocities differ due to the fact that the distance along the wavefront normal is different than the distance along the ray over the same time interval.

As shown in Exercise 8.4, the relationship between the magnitudes of the ray velocity, $V = |\dot{\mathbf{x}}|$, and phase velocity, $v = 1/|\mathbf{p}|$, is given by

$$V = \frac{v}{\mathbf{n} \cdot \mathbf{t}}, \tag{8.4.13}$$

where $\mathbf{n}$ and $\mathbf{t}$ are unit vectors normal to the wavefront and tangent to the ray, respectively.

Note that, in view of vector algebra, expression (8.4.13) shows that the phase-velocity vector is the projection of the ray-velocity vector onto the wavefront normal.[7] This means that, in general, the magnitude of ray velocity is always greater than, or equal to, the magnitude of the corresponding phase velocity.

Using the definition of the scalar product and the fact that $|\mathbf{n}| = |\mathbf{t}| = 1$, we can rewrite expression (8.4.13) as

$$V = \frac{v}{\cos(\theta - \vartheta)}. \tag{8.4.14}$$

Expression (8.4.14) conveniently involves all four entities discussed in this chapter, namely, ray velocity, V, phase velocity, v, ray angle, θ, and phase angle, ϑ.

[7] Readers interested in this formulation might refer to Auld, B.A., (1973) Acoustic fields and waves in solids: John Wiley and Sons, Vol. I, p. 222 and p. 227, to Born, M. and Wolf, E., (1999) Principles of optics (7*th* edition): Cambridge University Press, p. 794, and to Epstein, M. and Śniatycki, J. (1992) Fermat's principle in elastodynamics: Journal of Elasticity, **27**, 45–56.

8.5. Example: Elliptical Anisotropy and Linear Inhomogeneity

8.5.1. Introductory comments

[8] In this section, we study Hamilton's ray equations for a particular case of a wave that exhibits an elliptical velocity dependence with direction and a linear velocity dependence along one axis. This assumption, in the context of a two-dimensional continuum, allows us to illustrate the meaning of Hamilton's ray equations by considering analytic expressions for rays and traveltimes. Also, the same case is treated in Section 14.3 in the context of Lagrange's ray equations. Thus, our examination of Sections 8.5 and 14.3 allows us to investigate the same physical problem using the two different approaches that are available to study seismic ray theory.

We wish to emphasize that, as discussed in Section 6.11.3, the elliptical velocity dependence refers to the fact that infinitesimal wavefronts generated by a point source are elliptical. This behaviour of wavefronts is related to the properties of the continuum in which the given wave propagates. As shown in Section 7.3, three types of waves can propagate in anisotropic continua. In general, in a given continuum, each of the three waves exhibits a different infinitesimal wavefront. Thus, although in a particular continuum one of the three waves might exhibit an elliptical wavefront, the other two waves, in general, do not exhibit elliptical wavefronts. Often, for brevity, we refer to the elliptical velocity dependence as the elliptical anisotropy. However, it should be clear that elliptical anisotropy refers to the response of a given wave to the properties of the continuum, not to the material symmetry of the continuum itself.

---

[8] This section is based on the work that was published by Rogister, Y. and Slawinski, M.A., (2005) Analytic solution of ray-tracing equations for a linearly inhomogeneous and elliptically anisotropic velocity model: Geophysics, **70**, D37–D41. The same example—with a different emphasis and with three figures—is discussed by Bóna, A. and Slawinski, M.A., (2021) Wavefronts and rays as characteristics and asymptotics (*3rd* edition): World Scientific, Example 3.1.

8.5.2. Eikonal equation

As shown in expression (6.11.11), considering the xz-plane, we can write the phase velocity of a wave subjected to elliptical anisotropy as

$$v(\vartheta) = \sqrt{v_x^2 \sin^2\vartheta + v_z^2 \cos^2\vartheta}\,,$$

where v_x and v_z are the magnitudes of phase velocity along the x-axis and z-axis, respectively, and ϑ is the phase angle measured from the z-axis. For convenience of notation, we define parameter

$$\chi := \frac{v_x^2 - v_z^2}{2v_z^2}\,, \tag{8.5.1}$$

which is a dimensionless quantity that vanishes in the isotropic case. Using this definition, we can solve for

$$v_x^2 = v_z^2 (1+2\chi)\,,$$

and rewrite the expression for the phase velocity as

$$v(\vartheta) = v_z \sqrt{(1+2\chi)\sin^2\vartheta + \cos^2\vartheta}\,.$$

If the wave is also subjected to the linear increase of velocity along the z-axis, we can write

$$v(\vartheta, z) = (a+bz)\sqrt{(1+2\chi)\sin^2\vartheta + \cos^2\vartheta}\,, \tag{8.5.2}$$

where a and b are constants whose units are the units of velocity and the reciprocal of time, respectively. We refer to the velocity model described by this expression as the $ab\chi$ model.

Note that the meaning of the term "model" used herein, although consistent with the common use in seismology, is not the same as the meaning of models referred to in the footnote on page 8. Therein, the models constitute a complete physical picture in the context of a mathematical theory.

We wish to formulate the eikonal equation corresponding to equation (8.5.2). Following equation (8.1.1), we can write it as

$$p^2 := p_x^2 + p_z^2 = \frac{1}{(a+bz)^2\left[(1+2\chi)\sin^2\vartheta + \cos^2\vartheta\right]}\,.$$

To express ϑ in terms of vector $\mathbf{p} = [p_x, p_z]$, we can write

$$\vartheta = \arctan\frac{p_x}{p_z}, \tag{8.5.3}$$

which is equivalent to expression (6.11.9). Inserting the expression for ϑ into the above equation and using trigonometric identities, we get

$$p_x^2 + p_z^2 = \frac{p_x^2 + p_z^2}{(a+bz)^2\left[(1+2\chi)\,p_x^2 + p_z^2\right]}.$$

Simplifying, we obtain

$$(a+bz)^2\left[(1+2\chi)\,p_x^2 + p_z^2\right] = 1. \tag{8.5.4}$$

This is the eikonal equation that corresponds to elliptical anisotropy and linear inhomogeneity. Since the right-hand side is equal to unity, this expression is also the level set of function (8.1.4).

To avoid any confusion about the meaning of p_x, p_z, v_x and v_z, we refer the reader to Notation 6.11.1 on page 273.

To see that we are dealing with a differential equation, let us take a look at eikonal equation (8.5.4) and, in view of definition (7.3.2), rewrite it as

$$(a+bz)^2\left[(1+2\chi)\left[\frac{\partial\psi(x,z)}{\partial x}\right]^2 + \left[\frac{\partial\psi(x,z)}{\partial z}\right]^2\right] = 1. \tag{8.5.5}$$

We are looking for function $\psi(x,z)$. Rather than attempting to solve this nonlinear partial differential equation, we study the system of ordinary differential equations that are the characteristic equations of equation (8.5.4). The solution of this system provides us with information about the physical phenomenon that is governed by eikonal equation (8.5.4).

8.5.3. Hamilton's ray equations

Using expression (8.5.4), we write our ray-theory Hamiltonian as

$$\mathscr{H} = \frac{(a+bz)^2\left[(1+2\chi)\,p_x^2 + p_z^2\right]}{2},$$

which, as we can see, is a dimensionless quantity. Following equations (8.2.7), we write our Hamilton's ray equations as

$$\left\{\begin{array}{l} \dot{x}=\dfrac{\partial \mathscr{H}(z,p_x,p_z)}{\partial p_x}=(a+bz)^2(1+2\chi)\,p_x \\ \dot{z}=\dfrac{\partial \mathscr{H}(z,p_x,p_z)}{\partial p_z}=(a+bz)^2\,p_z \\ \dot{p}_x=-\dfrac{\partial \mathscr{H}(z,p_x,p_z)}{\partial x}=0 \\ \dot{p}_z=-\dfrac{\partial \mathscr{H}(z,p_x,p_z)}{\partial z}=-b(a+bz)\left[(1+2\chi)\,p_x^2+p_z^2\right] \end{array}\right. . \tag{8.5.6}$$

These are the characteristic equations of equation (8.5.4).

Since we can write equation (8.5.4) as

$$(1+2\chi)\,p_x^2+p_z^2=\frac{1}{(a+bz)^2},$$

we can rewrite the last Hamilton's ray equation as

$$\dot{p}_z=-\frac{b}{a+bz}.$$

Since $\dot{p}_x=0$, it immediately follows that $p_x(t)=\mathfrak{p}$, where $\mathfrak{p}$ denotes a constant. Now, we can write the remaining three Hamilton's ray equations as a system of ordinary differential equations to be solved for x, z and p_z. These equations are

$$\frac{\mathrm{d}x(t)}{\mathrm{d}t}=[a+bz(t)]^2(1+2\chi)\,\mathfrak{p}, \tag{8.5.7}$$

$$\frac{\mathrm{d}z(t)}{\mathrm{d}t}=[a+bz(t)]^2\,p_z(t) \tag{8.5.8}$$

and

$$\frac{\mathrm{d}p_z(t)}{\mathrm{d}t}=-\frac{b}{a+bz(t)}. \tag{8.5.9}$$

8.5.4. Initial conditions

To complete this system of differential equations, we need additional constraints. We choose to use the initial conditions, which correspond to the values of unknowns at the initial time. In other words, we need $x(t)$, $z(t)$, $p_x(t)$ and $p_z(t)$ at $t=0$. While we already have $p_x(0)=\mathfrak{p}$, let us set

$x(0)=0$ and $z(0)=0$ and $p_z(0)=p_{z0}$. The initial condition for $p_z(t)$ is not independent from the initial condition for $p_x(t)$. They are related by eikonal equation (8.5.4). Solving this equation for p_z at $t=0$, which corresponds to $z=0$, we get

$$p_z(0)=\sqrt{\frac{1}{a^2}-(1+2\chi)[p_x(0)]^2}.$$

Since $p_x(0)=\mathfrak{p}$, the initial condition for p_z that obeys the eikonal equation is

$$p_z(0)=\sqrt{\frac{1}{a^2}-(1+2\chi)\mathfrak{p}^2}. \tag{8.5.10}$$

System (8.5.6) accompanied by the initial conditions has a clear meaning in the context of ray theory. We discuss it in the next section.

8.5.5. Physical interpretation of equations and conditions

Ray velocity

The first two equations of system (8.5.6) define the vector field in the xz-plane. Herein, solution $[x(t),z(t)]$ describes the path of a signal under elliptical velocity dependence with direction and a linear velocity dependence along the z-axis; this path is the ray. Below, we discuss the physical information contained in the equations themselves.

Considering the first two equations of system (8.5.6), we can express the magnitude of the velocity of the signal along the ray as

$$V=\sqrt{\left(\frac{\mathrm{d}x}{\mathrm{d}t}\right)^2+\left(\frac{\mathrm{d}z}{\mathrm{d}t}\right)^2},$$

where V is referred to as the ray velocity. We can explicitly write

$$V=\sqrt{\left([a+bz(t)]^2(1+2\chi)\mathfrak{p}\right)^2+\left([a+bz(t)]^2p_z(t)\right)^2},$$

which we rewrite as

$$V=[a+bz(t)]^2\sqrt{(1+2\chi)^2\mathfrak{p}^2+p_z(t)^2}. \tag{8.5.11}$$

This expression relates the ray velocity to the wavefront slowness.

Ray orientation

Considering the first two equations of system (8.5.6), we can express the direction of a ray as

$$\theta = \arctan \frac{\mathrm{d}x}{\mathrm{d}z},$$

where θ is measured from the z-axis and is referred to as the ray angle. Herein, by referring to system (8.5.6), we write

$$\theta = \arctan \frac{\mathrm{d}x}{\mathrm{d}z} = \arctan \frac{\dot{x}}{\dot{z}} = \arctan \left[(1+2\chi) \frac{p_x}{p_z} \right],$$

which we can rewrite as

$$\tan\theta = (1+2\chi) \frac{p_x}{p_z}.$$

Invoking expression (6.11.9), we can write

$$\tan\theta = (1+2\chi) \tan\vartheta, \tag{8.5.12}$$

which explicitly relates ray and phase angles.

Wavefront orientation

To gain more understanding of expression (6.11.9), which was already discussed between pages 272 and 275, we examine the last two equations of system (8.5.6).

We recall that contours of $\psi(x,z)$ correspond to wavefronts at given instants of time. Hence, in view of definition (7.3.2), $p_x := \partial\psi/\partial x$ and $p_z = \partial\psi/\partial z$ at a given point, $(x(t), z(t))$, are components of slowness with which a wavefront propagates at this point. Also, vector $\mathbf{p} = [p_x, p_z]$ at $(x(t), z(t))$ is normal to the wavefront at this point. In view of expression (8.5.3), we write

$$\vartheta = \arctan \frac{p_x}{p_z},$$

where ϑ is the phase angle, which gives the direction of wavefront propagation.

Initial conditions

Now, let us examine the physical meaning of the initial conditions. Setting

$$[x(0), z(0)] = [0,0],$$

we fix the origin of the ray at the initial time. In other words, we locate the point source at the origin. In view of continuity of wavefronts and considering the inhomogeneity along the z-axis only, we know that

$$\mathfrak{p} = \frac{\sin\vartheta}{v(\vartheta, z)}.$$

Considering $z = 0$ and using expression (8.5.2), we can write

$$p_x(0) = \mathfrak{p} = \frac{\sin\vartheta_0}{a\sqrt{(1+2\chi)\sin^2\vartheta_0 + \cos^2\vartheta_0}}, \tag{8.5.13}$$

where ϑ_0 denotes the take-off phase angle; in other words, the direction of the wavefront at the source.

8.5.6. Solution of Hamilton's ray equations

General solution

We wish to solve system (8.5.6). Since we already know that $p_x = \mathfrak{p}$, we must solve equations (8.5.7), (8.5.8) and (8.5.9) for $x(t)$, $z(t)$ and $p_z(t)$.

To do so, let us write the second equation of system (8.5.6) as

$$p_z = \frac{\frac{\mathrm{d}z}{\mathrm{d}t}}{[a+bz]^2}.$$

Differentiating with respect to t, we get

$$\frac{\mathrm{d}p_z}{\mathrm{d}t} = \frac{\frac{\mathrm{d}^2 z}{\mathrm{d}t^2}(a+bz) - 2b\left(\frac{\mathrm{d}z}{\mathrm{d}t}\right)^2}{(a+bz)^3}.$$

Equating this result with the fourth equation of system (8.5.6) and rearranging, we get

$$(a+bz)\frac{\mathrm{d}^2 z}{\mathrm{d}t^2} - 2b\left(\frac{\mathrm{d}z}{\mathrm{d}t}\right)^2 + b(a+bz)^2 = 0.$$

Letting $(a+bz)^2 = y$, we get

$$\frac{1}{2b}\frac{\mathrm{d}^2 y}{\mathrm{d}t^2} - \frac{3}{4yb}\left(\frac{\mathrm{d}y}{\mathrm{d}t}\right)^2 + by = 0.$$

Letting $y = a^2 \exp u$, we get

$$\frac{\mathrm{d}^2 u}{\mathrm{d}t^2} - \frac{1}{2}\left(\frac{\mathrm{d}u}{\mathrm{d}t}\right)^2 + 2b^2 = 0.$$

Letting $\mathrm{d}u/\mathrm{d}t = q$, we get

$$\frac{\mathrm{d}q}{\mathrm{d}t} - \frac{1}{2}q^2 + 2b^2 = 0.$$

We can rewrite this equation as

$$\mathrm{d}t = \frac{\mathrm{d}q}{\frac{1}{2}q^2 - 2b^2}.$$

Integrating both sides of this equation, we get

$$t + A_1 = -\frac{1}{b}\tanh^{-1}\frac{q}{2b},$$

where $\tanh^{-1}$ stands for the inverse function of $\tanh$, which could also be written as atanh or arctanh, and A_1 is an integration constant. Solving for q, we get

$$q = 2b\tanh\left(-b(t+A_1)\right).$$

To obtain u, we integrate and get

$$u = -2\ln\left(\cosh\left(-b(t+A_1)\right)\right) + A_2,$$

where A_2 is an integration constant. Hence,

$$y = \frac{a^2 \exp[A_2]}{\cosh^2\left(-b(t+A_1)\right)}. \tag{8.5.14}$$

Since $z = \left(\sqrt{y} - a\right)/b$, we have the solution of the second equation of system (8.5.6), namely,

$$z(t) = \frac{1}{b}\frac{a\exp\left(\frac{A_2}{2}\right)}{\cosh\left(-b\left(t + A_1\right)\right)} - \frac{a}{b}.$$

Also—in view of the second equation of system (8.5.6)—we have $p_z = \dot{z}/y$. Thus, we have the solution of the fourth equation of system (8.5.6). This solution is

$$p_z(t) = -\frac{\sinh\left(-b\left(t + A_1\right)\right)}{a\exp\left(\frac{A_2}{2}\right)}.$$

We can write the remaining equation of system (8.5.6) as

$$\frac{\mathrm{d}x}{\mathrm{d}t} = y(1 + 2\chi)\,\mathfrak{p},$$

where y is given by expression (8.5.14). Integrating, we get

$$x(t) = \frac{\mathfrak{p}(1 + 2\chi)\,a^2 \exp A_2}{b}\tanh\left(b\left(t + A_1\right)\right) + A_3,$$

where A_3 is an integration constant. Thus, we can concisely write the general solution of system (8.5.6) as

$$\begin{cases} x(t) = \dfrac{\mathfrak{p}(1 + 2\chi)\,a^2 \exp A_2}{b}\tanh\left(b\left(t + A_1\right)\right) + A_3 \\ z(t) = \dfrac{1}{b}\dfrac{a\exp\left(\frac{A_2}{2}\right)}{\cosh\left(b\left(t + A_1\right)\right)} - \dfrac{a}{b} \\ p_x(t) = \mathfrak{p} \\ p_z(t) = -\dfrac{\sinh\left(-b\left(t + A_1\right)\right)}{a\exp\left(\frac{A_2}{2}\right)} \end{cases}, \qquad (8.5.15)$$

where, in view of its expression in system (8.5.6), $p_x(t)$ is constant along any ray; its value—for a given ray—is $\mathfrak{p}$.

Integration constants

Now, we can find the three integration constants. At $t = 0$, using the initial conditions discussed in Section 8.5.4, we can rewrite the solutions stated in the first two expressions of set (8.5.15) as

$$x(0) = \frac{\mathfrak{p}(1+2\chi)a^2 \exp A_2}{b} \tanh(bA_1) + A_3 = 0,$$

$$z(0) = \frac{1}{b}\frac{a \exp\left(\frac{A_2}{2}\right)}{\cosh(bA_1)} - \frac{a}{b} = 0.$$

Also, considering $z = 0$ and using the second equation of system (8.5.6) combined with solution $z(t)$ given in set (8.5.15) and evaluating at $t = 0$, we can write

$$p_z(0) = \frac{\dot{z}(0)}{a^2} = -\frac{\sinh(bA_1)}{a \exp\left(\frac{A_2}{2}\right)} = \sqrt{\frac{1}{a^2} - (1+2\chi)\mathfrak{p}^2},$$

where the right-hand side is given in expression (8.5.10).

Considering the last two equations, we have a system of two equations in two unknowns, A_1 and A_2. Solving, we obtain

$$A_1 = -\frac{1}{b}\tanh^{-1}\sqrt{1-(1+2\chi)\mathfrak{p}^2 a^2},$$

$$A_2 = -\ln\left[(1+2\chi)\mathfrak{p}^2 a^2\right].$$

Inserting A_1 and A_2 into the equation for $x(0)$, we obtain

$$A_3 = \frac{\sqrt{1-(1+2\chi)\mathfrak{p}^2 a^2}}{\mathfrak{p} b}.$$

Examining A_1, A_2 and A_3, we see that the units of A_1 are the units of time, A_2 is dimensionless and the units of A_3 are the units of distance. This is consistent with positions of the A_1, A_2 and A_3 in system (8.5.15).

Unique solution

Having found A_1, A_2 and A_3, we can rewrite solutions (8.5.15) as

$$\begin{cases} x(t) = \dfrac{1}{\mathfrak{p}b}\left[\tanh\left(bt - \tanh^{-1}\sqrt{1-(1+2\chi)\mathfrak{p}^2a^2}\right) + \sqrt{1-(1+2\chi)\mathfrak{p}^2a^2}\right] \\ z(t) = \dfrac{a}{b}\left[\dfrac{1}{\mathfrak{p}a\cosh\left(\tanh^{-1}\sqrt{1-(1+2\chi)\mathfrak{p}^2a^2} - bt\right)\sqrt{1+2\chi}} - 1\right] \\ p_x(t) = \mathfrak{p} \\ p_z(t) = \mathfrak{p}\sqrt{1+2\chi}\sinh\left(\tanh^{-1}\sqrt{1-(1+2\chi)\mathfrak{p}^2a^2} - bt\right) \end{cases}, \tag{8.5.16}$$

where, in view of expression (8.5.13), we have

$$\mathfrak{p} = \frac{\sin\vartheta_0}{a\sqrt{(1+2\chi)\sin^2\vartheta_0 + \cos^2\vartheta_0}}$$

with ϑ_0 being the take-off phase angle.

Thus—for a velocity model given by a, which is the velocity at $z = 0$, b, which describes the increase of velocity along the z-axis, and χ, which describes the elliptical velocity dependence with direction—we can choose the phase take-off angle, ϑ_0, and, using the first two expressions of solutions (8.5.16), obtain the ray along which the signal generated at $(0,0)$ propagates.

Geometrical interpretation

Examination of the first two expressions of solutions (8.5.16) allows us to learn about the shape of rays for the $ab\chi$ model. We can write each of these expressions as

$$\mathfrak{p}bx(t) - \sqrt{1-\mathfrak{p}^2a^2(1+2\chi)} = \tanh\left(bt - \tanh^{-1}\sqrt{1-(1+2\chi)\mathfrak{p}^2a^2}\right)$$

and

$$\left[\frac{b}{a}z(t)+1\right]\mathfrak{p}a\sqrt{1+2\chi}=\frac{1}{\cosh\left(bt-\tanh^{-1}\sqrt{1-(1+2\chi)\mathfrak{p}^2a^2}\right)},$$

respectively. Squaring these two equations, adding them together, and using standard identities, we obtain

$$\frac{\left(x-\frac{\sqrt{1-\mathfrak{p}^2a^2(1+2\chi)}}{\mathfrak{p}b}\right)^2}{\left(\frac{1}{\mathfrak{p}b}\right)^2}+\frac{\left(z+\frac{a}{b}\right)^2}{\left(\frac{1}{\mathfrak{p}b\sqrt{1+2\chi}}\right)^2}=1. \qquad (8.5.17)$$

This is the equation of an ellipse with a centre on the line given by $z=-a/b$. In other words, in the $ab\chi$ model, rays are elliptical arcs. In view of $v(z)=a+bz$, we conclude that the centre of the ellipse corresponds to the level where the velocity vanishes. This equation is identical to the one that we obtain in Section 14.3.2 using Lagrange's, rather than Hamilton's, ray equations.

8.5.7. Solution of eikonal equation

Using solutions (8.5.16), we can obtain the graph of the solution of the original partial differential equation, namely, eikonal equation (8.5.4), as a parametric plot $[x(t),z(t),t]$ for all $\mathfrak{p}$ that are consistent with the original equation. In the present case, we set $\vartheta_0\in(-\pi/2,\pi/2)$ to get, following expression (8.5.13),

$$\mathfrak{p}\in\left(-1/\left[a\sqrt{1+2\chi}\right],1/\left[a\sqrt{1+2\chi}\right]\right).$$

In the case of the present example, we can also obtain an explicit analytic form of the solution of the original partial differential equation, namely, eikonal equation (8.5.4). In other words, we can use our results to obtain $\psi(x,z)$. To do so, we proceed in the following way. Since $t=\psi(x,z)$, let us solve the first equation of set (8.5.16) for t. We get

$$t(x;\mathfrak{p})=\frac{\tanh^{-1}\left[\mathfrak{p}bx-\sqrt{1-\mathfrak{p}^2a^2(1+2\chi)}\right]+\tanh^{-1}\sqrt{1-\mathfrak{p}^2a^2(1+2\chi)}}{b}. \qquad (8.5.18)$$

To express t in terms of x and z, and the parameters of a given $ab\chi$ model, we solve equation (8.5.17) for $\mathfrak{p}$. We get

$$\mathfrak{p}(x,z) = \frac{2x}{\sqrt{\left[x^2+(1+2\chi)z^2\right]\left[(2a+bz)^2(1+2\chi)+b^2x^2\right]}}; \tag{8.5.19}$$

this value remains constant along a given ray. Expression (8.5.18) with $\mathfrak{p}$ given by expression (8.5.19) is the solution of equation (8.5.4). In the context of expressions (8.5.18) and (8.5.19), we can write this solution as $\psi = t(x,z)$.

We can also obtain the expression for t by solving the second equation of set (8.5.16), as shown in Exercise 8.10, where it appears as expression (8.7.17). Traveltime expression (8.5.18), stated above, is valid for the entirety of each ray. Traveltime expression (8.7.17), stated below, is valid only for the downward segments. Both traveltime expressions are valid as long as the corresponding coordinates, x and z, respectively, are increasing; coordinate x increases for the entire ray, but coordinate z increases only for the downward segment of the ray. This distinction of the validity of the traveltime expressions appear again, and are discussed in more detail, in Section 14.3.3, in the context of integration along the x-axis and the z-axis.

REMARK 8.5.1. Neither traveltime expression (8.5.18) nor (8.7.17) is valid for $x = 0$. At that point, the inverse hyperbolic functions tend to infinity. The value along $x = 0$—between 0 and z—can be obtained using[9,10]

$$t = \frac{1}{b}\ln\left(\frac{a+bz}{a}\,\frac{1+\sqrt{1-a^2\mathfrak{p}^2(1+2\chi)}}{1+\sqrt{1-(a+bz)^2\mathfrak{p}^2(1+2\chi)}}\right), \tag{8.5.20}$$

which—for $x = 0$ and, hence, $\mathfrak{p} = 0$—reduces to[11]

$$t = \frac{1}{b}\ln\frac{a+bz}{a}; \tag{8.5.21}$$

[9] Slawinski, M.A., Wheaton, C.J., and Powojowski, M. (2004) VSP traveltime inversion for linear inhomogeneity and elliptical anisotropy: Geophysics, **69**(2), 373–377, equation (11).

[10] In contrast to expression (8.5.18), expression (8.5.20) is valid only for the downward segment of a ray, hence, it fails for $x \gg 0$.

[11] Bóna, A. and Slawinski, M.A., (2021) Wavefronts and rays as characteristics and asymptotics (3*rd* edition): World Scientific, Example 3.1.

for an interval between z_1 and z_2,

$$t = \frac{1}{b} \ln \frac{a + b z_2}{a + b z_1}, \tag{8.5.22}$$

which results from

$$t = \int_{z_1}^{z_2} \frac{\mathrm{d}z}{a + b z}. \tag{8.5.23}$$

Expression (8.5.22) is equivalent to expression (8.5.21), with a replaced by $a + b z_1$, and z by $z_2 - z_1$. The same replacement in expressions (8.5.19) and (8.5.20) results in traveltime between z_1 and z_2, for $x \neq 0$.

Traveltimes discussed in Section 8.5.7 correspond to average velocities. To justify this statement, let us consider the following definition.

DEFINITION 8.5.2. Average velocity[12] is the ratio of distance along a ray to the traveltime along this ray,

$$\overline{V} := \frac{\int_0^t V(\tau)\,\mathrm{d}\tau}{\int_0^t \mathrm{d}\tau} = \frac{\int_0^z \mathrm{d}\zeta}{\int_0^z \frac{\mathrm{d}\zeta}{V(\zeta)}}, \tag{8.5.24}$$

where V is the ray velocity.

In this section, the traveltime expressions—notably, expressions (8.5.18) and (8.5.20)—are solutions of Hamilton's ray equations (8.5.6) and, hence, result in traveltimes along a ray, in accordance with Definition 8.5.2.

To illustrate it, we insert expression (8.5.11)—for $x = 0$—in definition (8.5.24) to obtain[13]

$$\overline{V} = \frac{\int_0^z \mathrm{d}\zeta}{\int_0^z \frac{\mathrm{d}\zeta}{a + b\zeta}} = \frac{b z}{\ln \frac{a + b z}{a}},$$

which is consistent with expression (8.5.21), as expected.

[12] Sheriff, R.E., (2002) Encyclopedic dictionary of applied geophysics (4*th* edition): Society of Exploration Geophysicists.

[13] Slawinski, R.A. and Slawinski, M.A. (1999) On raytracing in constant velocity-gradient media: Calculus approach: Canadian Journal of Exploration Geophysics, **35**, No. 1/2, 24–27, Appendix 1.

The magnitude of the average velocity at a given point is the speed of a signal along a ray, at that point. Except for isotropic media, it is not an intrinsic propagation speed within the medium; in general, speeds along distinct rays are different at the same point. This is also true for the *root-mean-square velocity*,

$$V_{\mathrm{rms}} := \sqrt{\frac{\int_0^t \overline{V}^2(\tau)\,\mathrm{d}\tau}{\int_0^t \mathrm{d}\tau}} = \sqrt{\frac{\int_0^z \overline{V}\,\mathrm{d}\zeta}{\int_0^z \frac{\mathrm{d}\zeta}{\overline{V}(\zeta)}}}, \tag{8.5.25}$$

since $\overline{V}$ is the average velocity. For an $ab\chi$ model, along $x = 0$,[14]

$$V_{\mathrm{rms}} = \sqrt{\frac{2abz + b^2z^2}{2\ln(a+bz)}}.$$

For the $ab\chi$ models of layered media, using the values of the root-mean-square velocity from expression (8.5.25) in[15]

$$\sqrt{\frac{V_{\mathrm{rms}_2}^2 t_2 - V_{\mathrm{rms}_1}^2 t_1}{t_2 - t_1}} \approx V_i,$$

where t_1 and t_2 correspond to two instants along the same ray, one might obtain approximate interval velocities between two points. Such velocities represent speeds of a signal—within a given interval—propagating along $x = 0$.

Mathematical accuracy of this approximation is a function of the orientation of the ray, $\Delta t := t_2 - t_1$, and the strength of anisotropy, which—in the context of Section 8.5—can be examined in terms of χ. Regardless of the mathematical accuracy, one needs to be cautious in interpreting the results as corresponding to physical properties of a material within a given interval, since the interval velocity is not formulated based on information

[14] Slawinski, R.A. and Slawinski, M.A. (1999) On raytracing in constant velocity-gradient media: Calculus approach: Canadian Journal of Exploration Geophysics, **35**, No. 1/2, 24–27, Appendix 1.

[15] Dix, C.H., (1955) Seismic velocities from surface measurements: Geophysics **20**, 60–86.

from that interval, but from a total traveltime. In other words, interval velocity must agree only with the average velocity, from which it is derived; it might not represent local properties in the interval to which it is assigned.

ASSUMPTION 8.5.3. In Remark 8.5.1, we assume that the arguments of natural logarithms are positive.

8.5.8. Physical interpretation of solutions

In Section 8.5, we discuss the rays and traveltimes, which are obtained from the solutions of Hamilton's ray equations (8.5.6). We also discuss wavefronts, which are obtained from the solution of eikonal equation (8.5.4). The physical meaning of these solutions is obvious.

Furthermore, we can use p_z, in solution (8.5.16), to express the magnitude of the slowness of the wavefront propagation. Since the wavefront slowness is the reciprocal of the wavefront velocity, we write

$$v = \frac{1}{\sqrt{\mathbf{p} \cdot \mathbf{p}}} = \frac{1}{\sqrt{[p_x, p_z] \cdot [p_x, p_z]}} = \frac{1}{\sqrt{\mathfrak{p}^2 + p_z^2}},$$

where v is also referred to as the phase velocity.

8.6. Example: Isotropy and Inhomogeneity

8.6.1. Parametric form

In most cases, one would solve Hamilton's ray equations given in expression (8.2.7) by numerical methods. Since Hamilton's ray equations are first-order linear ordinary differential equations, the solution can be obtained using standard computer tools.

As shown in Section (8.5), in particular cases system (8.2.7) allows us to analytically, rather than numerically, study ray theory in the context of anisotropic inhomogeneous continua. To gain further familiarity with this system, let us now consider a formulation for isotropic inhomogeneous continua, where eikonal equation (8.1.1) reduces to

$$p^2 = \frac{1}{v^2(\mathbf{x})}, \tag{8.6.1}$$

which is eikonal equation (6.11.22).

To study ray equations in isotropic inhomogeneous continua, let us choose function (8.1.3), which becomes

$$F(\mathbf{x}) = p^2 - \frac{1}{v^2(\mathbf{x})}. \tag{8.6.2}$$

Using system (8.1.13), we can write the corresponding characteristic equations as

$$\left\{ \begin{array}{l} \dfrac{\mathrm{d}x_i}{\mathrm{d}s} = 2\zeta p_i \\ \\ \dfrac{\mathrm{d}p_i}{\mathrm{d}s} = -2\zeta \dfrac{1}{v^3} \dfrac{\partial v}{\partial x_i} \end{array} \right. , \qquad i \in \{1,2,3\}. \tag{8.6.3}$$

Also, let us choose scaling factor ζ so that s is the arclength parameter. As shown in Exercise 8.6, we obtain the arclength parametrization of system (8.6.3) by letting $\zeta = v/2$. Furthermore, as shown in Exercise 8.7, system (8.6.3) can be restated as a single expression

$$\frac{\mathrm{d}}{\mathrm{d}s}\left[\frac{1}{v(\mathbf{x})}\frac{\mathrm{d}\mathbf{x}}{\mathrm{d}s}\right] = -\frac{\nabla v(\mathbf{x})}{v^2(\mathbf{x})}, \tag{8.6.4}$$

where $\mathbf{x} = [x_1, x_2, x_3]$.

Equation (8.6.4) relates the properties of the continuum, which are given by the phase-velocity function $v(\mathbf{x})$, to the ray $\mathbf{x}(s)$, which is described by arclength parameter s.

8.6.2. Explicit form

Consider a three-dimensional isotropic inhomogeneous continuum where $\mathbf{x} = [x, y, z]$. Expression (8.6.4) can be explicitly written as three parametric equations for $x(s)$, $y(s)$ and $z(s)$, namely,

$$\begin{aligned} \frac{\mathrm{d}}{\mathrm{d}s}\left(\frac{1}{v(\mathbf{x})}\frac{\mathrm{d}x}{\mathrm{d}s}\right) &= -\frac{1}{v^2(\mathbf{x})}\frac{\partial v(\mathbf{x})}{\partial x}, \\ \frac{\mathrm{d}}{\mathrm{d}s}\left(\frac{1}{v(\mathbf{x})}\frac{\mathrm{d}y}{\mathrm{d}s}\right) &= -\frac{1}{v^2(\mathbf{x})}\frac{\partial v(\mathbf{x})}{\partial y}, \\ \frac{\mathrm{d}}{\mathrm{d}s}\left(\frac{1}{v(\mathbf{x})}\frac{\mathrm{d}z}{\mathrm{d}s}\right) &= -\frac{1}{v^2(\mathbf{x})}\frac{\partial v(\mathbf{x})}{\partial z}, \end{aligned} \tag{8.6.5}$$

where s is the arclength parameter along the ray. Consequently, all three equations are related by $\mathrm{d}s = \sqrt{(\mathrm{d}x)^2 + (\mathrm{d}y)^2 + (\mathrm{d}z)^2}$, where x, y and z are the orthonormal coordinates. Consequently, as shown in Exercise 8.11, instead of using the parametric form, under certain conditions related to the behaviour of the curve $\mathbf{x}(s)$, we can write equations (8.6.5) as two equations for $x(z)$ and $y(z)$, namely,

$$\frac{\mathrm{d}}{\mathrm{d}z}\left[\frac{1}{v(\mathbf{x})}\frac{\frac{\mathrm{d}x}{\mathrm{d}z}}{\sqrt{\left(\frac{\mathrm{d}x}{\mathrm{d}z}\right)^2+\left(\frac{\mathrm{d}y}{\mathrm{d}z}\right)^2+1}}\right] = -\frac{1}{v^2(\mathbf{x})}\frac{\partial v(\mathbf{x})}{\partial x}\sqrt{\left(\frac{\mathrm{d}x}{\mathrm{d}z}\right)^2+\left(\frac{\mathrm{d}y}{\mathrm{d}z}\right)^2+1} \tag{8.6.6}$$

and

$$\frac{\mathrm{d}}{\mathrm{d}z}\left[\frac{1}{v(\mathbf{x})}\frac{\frac{\mathrm{d}y}{\mathrm{d}z}}{\sqrt{\left(\frac{\mathrm{d}x}{\mathrm{d}z}\right)^2+\left(\frac{\mathrm{d}y}{\mathrm{d}z}\right)^2+1}}\right] = -\frac{1}{v^2(\mathbf{x})}\frac{\partial v(\mathbf{x})}{\partial y}\sqrt{\left(\frac{\mathrm{d}x}{\mathrm{d}z}\right)^2+\left(\frac{\mathrm{d}y}{\mathrm{d}z}\right)^2+1}, \tag{8.6.7}$$

which form a system of explicit equations for isotropic inhomogeneous continua.

If the continuum exhibits only vertical inhomogeneity, $v = v(z)$, the right-hand sides of equations (8.6.6) and (8.6.7) vanish and, for the resulting equations, we can obtain an analytic solution, as shown in Exercise 8.12. If, however, the properties of the medium vary along the x-axis and the y-axis, we must often resort to numerical methods to obtain a solution.[16]

Closing Remarks

By solving the eikonal equation using the method of characteristics, we obtain Hamilton's ray equations whose solutions give rays. Hamilton's ray equations are rooted in the high-frequency approximation and the trial

[16]Readers interested in numerical techniques to solve these differential equations might refer to Červený, V., and Ravindra, R., (1971) Theory of seismic head waves: University of Toronto Press, pp. 25–26.

solutions discussed in Chapters 6 and 7, and the resulting rays are given by function $\mathbf{x} = \mathbf{x}(t)$. We can study the entire ray theory in the context of Hamilton's ray equations, which is the most rigorous method for studying seismic rays.

In Chapter 11, however, we explore another formulation of rays using the approach that transforms Hamilton's six first-order equations into Lagrange's three second-order equations. Also, this Lagrangian formulation coincides with the variational approach to the study of ray theory, which we discuss in *Part* 3. By investigating both of these approaches, we gain additional physical insight into ray theory, as well as additional knowledge of useful mathematical tools.

In general, ray theory is related to the Wentzel-Kramers-Brillouin-Jeffreys (WKBJ) method for solving differential equations. The WKBJ method is also used in other physical theories, for instance, in quantum mechanics.[17] Ray theory is an approximation of wave theory as classical mechanics is an approximation of quantum mechanics. The high-frequency approximation is analogous to assuming the action, discussed in Section 13.2.2, to be infinitely divisible, as is the case in classical mechanics. This is not the case in quantum mechanics due to the existence of Planck's constant, which is the fundamental unit, or quantum, of action.[18]

[17]Readers interested in the WKBJ method might refer to Aki, K. and Richards, P.G., (2002) Quantitative seismology (*2nd* edition): University Science Books, pp. 434–437, (Box 9.6), and to Woodhouse, N.M.J., (1992) Geometric quantization (*2nd* edition): Oxford Science Publications, pp. 197–201 and pp. 236–249.

[18]Readers interested in the association of the geometrical optics and quantum mechanics might refer to Goldstein, H., (1950/1980) Classical mechanics: Addison-Wesley Publishing Co., pp. 484–492, and to Batterman, R.W., (2002) The devil in the details: Asymptotic reasoning in explanation, reduction, and emergence: Oxford University Press, Chapters 6 and 7.

8.7. Exercises

EXERCISE 8.1. Consider a three-dimensional isotropic inhomogeneous continuum. Using Hamilton's ray equations (8.2.7), show that, in isotropic continua, rays are orthogonal to wavefronts.

SOLUTION 8.1. Following expression (8.2.8), we can explicitly write Hamiltonian $\mathscr{H}(\mathbf{x},\mathbf{p})$, in a three-dimensional isotropic inhomogeneous continuum, as

$$\mathscr{H}(\mathbf{x},\mathbf{p}) = \frac{1}{2}p^2v^2(\mathbf{x}) = \frac{1}{2}[p_1,p_2,p_3]\cdot[p_1,p_2,p_3]\,v^2(x_1,x_2,x_3)\,. \quad (8.7.1)$$

The corresponding Hamilton's ray equations (8.2.7) are

$$\begin{cases} \dot{x}_1 = p_1v^2 \\ \dot{x}_2 = p_2v^2 \\ \dot{x}_3 = p_3v^2 \\ \dot{p}_1 = -p^2v\dfrac{\partial v}{\partial x_1} \\ \dot{p}_2 = -p^2v\dfrac{\partial v}{\partial x_2} \\ \dot{p}_3 = -p^2v\dfrac{\partial v}{\partial x_3} \end{cases}. \quad (8.7.2)$$

Recalling definition (7.3.2), we can write the first three equations of system (8.7.2) as

$$[\dot{x}_1,\dot{x}_2,\dot{x}_3] = v^2\left[\frac{\partial\psi}{\partial x_1},\frac{\partial\psi}{\partial x_2},\frac{\partial\psi}{\partial x_3}\right]. \quad (8.7.3)$$

The left-hand side of equation (8.7.3) is a vector tangent to the curve $\mathbf{x}(t)$, while the right-hand side is the gradient of function $\psi(\mathbf{x})$, scaled by v^2. For a given point of the continuum, we can write equation (8.7.3) as

$$\mathbf{t}|_{x_1,x_2,x_3} = v^2(\nabla\psi)|_{x_1,x_2,x_3}\,.$$

This means that vector $\mathbf{t}$, which is tangent to curve $\mathbf{x}(t)$, is parallel to the gradient of the eikonal function, $\nabla\psi(\mathbf{x})$. Since curve $\mathbf{x}(t)$ corresponds to the ray and the level sets of the eikonal function correspond to the wavefronts, by the properties of the gradient operator, the rays in an isotropic inhomogeneous continuum are orthogonal to the wavefronts.

EXERCISE 8.2. Derive expression (8.4.12) using level-set function (8.1.4) and characteristic equations (8.1.13).

SOLUTION 8.2. In view of expression (8.4.10), the ray angle can be stated as

$$\tan\theta = \frac{\dfrac{\mathrm{d}x_1}{\mathrm{d}s}}{\dfrac{\mathrm{d}x_3}{\mathrm{d}s}}, \tag{8.7.4}$$

where s defines the parametrization of the ray $\mathbf{x}(s)$, and $\mathrm{d}x_i/\mathrm{d}s$ are the components of the vector tangent to the ray. Since expression (8.7.4) is given as a ratio, the actual parametrization has no effect on the ray angle. Consider a given point in an anisotropic continuum and the level-set function given by expression (8.1.4), namely,

$$F(\mathbf{p}) = p^2 v^2(\mathbf{p}) = 1. \tag{8.7.5}$$

At a given point, expression (8.7.5) is not a function of $\mathbf{x}$, and, hence, $\partial F/\partial x_i = 0$. Thus, characteristic equations (8.1.13), are reduced to

$$\frac{\mathrm{d}x_i}{\mathrm{d}s} = \zeta\frac{\partial F}{\partial p_i} = 2\zeta\left(p_i v^2 + p^2 v\frac{\partial v}{\partial p_i}\right), \qquad i \in \{1,2\}.$$

Following expression (8.7.4), we can write the ray angle as

$$\tan\theta = \frac{2\zeta\left(p_1 v^2 + p^2 v\dfrac{\partial v}{\partial p_1}\right)}{2\zeta\left(p_3 v^2 + p^2 v\dfrac{\partial v}{\partial p_3}\right)} = \frac{p_1 v^2 + p^2 v\dfrac{\partial v}{\partial p_1}}{p_3 v^2 + p^2 v\dfrac{\partial v}{\partial p_3}}. \tag{8.7.6}$$

We wish to express the quantities on the right-hand side of expression (8.7.6) in terms of the phase angle, ϑ. Recalling expression (8.4.5), we can write the differential operator in the numerator as

$$\frac{\partial}{\partial p_1} = \frac{\partial \vartheta}{\partial p_1}\frac{\partial}{\partial \vartheta} = \frac{\partial \arctan \frac{p_1}{p_3}}{\partial p_1}\frac{\partial}{\partial \vartheta}$$

$$= \frac{\frac{1}{p_3}}{1 + \left(\frac{p_1}{p_3}\right)^2}\frac{\partial}{\partial \vartheta} = \frac{p_3}{p^2}\frac{\partial}{\partial \vartheta}.$$

Similarly, we obtain the differential operator in the denominator, which is

$$\frac{\partial}{\partial p_3} = -\frac{p_1}{p^2}\frac{\partial}{\partial \vartheta}.$$

Using these differential operators in expression (8.7.6), we can rewrite it as

$$\tan\theta = \frac{p_1 v^2 + p_3 v \frac{\partial v}{\partial \vartheta}}{p_3 v^2 - p_1 v \frac{\partial v}{\partial \vartheta}} = \frac{p_1 + p_3 \frac{1}{v}\frac{\partial v}{\partial \vartheta}}{p_3 - p_1 \frac{1}{v}\frac{\partial v}{\partial \vartheta}}.$$

Again, recalling expression (8.4.5), we divide both the numerator and the denominator by p_3 to obtain

$$\tan\theta = \frac{\frac{p_1}{p_3} + \frac{1}{v}\frac{\partial v}{\partial \vartheta}}{1 - \frac{p_1}{p_3}\frac{1}{v}\frac{\partial v}{\partial \vartheta}} = \frac{\tan\vartheta + \frac{1}{v}\frac{\partial v}{\partial \vartheta}}{1 - \frac{\tan\vartheta}{v}\frac{\partial v}{\partial \vartheta}},$$

which is expression (8.4.12), as required.

EXERCISE 8.3. Derive expression (8.4.12) using level-set function (8.1.3) and characteristic equations (8.1.13).

SOLUTION 8.3. Recall expression (8.7.4). Consider a given point in an anisotropic continuum and the level-set function that is given by expression (8.1.3), namely,

$$F(\mathbf{p}) = p^2 - \frac{1}{v^2(\mathbf{p})} = 0. \tag{8.7.7}$$

At a given point, expression (8.7.7) is not a function of $\mathbf{x}$, and, hence, $\partial F/\partial x_i = 0$. Thus, characteristic equations (8.1.13), are reduced to

$$\frac{\mathrm{d}x_i}{\mathrm{d}s} = \zeta \frac{\partial F}{\partial p_i} = 2\zeta \left(p_i + \frac{1}{v^3} \frac{\partial v}{\partial p_i} \right), \qquad i \in \{1,2\}.$$

Following expression (8.7.4), we can write the ray angle as

$$\tan\theta = \frac{p_1 + \dfrac{1}{v^3}\dfrac{\partial v}{\partial p_1}}{p_3 + \dfrac{1}{v^3}\dfrac{\partial v}{\partial p_3}}. \qquad (8.7.8)$$

We wish to express the quantities on the right-hand side of expression (8.7.8) in terms of the phase angle, ϑ. In view of expression (8.4.5), we consider the differential operator in the numerator, namely,

$$\frac{\partial}{\partial p_1} = \frac{\partial \vartheta}{\partial p_1} \frac{\partial}{\partial \vartheta} = \frac{\partial \arctan \dfrac{p_1}{p_3}}{\partial p_1} \frac{\partial}{\partial \vartheta}$$
$$= \frac{\dfrac{1}{p_3}}{1 + \left(\dfrac{p_1}{p_3} \right)^2} \frac{\partial}{\partial \vartheta} = \frac{p_3}{p_1^2 + p_3^2} \frac{\partial}{\partial \vartheta}.$$

Considering the phase-slowness vector given by $\mathbf{p} = [p_1, p_3]$, we can write $p^2 = \mathbf{p} \cdot \mathbf{p}$. Hence, the differential operator becomes

$$\frac{\partial}{\partial p_1} = \frac{p_3}{p^2} \frac{\partial}{\partial \vartheta}.$$

Similarly, we obtain the differential operator in the denominator, which is

$$\frac{\partial}{\partial p_3} = -\frac{p_1}{p^2} \frac{\partial}{\partial \vartheta}.$$

Using these differential operators in expression (8.7.8), we can rewrite it as

$$\tan\theta = \frac{p_1 + \dfrac{1}{v^3}\dfrac{p_3}{p^2}\dfrac{\partial v}{\partial \vartheta}}{p_3 - \dfrac{1}{v^3}\dfrac{p_1}{p^2}\dfrac{\partial v}{\partial \vartheta}}.$$

Following eikonal equation (7.3.8), we can state $p^2v^2 = 1$, and, hence, we can write

$$\tan\theta = \frac{p_1 + \frac{p_3}{v}\frac{\partial v}{\partial \vartheta}}{p_3 - \frac{p_1}{v}\frac{\partial v}{\partial \vartheta}}.$$

Again, recalling expression (8.4.5) and dividing both numerator and denominator by p_3 , we obtain

$$\tan\theta = \frac{\frac{p_1}{p_3} + \frac{1}{v}\frac{\partial v}{\partial \vartheta}}{1 - \frac{\frac{p_1}{p_3}}{v}\frac{\partial v}{\partial \vartheta}} = \frac{\tan\vartheta + \frac{1}{v(\vartheta)}\frac{\partial v(\vartheta)}{\partial \vartheta}}{1 - \frac{\tan\vartheta}{v(\vartheta)}\frac{\partial v(\vartheta)}{\partial \vartheta}},$$

which is expression (8.4.12), as required.

EXERCISE 8.4. Using expressions (8.4.9) and (8.4.12), derive expression (8.4.13).

SOLUTION 8.4. Using algebraic manipulation, we can write expression (8.4.12), namely,

$$\tan\theta = \frac{\tan\vartheta + \frac{1}{v}\frac{\partial v}{\partial \vartheta}}{1 - \frac{\tan\vartheta}{v}\frac{\partial v}{\partial \vartheta}},$$

as

$$\frac{\partial v}{\partial \vartheta} = v\frac{\tan\theta - \tan\vartheta}{1 + \tan\theta\tan\vartheta}.$$

Recognizing the trigonometric identity, we can rewrite it as

$$\frac{\partial v}{\partial \vartheta} = v\tan(\theta - \vartheta). \tag{8.7.9}$$

Consider expression (8.4.9). In view of expression (8.7.9), we can write

$$\begin{aligned} V &= \sqrt{v^2 + \left(\frac{\partial v}{\partial \vartheta}\right)^2} \\ &= \sqrt{v^2 + v^2\tan^2(\theta - \vartheta)} \\ &= v\sqrt{1 + \tan^2(\theta - \vartheta)}. \end{aligned}$$

Using trigonometric identities, we obtain

$$V=\frac{v}{\cos\left(\theta-\vartheta\right)}, \tag{8.7.10}$$

which, notably, is expression (8.4.14). The argument of the cosine function is the angle between the ray-velocity vector, $\mathbf{V}$, and the phase-velocity vector, $\mathbf{v}$. As defined in expression (8.4.13), let $\mathbf{t}$ be the unit vector tangent to the ray, and $\mathbf{n}$ be the unit vector normal to the wavefront. Hence, $\theta-\vartheta$ is the angle between $\mathbf{n}$ and $\mathbf{t}$. Thus, we can immediately rewrite expression (8.7.10) as

$$V=\frac{v}{\mathbf{n}\cdot\mathbf{t}},$$

which is expression (8.4.13), as required.

EXERCISE 8.5. Derive a particular case of expression (8.4.12) that corresponds to the elliptical velocity dependence.

SOLUTION 8.5. Inserting expression (6.11.11) into expression (8.4.9), we can write the magnitude of the ray-velocity vector as

$$V\left(\vartheta\right)=\sqrt{\frac{v_x^4\tan^2\vartheta+v_z^4}{v_x^2\tan^2\vartheta+v_z^2}}. \tag{8.7.11}$$

This is the magnitude of ray velocity in terms of the phase velocity as a function of the phase angle for the case of elliptical velocity dependence. Also, inserting expression (6.11.11) into expression (8.4.12), we obtain

$$\tan\theta=\left(\frac{v_x}{v_z}\right)^2\tan\vartheta, \tag{8.7.12}$$

which is the relation between the phase angle and the ray angle for elliptical velocity dependence. Expression (8.7.12) is analogous to expression (9.4.14), which corresponds to *SH* waves in transversely isotropic continua. Inserting expression (8.7.12) into expression (8.7.11), we can write the magnitude of the ray-velocity vector in terms of ray-related quantities, namely,

$$V\left(\theta\right)=V_z\sqrt{\frac{\tan^2\theta+1}{\left(\frac{V_z}{V_x}\right)^2\tan^2\theta+1}}, \tag{8.7.13}$$

where V_x and V_z are the magnitudes of the ray-velocity vector along the x-axis and z-axis, respectively. Herein, we use the fact that, along the axes of the ellipse, the magnitudes of the phase velocity and the ray velocity coincide.

REMARK 8.7.1. Characteristic equations (8.1.13) can be parametrized by choosing various expressions for scaling factor ζ. Two typical examples are shown in Exercises 8.7 and 8.8, below. In both cases, we invoke function (8.1.3) and consider isotropic inhomogeneous continua. Hence, characteristic equations (8.1.13) become equations (8.6.3).

EXERCISE 8.6. [19]Show that the arclength parametrization of system (8.6.3) requires $\zeta = v/2$.

SOLUTION 8.6. In general, if $\mathbf{x} = \mathbf{x}(s)$, using definition (8.1.2), we can write

$$\frac{\mathrm{d}\psi(\mathbf{x})}{\mathrm{d}s} = \sum_{i=1}^{3} \frac{\partial \psi}{\partial x_i} \frac{\mathrm{d}x_i}{\mathrm{d}s} = \sum_{i=1}^{3} p_i \frac{\mathrm{d}x_i}{\mathrm{d}s},$$

which, in view of characteristic equations (8.6.3), we can rewrite as

$$\frac{\mathrm{d}\psi(\mathbf{x})}{\mathrm{d}s} = \sum_{i=1}^{3} 2\zeta p_i p_i,$$

which we can immediately restate as

$$\frac{\mathrm{d}\psi(\mathbf{x})}{\mathrm{d}s} = 2\zeta p^2. \tag{8.7.14}$$

If s is the arclength parameter, then

$$\mathrm{d}s = \sqrt{(\mathrm{d}x)^2 + (\mathrm{d}y)^2 + (\mathrm{d}z)^2},$$

and, hence,

$$\frac{\mathrm{d}s}{\mathrm{d}t} = \sqrt{\left(\frac{\mathrm{d}x}{\mathrm{d}t}\right)^2 + \left(\frac{\mathrm{d}y}{\mathrm{d}t}\right)^2 + \left(\frac{\mathrm{d}z}{\mathrm{d}t}\right)^2}, \tag{8.7.15}$$

[19] See also Section 8.6.1.

where t stands for traveltime. Combining expressions (8.7.14) and (8.7.15), we obtain

$$\frac{\mathrm{d}\psi(\mathbf{x})}{\mathrm{d}t}=\frac{\mathrm{d}\psi(\mathbf{x})}{\mathrm{d}s}\frac{\mathrm{d}s}{\mathrm{d}t}=2\zeta p^2\sqrt{\left(\frac{\mathrm{d}x}{\mathrm{d}t}\right)^2+\left(\frac{\mathrm{d}y}{\mathrm{d}t}\right)^2+\left(\frac{\mathrm{d}z}{\mathrm{d}t}\right)^2}. \qquad (8.7.16)$$

In view of condition (8.2.1) and since the square root gives the magnitude of velocity, we can rewrite equation (8.7.16) as

$$1=2\zeta p^2 v.$$

Herein, since system (8.6.3) refers to the isotropic case—where the ray and wavefront velocities are equal to one another—we denote this velocity by v, even though, in general, the above square root corresponds to the ray velocity. Solving for ζ, where—in view of equation (8.6.1)—we use $p^2v^2=1$, we get

$$\zeta=\frac{1}{2p^2v}=\frac{v}{2p^2v^2}=\frac{v}{2},$$

as required.

EXERCISE 8.7. [20]Letting $\zeta=v/2$, show that characteristic equations (8.6.3) can be reduced to equation (8.6.4), namely,

$$\frac{\mathrm{d}}{\mathrm{d}s}\left[\frac{1}{v(\mathbf{x})}\frac{\mathrm{d}\mathbf{x}}{\mathrm{d}s}\right]=-\frac{\nabla v(\mathbf{x})}{v^2(\mathbf{x})}.$$

SOLUTION 8.7. If $\zeta=v/2$, characteristic equations (8.6.3) become

$$\left\{\begin{array}{l}\dfrac{\mathrm{d}x_i}{\mathrm{d}s}=vp_i\\[2ex]\dfrac{\mathrm{d}p_i}{\mathrm{d}s}=-\dfrac{1}{v^2}\dfrac{\partial v}{\partial x_i}\end{array}\right., \qquad i\in\{1,2,3\}.$$

The first equation of this system can be rewritten as

$$p_i=\frac{1}{v}\frac{\mathrm{d}x_i}{\mathrm{d}s}, \qquad i\in\{1,2,3\}.$$

[20]See also Section 8.6.1.

Hence, the second equation can be stated as

$$\frac{\mathrm{d}p_i}{\mathrm{d}s} = \frac{\mathrm{d}}{\mathrm{d}s}\left(\frac{1}{v}\frac{\mathrm{d}x_i}{\mathrm{d}s}\right) = -\frac{1}{v^2}\frac{\partial v}{\partial x_i}, \qquad i \in \{1,2,3\}.$$

Thus, the system of characteristic equations can be written as a single expression

$$\frac{\mathrm{d}}{\mathrm{d}s}\left(\frac{1}{v(\mathbf{x})}\frac{\mathrm{d}\mathbf{x}}{\mathrm{d}s}\right) = -\frac{\nabla v(\mathbf{x})}{v^2(\mathbf{x})},$$

where $\mathbf{x} = [x_1, x_2, x_3]$, which is equation (8.6.4), as required.

EXERCISE 8.8. [21]Letting $\zeta = v^2/2$, show that characteristic equations (8.6.3) can be written as a system of equations given by

$$\begin{cases} \dot{x}_i = v^2 p_i \\ \\ \dot{p}_i = -\dfrac{\partial}{\partial x_i}\ln v \end{cases}, \qquad i \in \{1,2,3\}.$$

SOLUTION 8.8. As shown in Section 8.2, using function (8.1.3) and letting $\zeta = v^2/2$ results in the time parametrization of characteristic equations (8.1.13). Hence, characteristic equations (8.6.3) can be written as

$$\begin{cases} \dot{x}_i := \dfrac{\mathrm{d}x_i}{\mathrm{d}t} = v^2 p_i \\ \\ \dot{p}_i := \dfrac{\mathrm{d}p_i}{\mathrm{d}t} = -\dfrac{1}{v}\dfrac{\partial v}{\partial x_i} \end{cases}, \qquad i \in \{1,2,3\}.$$

Following the chain rule, we can restate the second equation of this system to obtain

$$\begin{cases} \dot{x}_i = v^2 p_i \\ \\ \dot{p}_i = -\dfrac{\partial}{\partial x_i}\ln v \end{cases}, \qquad i \in \{1,2,3\},$$

as required.

[21] See also Section 8.2.2 and Exercise 13.3.

REMARK 8.7.2. Lemma 8.1.1 shows that both functions (8.1.3) and (8.1.4) yield the same characteristics. Thus, in a seismological context, both functions result in the same rays. In view of Exercise 8.8, Exercise 8.9 illustrates this property for isotropic inhomogeneous continua.

EXERCISE 8.9. [22]Using characteristic equations (8.2.4) and considering functions (8.1.4), show that, for isotropic inhomogeneous continua, we obtain the system of equations

$$\begin{cases} \dot{x}_i = v^2 p_i \\ \\ \dot{p}_i = -\dfrac{\partial}{\partial x_i} \ln v \end{cases}, \qquad i \in \{1,2,3\}.$$

SOLUTION 8.9. Considering functions (8.1.4) for isotropic inhomogeneous continua, characteristic equations (8.2.4), which are parametrized in terms of time, become

$$\begin{cases} \dot{x}_i = v^2 p_i \\ \\ \dot{p}_i = -p^2 v \dfrac{\partial v}{\partial x_i} \end{cases}, \qquad i \in \{1,2,3\}.$$

Since $p^2 v^2 = 1$, we can write

$$\begin{cases} \dot{x}_i = v^2 p_i \\ \\ \dot{p}_i = -\dfrac{1}{v}\dfrac{\partial v}{\partial x_i} = -\dfrac{\partial}{\partial x_i} \ln v \end{cases}, \qquad i \in \{1,2,3\},$$

which is also the solution of Exercise 8.8.

EXERCISE 8.10. Solve the second equation of set (8.5.16) for t.

[22] See also Exercise 13.3.

SOLUTION 8.10. Consider the second equation of set (8.5.16), namely,

$$z(t) = \frac{a}{b}\left[\frac{1}{\mathfrak{p}a\cosh\left(\tanh^{-1}\sqrt{1-(1+2\chi)\mathfrak{p}^2a^2}-bt\right)\sqrt{1+2\chi}}-1\right].$$

We can rearrange this equation to get

$$\cosh\left(\tanh^{-1}\sqrt{1-(1+2\chi)\mathfrak{p}^2a^2}-bt\right) = \frac{1}{\mathfrak{p}(a+bz)\sqrt{1+2\chi}}.$$

Applying the inverse function, $\cosh^{-1}$, to both sides, we write

$$\tanh^{-1}\sqrt{1-(1+2\chi)\mathfrak{p}^2a^2}-bt = \cosh^{-1}\frac{1}{\mathfrak{p}(a+bz)\sqrt{1+2\chi}}.$$

Solving for t, we obtain

$$t(z;\mathfrak{p}) = \frac{1}{b}\left(\tanh^{-1}\sqrt{1-(1+2\chi)\mathfrak{p}^2a^2}-\cosh^{-1}\frac{1}{\mathfrak{p}(a+bz)\sqrt{1+2\chi}}\right), \tag{8.7.17}$$

as required.

REMARK 8.7.3. Expression (8.7.17) is valid only for the downward segment of the ray, in contrast to expression (8.5.18), which is valid for the entire ray, as discussed on page 375. Both equations, however, have a singularity for $\mathfrak{p} = 0$, which corresponds $x = 0$; at that point, both inverse hyperbolic functions tend to infinity.

EXERCISE 8.11. Formally, show the steps leading from set (8.6.5) to equations (8.6.6) and (8.6.7).

SOLUTION 8.11. The first two equations can be written as

$$\frac{\mathrm{d}}{\mathrm{d}s}\left(\frac{1}{v(\mathbf{x})}\frac{\mathrm{d}x}{\mathrm{d}s}\right) = -\frac{1}{v^2(\mathbf{x})}\frac{\partial v(\mathbf{x})}{\partial x},$$

$$\frac{\mathrm{d}}{\mathrm{d}s}\left(\frac{1}{v(\mathbf{x})}\frac{\mathrm{d}y}{\mathrm{d}s}\right) = -\frac{1}{v^2(\mathbf{x})}\frac{\partial v(\mathbf{x})}{\partial y},$$

which leads to

$$\frac{\mathrm{d}s}{\mathrm{d}z}\frac{\mathrm{d}}{\mathrm{d}s}\left(\frac{1}{v(\mathbf{x})}\frac{\mathrm{d}x}{\mathrm{d}s}\frac{\mathrm{d}s}{\mathrm{d}z}\frac{\mathrm{d}z}{\mathrm{d}s}\right)=-\frac{1}{v^2(\mathbf{x})}\frac{\partial v(\mathbf{x})}{\partial x}\frac{\mathrm{d}s}{\mathrm{d}z},$$

$$\frac{\mathrm{d}s}{\mathrm{d}z}\frac{\mathrm{d}}{\mathrm{d}s}\left(\frac{1}{v(\mathbf{x})}\frac{\mathrm{d}y}{\mathrm{d}s}\frac{\mathrm{d}s}{\mathrm{d}z}\frac{\mathrm{d}z}{\mathrm{d}s}\right)=-\frac{1}{v^2(\mathbf{x})}\frac{\partial v(\mathbf{x})}{\partial y}\frac{\mathrm{d}s}{\mathrm{d}z},$$

where we multiplied both sides of the equations by ds/dz, and we multiplied the factors inside the parentheses by unity in the form $(\mathrm{d}s/\mathrm{d}z)(\mathrm{d}z/\mathrm{d}s)$. The two equations can be immediately restated as

$$\frac{\mathrm{d}}{\mathrm{d}z}\left(\frac{1}{v(\mathbf{x})}\frac{\mathrm{d}x}{\mathrm{d}z}\frac{\mathrm{d}z}{\mathrm{d}s}\right)=-\frac{1}{v^2(\mathbf{x})}\frac{\partial v(\mathbf{x})}{\partial x}\frac{\mathrm{d}s}{\mathrm{d}z},$$

$$\frac{\mathrm{d}}{\mathrm{d}z}\left(\frac{1}{v(\mathbf{x})}\frac{\mathrm{d}y}{\mathrm{d}z}\frac{\mathrm{d}z}{\mathrm{d}s}\right)=-\frac{1}{v^2(\mathbf{x})}\frac{\partial v(\mathbf{x})}{\partial y}\frac{\mathrm{d}s}{\mathrm{d}z}. \tag{8.7.18}$$

We assume the invertibilty of function $z=z(s)$, which allows us to write $s=s(z)$. Furthermore, we assume that the behaviour of the space curve $[x(s),y(s),z(s)]$ allows us to express it as $[x(z),y(z)]$. Consequently, from formal operations, we get

$$\frac{\mathrm{d}s}{\mathrm{d}z}\equiv\frac{\mathrm{d}s(x(z),y(z),z)}{\mathrm{d}z}=\frac{\sqrt{[\mathrm{d}x(z)]^2+[\mathrm{d}y(z)]^2+[\mathrm{d}z(z)]^2}}{\mathrm{d}z}$$

$$=\sqrt{\left(\frac{\mathrm{d}x}{\mathrm{d}z}\right)^2+\left(\frac{\mathrm{d}y}{\mathrm{d}z}\right)^2+\left(\frac{\mathrm{d}z}{\mathrm{d}z}\right)^2}.$$

Thus, since $\mathrm{d}s/\mathrm{d}z=1/(\mathrm{d}z/\mathrm{d}s)$, equations (8.7.18) can be stated as

$$\frac{\mathrm{d}}{\mathrm{d}z}\left(\frac{1}{v(\mathbf{x})}\frac{\frac{\mathrm{d}x}{\mathrm{d}z}}{\sqrt{\left(\frac{\mathrm{d}x}{\mathrm{d}z}\right)^2+\left(\frac{\mathrm{d}y}{\mathrm{d}z}\right)^2+1}}\right)$$

$$=-\frac{1}{v^2(\mathbf{x})}\frac{\partial v(\mathbf{x})}{\partial x}\sqrt{\left(\frac{\mathrm{d}x}{\mathrm{d}z}\right)^2+\left(\frac{\mathrm{d}y}{\mathrm{d}z}\right)^2+1}$$

and

$$\frac{\mathrm{d}}{\mathrm{d}z}\left(\frac{1}{v(\mathbf{x})}\frac{\frac{\mathrm{d}y}{\mathrm{d}z}}{\sqrt{\left(\frac{\mathrm{d}x}{\mathrm{d}z}\right)^2+\left(\frac{\mathrm{d}y}{\mathrm{d}z}\right)^2+1}}\right)$$
$$=-\frac{1}{v^2(\mathbf{x})}\frac{\partial v(\mathbf{x})}{\partial y}\sqrt{\left(\frac{\mathrm{d}x}{\mathrm{d}z}\right)^2+\left(\frac{\mathrm{d}y}{\mathrm{d}z}\right)^2+1},$$

which—as required—are equations (8.6.6) and (8.6.7), respectively.

EXERCISE 8.12. Solve ray equations (8.6.6) and (8.6.7) for a vertically inhomogeneous continuum, where $v = v(z)$.

SOLUTION 8.12. Since $v = v(z)$, the right-hand sides of equations (8.6.6) and (8.6.7) vanish. Consequently, we obtain

$$\frac{\mathrm{d}}{\mathrm{d}z}\left[\frac{1}{v(z)}\frac{\frac{\mathrm{d}x}{\mathrm{d}z}}{\sqrt{\left(\frac{\mathrm{d}x}{\mathrm{d}z}\right)^2+\left(\frac{\mathrm{d}y}{\mathrm{d}z}\right)^2+1}}\right]=0 \tag{8.7.19}$$

and

$$\frac{\mathrm{d}}{\mathrm{d}z}\left[\frac{1}{v(z)}\frac{\frac{\mathrm{d}y}{\mathrm{d}z}}{\sqrt{\left(\frac{\mathrm{d}x}{\mathrm{d}z}\right)^2+\left(\frac{\mathrm{d}y}{\mathrm{d}z}\right)^2+1}}\right]=0.$$

Since the velocity gradient is present only along the z-axis, the ray is contained in a single vertical plane. Thus, with no loss of generality, we can assume that a given ray is contained in the xz-plane and, hence, consider only equation (8.7.19). In view of the vanishing of the total derivative, equation (8.7.19) can be restated as

$$\frac{1}{v(z)}\frac{\frac{\mathrm{d}x}{\mathrm{d}z}}{\sqrt{\left(\frac{\mathrm{d}x}{\mathrm{d}z}\right)^2+1}}=\mathfrak{p}, \tag{8.7.20}$$

where $\mathfrak{p}$ is a constant. Equation (8.7.20), can be rewritten as

$$\left(\frac{\mathrm{d}x}{\mathrm{d}z}\right)^2=\mathfrak{p}^2v^2\left[\left(\frac{\mathrm{d}x}{\mathrm{d}z}\right)^2+1\right].$$

Solving for $\mathrm{d}x/\mathrm{d}z$, we obtain

$$\frac{\mathrm{d}x}{\mathrm{d}z}=\frac{\mathfrak{p}v}{\sqrt{1-\mathfrak{p}^2v^2}}, \tag{8.7.21}$$

and, hence we can state the solution as

$$x(z)=\int_{z_0}^{z}\frac{\mathfrak{p}v(\xi)}{\sqrt{1-\mathfrak{p}^2v^2(\xi)}}\mathrm{d}\xi,$$

where ξ is the integration variable. This is a standard expression for a ray in vertically inhomogeneous continua, where, as shown in Exercise 8.13, $\mathfrak{p}=\sin\theta/v(z)$.

EXERCISE 8.13. Consider equation (8.7.20). Show that $\mathfrak{p}=\sin\theta/v(z)$.

SOLUTION 8.13. Since $\mathrm{d}x/\mathrm{d}z=\tan\theta$, following standard trigonometric identities, we can write equation (8.7.20) as

$$\mathfrak{p}=\frac{1}{v(z)}\frac{\frac{\mathrm{d}x}{\mathrm{d}z}}{\sqrt{\left(\frac{\mathrm{d}x}{\mathrm{d}z}\right)^2+1}}=\frac{1}{v(z)}\frac{\tan\theta}{\sqrt{\tan^2\theta+1}}=\frac{\frac{\sin\theta}{\cos\theta}}{v(z)\sec\theta}=\frac{\sin\theta}{v(z)}.$$

EXERCISE 8.14. Consider a one-dimensional homogeneous continuum. Show that solution $x(t)$ of Hamilton's ray equations (8.2.7) corresponds to coordinates (6.5.4), which can be written as

$$x(t)=x_0\pm vt.$$

SOLUTION 8.14. For a one-dimensional case, letting $x_1 \equiv x$ and $p_1 \equiv p$, we can write Hamilton's ray equations (8.2.7) as

$$\begin{cases} \dot{x} = pv^2 + p^2 v \dfrac{\partial v}{\partial p} \\ \\ \dot{p} = -p^2 v \dfrac{\partial v}{\partial x} \end{cases},$$

To study solution $x(t)$, we consider the first equation. In elasticity theory, a one-dimensional continuum must be isotropic, hence, $\partial v / \partial p = 0$. Thus, we obtain

$$\dot{x} = pv^2 .$$

Since, in the one-dimensional case, p is the magnitude of the phase-slowness vector, we can write

$$\dot{x} = \frac{1}{p} p^2 v^2 .$$

In view of eikonal equation (7.3.8) and since $v = \pm 1/p$, we can write

$$\dot{x} := \frac{\mathrm{d}x}{\mathrm{d}t} = \pm v .$$

Solving for $\mathrm{d}x$, we obtain

$$\mathrm{d}x = \pm v \mathrm{d}t .$$

Integrating both sides, we obtain

$$x(t) = x_0 \pm vt ,$$

as required and where x_0 is the integration constant.

EXERCISE 8.15. Following ray equation (8.6.4), show that rays are straight lines in homogeneous continua.[23]

[23] Readers interested in an insightful explanation of the straight-line appearance of the optical rays by the theory of quantum electrodynamics might refer to Feynman, R.P., (1985/2006) QED: The strange theory of light and matter: Princeton University Press, pp. 53–56, where we read that

> the idea that light goes in a straight line is a convenient approximation to describe what happens in the world that is familiar to us.

SOLUTION 8.15. For homogeneous continua, v is constant and, hence, the right-hand side of ray equation (8.6.4) vanishes. Thus, we obtain

$$\frac{\mathrm{d}}{\mathrm{d}s}\left(\frac{1}{v}\frac{\mathrm{d}\mathbf{x}}{\mathrm{d}s}\right)=\mathbf{0}.$$

The vanishing of the total derivative implies that the term in parentheses can be written as

$$\frac{1}{v}\frac{\mathrm{d}\mathbf{x}}{\mathrm{d}s}=\mathbf{C},$$

where $\mathbf{C}$ denotes a constant vector. Rearranging and integrating gives

$$\mathbf{x}=\mathbf{a}s+\mathbf{b},$$

which is an equation of a straight line, where $\mathbf{a}:=\mathbf{C}v$.

CHAPTER 9

Christoffel's Equations

> Mathematical applications to physics occur in at least two aspects. Mathematics is of course the principal tool for solving technical analytical problems, but increasingly it is also a principal guide in our understanding of the basic structure and concepts involved.[1]
>
> *Theodore Frankel (1997)*
> *The geometry of physics*

Preliminary Remarks

In Chapter 7, where we studied the equations of motion in anisotropic continua, we noted that waves propagate therein with three distinct phase velocities. In Chapters 7 and 8, we denoted each of these velocities by

[1] Readers interested in philosophical aspects of this statement might refer to Steiner, M., (1998) The applicability of mathematics as a philosophical problem: Harvard University Press, pp. 1–11.

Also, we might remind ourselves that

> La physique a bâti ses propres critères de validation: n'est valide à ses yeux que ce qui est confirmé par l'expèrience; ni la cohèrence logique ni l'élégance mathématique ne suffisent à elles seules à étayer une théorie physique, même si elles sont souvant de précieux indicateurs.*
>
> *Etienne Klein (1991) Conversations avec le Sphinx: Les paradoxes en physique*
>
> *Physics has established its own criteria of validation: In its eyes, only things confirmed by experiments are valid; neither logical consistency nor mathematical elegance suffice by themselves to support a physical theory, even though they are often valuable indicators.

$v = v(\mathbf{x}, \mathbf{p})$, which is a function of both position and direction. Such a formulation allowed us to derive general forms of the equations governing ray theory in anisotropic inhomogeneous continua, namely, the eikonal equation and Hamilton's ray equations. In this chapter, we wish to derive explicit expressions for these three velocities in terms of the properties of a given continuum, namely, its mass density and elasticity parameters.

We begin this chapter by writing Christoffel's equations, derived in Chapter 7, explicitly in terms of mass density and elasticity parameters. Based on the solvability of these equations, we are then able to formulate the expressions for the three wave velocities, as well as for the associated displacement directions. Using these expressions, we study two specific cases—the three waves that propagate along the symmetry axis in a monoclinic continuum and the three waves that propagate in an arbitrary direction in a transversely isotropic continuum.[2] The chapter concludes with a discussion of the three corresponding phase-slowness surfaces and their intersections.

9.1. Explicit form of Christoffel's Equations

We wish to study Christoffel's equations, shown in expression (7.3.3), namely,

$$\sum_{k=1}^{3}\left(\sum_{j=1}^{3}\sum_{\ell=1}^{3} c_{ijk\ell}(\mathbf{x})\, p_j p_\ell - \rho(\mathbf{x})\,\delta_{ik}\right) A_k(\mathbf{x}) = \mathbf{0}, \qquad i \in \{1,2,3\}, \tag{9.1.1}$$

in the context of a specific continuum. In other words, we wish to rewrite equations (9.1.1) in a way that allows us to conveniently insert the elasticity parameters of a continuum exhibiting a particular symmetry, as discussed in Chapter 5.

Expressing the phase slowness as the reciprocal of the phase velocity, namely,

$$p^2 = \frac{1}{v^2}, \tag{9.1.2}$$

[2] Readers interested in an insightful formulation of Christoffel's equations and waves in a transversely isotropic continuum, which is exemplified by a hexagonal crystal, might also refer to Newnham, R.E., (2005) Properties of materials: Anisotropy, symmetry, structure: Oxford University Press, pp. 249–255.

and letting $n_i^2 = p_i^2/p^2$, where $p^2 := \mathbf{p}\cdot\mathbf{p}$, be the squared components of the unit vector normal to the wavefront, we can rewrite equations (9.1.1) as

$$p^2 \sum_{k=1}^{3} \left(\sum_{j=1}^{3} \sum_{\ell=1}^{3} c_{ijk\ell}(\mathbf{x}) n_j n_\ell - \rho(\mathbf{x}) v^2 \delta_{ik} \right) A_k(\mathbf{x}) = \mathbf{0}, \qquad i \in \{1,2,3\}. \tag{9.1.3}$$

We can state equations (9.1.3) in matrix notation as

$$p^2 \left[\Gamma(\mathbf{x},\mathbf{n}) - \rho(\mathbf{x}) v^2 \mathbf{I} \right] \mathbf{A}(\mathbf{x}) = \mathbf{0}, \tag{9.1.4}$$

where

$$\Gamma(\mathbf{x},\mathbf{n}) := \begin{bmatrix} \sum_{j=1}^{3}\sum_{\ell=1}^{3} c_{1j1l}(\mathbf{x}) n_j n_\ell & \sum_{j=1}^{3}\sum_{\ell=1}^{3} c_{1j2l}(\mathbf{x}) n_j n_\ell & \sum_{j=1}^{3}\sum_{\ell=1}^{3} c_{1j3l}(\mathbf{x}) n_j n_\ell \\ \sum_{j=1}^{3}\sum_{\ell=1}^{3} c_{2j1l}(\mathbf{x}) n_j n_\ell & \sum_{j=1}^{3}\sum_{\ell=1}^{3} c_{2j2l}(\mathbf{x}) n_j n_\ell & \sum_{j=1}^{3}\sum_{\ell=1}^{3} c_{2j3l}(\mathbf{x}) n_j n_\ell \\ \sum_{j=1}^{3}\sum_{\ell=1}^{3} c_{3j1l}(\mathbf{x}) n_j n_\ell & \sum_{j=1}^{3}\sum_{\ell=1}^{3} c_{3j2l}(\mathbf{x}) n_j n_\ell & \sum_{j=1}^{3}\sum_{\ell=1}^{3} c_{3j3l}(\mathbf{x}) n_j n_\ell \end{bmatrix}, \tag{9.1.5}$$

which is equivalent to Christoffel's matrix (7.3.7).[3] Using formula (3.2.5), we can state the entries of matrix $\Gamma(\mathbf{x},\mathbf{n})$ in terms of the elasticity parameters $C_{mn}(\mathbf{x})$, to obtain

$$\begin{aligned}
\Gamma_{11} &= C_{11}n_1^2 + C_{66}n_2^2 + C_{55}n_3^2 + 2\left(C_{16}n_1n_2 + C_{56}n_2n_3 + C_{15}n_1n_3\right),\\
\Gamma_{22} &= C_{66}n_1^2 + C_{22}n_2^2 + C_{44}n_3^2 + 2\left(C_{26}n_1n_2 + C_{24}n_2n_3 + C_{46}n_1n_3\right),\\
\Gamma_{33} &= C_{55}n_1^2 + C_{44}n_2^2 + C_{33}n_3^2 + 2\left(C_{45}n_1n_2 + C_{34}n_2n_3 + C_{35}n_1n_3\right),\\
\Gamma_{12} &= \Gamma_{21}\\
&= C_{16}n_1^2 + C_{26}n_2^2 + C_{45}n_3^2\\
&\quad + \left(C_{12}+C_{66}\right)n_1n_2 + \left(C_{25}+C_{46}\right)n_2n_3 + \left(C_{14}+C_{56}\right)n_1n_3,\\
\Gamma_{13} &= \Gamma_{31} \qquad (9.1.7)\\
&= C_{15}n_1^2 + C_{46}n_2^2 + C_{35}n_3^2\\
&\quad + \left(C_{14}+C_{56}\right)n_1n_2 + \left(C_{36}+C_{45}\right)n_2n_3 + \left(C_{13}+C_{55}\right)n_1n_3,\\
\Gamma_{23} &= \Gamma_{32}\\
&= C_{56}n_1^2 + C_{24}n_2^2 + C_{34}n_3^2\\
&\quad + \left(C_{25}+C_{46}\right)n_1n_2 + \left(C_{23}+C_{44}\right)n_2n_3 + \left(C_{36}+C_{45}\right)n_1n_3,
\end{aligned}$$

[3] It is common to divide the elasticity parameters by mass density and, hence, to write Christoffel's equations (9.1.1) as

$$\sum_{k=1}^{3}\left(\sum_{j=1}^{3}\sum_{\ell=1}^{3}\frac{c_{ijk\ell}(\mathbf{x})}{\rho(\mathbf{x})}p_jp_\ell - \delta_{ik}\right)A_k(\mathbf{x}) = \mathbf{0}, \qquad i \in \{1,2,3\},$$

where, as we see in view of Exercise 2.7, the $c_{ijk\ell}/\rho$ have units of velocity squared. The corresponding solvability condition can be written as

$$\det\left[\Gamma_{ik}(\mathbf{x},\mathbf{p}) - \delta_{ik}\right] = 0, \qquad i,k \in \{1,2,3\}, \qquad (9.1.6)$$

where the entries of matrix $\Gamma(\mathbf{x},\mathbf{p})$ are

$$\Gamma_{ik}(\mathbf{x},\mathbf{p}) := \sum_{j=1}^{3}\sum_{\ell=1}^{3}\frac{c_{ijk\ell}(\mathbf{x})}{\rho(\mathbf{x})}p_jp_\ell, \qquad i,k \in \{1,2,3\}.$$

Each of the three eigenvalues of $\Gamma(\mathbf{x},\mathbf{p})$, namely, $G_i(\mathbf{x},\mathbf{p})$, where $i \in \{1,2,3\}$, results in an eikonal equation, which we can write as

$$G_i(\mathbf{x},\mathbf{p}) = 1, \qquad i \in \{1,2,3\},$$

and which is equivalent to equation (7.3.8).

where, for convenience of notation, we do not explicitly write $\Gamma_{rs}(\mathbf{x},\mathbf{n})$ and $C_{mn}(\mathbf{x})$. Thus, using the elasticity matrices formulated in Chapter 5, expressions (9.1.7) allow us to state Christoffel's equations for a given continuum. Hence, we can conveniently study behaviour of the continuum in terms of its properties, namely, its mass density and elasticity parameters.

The four-subscript notation, $c_{ijk\ell}$, contains intrinsic patterns lost in the two-subscript notation, C_{mn}. As an example, let us write Γ_{13} of expression (9.1.7) using the four-subscript notation.

$$\Gamma_{13} = c_{\mathbf{1}1\mathbf{1}3}n_{\mathbf{1}}^2 + c_{\mathbf{2}3\mathbf{2}1}n_{\mathbf{2}}^2 + c_{\mathbf{3}3\mathbf{3}1}n_{\mathbf{3}}^2 + \left(c_{\mathbf{1}1\mathbf{2}3} + c_{\mathbf{1}3\mathbf{2}1}\right)n_{\mathbf{1}}n_{\mathbf{2}}$$
$$+ \left(c_{\mathbf{2}1\mathbf{3}3} + c_{\mathbf{2}3\mathbf{3}1}\right)n_{\mathbf{2}}n_{\mathbf{3}} + \left(c_{\mathbf{1}1\mathbf{3}3} + c_{\mathbf{1}3\mathbf{3}1}\right)n_{\mathbf{1}}n_{\mathbf{3}},$$

where the bold font emphasizes the pattern: indices of $\mathbf{n}$ correspond to i and k, in $c_{ijk\ell}$.

System (9.1.4) is a homogeneous system of linear equations. In general, such a system has either only the trivial solution, namely, $\mathbf{A} = \mathbf{0}$, or infinitely many solutions in addition to the trivial solution. A necessary and sufficient condition for a system of n homogeneous equations in n unknowns to have nontrivial solutions is the vanishing of the determinant of the coefficient matrix.

We wish to examine the solvability of system (9.1.4). Since, for physically meaningful solutions, we require $p^2 \neq 0$, as discussed in Section 7.3, system (9.1.4) can be written as

$$\left[\Gamma(\mathbf{x},\mathbf{n}) - \rho(\mathbf{x})v^2\mathbf{I}\right]\mathbf{A}(\mathbf{x}) = \mathbf{0}. \tag{9.1.8}$$

Hence, we can write the solvability condition of system (9.1.8) as

$$\det\begin{bmatrix} \Gamma_{11}(\mathbf{x},\mathbf{n}) - \rho(\mathbf{x})v^2 & \Gamma_{12}(\mathbf{x},\mathbf{n}) & \Gamma_{13}(\mathbf{x},\mathbf{n}) \\ \Gamma_{12}(\mathbf{x},\mathbf{n}) & \Gamma_{22}(\mathbf{x},\mathbf{n}) - \rho(\mathbf{x})v^2 & \Gamma_{23}(\mathbf{x},\mathbf{n}) \\ \Gamma_{13}(\mathbf{x},\mathbf{n}) & \Gamma_{23}(\mathbf{x},\mathbf{n}) & \Gamma_{33}(\mathbf{x},\mathbf{n}) - \rho(\mathbf{x})v^2 \end{bmatrix} = 0. \tag{9.1.9}$$

Determinantal equation (9.1.9) is an eigenvalue equation. We wish to learn about the associated eigenvalues and eigenvectors. Γ is a symmetric matrix as shown in Exercise 7.5. Consequently, we can invoke explicitly two theorems of linear algebra.[4]

THEOREM 9.1.1. *Since Γ is symmetric, the corresponding eigenvalues are real.*

THEOREM 9.1.2. *Since Γ is symmetric, the corresponding eigenvectors are orthogonal to each other.*

In view of Theorem 9.1.1, the determinantal equation, stated in expression (9.1.9), has three real roots—the eigenvalues ρv_i^2, where $i = 1,2,3$. Furthermore, in view of Theorem 9.1.2, the three corresponding eigenvectors are orthogonal to each other.

To recognize the physical meaning of the eigenvalues and eigenvectors of system (9.1.8), consider trial solution (7.2.1), which led to Christoffel's equations and which can be written as

$$\mathbf{u}(\mathbf{x},t) = \mathbf{A}(\mathbf{x}) f\{v_0[\psi(\mathbf{x}) - t]\}. \tag{9.1.10}$$

Examining expression (9.1.10) and in view of definition (7.3.2), namely, $p_j := \partial\psi/\partial x_j$, and expression (9.1.2), namely, $p^2 = 1/v^2$, we see that the three eigenvalues correspond to three distinct phase velocities, which are measured normal to the wavefront of a given wave. In view of Theorem 9.1.1, these velocities are real.

Also, as stated in trial solution (7.2.1), $\mathbf{A}(\mathbf{x})$ is the displacement vector. Hence, each eigenvector corresponds to the displacements of the continuum associated with the propagation of a given wave. In view of Theorem 9.1.2, each wave exhibits the displacement vector that is orthogonal to the displacement vectors of the other two waves.

The three displacement vectors are orthogonal to each other at a given point of the continuum only if all three corresponding wavefronts exhibit the same direction at that point. In seismological studies, if we place a receiver in an inhomogeneous continuum at a certain distance from the source—where, in general, the three wavefront normals do not coincide—the three recorded displacement directions are not orthogonal to each other

---

[4] For proofs of Theorem 9.1.1 and Theorem 9.1.2, interested readers might refer to Anton, H., (1973) Elementary linear algebra: John Wiley & Sons, p. 289 and p. 399, and pp. 286–287, respectively.

since each displacement vector corresponds to a wavefront that exhibits a different orientation than the two other wavefronts.

Examining matrix (9.1.5), we can also conclude that, for a given wave in a continuum defined by stress-strain equations (7.1.2), the magnitude of the phase velocity, at a given point, depends only on the elasticity parameters and mass density at that point and is a function of the direction of propagation. Hence, given the properties of the continuum, at each point, we can uniquely determine the magnitude of phase velocity for every direction.

The corresponding displacement direction depends on the same quantities and can be also uniquely determined at a given point of an anisotropic continuum. This is not the case in isotropic continua, where the displacement direction of S waves, although contained in the plane orthogonal to the phase-slowness vector, $\mathbf{p}$, cannot be uniquely determined, as shown in Exercise 9.2.

For the remainder of this chapter, we focus our attention on a given point of the continuum. Hence, for convenience of notation, we write $\rho(\mathbf{x}) \equiv \rho$ and $C_{mn}(\mathbf{x}) \equiv C_{mn}$.

9.2. Christoffel's Equations and Anisotropic Continua

9.2.1. Introductory comments

We wish to study equation (9.1.9), which provides us with the phase velocities of the three waves within an anisotropic continuum. We also wish to examine the eigenvectors of the corresponding matrix Γ, which are the displacement vectors.

Explicit expressions for these velocities in a generally anisotropic continuum can be obtained by inserting entries (9.1.7) into equation (9.1.9). Thus, we obtain three phase velocities, which are functions of both the properties of the continuum—given by its mass density, ρ, and the elasticity parameters, C_{mn}—and the orientation of the wavefront—given by its unit normal, $\mathbf{n}$. Once the phase velocities are obtained, we can find the displacement directions that correspond to each of the three waves by using system (9.1.8).

Note that in the formulation discussed in Chapters 1 and 2, we assumed the displacements of material points associated with the propagation of the waves to be infinitesimal. This is justified by the fact that these displacements are many orders of magnitude smaller than the size of the continuum under investigation, as well as, several orders of magnitude smaller than the wavelength of a given wave. Nevertheless, seismic receivers measure the direction and the amplitude of these displacements, thereby providing us with important information for our study of the properties of the materials through which waves propagate. These measurements are discussed in this chapter and in Chapter 10, respectively.

To illustrate explicit expressions for phase velocities and displacement directions, we consider two particular cases. In the case of a monoclinic continuum, we investigate velocities and displacements for the three waves that are associated with the propagation along the symmetry axis. Notably, this formulation also allows us to illustrate the condition of the natural coordinate system, discussed in Section 5.1. In the case of a transversely isotropic continuum, we investigate velocities and displacements of the three waves for an arbitrary direction of propagation. Notably, this formulation allows us to show that, in general, for anisotropic continua, the displacement direction is neither parallel nor orthogonal to the direction of propagation, as is the case for isotropic continua.

9.2.2. Monoclinic continua

Christoffel's equations along symmetry axis

Consider a monoclinic continuum and let the x_3-axis coincide with the normal to the symmetry plane. In other words, let the x_3-axis be the symmetry axis. Such a continuum is described by elasticity matrix (5.6.4).

Consider a propagation along the x_3-axis. Hence, $n_1 = n_2 = 0$, and the unit vector normal to the wavefront is $\mathbf{n} = [0, 0, 1]$. Following entries (9.1.7) and in view of elasticity matrix (5.6.4), we note that system (9.1.8) becomes

$$\begin{bmatrix} C_{55} - \rho v^2 & C_{45} & 0 \\ C_{45} & C_{44} - \rho v^2 & 0 \\ 0 & 0 & C_{33} - \rho v^2 \end{bmatrix} \begin{bmatrix} A_1 \\ A_2 \\ A_3 \end{bmatrix} = \begin{bmatrix} 0 \\ 0 \\ 0 \end{bmatrix}. \tag{9.2.1}$$

System (9.2.1) can be rewritten as

$$\begin{bmatrix} C_{55}-\rho v^2 & C_{45} \\ C_{45} & C_{44}-\rho v^2 \end{bmatrix} \begin{bmatrix} A_1 \\ A_2 \end{bmatrix} = \begin{bmatrix} 0 \\ 0 \end{bmatrix}, \tag{9.2.2}$$

and

$$\left[C_{33}-\rho v^2\right] A_3 = 0 . \tag{9.2.3}$$

The displacement vectors associated with equations (9.2.2) are contained in the x_1x_2-plane. The displacement vector associated with equation (9.2.3) coincides with the x_3-axis. Hence, the displacement directions associated with equations (9.2.2) are orthogonal to the direction of propagation, while the displacement direction associated with equation (9.2.3) is parallel to the direction of propagation. We refer to the waves whose displacement directions are either orthogonal or parallel to the direction of propagation as the pure-mode waves, and denote them by S or P, respectively.[5]

Note that this monoclinic example illustrates the fact that, along the symmetry axes, all waves propagate as pure-mode waves.

Phase velocities along symmetry axis

To obtain the phase velocity along the symmetry axis, consider equations (9.2.2). The solvability condition is

$$\det \begin{bmatrix} C_{55}-\rho v^2 & C_{45} \\ C_{45} & C_{44}-\rho v^2 \end{bmatrix} = 0 .$$

Thus, we obtain the determinantal equation, namely,

$$\rho^2 \left(v^2\right)^2 - \left[\left(C_{44}+C_{55}\right)\rho\right]\left(v^2\right) - \left(C_{45}^2 - C_{44}C_{55}\right) = 0 ,$$

and, hence, the velocities of the S waves are

$$v_{S_1} = \sqrt{\frac{\left(C_{44}+C_{55}\right)+\sqrt{\left(C_{44}-C_{55}\right)^2+4C_{45}^2}}{2\rho}}, \tag{9.2.4}$$

[5] Readers interested in an insightful description of the pure-mode waves might also refer to Newnham, R.E., (2005) Properties of materials: Anisotropy, symmetry, structure: Oxford University Press, pp. 256 – 258.

and

$$v_{S_2} = \sqrt{\frac{(C_{44}+C_{55}) - \sqrt{(C_{44}-C_{55})^2 + 4C_{45}^2}}{2\rho}}.$$

Also, consider equation (9.2.3). A nontrivial solution requires that $A_3 \neq 0$. Thus, the velocity of the P wave is

$$v_P = \sqrt{\frac{C_{33}}{\rho}}.$$

Displacement directions along symmetry axis

In view of equation (9.2.3), the P-wave displacement vector is parallel to the x_3-axis. Considering a three dimensional continuum, we can write this displacement vector as

$$\mathbf{A}_P = \begin{bmatrix} A_1 \\ A_2 \\ A_3 \end{bmatrix} = a \begin{bmatrix} 0 \\ 0 \\ 1 \end{bmatrix},$$

where a is a nonzero constant.

Now, we wish to find the orientations of the displacement vectors of the S waves. In view of equations (9.2.2), these vectors are contained in the x_1x_2-plane. Inserting eigenvalue (9.2.4) into equations (9.2.2), we obtain

$$\begin{bmatrix} \dfrac{C_{55} - C_{44} - \sqrt{(C_{44}-C_{55})^2 + 4C_{45}^2}}{2} & C_{45} \\ C_{45} & \dfrac{C_{44} - C_{55} - \sqrt{(C_{44}-C_{55})^2 + 4C_{45}^2}}{2} \end{bmatrix} \begin{bmatrix} A_1 \\ A_2 \end{bmatrix} = \begin{bmatrix} 0 \\ 0 \end{bmatrix}. \tag{9.2.5}$$

In view of a three-dimensional continuum, we can write the nontrivial solution of system (9.2.5) as the displacement vector given by

$$\mathbf{A}_{S_1} = \begin{bmatrix} A_1 \\ A_2 \\ A_3 \end{bmatrix} = b \begin{bmatrix} \dfrac{C_{55} - C_{44} + \sqrt{(C_{44}-C_{55})^2 + 4C_{45}^2}}{2C_{45}} \\ 1 \\ 0 \end{bmatrix}, \tag{9.2.6}$$

where b is a nonzero constant. Hence, the angle that this vector makes with a coordinate axis in the x_1x_2-plane is

$$\tan\Theta = \frac{A_1}{A_2} = \frac{C_{55} - C_{44} + \sqrt{(C_{44} - C_{55})^2 + 4C_{45}^2}}{2C_{45}}. \tag{9.2.7}$$

We can also find, in an analogous manner, the displacement vector that corresponds to the other S wave. It is given by

$$\mathbf{A}_{S_2} = \begin{bmatrix} A_1 \\ A_2 \\ A_3 \end{bmatrix} = c \begin{bmatrix} -1 \\ \frac{C_{55} - C_{44} + \sqrt{(C_{44} - C_{55})^2 + 4C_{45}^2}}{2C_{45}} \\ 0 \end{bmatrix},$$

where c is a nonzero constant. We recognize that eigenvectors $\mathbf{A}_P$, $\mathbf{A}_{S_1}$ and $\mathbf{A}_{S_2}$ are linearly independent. Thus, as expected, by Theorem 9.1.2, the three displacement directions are orthogonal to each other, since

$$\mathbf{A}_P \cdot \mathbf{A}_{S_1} = \mathbf{A}_P \cdot \mathbf{A}_{S_2} = \mathbf{A}_{S_1} \cdot \mathbf{A}_{S_2} = 0.$$

Furthermore, in this particular case of the waves propagating along the symmetry axis, the displacement vectors are either parallel or orthogonal to the wavefront normal, $\mathbf{n}$.

In general, in anisotropic continua, the wavefront normal, $\mathbf{n}$, is neither parallel nor orthogonal to the displacement vector. However, in any anisotropic continuum, there exist at least three directions of propagation where the wavefront normal is either parallel or orthogonal to the displacement direction.[6] Such directions are called the pure-mode directions. As illustrated herein, symmetry axes are pure-mode directions.

Natural coordinate systems

In Section 5.6.3, we use the natural coordinate system to describe a monoclinic continuum using the smallest number of nonzero elasticity parameters. The relation between the natural coordinate system and pure-mode directions is stated by the following proposition.

---

[6] Readers interested in further description and additional references might refer to Helbig, K., (1994) Foundations of anisotropy for exploration seismics: Pergamon, p. 166.

PROPOSITION 9.2.1. *Given a propagation along a pure-mode direction, the coordinate system whose axes coincide with the displacement directions of the three waves is a natural coordinate system.*

To elucidate Proposition 9.2.1, consider expression (9.2.7). Invoking the trigonometric identity given by

$$\tan(2\Theta) = \frac{2\tan\Theta}{1-\tan^2\Theta},$$

we can restate expression (9.2.7) as

$$\tan(2\Theta) = \frac{2C_{45}}{C_{44}-C_{55}}. \tag{9.2.8}$$

Expression (9.2.8) is precisely expression (5.6.6), which allows us to express elasticity matrix (5.6.4) in a natural coordinate system to obtain matrix (5.6.7). To further illustrate this result, we notice that, using elasticity matrix (5.6.7), equations (9.2.1) become

$$\begin{bmatrix} \hat{C}_{55}-\rho v^2 & 0 & 0 \\ 0 & \hat{C}_{44}-\rho v^2 & 0 \\ 0 & 0 & \hat{C}_{33}-\rho v^2 \end{bmatrix} \begin{bmatrix} A_1 \\ A_2 \\ A_3 \end{bmatrix} = \begin{bmatrix} 0 \\ 0 \\ 0 \end{bmatrix}, \tag{9.2.9}$$

where all three displacement directions are along the axes of the natural coordinate system, as expected.

Square submatrix

$$\begin{bmatrix} \hat{C}_{55}-\rho v^2 & 0 \\ 0 & \hat{C}_{44}-\rho v^2 \end{bmatrix},$$

in equation (9.2.9), is the diagonal form of the square matrix shown in equation (9.2.2). In terms of a natural coordinate system, such a diagonalization is also obtained using equation (5.6.8), which in the present case, we can write as

$$\begin{bmatrix} \hat{C}_{55}-\rho v^2 & 0 \\ 0 & \hat{C}_{44}-\rho v^2 \end{bmatrix}$$
$$= \begin{bmatrix} \cos\Theta & \sin\Theta \\ -\sin\Theta & \cos\Theta \end{bmatrix} \begin{bmatrix} C_{55}-\rho v^2 & C_{45} \\ C_{45} & C_{44}-\rho v^2 \end{bmatrix} \begin{bmatrix} \cos\Theta & -\sin\Theta \\ \sin\Theta & \cos\Theta \end{bmatrix},$$

where C_{44} and C_{55} are entries of matrix (5.6.4), while $\hat{C}_{44}$ and $\hat{C}_{55}$ are entries of matrix (5.6.7). Also, examining systems (9.2.1) and (9.2.9), we see that the third equation remains unchanged; hence, $C_{33} = \hat{C}_{33}$. This results from the fact that to obtain a natural coordinate system, the original coordinate system is rotated by angle Θ about the x_3-axis, whose orientation remains unchanged.

9.2.3. Transversely isotropic continua

Christoffel's equations

In seismological studies, transverse isotropy plays an important role. For instance, transverse isotropy can be conveniently used to describe layered media.

Consider a transversely isotropic continuum and let the x_3-axis coincide with the normal to the plane of transverse isotropy. In other words, let the x_3-axis be the rotation-symmetry axis. Such a continuum is described by elasticity matrix (5.10.3). For notational convenience, letting

$$\frac{C_{11} - C_{12}}{2} = C_{66}, \tag{9.2.10}$$

in matrix (5.10.3), we can write the entries of matrix Γ, given by expressions (9.1.7), as

$$\begin{aligned}
\Gamma_{11} &= n_1^2 C_{11} + n_2^2 C_{66} + n_3^2 C_{44}, \\
\Gamma_{22} &= n_1^2 C_{66} + n_2^2 C_{11} + n_3^2 C_{44}, \\
\Gamma_{33} &= \left(n_1^2 + n_2^2\right) C_{44} + n_3^2 C_{33}, \\
\Gamma_{12} = \Gamma_{21} &= n_1 n_2 \left(C_{11} - C_{66}\right), \\
\Gamma_{13} = \Gamma_{31} &= n_1 n_3 \left(C_{13} + C_{44}\right), \\
\Gamma_{23} = \Gamma_{32} &= n_2 n_3 \left(C_{13} + C_{44}\right).
\end{aligned} \tag{9.2.11}$$

Note that, in view of expression (9.2.10), $C_{12} = C_{11} - 2C_{66}$. Thus, we could also write $\Gamma_{12} = \Gamma_{21} = n_1 n_2 \left(C_{12} + C_{66}\right)$, which is consistent with the pattern of the last two lines of set (9.2.11). However, in this chapter, we choose to describe a transversely isotropic continuum using C_{11}, C_{13}, C_{33}, C_{44} and C_{66}.

Thus, Christoffel's equations for a transversely isotropic continuum are given by system (9.1.8) with entries (9.2.11). Note that, in view of transverse isotropy, with no loss of generality, we can set either $n_1 = 0$ or $n_2 = 0$.

Phase velocities in transverse-isotropy plane

In this section, we wish to obtain three distinct phase-velocity expressions for the pure-mode waves in a transversely isotropic continuum to identify the general expressions derived in the following section. All waves that propagate along the rotation-symmetry axis, as well as the waves that propagate within the plane of transverse isotropy, are pure-mode waves. However, along the rotation-symmetry axis, the displacement directions of the S waves are subject to the same elastic properties, and their phase-velocity expressions are not distinct. Consequently, to obtain three distinct velocities, we consider the propagation in the plane of transverse isotropy, where $n_3 = 0$.

Furthermore, in view of transverse isotropy, we can consider the propagation in any direction in this plane. We choose the propagation along the x_1-axis and, hence, we set $n_2 = 0$. Consequently, $n_1^2 = 1$. Thus, entries (9.2.11) become $\Gamma_{11} = C_{11}$, $\Gamma_{22} = C_{66}$, $\Gamma_{33} = C_{44}$ and $\Gamma_{12} = \Gamma_{13} = \Gamma_{23} = 0$. Hence, for the propagation along the x_1-axis, system (9.1.8) becomes

$$\begin{bmatrix} C_{11} - \rho v^2 & 0 & 0 \\ 0 & C_{66} - \rho v^2 & 0 \\ 0 & 0 & C_{44} - \rho v^2 \end{bmatrix} \begin{bmatrix} A_1 \\ A_2 \\ A_3 \end{bmatrix} = \begin{bmatrix} 0 \\ 0 \\ 0 \end{bmatrix}. \tag{9.2.12}$$

By examining system (9.2.12), we recognize that all equations are independent of each other and, as expected, all three waves propagate as pure-mode waves. To consider a P wave propagating along the x_1-axis, we set the displacement amplitude along the x_1-axis to unity. Hence, the corresponding vector is $\mathbf{A}_P = [1, 0, 0]^T$. This immediately results in the expression for the P-wave velocity along the x_1-axis, namely,

$$v_P = \pm\sqrt{\frac{C_{11}}{\rho}}. \tag{9.2.13}$$

To consider an S wave propagating along the x_1-axis, we set to unity the displacement amplitude along the axis orthogonal to the x_1-axis and contained in the x_1x_2-plane. We view the x_1x_2-plane as a horizontal plane and, therefore, we refer to this wave as an SH wave . Hence, the corresponding vector is $\mathbf{A}_{SH} = [0, 1, 0]^T$. This immediately results in the expression for the SH-wave velocity along the x_1-axis, namely,

$$v_{SH} = \pm\sqrt{\frac{C_{66}}{\rho}}. \tag{9.2.14}$$

To consider the other S wave propagating along the x_1-axis, we set to unity the displacement amplitude along the axis orthogonal to the x_1x_2-plane. We refer to this wave as an SV wave. Hence, the corresponding vector is $\mathbf{A}_{SV} = [0, 0, 1]^T$. This immediately results in the expression for the SV-wave velocity along the x_1-axis, namely,

$$v_{SV} = \pm\sqrt{\frac{C_{44}}{\rho}}. \tag{9.2.15}$$

Expressions (9.2.13), (9.2.14) and (9.2.15) are distinct from each other. Hence, we can use these expressions to identify general expressions for wave velocities, which are derived below.

Phase velocities in arbitrary directions

We wish to obtain phase-velocity expressions for the three waves propagating in arbitrary directions within a transversely isotropic continuum. Using entries (9.2.11), we write determinant (9.1.9) as

$$\begin{aligned}\det\left[\Gamma - \rho v^2 \mathbf{I}\right] = {} & \left[C_{66}\left(n_1^2 + n_2^2\right) + C_{44} n_3^2 - \rho v^2\right] \\ & \{-C_{13}^2\left(n_1^2 + n_2^2\right) n_3^2 - 2C_{13}C_{44}\left(n_1^2 + n_2^2\right) n_3^2 \\ & + C_{33}C_{44} n_3^4 - C_{44}\left(n_1^2 + n_2^2\right)\rho v^2 - C_{33} n_3^2 \rho v^2 - C_{44} n_3^2 \rho v^2 \\ & + C_{11}\left(n_1^2 + n_2^2\right)\left[C_{44}\left(n_1^2 + n_2^2\right) + C_{33} n_3^2 - \rho v^2\right] + \rho^2 v^4\}.\end{aligned}$$

Using a property of the components of the unit vector, namely, $n_1^2+n_2^2=1-n_3^2$, we can write this determinant as a function of a single component, namely, n_3.

Rearranging, we write this determinant as the product of a quadratic expression in v and a biquadratic expression in v, namely,

$$\begin{aligned}\det\left[\Gamma-\rho v^2\mathbf{I}\right] &= \left[C_{66}\left(1-n_3^2\right)+C_{44}n_3^2-\rho v^2\right] \\ &\left\{\left[C_{33}C_{44}n_3^4-\left[2C_{13}C_{44}-C_{11}C_{33}+C_{13}^2\right]n_3^2\left(1-n_3^2\right)\right.\right. \\ &\left.+C_{11}C_{44}\left(1-n_3^2\right)^2\right]+\left[(C_{11}-C_{33})n_3^2\right. \\ &\left.\left.-(C_{11}+C_{44})\right]\rho v^2+\rho^2v^4\right\}.\end{aligned} \tag{9.2.16}$$

Note that determinant (9.2.16) depends only on n_3, which is the orientation of the wavefront normal, $\mathbf{n}$, with respect to the x_3-axis, which is the rotation-symmetry axis. The absence of n_1 and n_2 illustrates the fact that to study the properties of a transversely isotropic continuum, we can use an arbitrary plane that contains the rotation-symmetry axis.

Following equation (9.1.9) and, hence, setting expression (9.2.16) to zero, we immediately obtain the equation to be solved for the three velocities.

Solving the quadratic equation, which is the first term in brackets on the right-hand side of expression (9.2.16) set to zero, and considering only the positive root, we obtain

$$v(\mathbf{n})=\sqrt{\frac{C_{66}\left(1-n_3^2\right)+C_{44}n_3^2}{\rho}}. \tag{9.2.17}$$

Setting $n_3=0$ and comparing to expressions (9.2.13), (9.2.14) and (9.2.15), we recognize expression (9.2.17) as corresponding to expression (9.2.14). Thus, we denote it as

$$v_{SH}(\mathbf{n})=\sqrt{\frac{C_{66}\left(1-n_3^2\right)+C_{44}n_3^2}{\rho}}. \tag{9.2.18}$$

Solving the biquadratic equation, which is the term in braces in expression (9.2.16) set to zero, we obtain two solutions. Again, setting $n_3=0$, we recognize them as corresponding to expressions (9.2.13) and (9.2.15). We denote them as v_{qP} and v_{qSV} , respectively. Following algebraic simplifications and considering only the positive roots, we write these two solutions as

$$v_{qP}(\mathbf{n}) = \sqrt{\frac{(C_{33} - C_{11})\, n_3^2 + C_{11} + C_{44} + \sqrt{\Delta}}{2\rho}} \tag{9.2.19}$$

and

$$v_{qSV}(\mathbf{n}) = \sqrt{\frac{(C_{33} - C_{11})\, n_3^2 + C_{11} + C_{44} - \sqrt{\Delta}}{2\rho}}, \tag{9.2.20}$$

where the discriminant, Δ, is

$$\Delta \equiv \left[(C_{11} - C_{33})n_3^2 - C_{11} - C_{44}\right]^2 - 4\left[C_{33}C_{44}n_3^4 - \left[2C_{13}C_{44} - C_{11}C_{33} + C_{13}^2\right]n_3^2\left(1 - n_3^2\right) + C_{11}C_{44}\left(1 - n_3^2\right)^2\right]. \tag{9.2.21}$$

Since

$$n_3 = \cos\vartheta\,, \tag{9.2.22}$$

where ϑ is the phase angle, velocity expressions (9.2.18), (9.2.19) and (9.2.20) can be immediately stated in terms of the phase angle.

Let us examine discriminant (9.2.21). To do so, we rewrite it as

$$\Delta \equiv \left[(C_{11} - C_{44})\left(1 - n_3^2\right) - (C_{33} - C_{44})\, n_3^2\right]^2 + 4\,(C_{44} + C_{13})^2\, n_3^2\left(1 - n_3^2\right). \tag{9.2.23}$$

We see that Δ is a nonnegative quantity; hence, expressions (9.2.19) and (9.2.20) are real, as expected in view of the symmetry of matrix Γ. Expressions (9.2.19) and (9.2.20) are distinct from one another if $\Delta \neq 0$. Examining expression (9.2.23), we see that $\Delta = 0$ for the following cases: $n_3 = 0$ and $C_{11} = C_{44}$, $n_3 = \pm 1$ and $C_{33} = C_{44}$, as well as particular values of n_3 if $C_{44} = -C_{13}$. Since these equalities lead to results that are uncommon in the theory of elasticity, we assume that $C_{11} \neq C_{44}$, $C_{33} \neq C_{44}$, $C_{44} \neq -C_{13}$. Hence, we consider Δ as a strictly positive quantity for all values of n_3 , except in a brief discussion of $\Delta = 0$, at the end of Section 9.3.3.

To gain further insight into the phase-velocity expressions, let us consider discriminant (9.2.23) for the case of $n_3 = 0$ and $n_3 = 1$ —the horizontal and vertical propagation directions, respectively. In the former case,

$$\Delta = (C_{11} - C_{44})^2 ; \tag{9.2.24}$$

in the latter case,

$$\Delta = (C_{33} - C_{44})^2 . \tag{9.2.25}$$

[7] Using these results to examine the phase velocity of the qSV wave, which is given in expression (9.2.20), we conclude that $v_{qSV} = \sqrt{C_{44}/\rho}$ for both the horizontal and vertical propagation directions. Also, for the vertical propagation, $v_{qSV} = v_{SH}$, as we see by setting $n_3 = 1$ in expression (9.2.18).

As commented on page 412, the equality of v_{qSV} and v_{SH} along the rotation-symmetry axis is related to the displacement directions of both S waves being subject to the same elastic properties. In general, expressions for v_{SH} and v_{qSV} are functions of the material properties; namely, the elasticity parameters and mass density. The elasticity parameters are associated with directions, which includes both the propagation and displacement directions. If the phase velocity were only a function of the displacement direction, it would follow that the behaviour of the SH wave would be isotropic, since its displacement direction remains in the same plane for all propagation directions.

---

[7] A value of C_{44} is constrained by neither C_{11} nor C_{33} . Hence, according to expressions (9.2.24) and (9.2.25), for $n_3 = 0$ and $n_3 = 1$, $\sqrt{\Delta}$ can be negative in expressions (9.2.19) and (9.2.20).

To justify this statement, we write the components of a transversely isotropic tensor in a manner presented in matrix (5.2.18); its eigenvalues are $\lambda_1 = \lambda_2 = 2c_{1212}$, $\lambda_3 = \lambda_4 = 2c_{2323}$ and

$$\lambda_{5,6} = \frac{1}{2}\left(2c_{1111} + c_{3333} - 2c_{1212} \pm \sqrt{8c_{1133}^2 + (-2c_{1111} + c_{3333} + 2c_{1212})^2}\right) .$$

We see that among these eigenvalues—whose positivity is the sole constraint on the values of elasticity parameters—there is no eigenvalue involving c_{2323} with either c_{1111} or c_{3333} .

Let us examine consequences of $\sqrt{\Delta} < 0$. For $n_3 = 0$ and $\sqrt{\Delta} < 0$, the numerator of the radicand of expression (9.2.19) is $C_{11} + C_{44} - (C_{11} - C_{44}) = 2C_{44}$, which is equivalent to expression (9.2.20), for $n_3 = 0$ and $\sqrt{\Delta} > 0$. For $n_3 = 1$ and $\sqrt{\Delta} < 0$, the numerator of the radicand of expression (9.2.19) is $C_{33} - C_{11} + C_{11} + C_{44} - (C_{33} - C_{44}) = 2C_{44}$, which is equivalent to expression (9.2.20), for $n_3 = 1$ and $\sqrt{\Delta} > 0$. Thus, for $n_3 = 0$ and $n_3 = 1$, if $\sqrt{\Delta} < 0$, expression (9.2.19) becomes expression (9.2.20) and expression (9.2.20) becomes expression (9.2.19).

Displacement directions

To find the displacement directions of waves propagating in a transversely isotropic continuum, we consider, with no loss of generality, any plane that contains the rotation-symmetry axis. Letting this plane coincide with the x_1x_3-plane, we set $n_2 = 0$, and, hence, using entries (9.2.11), we can write the coefficient matrix of system (9.1.8) as

$$\begin{bmatrix} n_1^2 C_{11} + n_3^2 C_{44} - \rho v^2 & 0 & n_1 n_3 (C_{13} + C_{44}) \\ 0 & n_1^2 C_{66} + n_3^2 C_{44} - \rho v^2 & 0 \\ n_1 n_3 (C_{13} + C_{44}) & 0 & n_1^2 C_{44} + n_3^2 C_{33} - \rho v^2 \end{bmatrix}. \tag{9.2.26}$$

Considering equations (9.1.8) and in view of the coefficient matrix (9.2.26), we see that the second equation is not coupled with the remaining two. Hence, we can rewrite system (9.1.8) as

$$\left[n_1^2 C_{66} + n_3^2 C_{44} - \rho v^2 (\mathbf{n}) \right] A_2 = 0, \tag{9.2.27}$$

and

$$\begin{bmatrix} n_1^2 C_{11} + n_3^2 C_{44} - \rho v^2 (\mathbf{n}) & n_1 n_3 (C_{13} + C_{44}) \\ n_1 n_3 (C_{13} + C_{44}) & n_1^2 C_{44} + n_3^2 C_{33} - \rho v^2 (\mathbf{n}) \end{bmatrix} \begin{bmatrix} A_1 \\ A_3 \end{bmatrix} = \begin{bmatrix} 0 \\ 0 \end{bmatrix}. \tag{9.2.28}$$

Note the decoupling of a 3×3 matrix into a 1×1 matrix and a 2×2 matrix, where, the former corresponds to the *SH* waves while the latter corresponds to the *qP* and the *qSV* waves. This decoupling of mathematical entities has a physical reason. The displacement vector associated with equation (9.2.27) is parallel to the x_2-axis, while the displacement vectors associated with equations (9.2.28) are contained in the x_1x_3-plane. Since the two sets of displacement vectors are orthogonal to one another and, hence, do not share any components, they do not affect one another.

Let us investigate the displacement vector associated with equation (9.2.27). The trivial solution is $A_2 = 0$. To find a nontrivial solution, we consider a nonzero vector. This displacement vector, $\mathbf{A} = [0, A_2, 0]$, is parallel to the x_2-axis and, hence, it is orthogonal to the propagation plane. Such a displacement must result from the propagation of a pure *SH* wave.

We can verify that expression (9.2.18), which can be written as

$$\rho v_{SH}^2(\mathbf{n}) = n_1^2 C_{66} + n_3^2 C_{44}, \tag{9.2.29}$$

corresponds to the SH wave. Inserting expression (9.2.29) into (9.2.27), we notice that the term in brackets vanishes, as expected. In accordance with the theory of linear equations, this also means that any value of A_2 satisfies equation (9.2.27). In other words, this equation constrains the orientation, but not the magnitude, of the displacement vector.

Now, we focus our attention on the displacement vectors associated with the remaining two equations, which are stated in system (9.2.28) and correspond to the qP and qSV waves. The trivial solution is $A_1 = A_3 = 0$. To find a nontrivial solution, we consider a nonzero displacement vector, $\mathbf{A} = [A_1, 0, A_3]$, which is contained in the x_1x_3-plane.

System (9.2.28) allows us to show that, in general, in anisotropic continua, the displacement direction is neither parallel nor orthogonal to the direction of propagation. To do so, we find the angle that the displacement vector makes with the x_3-axis. This angle is given by

$$\phi = \arctan\frac{A_1}{A_3}. \tag{9.2.30}$$

Using the second equation of system (9.2.28), we obtain

$$\frac{A_1}{A_3} = \frac{\rho v^2(\mathbf{n}) - n_1^2 C_{44} - n_3^2 C_{33}}{n_1 n_3 (C_{13} + C_{44})}. \tag{9.2.31}$$

Note that the same value of the displacement angle is obtained if we use the first equation of system (9.2.28), as illustrated in Exercise 9.6.

Since $\mathbf{n}^2 = n_1^2 + n_3^2 = 1$ and n_3 is given by expression (9.2.22), we can write expression (9.2.30) as

$$\phi = \arctan\frac{\rho v^2(\vartheta) - C_{44}\sin^2\vartheta - C_{33}\cos^2\vartheta}{(C_{13} + C_{44})\sin\vartheta\cos\vartheta}, \tag{9.2.32}$$

where $v(\vartheta)$ is given by expressions (9.2.19) or (9.2.20), together with expression (9.2.22), for the qP or qSV waves, respectively. In other words, if we wish to find the displacement direction associated with the qP wave, we insert expressions (9.2.19) and (9.2.22) into expression (9.2.32). If we wish to find the displacement direction associated with the qSV wave, we insert expressions (9.2.20) and (9.2.22) into expression (9.2.32).

Examining expression (9.2.32), we see that, in general, ϕ and ϑ are neither equal to one another nor differ by precisely $\pi/2$; this is shown below in Figure 9.4.1 in Exercise 9.12. Hence, in general, in anisotropic continua, waves do not propagate as pure-mode waves. However, in many geological materials, the angle between ϕ and ϑ is not much different from 0 or $\pi/2$; this is the reason for our referring to these waves as *quasiP* or *quasiS*, respectively.

Note that, as expected from the theory of linear equations, in spite of having determined the orientation of the displacement vector, we still have infinitely many nontrivial solutions given by $A_1 = s$ and $A_3 = ms$, where s is a nonzero parameter and m is the right-hand side of equation (9.2.31). In other words, we find the orientation but not the magnitude of the displacement vectors.

9.3. Phase-slowness Surfaces

9.3.1. Introductory comments

Let us consider a point within a continuum and the phase-slowness vectors emanating, in every direction, from this point. The phase-slowness surface is a surface that contains the endpoints of these phase-slowness vectors. In general, in view of three distinct velocities, there are three distinct sheets of the phase-slowness surface.

Phase-slowness surfaces are used in formulating and applying seismic theory associated with anisotropic continua, as shown in Chapter 10. They possess important topological properties. For the elasticity parameters used to describe geological materials, the two outer sheets of the phase-slowness surface intersect. In other words, the magnitudes of the phase velocity of the two slower waves coincide for certain propagation directions.

9.3.2. Convexity of innermost sheet

For the elastic continua, the phase-slowness surface—which, for the transversely isotropic case, results from the bicubic equation given by expression (9.2.16)—is of degree 6. Consequently, any straight line can intersect the surface in, at most, six points. Since the line intersecting the innermost sheet of the phase-slowness surface must intersect the two outer sheets twice, the innermost sheet can be intersected at, at most, two points. This results in the following theorem.[8]

THEOREM 9.3.1. *In elastic continua, if the innermost sheet of the phase-slowness surface is detached, it is convex.*

There are particular cases in which the innermost phase-slowness sheet is not detached, such as the case of $C_{44} = -C_{13}$, mentioned above.[9] Nevertheless, Theorem 9.3.1 is still valid, because the detachment is not necessary for the validity of the theorem. In other words, in elastic continua, the innermost sheet of the phase-slowness surface is convex.[10]

9.3.3. Intersection points

S_1 and S_2 waves are the two slower waves. As stated above, along certain directions, the velocities of these waves must be the same. We wish to find these directions for the S waves propagating in transversely isotropic continua, namely, the intersections of the SH and qSV phase-slowness sheets.

In a transversely isotropic continuum, discussed herein, we consider a cross-section of the phase-slowness surface in the x_1x_3-plane. In view of phase-velocity expressions (9.2.18), (9.2.19) and (9.2.20), and using expression (9.2.22), the corresponding phase-slowness curves can be generated as a polar plot with the radius given by the reciprocal of the phase-velocity magnitude.

Note that the intersection points of the phase-slowness curves in the x_1x_3-plane correspond to intersection lines of the phase-slowness sheets in the $x_1x_2x_3$-space. In view of the rotation symmetry about the x_3-axis, these lines are circles that are parallel to the x_1x_2-plane.

[8] Interested readers might refer to Musgrave, M.J.P., (1970) Crystal acoustics: Introduction to the study of elastic waves and vibrations in crystals: Holden-Day, pp. 91–92.

[9] For other examples of a nondetached innermost sheets, readers might refer to Auld, B.A., (1973) Acoustic fields and waves in solids: John Wiley and Sons, Vol. I, p. 406, and to Helbig, K., (1994) Foundations of anisotropy for exploration seismics: Pergamon, p. 205.

[10] Interested readers might refer to Bucataru, I. and Slawinski, M.A. (2009) On convexity and detachment of innermost wavefront-slowness surface: Geophysics **74**(5), WB63-WB66.

Consider determinant (9.2.16). In view of the fact that the quadratic expression in v^2 contains SH waves while the biquadratic expression in v^2 contains qSV waves, at the intersection points the solution of the quadratic equation must satisfy the biquadratic equation for values of $n_3 \in [-1,1]$. Thus, inserting v^2—given by expression (9.2.18)— into the biquadratic part of equation (9.2.16)—shown in braces—and simplifying, we obtain

$$\left(n_3^2-1\right)\left\{\left(C_{66}-C_{11}\right)\left(C_{44}-C_{66}\right)\right.$$
$$\left.+\left[\left(C_{13}+C_{44}\right)^2-\left(C_{11}-C_{66}\right)\left(C_{33}-2C_{44}+C_{66}\right)\right]n_3^2\right\}=0,$$

which is an expression of the form

$$\left(n_3^2-1\right)\left(A+Bn_3^2\right)=0, \tag{9.3.1}$$

where

$$A:=\left(C_{66}-C_{11}\right)\left(C_{44}-C_{66}\right),$$

and

$$B:=\left(C_{13}+C_{44}\right)^2-\left(C_{11}-C_{66}\right)\left(C_{33}-2C_{44}+C_{66}\right).$$

Hence, immediate solutions of equation (9.3.1) are given by

$$n_3=\pm 1,$$

which correspond to the propagation along the rotation-symmetry axis. Setting $n_3 = \pm 1$ in expressions (9.2.18) and (9.2.20), we can verify that these are the velocities of SH and qSV that are equal to one another.

The remaining solutions of equation (9.3.1) depend on the values of A and B, namely, on the properties of a given continuum given by its elasticity parameters, C_{mn}. In general, we get four distinct cases, namely,

- if $B = 0$, and $A \neq 0$, there are no additional solutions and the magnitudes of the velocity coincide only for the propagation along the rotation-symmetry axis.
- if $B \neq 0$, and $A/B > 0$, there are no additional solutions and the magnitudes of the velocity coincide only for the propagation along the rotation-symmetry axis. Also, except at those two points, the qSV-wave velocity is greater than the SH-wave velocity.

- if $B \neq 0$, and $A/B \leqslant 0$, there is an additional solution given by

$$n_3 = \pm\sqrt{\frac{(C_{11}-C_{66})(C_{44}-C_{66})}{(C_{13}+C_{44})^2-(C_{11}-C_{66})(C_{33}-2C_{44}+C_{66})}}. \tag{9.3.2}$$

- if $A = B = 0$, all values of n_3 are the solutions and, hence, the magnitudes of the SH-wave and the qSV-wave velocities coincide for all directions. This is the case for isotropic continua.

In a seismological context, expression (9.3.2) is of particular interest, because, in connection with expression (9.2.22), namely, $n_3 = \cos\vartheta$, it gives the value of the phase angle at which the intersection points occur, as shown in Exercise 9.3. The equality of the two shear-wave phase velocities results from the equality of two eigenvalues. Consequently, the two corresponding eigenvectors, and, hence, the displacement-vector directions, are not uniquely determined. This is also the case for S waves in isotropic continua, as stated in Remark 9.4.5, which follows Exercise 9.2.

In above derivations, we have excluded three particular cases, as stated on page 415. Let us comment on the most interesting one: $C_{44} = -C_{13}$. If $C_{44} = -C_{13}$, expressions (9.2.19) and (9.2.20) become

$$v_{qP}(\mathbf{n}) = \sqrt{\frac{(C_{33}-C_{11})n_3^2+C_{11}+C_{44}+\left|(C_{11}-C_{44})\left(1-n_3^2\right)-(C_{33}-C_{44})n_3^2\right|}{2\rho}} \tag{9.3.3}$$

and

$$v_{qSV}(\mathbf{n}) = \sqrt{\frac{(C_{33}-C_{11})n_3^2+C_{11}+C_{44}-\left|(C_{11}-C_{44})\left(1-n_3^2\right)-(C_{33}-C_{44})n_3^2\right|}{2\rho}}. \tag{9.3.4}$$

Expressions (9.3.3) and (9.3.4) are equal to one another for particular values of n_3. To find these values, we set the right-hand side of expression (9.2.23) to zero, and get

$$\mathbf{n}_3 = \pm\sqrt{\frac{C_{11} - C_{44}}{C_{11} + C_{33} - 2C_{44}}}. \qquad (9.3.5)$$

At points corresponding to the values given by expression (9.3.5), the sheets of the phase-slowness surfaces that correspond to the qP and qSV waves are not smooth, due to the change of sign within the absolute values in expressions (9.3.3) and (9.3.4). Also, since n_3 is a component of a unit vector, we require that $|n_3| \in (0,1)$, and thus expression (9.3.5) imposes constraints on elasticity parameters. To investigate these constraints, we rewrite this expression as

$$n_3 = \pm\frac{1}{\sqrt{1 + \frac{C_{33} - C_{44}}{C_{11} - C_{44}}}};$$

thus, we require the fraction in the denominator to be positive. Hence, we must assume that C_{33} and C_{11} are both greater, or both smaller, than C_{44}. A transversely isotropic continuum with these constraints has the innermost sheet of the phase-slowness surface that is not detached from the other sheets. The displacement direction associated with this sheet means that the sheet corresponds to the qP wave. This correspondence is expected in view of expression (9.3.3), whose value is greater than the one of expression (9.3.4) for all n_3, except for the points given by expression (9.3.5).[11]

Closing Remarks

Explicit velocity and displacement-angle expressions allow us to study wave phenomena in the context of specific materials. In particular, these expressions can be used in formulating inverse problems where the elasticity parameters are calculated based on the traveltime and displacement-angle information, which are obtained from experimental measurements.

[11] Readers interested in the formulation of the displacement vectors for the waves in the transversely isotropic continuum discussed herein might refer to Bucataru, I. and Slawinski, M.A. (2009) On convexity and detachment of innermost wavefront-slowness surface: Geophysics **74**(5), WB63-WB66.

Studying anisotropic materials, we need to consider three types of angles, namely, the phase angles, discussed in Chapters 6 and 7, as well as the ray angles and the displacement angles, discussed herein. As illustrated in Exercise 9.12, all three angles are related by analytic expressions. However, each angle plays a distinct role in theoretical formulations and the analysis of experimental measurements.

9.4. Exercises

EXERCISE 9.1. [12] Prove the following theorem.

THEOREM 9.4.1. *The solvability condition stated in equation (9.1.6) remains the same if $c_{ijk\ell}$ and $\mathbf{p}$ are transformed in the following manner.*

Consider tensor $c_{ijk\ell} \in \mathbb{R}^d$ and vector $\mathbf{p} \in \mathbb{R}^{\mathbf{d}}$. Also, consider an orthogonal transformation, $A \in \mathbb{R}^{d\times d}$. It follows that matrices

$$\left[\sum_{j=1}^{d}\sum_{\ell=1}^{d} c_{ijk\ell}\,\hat{p}_j\hat{p}_\ell\right]_{1\leqslant i,k\leqslant d} \in \mathbb{R}^{d\times d} \tag{9.4.1}$$

and

$$\left[\sum_{j=1}^{d}\sum_{\ell=1}^{d} \check{c}_{ijk\ell}\,p_j p_\ell\right]_{1\leqslant i,k\leqslant d} \in \mathbb{R}^{d\times d}, \tag{9.4.2}$$

where

$$\hat{p}_i := \sum_{j=1}^{d} A_{ij}\,p_j \qquad \text{and} \qquad \check{c}_{ijk\ell} := \sum_{m=1}^{d}\sum_{n=1}^{d}\sum_{o=1}^{d}\sum_{q=1}^{d} A_{mi}A_{nj}A_{ok}A_{q\ell}\,c_{mnoq},$$

are similar to one another and, consequently, have the same spectra.

[12]This exercise is based on Bos, L., Slawinski, M.A., Stanoev, T., Vianello, M. (2019) On orthogonal transformations of Christoffel equations: arXiv 1901.03926 [physics.geo-ph] and International Journal on Geomathematics, which also contains a numerical example. This theorem is used by Ivanov, Y., and Stovas, A. (2019) 3D mapping of kinematic attributes in anisotropic media: Geophysics **84**(3), C159–C170.

NOTATION 9.4.2. Herein, $\hat{}$ and $\check{}$ mean that $\hat{p}_i$ and $\check{c}_{ijk\ell}$ are transformed by $\mathbf{A}$ and $\mathbf{A}^T$, respectively.

SOLUTION 9.1. Let us consider the following proof.

PROOF. The fourth-rank tensor, $c_{ijk\ell} \in \mathbb{R}^d$, can be viewed as a $d \times d$ matrix, whose entries are $d \times d$ matrices,

$$\mathbf{C} = [C_{ik}]_{1 \leqslant i,k \leqslant d} \in \left(\mathbb{R}^{d \times d}\right)^{d \times d},$$

with $C_{ik} \in \mathbb{R}^{d \times d}$ and $(C_{ik})_{j\ell} := c_{ijk\ell}$. Thus, matrix (9.4.1) can be written as

$$\begin{aligned}\left[\hat{\mathbf{p}}^T C_{ik}\,\hat{\mathbf{p}}\right]_{1 \leqslant i,k \leqslant d} &= \left[(\mathbf{A}\,\mathbf{p})^T C_{ik}\,(\mathbf{A}\,\mathbf{p})\right]_{1 \leqslant i,k \leqslant d} \\ &= \left[\mathbf{p}^T \left(\mathbf{A}^T C_{ik}\,\mathbf{A}\right)\mathbf{p}\right]_{1 \leqslant i,k \leqslant d} \in \mathbb{R}^{d \times d},\end{aligned} \tag{9.4.3}$$

where ^{T} denotes the transpose.

We claim that matrix (9.4.2) can be written as

$$\mathbf{A}^T \left[\mathbf{p}^T \left(\mathbf{A}^T C_{ik}\,\mathbf{A}\right)\mathbf{p}\right]_{1 \leqslant i,k \leqslant d} A \in \mathbb{R}^{d \times d}. \tag{9.4.4}$$

To see this, we let matrix (9.4.4) be

$$\mathbf{M} = \mathbf{A}^T \mathbf{X} \mathbf{A},$$

where $\mathbf{X} := \left[\mathbf{p}^T \left(\mathbf{A}^T C_{ik}\,\mathbf{A}\right)\mathbf{p}\right]_{1 \leqslant i,k \leqslant d}$, to write

$$M_{ik} = \sum_{m=1}^{d} \sum_{o=1}^{d} (A^T)_{im} X_{mo} A_{ok} = \sum_{m=1}^{d} \sum_{o=1}^{d} A_{mi} X_{mo} A_{ok}.$$

Defining $\mathbf{Y} := \mathbf{A}^T C_{mo}\,\mathbf{A}$, we have

$$X_{mo} = \sum_{j=1}^{d} \sum_{\ell=1}^{d} Y_{j\ell}\, p_j\, p_\ell,$$

where

$$Y_{j\ell} = \sum_{n=1}^{d} \sum_{q=1}^{d} (A^T)_{jn} (C_{mo})_{nq} A_{q\ell} = \sum_{n=1}^{d} \sum_{q=1}^{d} A_{nj} A_{q\ell}\, c_{mnoq}.$$

Hence,

$$X_{mo}=\sum_{j=1}^{d}\sum_{\ell=1}^{d}\left(\sum_{n=1}^{d}\sum_{q=1}^{d}A_{nj}A_{q\ell}\,c_{mnoq}\right)p_jp_\ell$$

and, in turn,

$$\begin{aligned}M_{ik}&=\sum_{j=1}^{d}\sum_{\ell=1}^{d}\left(\sum_{m=1}^{d}\sum_{n=1}^{d}\sum_{o=1}^{d}\sum_{q=1}^{d}A_{mi}A_{nj}A_{ok}A_{q\ell}\,c_{mnoq}\right)p_jp_\ell\\&=\sum_{j=1}^{d}\sum_{\ell=1}^{d}\check{c}_{ijk\ell}\,p_jp_\ell\,,\qquad i,k=1,\ldots,d\,,\end{aligned}$$

which is matrix (9.4.2), as required. □

COROLLARY 9.4.3. *From Theorem 9.4.1—and the aforementioned fact that the similar matrices share the same spectrum, as well as the fact that the similarity of matrices is not affected by subtracting from them the identity matrices—it follows that*

$$\det\left[\sum_{j=1}^{3}\sum_{\ell=1}^{3}c_{ijk\ell}\,\hat{p}_j\hat{p}_\ell-\delta_{ik}\right]=\det\left[\sum_{j=1}^{3}\sum_{\ell=1}^{3}\check{c}_{ijk\ell}\,p_jp_\ell-\delta_{ik}\right],\qquad i,k=1,2,3\,,$$

and, hence, equations

$$\det\left[\sum_{j=1}^{3}\sum_{\ell=1}^{3}c_{ijk\ell}\,\hat{p}_j\hat{p}_\ell-\delta_{ik}\right]=0\,,\qquad i,k=1,2,3\,,\tag{9.4.5}$$

and

$$\det\left[\sum_{j=1}^{3}\sum_{\ell=1}^{3}\check{c}_{ijk\ell}\,p_jp_\ell-\delta_{ik}\right]=0\,,\qquad i,k=1,2,3\,,\tag{9.4.6}$$

which are tantamount to the solvability condition stated in equation (9.1.6), are equivalent to one another, as required.

REMARK 9.4.4. Corollary 9.4.3 is valid even without requiring the index symmetries of Hookean solids. Furthermore, Theorem 9.4.1 and Corollary 9.4.3 are valid for arbitrary vectors and fourth-rank tensors, even though, in this corollary, they are related by Christoffel's equations. However, the equivalence of equations (9.4.5) and (9.4.6) does not imply their equivalence to

$$\det\left[\sum_{j=1}^{3}\sum_{\ell=1}^{3} c_{ijk\ell}\, p_j p_\ell - \delta_{ik}\right] = 0\,, \qquad i,k = 1,2,3\,.$$

EXERCISE 9.2. [13]Formulate and solve equation (9.1.9) for isotropic continua.

SOLUTION 9.2. Since isotropy implies directional invariance, with no loss of generality, consider propagation along the x_3-axis and, hence, let $n_1 = n_2 = 0$ and $n_3 = 1$. Considering elasticity matrix (5.12.4) and following entries (9.1.7), we can write equation (9.1.9) as

$$\det\begin{bmatrix} \mu - \rho v^2 & 0 & 0 \\ 0 & \mu - \rho v^2 & 0 \\ 0 & 0 & \lambda + 2\mu - \rho v^2 \end{bmatrix} = 0\,, \tag{9.4.7}$$

to obtain

$$\left(\rho v^2 - \mu\right)^2 \left[\rho v^2 - (\lambda + 2\mu)\right] = 0\,. \tag{9.4.8}$$

Hence, the solutions are $v_1 = v_2 = \sqrt{\mu/\rho}$ and $v_3 = \sqrt{(\lambda+2\mu)/\rho}$, as expected in view of equations (6.1.17) and (6.1.13), respectively.

REMARK 9.4.5. The first two solutions in Exercise 9.2 correspond to the S waves since we can write the corresponding displacement directions as vectors $\mathbf{A} = [1,0,0]^T$ and $\mathbf{A} = [0,1,0]^T$, which are orthogonal to the direction of propagation, $\mathbf{n} = [0,0,1]^T$. The third solution corresponds to the P waves since we can write the corresponding displacement direction as vector $\mathbf{A} = [0,0,1]^T$, which is parallel to the direction of propagation. In view of the double root in equation (9.4.8), there are only two eigenspaces associated with matrix Γ for an isotropic case, in contrast to the anisotropic case, where there are three eigenspaces. Exercise 9.2 shows that in isotropic continua the displacement directions of S waves are contained in the plane that is orthogonal to the direction of propagation. However, these displacement directions cannot be determined uniquely, as is the case for anisotropic continua.

[13]See also Section 9.3.3.

EXERCISE 9.3. Given the values of the elasticity parameters of the Green-river shale[14], namely,

$$\begin{cases} C_{11} = 3.13 \times 10^{10} N/m^2 \\ C_{13} = 0.34 \times 10^{10} N/m^2 \\ C_{33} = 2.25 \times 10^{10} N/m^2 \\ C_{44} = 0.65 \times 10^{10} N/m^2 \\ C_{66} = 0.88 \times 10^{10} N/m^2 \end{cases}, \tag{9.4.9}$$

find the intersection points for the SH and qSV waves.

SOLUTION 9.3. Since $C_{44} \neq -C_{13}$, we can use expression (9.3.2), which refers to the SH and qSV waves only—herein, the qP wave is detached, as discussed on page 415. Combining expression (9.2.22) and (9.3.2), we obtain

$$\vartheta = \arccos \sqrt{\frac{(C_{11} - C_{66})(C_{44} - C_{66})}{(C_{13} + C_{44})^2 - (C_{11} - C_{66})(C_{33} - 2C_{44} + C_{66})}} \approx 66^\circ .$$

The intersection points for the SH and qSV waves occur at $\vartheta \approx 66^\circ$.

EXERCISE 9.4. In view of Section 9.3.3, show that for isotropic continua, SH-wave velocity and SV-wave velocity coincide with one another for all directions.

SOLUTION 9.4. As shown in elasticity matrix (5.12.4), for an isotropic continuum, we have

$$C_{11} = C_{22} = C_{33} = \lambda + 2\mu ,$$

$$C_{13} = \lambda ,$$

$$C_{44} = C_{66} = \mu ,$$

where λ and μ are Lamé's parameters. Thus,

$$\begin{aligned} A &= (C_{11} - C_{66})(C_{44} - C_{66}) = (\lambda + \mu)(\mu - \mu) \\ &= 0 , \end{aligned}$$

[14]These values are taken from Thomsen, L., (1986) Weak elastic anisotropy: Geophysics, **51**, 1954–1966.

and

$$B = (C_{13} + C_{44})^2 - (C_{11} - C_{66})(C_{33} - 2C_{44} + C_{66}) = (\lambda + \mu)^2 - (\lambda + \mu)^2$$
$$= 0 .$$

As stated in Section 9.3.3, if $A = B = 0$, the phase-slowness curves coincide for all directions.

EXERCISE 9.5. Using expression (9.2.32), namely,

$$\phi = \arctan \frac{\rho v^2(\vartheta) - C_{44}\sin^2\vartheta - C_{33}\cos^2\vartheta}{(C_{13} + C_{44})\sin\vartheta\cos\vartheta}, \tag{9.4.10}$$

show that, for P waves in isotropic continua, the phase angle, ϑ, and the displacement angle, ϕ, coincide.

SOLUTION 9.5. Considering the elasticity matrix for an isotropic continuum, namely, matrix (5.12.1), we see that

$$C_{13} = C_{11} - 2C_{44},$$

and

$$C_{11} = C_{33} .$$

Considering elasticity matrix (5.12.4) and expression (6.1.13), we can express the velocity of a P wave in an isotropic continuum as

$$v_P = \sqrt{\frac{C_{33}}{\rho}} .$$

Hence, expression (9.4.10) can be rewritten as

$$\phi = \arctan \frac{C_{11} - C_{44}\sin^2\vartheta - C_{11}\cos^2\vartheta}{(C_{11} - C_{44})\sin\vartheta\cos\vartheta} .$$

Rearranging and using standard trigonometric identities, we obtain

$$\phi = \arctan \frac{(C_{11} - C_{44})\sin^2\vartheta}{(C_{11} - C_{44})\sin\vartheta\cos\vartheta} = \arctan(\tan\vartheta) .$$

Hence, $\phi = \vartheta$, as required and as expected from our discussion in Section 6.2.

EXERCISE 9.6. Expression (9.4.10) is obtained using the second equation of system (9.2.28). Verify that using the first equation of this system to obtain A_1/A_3, we get the same result as shown in Exercise 9.5.

SOLUTION 9.6. Using the first equation of system (9.2.28), we can write expression (9.2.30) as

$$\phi = \arctan\frac{A_1}{A_3} = \arctan\frac{(C_{13}+C_{44})\sin\vartheta\cos\vartheta}{\rho v^2(\vartheta) - C_{11}\sin^2\vartheta - C_{44}\cos^2\vartheta}. \qquad (9.4.11)$$

In view of the isotropic-case expressions, stated in Exercise 9.5, we can rewrite expression (9.4.11) as

$$\phi = \arctan\frac{(C_{11}-C_{44})\sin\vartheta\cos\vartheta}{C_{11} - C_{11}\sin^2\vartheta - C_{44}\cos^2\vartheta} = \arctan(\tan\vartheta).$$

Hence, $\phi = \vartheta$, as required.

EXERCISE 9.7. Using expression (9.4.10), show that, for S waves in isotropic continua, the phase angle, ϑ, and the displacement angle, ϕ, differ by $\pi/2$, which implies that the propagation and displacement directions are orthogonal to one another.

SOLUTION 9.7. Considering the elasticity matrix for an isotropic continuum, namely, matrix (5.12.1), we see that

$$C_{13} = C_{11} - 2C_{44},$$

and

$$C_{11} = C_{33}.$$

Considering elasticity matrix (5.12.4) and expression (6.1.17), we can express the velocity of an S wave in an isotropic continuum as

$$v_S = \sqrt{\frac{C_{44}}{\rho}}.$$

Hence, in a manner analogous to the one used to obtain the solution of Exercise 9.5, expression (9.4.10) becomes

$$\phi = \arctan(-\cot\vartheta) = -\arctan(\cot\vartheta).$$

Using properties of the inverse trigonometric functions, we can rewrite this expression as

$$\phi = \arctan(\tan\vartheta) - \frac{\pi}{2} = \vartheta - \frac{\pi}{2},$$

as required and as expected from our discussion in Section 6.2.

EXERCISE 9.8. Using determinant (9.2.16) obtain expressions (9.2.13), (9.2.14) and (9.2.15).

SOLUTION 9.8. Consider the determinantal expression (9.2.16), namely,

$$\det\left[\Gamma - \rho(\mathbf{x})v^2\mathbf{I}\right] = \left[C_{66}\left(1-n_3^2\right) + C_{44}n_3^2 - \rho v^2\right]$$
$$\left\{\left[C_{33}C_{44}n_3^4 - \left[2C_{13}C_{44} - C_{11}C_{33} + C_{13}^2\right]n_3^2\left(1-n_3^2\right) + C_{11}C_{44}\left(1-n_3^2\right)^2\right]\right.$$
$$\left. + \left[\left(C_{11}-C_{33}\right)n_3^2 - \left(C_{11}+C_{44}\right)\right]\rho v^2 + \rho^2 v^4\right\}.$$

To consider propagation in the plane of transverse isotropy, we let $n_3 = 0$ to obtain

$$\det\left[\Gamma - \rho(\mathbf{x})v^2\mathbf{I}\right] = \left(C_{66} - \rho v^2\right)\left[\rho^2 v^4 - \left(C_{11}+C_{44}\right)\rho v^2 + C_{11}C_{44}\right]. \tag{9.4.12}$$

Setting expression (9.4.12) to zero, we obtain expressions (9.2.13), (9.2.14) and (9.2.15), as required.

EXERCISE 9.9. [15]Show that SH waves in transversely isotropic continua exhibit elliptical velocity dependence.

SOLUTION 9.9. Consider expression (9.2.18). Recalling expression (9.2.22) and using trigonometric identities, we can write

$$v_{SH}(\vartheta) = \sqrt{\frac{C_{66}}{\rho}\sin^2\vartheta + \frac{C_{44}}{\rho}\cos^2\vartheta}.$$

Setting $\vartheta = 0$, we get $v_{SH}(0) = \sqrt{C_{44}/\rho}$, while setting $\vartheta = \pi/2$, we get $v_{SH}(\pi/2) = \sqrt{C_{66}/\rho}$, which can be denoted as v_z and v_x, respectively. Thus, we can write

$$v_{SH}(\vartheta) = \sqrt{v_x^2\sin^2\vartheta + v_z^2\cos^2\vartheta},$$

[15]See also Section 6.11.3.

which is expression (6.11.11), giving the magnitude of phase velocity for the case of elliptical velocity dependence.

EXERCISE 9.10. Formulate Hamiltonian $\mathscr{H}$ that corresponds to SH waves in a transversely isotropic continuum.

SOLUTION 9.10. In view of expression (8.2.8) and considering a given point of the continuum, we can write the corresponding ray-theory Hamiltonian as

$$\mathscr{H}(\mathbf{p}) = \frac{1}{2}p^2v^2(\mathbf{p}).$$

Considering the SH-wave velocity given by expression (9.2.18), namely,

$$v_{SH}^2(\mathbf{n}) = \frac{C_{66}\left(1-n_3^2\right)+C_{44}n_3^2}{\rho},$$

and since $n_i^2 = p_i^2/p^2$ and $n_1^2 = 1 - n_3^2$, we can write

$$v_{SH}^2(\mathbf{p}) = \frac{C_{66}\dfrac{p_1^2}{p^2}+C_{44}\dfrac{p_3^2}{p^2}}{\rho} = \frac{1}{p^2}\frac{C_{66}p_1^2+C_{44}p_3^2}{\rho}.$$

Hence, we can write

$$\mathscr{H}_{SH}(\mathbf{p}) = \frac{1}{2}\frac{C_{66}p_1^2+C_{44}p_3^2}{\rho}. \tag{9.4.13}$$

EXERCISE 9.11. Using Legendre's transformation and expression (9.4.13), find the corresponding relation between the phase and the ray angles for SH waves in a transversely isotropic continuum.

SOLUTION 9.11. As shown in expression (8.7.4), the ray angle is given by

$$\tan\theta = \frac{\dfrac{\mathrm{d}x_1}{\mathrm{d}s}}{\dfrac{\mathrm{d}x_3}{\mathrm{d}s}}.$$

Using time parametrization, we can immediately restate this expression as

$$\tan\theta = \frac{\dfrac{\mathrm{d}x_1}{\mathrm{d}t}}{\dfrac{\mathrm{d}x_3}{\mathrm{d}t}} \equiv \frac{\dot{x}_1}{\dot{x}_3},$$

where t denotes time. In view of transformation (B.3.7), we can write

$$\dot{x}_i = \frac{\partial \mathcal{H}}{\partial p_i}.$$

Thus, using expression (9.4.13), we obtain

$$\dot{x}_1 = \frac{1}{2}\frac{\partial}{\partial p_1}\frac{C_{66}p_1^2 + C_{44}p_3^2}{\rho} = \frac{C_{66}}{\rho}p_1,$$

and

$$\dot{x}_3 = \frac{1}{2}\frac{\partial}{\partial p_3}\frac{C_{66}p_1^2 + C_{44}p_3^2}{\rho} = \frac{C_{44}}{\rho}p_3.$$

Hence, we can write

$$\tan\theta = \frac{\dot{x}_1}{\dot{x}_3} = \frac{C_{66}}{C_{44}}\frac{p_1}{p_3}.$$

Recalling expression (8.4.5), we can restate the above expression in terms of the phase angle, as

$$\tan\theta = \frac{C_{66}}{C_{44}}\tan\vartheta. \tag{9.4.14}$$

REMARK 9.4.6. Expression (9.4.14) allows us to explicitly express the phase angle as a function of the ray angle and vice versa, in terms of the properties of the continuum given by its elasticity parameters. An explicit, closed-form expression of the phase angle in terms of the ray angle is possible only for elliptical velocity dependence.

EXERCISE 9.12. Using expression (9.2.32), namely,

$$\phi = \arctan\frac{\rho v^2(\vartheta) - C_{44}\sin^2\vartheta - C_{33}\cos^2\vartheta}{(C_{13}+C_{44})\sin\vartheta\cos\vartheta}, \tag{9.4.15}$$

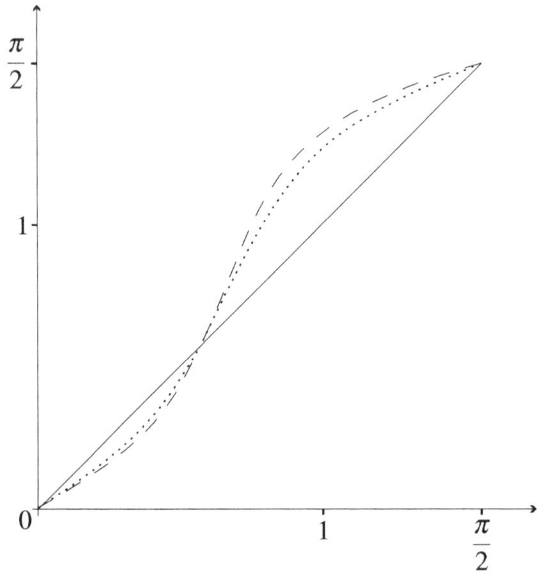

Fig. 9.4.1 Ray, phase and displacement angles; this figure refers to Exercise 9.12. The ray angle (dashed line) and the displacement angle (dotted line) are plotted as functions of the phase angle (solid line). The units of both axes are displayed in radians.

and expression (8.4.12), which can be rewritten as

$$\theta = \arctan \frac{\tan \vartheta + \dfrac{1}{v(\vartheta)} \dfrac{\partial v(\vartheta)}{\partial \vartheta}}{1 - \dfrac{\tan \vartheta}{v(\vartheta)} \dfrac{\partial v(\vartheta)}{\partial \vartheta}}, \tag{9.4.16}$$

as well as the elasticity parameters of the Green-river shale, shown in expressions (9.4.9), and its mass density, given by $\rho = 2310\ kg/m^3$, plot the displacement angles, ϕ, and the ray angle, θ, as a function of the phase angle, ϑ, for qP waves. Comment on the results.

SOLUTION 9.12. Inserting phase-velocity expression (9.2.19) and expression (9.2.22), into expressions (9.4.15) and (9.4.16), we generate the plot of the displacement and the ray angles, respectively. This plot is shown in

Figure 9.4.1. Examining this figure, we see that the values of the displacement angle are closer to the values of the ray angle than to the values of the phase angle.[16]

In general, the phase angles, the ray angles and the displacement angles are distinct. The three angles coincide along the pure-mode directions, in other words, along the directions for which the orientation of the ray coincides with the wavefront normal. These are the directions for which the values of $\theta = \vartheta$ satisfy equation (9.4.16). This is the case of propagation along the symmetry axes, which—for a transversely isotropic continuum with a vertical rotation-symmetry axis—correspond to 0 and $\pi/2$. There is also a third pure-mode direction, which—using the elasticity parameters of the Green-river shale—is obtained graphically in Figure 9.4.1. The existence of the third pure-mode direction is consistent with the statement on page 409, according to which—in any anisotropic continuum—there exist at least three directions of propagation where the wavefront normal is either parallel or orthogonal to the displacement direction. Along the pure-mode directions, the qP waves become P waves; hence, the displacements are along the propagation directions; in other words, the phase, the ray and the displacement angles are equal to each other along these directions.

Based on this physical argument, we conclude that the equality of any two among the aforementioned three angles implies the equality of the three angles; a statement that one might verify formally by using expressions relating these angles.

[16]Readers interested in relations among the phase angle, the ray angle and the displacement angle might also refer to Tsvankin, I., (2001) Seismic signatures and analysis of reflection data in anisotropic media: Pergamon, pp. 34 – 36.

CHAPTER 10

Reflection and Transmission

> A "perfect" scientific theory may be described as one which proceeds logically from a few simple hypotheses to conclusions which are in complete agreement with observation, to within the limits of accuracy of observation. [...] As accuracy of observation increases, a theory ceases to be perfect.[1]
>
> *John Lighton Synge (1937)*
> *Geometrical optics: An introduction to Hamilton's method*

Preliminary Remarks

Discussing ray theory in Chapter 7, we assumed the smoothness of functions describing mass density and elasticity parameters. Hence, the velocity function was smooth with respect to both position and direction. In other words, we assumed that the continuum was not separated by interfaces.

Certain seismic techniques do not require any *a priori* treatment of interfaces and, hence, smooth velocity functions suffice. For instance, for imaging seismic data, we might only need a background velocity field, which can be given by a smooth function. Other seismological studies, however, require an explicit treatment of interfaces. In particular, we need

[1] Readers interested in philosophical implications of this statement might refer to Steiner, M., (1998) The applicability of mathematics as a philosophical problem: Harvard University Press, pp. 58–59 and pp. 105–114.

to consider interfaces to study the phenomena of reflection and transmission. To study these phenomena, we invoke the principles of the continuity of phase, the equality of the sum of displacements and the equality of the traction components across the interface.

We begin this chapter with the derivation of relations among the incidence, reflection, and transmission angles for interfaces between two anisotropic continua. A specific case of elliptical velocity dependence is used to illustrate the general formulation. Then, we consider the amplitudes of the reflected and transmitted signals as functions of the angle of incidence. For a mathematical convenience, the explicit expressions are derived only for the case of *SH* waves in transversely isotropic continua.

10.1. Angles at Interface

10.1.1. Phase angles

Consider a three-dimensional continuum that is composed of parallel homogeneous layers of finite thickness. Let each layer be parallel to the x_1x_2-plane. We choose to view the $x_1x_2x_3$-coordinate system in such a way that we refer to the x_3-axis as the vertical axis. In other words, herein, we study phenomena associated with horizontal layers.

Recall Hamilton's ray equations (8.2.7), namely,

$$\left\{\begin{aligned} \dot{x}_i &= \frac{\partial \mathscr{H}}{\partial p_i} \\ \\ \dot{p}_i &= -\frac{\partial \mathscr{H}}{\partial x_i} \end{aligned}\right. , \qquad i \in \{1,2,3\}, \tag{10.1.1}$$

where Hamiltonian $\mathscr{H}$ is given by expression (8.2.8), namely,

$$\mathscr{H} = \frac{1}{2} p^2 v^2 (\mathbf{x}, \mathbf{p}). \tag{10.1.2}$$

Examining equations (10.1.1) and expression (10.1.2), in view of the horizontal layering, where the elastic properties remain unchanged along the x_1-axis and the x_2-axis, we see that

$$\dot{p}_i \equiv \frac{\mathrm{d}p_i}{\mathrm{d}t} = -\frac{\partial \mathscr{H}}{\partial x_i} = 0, \qquad i \in \{1,2\}.$$

Consequently, p_1 and p_2 are constant for a given solution curve $\mathbf{x}(t)$. In other words, the phase-slowness vector components that are parallel to the interfaces are conserved across these interfaces. We refer to this property as the continuity of phase.

The continuity of phase can be justified by a physical argument. The continuity of phase is tantamount to the continuity of wavefronts, which are the loci of constant phase. Equality of p_i, where $i \in \{1,2\}$, across the interface implies that although the orientation of vector $\mathbf{p}$ might change, its horizontal components must remain the same. In other words, the wavefronts are continuous across the interface. We can see this requirement as resulting from Huygens' principle and from the associated causality.

Let us consider propagation in the x_1x_3-plane. In other words, let $\mathbf{p} = [p_1, 0, p_3]$. We can write the horizontal component of the phase-slowness vector as

$$p_1 = |\mathbf{p}| \, n_1 ,$$

where $|\mathbf{p}|$ is the magnitude of the phase-slowness vector and n_1 is the horizontal component of the unit vector normal to the wavefront. Recalling expression (9.2.22) and using the fact that, in the x_1x_3-plane, $n_1 = \sqrt{1 - n_3^2}$, we obtain

$$p_1 = |\mathbf{p}| \sin \vartheta ,$$

where ϑ is the phase angle, which is measured between the wavefront normal and the vertical axis.

Since p_1 is conserved across the interfaces separating homogeneous horizontal layers, we denote this constant by $\mathfrak{p}$. Now, since the magnitudes of phase slowness and phase velocity are the reciprocals of one another, we can write conserved quantity $\mathfrak{p}$ as

$$\mathfrak{p} = \frac{\sin \vartheta}{v(\vartheta)} , \tag{10.1.3}$$

where $v(\vartheta)$ gives the magnitude of phase velocity as a function of the phase angle.

Expression (10.1.3) is a general statement of Snell's law[2] in the context of phase angle and phase velocity. It is valid across interfaces between generally anisotropic continua. Since $\mathfrak{p}$ is a conserved quantity for a given solution curve $\mathbf{x}(t)$, which corresponds to a ray, we refer to $\mathfrak{p}$ as ray parameter. We discuss it further, in the context of Hamilton's and Lagrange's ray equations, in Section 14.6.

The continuity of the horizontal phase-slowness components provides us with a convenient formulation to relate the angles of incidence, reflection and transmission.[3]

10.1.2. Ray angles

We wish to use the continuity of the phase-slowness components to derive the relation between the ray angles across the interface.

Examining Hamilton's ray equations, an analytic relation between the phase angles and the ray angles was derived in Section 8.4.3 and given by expression (8.4.12). As discussed on page 360, this expression also states that at any point of the phase-slowness curve, the corresponding ray direction is always normal to this curve. Herein, we use this geometrical property to formulate expressions relating incidence, reflection and transmission ray angles across an interface between two anisotropic continua.

---

[2] Readers interested in discussion of Snell's law in the context of quantum electrodynamics might refer to Feynman, R.P., (1985/2006) QED: The strange theory of light and matter: Princeton University Press, pp. 38–45, for reflection, and pp. 50–52, for refraction. On page 56 of that book, we read that

> the idea that light goes in a straight line is a convenient approximation to describe what happens in the world that is familiar to us; it's similar to the crude approximation that says when light reflects off a mirror, the angle of incidence is equal to the angle of reflection.

Readers interested in an insightful discussion of Snell's law, which includes both geometrical and mechanical formulations, might refer to Polya, G., (1954) Mathematics and plausible reasoning, Vol. I: Induction and analogy in mathematics: Princeton University Press, pp. 142–155. Readers interested in Snell's law as an example of causation and determinism might refer to Weinert, F., (2005) The scientist as philosopher: Springer-Verlag, pp. 208–209.

[3] Readers interested in a geometrical formulation of the relation among the incidence, reflection and transmission angles might refer to Auld, B.A., (1973) Acoustic fields and waves in solids: John Wiley and Sons, Vol. II, pp. 1–14.

Note that, while expression (10.1.3) is generally true for $\vartheta \in (-\pi, \pi)$, obtaining analytic expressions in terms of ray angles and ray velocities is not always possible. If we wish to obtain such expressions, we must restrict our studies to particular symmetries or use convenient approximations.[4] This is a consequence of restrictions imposed by Legendre's transformation, which is discussed in Appendix B. Nevertheless, the geometrical construction relating the phase and ray angles, which results from polar reciprocity, is possible at any given point of the phase-slowness surface of a generally anisotropic continuum.

In the following section, we consider a particular symmetry due to elliptical velocity dependence. Therein, we derive analytic expressions between the ray angles of incidence and transmission.

10.1.3. Example: Elliptical velocity dependence

Phase-slowness curves

Consider a two-dimensional continuum that is contained in the xz-plane. Let this continuum consist of two halfspaces, and let the interface coincide with the x-axis.

We wish to characterize each layer by the phase-slowness curve, which is expressed in terms of the horizontal and vertical velocities.

The two phase-slowness curves can be stated as

$$\begin{cases} f(p_x, p_z) = ({}_1v_x p_x)^2 + ({}_1v_z p_z)^2 = 1 \\ g(p_x, p_z) = ({}_2v_x p_x)^2 + ({}_2v_z p_z)^2 = 1 \end{cases}, \tag{10.1.4}$$

for the medium of incidence and transmission, respectively, where v_x and v_z specify the horizontal and vertical phase velocities, respectively.

[4] Readers interested in formulations using expressions based on the weak-anisotropy approximation might refer to Slawinski, M.A., Slawinski, R.A., Brown, R.J., and Parkin, J.M., (2000), A generalized form of Snell's law in anisotropic media. Geophysics, **65**, No. 2, 632–637.

Note that either expression of set (10.1.4) is the equation of an ellipse in the $p_x p_z$-plane, given by

$$\frac{p_x^2}{\left(\frac{1}{{}_m v_x}\right)^2}+\frac{p_z^2}{\left(\frac{1}{{}_m v_z}\right)^2}=1, \qquad m\in\{1,2\}, \tag{10.1.5}$$

where $m=1$ corresponds to the medium of incidence, while $m=2$ corresponds to the medium of transmission.

Conserved quantity in terms of phase angles and phase velocities

Since the continuum is homogeneous along the x-axis, we wish to obtain the quantity that is conserved across the interface in terms of the horizontal and vertical velocities.

In view of expression (10.1.3), we can write

$$\mathfrak{p}=p_x=\frac{\sin\vartheta}{v(\vartheta)}=\frac{\sin\vartheta_m}{\sqrt{{}_m v_x^2\sin^2\vartheta_m+{}_m v_z^2\cos^2\vartheta_m}}, \qquad m\in\{1,2\}, \tag{10.1.6}$$

where, in view of elliptical velocity dependence, $v(\vartheta)$ is given by expression (6.11.11).

Conserved quantity in terms of ray angles and ray velocities

We wish to express conserved quantity (10.1.6) in terms of the ray angle and the ray velocity.

In view of the symmetry of the ellipse, the values of the horizontal phase velocity and vertical phase velocity are equal to the corresponding values of the ray velocities, namely, $v_x=V_x$ and $v_z=V_z$. Hence, set (10.1.4) can be restated as

$$\begin{cases} f(p_x,p_z)=({}_1V_x p_x)^2+({}_1V_z p_z)^2=1 \\ g(p_x,p_z)=({}_2V_x p_x)^2+({}_2V_z p_z)^2=1 \end{cases}. \tag{10.1.7}$$

To find the angle of a normal to a phase-slowness curve, we can consider the phase-slowness curves as the level curves of functions f and g, and use the fact that the ray directions are normal to the phase-slowness curves. In view of the properties of the gradient operator and using, for instance, function f, we can write the unit vector normal to the phase-slowness curve as $\nabla_{\mathbf{p}} f / |\nabla_{\mathbf{p}} f|$, where $\nabla_{\mathbf{p}}$ is the gradient operator given by $[\partial/\partial p_x, \partial/\partial p_z]$. Now, using the scalar product, we obtain the angle between the vector normal to the phase-slowness surface and the vertical axis. This angle, which is the ray angle, is given by

$$\cos\theta_1 = \mathbf{e}_z \cdot \frac{\nabla_{\mathbf{p}} f}{|\nabla_{\mathbf{p}} f|} = \frac{\dfrac{\partial f}{\partial p_z}}{|\nabla_{\mathbf{p}} f|}$$

evaluated at (p_x, p_z), where $\mathbf{e}_z$ is the unit vector along the vertical axis. Thus, using expressions for f and g stated in set (10.1.7), we get the corresponding expression for a ray angle in elliptical velocity dependence, namely,

$$\cos\theta_m = \frac{{}_mV_z^2 p_z}{\sqrt{\left({}_mV_x^2 p_x\right)^2 + \left({}_mV_z^2 p_z\right)^2}}, \qquad m \in \{1,2\}. \tag{10.1.8}$$

To invoke the conserved quantity, $\mathfrak{p} = p_x$, we would like to explicitly solve equations (10.1.8) for p_x.

Using expressions of set (10.1.7), we can write

$$p_z = \frac{\sqrt{1 - \left({}_mV_x p_x\right)^2}}{{}_mV_z}, \qquad m \in \{1,2\}, \tag{10.1.9}$$

and, hence, inserting expressions (10.1.9) into equations (10.1.8), we get

$$\cos\theta_m = \frac{{}_mV_z\sqrt{1 - \left({}_mV_x p_x\right)^2}}{\sqrt{{}_mV_x^4 p_x^2 + {}_mV_z^2\left[1 - \left({}_mV_x p_x\right)^2\right]}}, \qquad m \in \{1,2\}. \tag{10.1.10}$$

Solving equations (10.1.10) for p_x, we obtain

$$p_x^2 = \frac{{}_mV_z^2 \sin^2\theta_m}{{}_mV_x^2\left({}_mV_z^2 \sin^2\theta_m + {}_mV_x^2 \cos^2\theta_m\right)}, \qquad m \in \{1,2\}.$$

Simplifying, we can write

$$p_x^2 = \frac{1}{{}_mV_x^2\left[\left(\frac{{}_mV_x}{{}_mV_z}\right)^2 \cot^2\theta_m + 1\right]}, \qquad m \in \{1,2\}.$$

Consequently, the conserved quantity, $\mathfrak{p} = p_x$, can be written as

$$\mathfrak{p} = \frac{1}{{}_mV_x\sqrt{\left(\frac{{}_mV_x}{{}_mV_z}\right)^2 \cot^2\theta_m + 1}}, \qquad m \in \{1,2\}, \tag{10.1.11}$$

which is conserved quantity (10.1.6) stated in terms of ray angles and ray velocities.

Note that we can write expression (10.1.6) as

$$\mathfrak{p} = \frac{1}{{}_mv_x\sqrt{\left(\frac{{}_mv_z}{{}_mv_x}\right)^2 \cot^2\vartheta_m + 1}}, \qquad m \in \{1,2\}, \tag{10.1.12}$$

which allows us to see the similarity of form between expressions (10.1.6) and (10.1.11). Notice, however, that in expression (10.1.11), we have V_x/V_z, while, in expression (10.1.12), we have v_z/v_x.

In general, as shown in Exercise 10.1, expressions (10.1.6) and (10.1.11) are equivalent to one another. For the isotropic case, as shown in Exercise 10.2, expressions (10.1.6) and (10.1.11) become identical.

Following expression (10.1.11) and in view of set (10.1.7), we can write

$${}_1V_x^2\left[\left(\frac{{}_1V_x}{{}_1V_z}\right)^2 \cot^2\theta_1 + 1\right] = {}_2V_x^2\left[\left(\frac{{}_2V_x}{{}_2V_z}\right)^2 \cot^2\theta_2 + 1\right], \tag{10.1.13}$$

where the subscripts 1 and 2 correspond to the medium of incidence and transmission, respectively. Equation (10.1.13) can be viewed as a statement of Snell's law for elliptical velocity dependence, expressed in terms of ray angles and ray velocities.

10.2. Amplitudes at Interface

10.2.1. Kinematic and dynamic boundary conditions

Introductory comments

In Section 10.1, we related the directions of waves across the interface. For this purpose, we used the continuity of phase. Herein, we relate the amplitudes of waves across the interface. For this purpose, we use the equality of the sum of displacements and the equality of the traction components across the interface, which we refer to as the kinematic and the dynamic boundary conditions, respectively.

In general, if a wave encounters an interface, it generates both reflected and transmitted waves. In this process, the energy of the incident wave is partially reflected and partially transmitted. The fractions of the incident-wave energy that are reflected and transmitted are functions of the direction of the incident wave and the material properties on either side of the interface. Since energy carried by a wave is directly proportional to the square of the amplitude of the displacement, which can be measured by a seismic receiver, we discuss reflection and transmission amplitudes.

The formulation presented in this section deals specifically with amplitudes of plane *SH* waves in the context of a plane interface between two transversely isotropic continua whose rotation-symmetry axes are normal to the interface. Also, these two continua are assumed to be in a welded contact, which implies that they cannot slip with respect to one another. *SH* waves are used because their elliptical velocity dependence lends itself to a convenient illustration of the physical concepts involved.

Displacement vectors

In a three-dimensional transversely isotropic continuum, where the rotation-symmetry axis is assumed to coincide with the x_3-axis, we consider an *SH* wave whose phase-slowness vector, $\mathbf{p}$, is contained in the x_1x_3-plane. Hence, this *SH* wave exhibits a displacement in the x_2-direction only, and, consequently, we can write its displacement vector as

$$\mathbf{u} = [0, u_2, 0] . \qquad (10.2.1)$$

Considering the oscillatory nature of waves and in view of expression (6.11.14), we can write the nonzero component of displacement as

$$u_2 = A\exp\left[i\omega\left(\mathbf{p}\cdot\mathbf{x}-t\right)\right], \tag{10.2.2}$$

where A denotes the amplitude of the displacement and $\exp[\cdot]$ is the phase factor.

Kinematic boundary conditions

The kinematic boundary conditions require the equality of the sum of displacements across the interface. This equality has the following physical meanings. The equality of displacements parallel to the interface implies that the materials cannot slip with respect to one another. The equality of the displacement normal to the interface implies that the materials cannot separate from one another or penetrate one another. These equalities are tantamount to the assumption of a welded contact.

In view of expression (10.2.2) and setting the amplitude of the incident signal to unity, we can write the kinematic boundary condition as

$$\begin{bmatrix}0\\1\\0\end{bmatrix}\exp\left[i\omega\left(\mathbf{p}^{i}\cdot\mathbf{x}-t\right)\right]+\begin{bmatrix}0\\A_r\\0\end{bmatrix}\exp\left[i\omega\left(\mathbf{p}^{r}\cdot\mathbf{x}-t\right)\right]$$
$$=\begin{bmatrix}0\\A_t\\0\end{bmatrix}\exp\left[i\omega\left(\mathbf{p}^{t}\cdot\mathbf{x}-t\right)\right],$$

where i, r and t, as superscripts or subscripts, refer to the incident, reflected and transmitted waves, respectively. We can immediately rewrite this kinematic boundary condition as

$$\exp\left[i\omega\left(\mathbf{p}^{i}\cdot\mathbf{x}-t\right)\right]+A_r\exp\left[i\omega\left(\mathbf{p}^{r}\cdot\mathbf{x}-t\right)\right]=A_t\exp\left[i\omega\left(\mathbf{p}^{t}\cdot\mathbf{x}-t\right)\right]. \tag{10.2.3}$$

Dynamic boundary conditions

The dynamic boundary conditions require the equality of the traction components across the interface, which is tantamount to the equality of the stress-tensor components. We deduce this equality from Newton's third law of motion. To do so, we require the welded contact in order for the adjacent points to remain in constant contact across the interface, and we view the continuum on either side of the interface as two distinct bodies acting on one another.[5]

Since the stress-tensor components are the components of the traction acting on a plane with a given orientation, let us recall the sign convention described on page 57.

> On a surface whose outward normal points in the positive direction of the corresponding coordinate axis, all traction components that act in the positive direction of a given axis are positive. On a surface whose outward normal points in the negative direction of the corresponding coordinate axis, all traction components that act in the negative direction of a given axis are positive.

For convenience, let us consider an interface coinciding with the x_1x_2-plane, and let traction $\mathbf{T}^{(\mathbf{n})}$ act on this plane in such a way that $T_1^{(\mathbf{n})}$, $T_2^{(\mathbf{n})}$ and $T_3^{(\mathbf{n})}$ are positive; **n** stands for the outward normal of the plane, it points in the positive direction of the x_3-axis. Thus, $T_1^{(\mathbf{n})}$, $T_2^{(\mathbf{n})}$ and $T_3^{(\mathbf{n})}$ point in the positive directions of the x_1-axis, the x_2-axis and the x_3-axis, respectively. Newton's third law of motion implies that $T_1^{(-\mathbf{n})}$, $T_2^{(-\mathbf{n})}$ and $T_3^{(-\mathbf{n})}$ are equal in magnitude to their counterparts on the other side of the interface, and they point in the negative directions of the corresponding axes. Since $\mathbf{T}^{(-\mathbf{n})}$ acts on a plane whose outward normal points in the negative direction of the x_3-axis, the three components of $\mathbf{T}^{(-\mathbf{n})}$ are positive. Hence, we can write $T_i^{(\mathbf{n})} = T_i^{(-\mathbf{n})}$. In other words, Newton's third law implies the equality of the traction components across the interface.

To express this result in terms of the stress-tensor components, let us recall the index convention described on page 57 and illustrated in Figure 2.5.1. Following this convention, both vectors $\mathbf{T}^{(\mathbf{n})}$ and $\mathbf{T}^{(-\mathbf{n})}$ are described by σ_{31}, σ_{32} and σ_{33} shown, respectively, by the solid arrows on the upper

[5] Readers interested in an insightful explanation of the requirement of the constant contact between two bodies acting on one another might refer to Schutz, B., (2003) Gravity from the ground up: Cambridge University Press, pp. 11–12.

horizontal plane and by the dashed arrows on the lower horizontal plane in Figure 2.5.1. Since we would have reached the same conclusion of equality if we had chosen a different orientation of the interface or different signs of the traction components, we have shown that Newton's third law implies the equality of the stress-tensor components, which are the dynamic boundary conditions. Let us use these conditions to discuss the displacement amplitudes at a given interface, say, the x_1x_2-plane.

For an interface coinciding with the x_1x_2-plane, $n_1 = n_2 = 0$. Thus, in view of expression (2.5.15) on page 63, we see that the corresponding traction components are identically zero. Consequently, their equality is satisfied trivially. As shown above, the equality of the nonzero components of traction implies the equality of the stress-tensor components given by

$$\begin{bmatrix} \sigma_{31}^{I} \\ \sigma_{32}^{I} \\ \sigma_{33}^{I} \end{bmatrix} = \begin{bmatrix} \sigma_{31}^{II} \\ \sigma_{32}^{II} \\ \sigma_{33}^{II} \end{bmatrix}, \tag{10.2.4}$$

where the superscript I indicates the medium of incidence and reflection, and the superscript II indicates the medium of transmission. Furthermore, in view of the symmetry of the stress tensor stated in Theorem 2.7.1 on page 72, we can write

$$\sigma_{13}^{I} = \sigma_{31}^{I} = \sigma_{13}^{II} = \sigma_{31}^{II}$$

and

$$\sigma_{23}^{I} = \sigma_{32}^{I} = \sigma_{23}^{II} = \sigma_{32}^{II}.$$

To study the amplitudes of reflected and transmitted waves in terms of the properties of the continua on either side of the interface, we wish to rewrite conditions (10.2.4) in terms of elasticity parameters and mass density. Recalling definition (1.4.6) and considering stress-strain equations (4.2.8) with the elasticity matrix for a transversely isotropic continuum, given by matrix (5.10.3), we can write

$$\sigma_{13} = C_{44}\left(\frac{\partial u_1}{\partial x_3} + \frac{\partial u_3}{\partial x_1}\right),$$

$$\sigma_{23} = C_{44}\left(\frac{\partial u_2}{\partial x_3} + \frac{\partial u_3}{\partial x_2}\right)$$

and

$$\sigma_{33} = C_{13}\frac{\partial u_1}{\partial x_1} + C_{13}\frac{\partial u_2}{\partial x_2} + C_{33}\frac{\partial u_3}{\partial x_3},$$

where—using the stress-tensor symmetry—we chose to write σ_{13} and σ_{23} rather than σ_{31} and σ_{32}. Considering displacement vector (10.2.1), where $u_1 = u_3 = 0$, we can write explicitly the three conditions as

$$\sigma_{13} = 0,$$

$$\sigma_{23} = C_{44}\frac{\partial u_2}{\partial x_3} \tag{10.2.5}$$

and

$$\sigma_{33} = C_{13}\frac{\partial u_2}{\partial x_2}.$$

Since $\sigma_{13} = 0$, the first condition is satisfied identically, and the remaining dynamic boundary conditions are

$$\sigma_{23}^{I} = \sigma_{23}^{II} \tag{10.2.6}$$

and

$$\sigma_{33}^{I} = \sigma_{33}^{II}. \tag{10.2.7}$$

Considering the displacement-vector components for the incident, reflected and transmitted waves in view of expression (10.2.5), we can write boundary condition (10.2.6) as

$$C_{44}^{I}\left(\frac{\partial u_2^i}{\partial x_3} + \frac{\partial u_2^r}{\partial x_3}\right) = C_{44}^{II}\frac{\partial u_2^t}{\partial x_3}, \tag{10.2.8}$$

where the left-hand side contains the contributions of both the incident and reflected waves. Invoking expression (10.2.2), we can write equation (10.2.8) as

$$C_{44}^{I}\left(\omega p_3^i \exp\left[i\omega\left(\mathbf{p}^i \cdot \mathbf{x} - t\right)\right] + \omega p_3^r A_r \exp\left[i\omega\left(\mathbf{p}^r \cdot \mathbf{x} - t\right)\right]\right)$$
$$= \omega p_3^t C_{44}^{II} A_t \exp\left[i\omega\left(\mathbf{p}^t \cdot \mathbf{x} - t\right)\right]. \tag{10.2.9}$$

In an analogous manner, we consider boundary condition (10.2.7) to obtain

$$C_{13}^{I}\left(\omega p_2^i \exp\left[i\omega\left(\mathbf{p}^i \cdot \mathbf{x} - t\right)\right] + \omega p_2^r A_r \exp\left[i\omega\left(\mathbf{p}^r \cdot \mathbf{x} - t\right)\right]\right)$$
$$= \omega p_2^t C_{13}^{II} A_t \exp\left[i\omega\left(\mathbf{p}^t \cdot \mathbf{x} - t\right)\right].$$

Since the phase-slowness vector, $\mathbf{p}$, is contained in the x_1x_3-plane, $p_2 = 0$. Hence, this condition is satisfied identically.

Thus, for *SH* waves propagating across the interface separating two transversely isotropic continua in welded contact, equation (10.2.3) is the only kinematic boundary condition and equation (10.2.9) is the only dynamic boundary condition. These equations form the system to be solved for the reflection and transmission amplitudes in terms of properties of continua on either side of the interface; notably, the mass density is implicitly present in these conditions since it is contained in expressions for the phase-slowness vectors, $\mathbf{p}$.

10.2.2. Reflection and transmission amplitudes

Derivation of expressions

We wish to obtain the values of the reflection amplitude, A_r, and the transmission amplitude, A_t. Thus, we need to solve the system composed of equations (10.2.3) and (10.2.9). Since these equations relate to a point on the interface, in view of the previous assumptions, we can make certain simplifications without further affecting the generality of the formulation.

Since we are considering the interface that coincides with the x_1x_2-plane, we set $x_3 = 0$. In view of the transversely isotropic continuum with the x_3-axis corresponding to the rotation-symmetry axis and our choice of the propagation in the x_1x_3-plane, the corresponding phase-slowness vector is $\mathbf{p} = [p_1, 0, p_3]$. Furthermore, the homogeneity of the continuum along the x_1-axis and the x_2-axis allows us to conveniently choose any incidence point on the interface; hence, we choose $(0,0,0)$. Also, at the instant of incidence, the incident, reflected and transmitted waves are considered at the boundary at the same time t. Moreover, considering monochromatic waves, the value of frequency, ω, is the same for the incident, reflected and transmitted waves. Thus, equations (10.2.3) and (10.2.9) simplify to

$$1 + A_r = A_t, \tag{10.2.10}$$

and

$$C_{44}^{I}\left(p_3^i + A_r p_3^r\right) = C_{44}^{II} p_3^t A_t, \tag{10.2.11}$$

respectively.

We can further simplify condition (10.2.11). In view of the phase-slowness curve being symmetric about the x_3-axis, the equality of the p_1 components for the incident and reflected waves implies that $p_3^i = -p_3^r$. In other words, the vertical components of the phase-slowness vectors for the incident and reflected waves exhibit the same magnitudes and opposite directions.

Hence, dynamic boundary condition (10.2.11) becomes

$$C_{44}^I p_3^i (1 - A_r) = C_{44}^{II} p_3^t A_t . \tag{10.2.12}$$

Now, it is convenient to explicitly include mass density in condition (10.2.12). Since p_3 is a vertical component of the phase-slowness vector, recalling expression (9.2.18), we can write

$$C_{44}^I \frac{\cos \vartheta_i}{\sqrt{\frac{C_{66}^I \sin^2 \vartheta_i + C_{44}^I \cos^2 \vartheta_i}{\rho_1}}} (1 - A_r) = C_{44}^{II} \frac{\cos \vartheta_t}{\sqrt{\frac{C_{66}^{II} \sin^2 \vartheta_t + C_{44}^{II} \cos^2 \vartheta_t}{\rho_2}}} A_t . \tag{10.2.13}$$

Equations (10.2.10) and (10.2.13) form a system of two equations to be solved for the two unknowns, namely, the reflection and transmission amplitudes. These solutions are

$$A_r(\vartheta) = \frac{\dfrac{\sqrt{\rho_1} C_{44}^I \cos \vartheta_i}{\sqrt{C_{66}^I \sin^2 \vartheta_i + C_{44}^I \cos^2 \vartheta_i}} - \dfrac{\sqrt{\rho_2} C_{44}^{II} \cos \vartheta_t}{\sqrt{C_{66}^{II} \sin^2 \vartheta_t + C_{44}^{II} \cos^2 \vartheta_t}}}{\dfrac{\sqrt{\rho_1} C_{44}^I \cos \vartheta_i}{\sqrt{C_{66}^I \sin^2 \vartheta_i + C_{44}^I \cos^2 \vartheta_i}} + \dfrac{\sqrt{\rho_2} C_{44}^{II} \cos \vartheta_t}{\sqrt{C_{66}^{II} \sin^2 \vartheta_t + C_{44}^{II} \cos^2 \vartheta_t}}}, \tag{10.2.14}$$

and

$$A_t(\vartheta) = \frac{2\dfrac{\sqrt{\rho_1} C_{44}^I \cos \vartheta_i}{\sqrt{C_{66}^I \sin^2 \vartheta_i + C_{44}^I \cos^2 \vartheta_i}}}{\dfrac{\sqrt{\rho_1} C_{44}^I \cos \vartheta_i}{\sqrt{C_{66}^I \sin^2 \vartheta_i + C_{44}^I \cos^2 \vartheta_i}} + \dfrac{\sqrt{\rho_2} C_{44}^{II} \cos \vartheta_t}{\sqrt{C_{66}^{II} \sin^2 \vartheta_t + C_{44}^{II} \cos^2 \vartheta_t}}}. \tag{10.2.15}$$

Expressions (10.2.14) and (10.2.15) give the reflection amplitude and the transmission amplitude, respectively, for *SH* waves in transversely isotropic continua with the rotation-symmetry axes normal to the interface. The reflection and transmission amplitudes depend on the values of the elasticity parameters and mass density on either side of the interface, and are functions of the phase angles of incidence and transmission.[6]

Interpretation of expressions

Examining expressions (10.2.14) and (10.2.15), we learn about the behaviour of the seismic signal in the context of its being transmitted through, or reflected from, the interface.

Depending on the values of elasticity parameters, mass densities and the incidence angle, the value of expression (10.2.14) can be either positive or negative. The positive sign implies that the direction of the displacement vectors for both the incident wave and the reflected wave is the same. The negative sign implies the reversal of the direction of the displacement vector. Also, while the amplitude of the incident wave is set to unity, the amplitude of the transmitted wave can be greater than unity. This is in agreement with balance of energy, as shown in Exercise 10.6.

If the values of elasticity parameters and mass densities are such that the magnitude of the velocity that is parallel to the interface is greater in the medium of transmission than in the medium of incidence, by examining expression (10.1.6), we conclude that once ϑ_i is large enough, $\sin\vartheta_t$ is greater than unity and, consequently, $\cos\vartheta_t = \sqrt{1-\sin^2\vartheta_t}$ is purely imaginary. Furthermore, examining expressions (10.2.14) and (10.2.15), we conclude that, in such a case, A_r and A_t are complex numbers.

Let us consider the transmitted wave. Returning to expression (10.2.2), we can write it as

$$\begin{aligned} u_2^t &= A_t \exp\left[i\omega\left(\left|\mathbf{p}^t\right|\cos\vartheta_t z + \left|\mathbf{p}^t\right|\sin\vartheta_t x - t\right)\right] \\ &= A_t \exp\left[i\omega\left(\left|\mathbf{p}^t\right| i\left|\cos\vartheta_t\right| z + \left|\mathbf{p}^t\right|\sin\vartheta_t x - t\right)\right] \\ &= A_t \exp\left(-\omega\left|\mathbf{p}^t\right|\left|\cos\vartheta_t\right| z\right)\exp\left[i\left(\left|\mathbf{p}^t\right|\sin\vartheta_t x - t\right)\right] . \end{aligned} \quad (10.2.16)$$

[6] Readers interested in an insightful explanation of the reflection and transmission amplitudes—as functions of the properties of the media and the incidence angle—in the context of the theory of quantum electrodynamics might refer to Feynman, R.P., (1985/2006) QED: The strange theory of light and matter: Princeton University Press, pp. 107–109.

Expression (10.2.16) describes a wave that propagates in the positive x-direction and decays exponentially in the positive z-direction. Such a wave is called evanescent. In such a case there is no energy transmitted across the interface. Also, in such a case, the corresponding magnitude of A_r is equal to unity, as shown in Exercise 10.7.

Since for evanescent waves there is no energy transmitted across the interface, let us focus our attention on the reflected wave. For evanescent waves, A_r is a complex number that we can write as

$$A_r(\vartheta) = |A_r| \exp(i\varkappa) , \tag{10.2.17}$$

where $|A_r|$ is the magnitude and $\varkappa$ is the angle in the complex plane. In view of expressions (10.2.2) and (10.2.17), as well as using the fact that $|A_r| = 1$, we can write the nonzero component of displacement of the reflected wave as

$$u_2^r = \exp(i\varkappa)\exp[i\omega(\mathbf{p}\cdot\mathbf{x}-t)] = \exp\{i[\varkappa+\omega(\mathbf{p}\cdot\mathbf{x}-t)]\} , \tag{10.2.18}$$

where $\exp\{\cdot\}$ is the phase factor. Consequently, examining expression (10.2.18) and following the sign convention used for the phase factor in expression (10.2.2), we see that if $\varkappa > 0$, the reflected wave is phase-delayed relative to the incident wave. This is the consequence of the fact that positive $\varkappa$ results in the phase factor being evaluated at an earlier time. In other words, $\exp[i(\varkappa - \omega t)]$ lags $\exp(-i\omega t)$ in time. Similarly, if $\varkappa < 0$, the reflected wave is phase-advanced.[7]

Expressions in terms of incidence phase angle

As shown in Section 10.1, the incidence and transmission angles can be expressed in terms of one another. Consequently, we wish to state expressions (10.2.14) and (10.2.15) in terms of the phase angle of incidence only.

Recall conserved quantity (10.1.3). Let the phase velocity be given by expression (9.2.18), and the phase angle be stated by expression (9.2.22).

[7] Readers interested in phase shifts might refer to Aki, K., and Richards, P.G., (2002) Quantitative seismology (2*nd* edition): University Science Books, pp. 149–157.

Thus, we can write

$$\frac{\sin\vartheta_i}{\sqrt{\dfrac{C_{66}^I\sin^2\vartheta_i + C_{44}^I\cos^2\vartheta_i}{\rho_1}}} = \frac{\sin\vartheta_t}{\sqrt{\dfrac{C_{66}^{II}\sin^2\vartheta_t + C_{44}^{II}\cos^2\vartheta_t}{\rho_2}}}. \tag{10.2.19}$$

Solving equation (10.2.19) for the angle of transmission, yields

$$\vartheta_t = \arcsin\sqrt{\frac{\rho_1 C_{44}^{II}\sin^2\vartheta_i}{\left[\rho_2\left(C_{66}^I - C_{44}^I\right) - \rho_1\left(C_{66}^{II} - C_{44}^{II}\right)\right]\sin^2\vartheta_i + \rho_2 C_{44}^I}}. \tag{10.2.20}$$

Hence, by inserting expression (10.2.20) into expressions (10.2.14) and (10.2.15), we can state the latter expressions in terms of the phase angle of incidence only.

Expressions in terms of incidence ray angle

It is often convenient to state expressions (10.2.14) and (10.2.15) in terms of the ray angle of incidence, rather than the phase angle of incidence. Following equation (9.4.14), we can express the phase angle in terms of the ray angle as

$$\vartheta_i = \arctan\left(\frac{C_{44}^I}{C_{66}^I}\tan\theta_i\right). \tag{10.2.21}$$

Consequently, by inserting expression (10.2.21) into expression (10.2.20), and inserting the resulting expression into expressions (10.2.14) and (10.2.15), we can state the latter expressions in terms of the ray angle of incidence only.

Turning points

Herein, we consider reflection amplitudes for a ray in a continuum where the velocity increases monotonically with depth. For this purpose—with no loss of generality—we can use the expression derived in Exercise 10.3, which corresponds to an isotropic case, namely,

$$A_r = \frac{\rho_1 v_1\cos\vartheta_i - \rho_2 v_2\cos\vartheta_t}{\rho_1 v_1\cos\vartheta_i + \rho_2 v_2\cos\vartheta_t},$$

where the values of mass density and velocity are positive real numbers. We can write

$$\cos\vartheta_i = \sqrt{1-\sin^2\vartheta_i}$$

and

$$\cos\vartheta_t = \sqrt{1-\sin^2\vartheta_t}$$

Invoking

$$\mathfrak{p} = \frac{\sin\vartheta_i}{v_1} = \frac{\sin\vartheta_t}{v_2},$$

we obtain

$$A_r = \frac{\rho_1 v_1\sqrt{1-\mathfrak{p}^2 v_1^2} - \rho_2 v_2\sqrt{1-\mathfrak{p}^2 v_2^2}}{\rho_1 v_1\sqrt{1-\mathfrak{p}^2 v_1^2} + \rho_2 v_2\sqrt{1-\mathfrak{p}^2 v_2^2}}.$$

Let $\rho_1 = \rho - \Delta\rho$, $\rho_2 = \rho + \Delta\rho$, $v_1 = v - \Delta v$ and $v_2 = v + \Delta v$. We get

$$A_r = \frac{(\rho-\Delta\rho)(v-\Delta v)\sqrt{1-\mathfrak{p}^2(v-\Delta v)^2}}{(\rho-\Delta\rho)(v-\Delta v)\sqrt{1-\mathfrak{p}^2(v-\Delta v)^2}+(\rho+\Delta\rho)(v+\Delta v)\sqrt{1-\mathfrak{p}^2(v+\Delta v)^2}}$$
$$-\frac{(\rho+\Delta\rho)(v+\Delta v)\sqrt{1-\mathfrak{p}^2(v+\Delta v)^2}}{(\rho-\Delta\rho)(v-\Delta v)\sqrt{1-\mathfrak{p}^2(v-\Delta v)^2}+(\rho+\Delta\rho)(v+\Delta v)\sqrt{1-\mathfrak{p}^2(v+\Delta v)^2}}.$$

Consider the root given by

$$\sqrt{1-\mathfrak{p}^2(v-\Delta v)^2} = \sqrt{1-\mathfrak{p}^2\left[v^2-2v\Delta v+(\Delta v)^2\right]}.$$

Assuming Δv is small, we ignore the second-order term and write

$$\sqrt{1-\mathfrak{p}^2(v-\Delta v)^2} \approx \sqrt{1-\mathfrak{p}^2(v^2-2v\Delta v)}.$$

As ϑ_i approaches $\pi/2$, $\mathfrak{p}$ tends to $1/v$. Thus, we can write

$$\sqrt{1-\mathfrak{p}^2(v-\Delta v)^2} \approx \sqrt{1-\frac{1}{v^2}\left[v^2-2v\Delta v\right]} = \sqrt{-2\frac{\Delta v}{v}}.$$

Similarly, consider the other root, we get

$$\sqrt{1-\mathfrak{p}^2(v+\Delta v)^2} \approx \sqrt{2\frac{\Delta v}{v}}.$$

Neglecting linear terms in $\Delta\rho$ and Δv, we obtain

$$A_r \approx \frac{\sqrt{2\frac{\Delta v}{v}} - \sqrt{-2\frac{\Delta v}{v}}}{\sqrt{2\frac{\Delta v}{v}} + \sqrt{-2\frac{\Delta v}{v}}}.$$

Rearranging, we write

$$A_r \approx \frac{\sqrt{2\frac{\Delta v}{v}} - i\sqrt{2\frac{\Delta v}{v}}}{\sqrt{2\frac{\Delta v}{v}} + i\sqrt{2\frac{\Delta v}{v}}}.$$

Simplifying, we get

$$A_r \approx \frac{1-i}{1+i}.$$

Multiplying both the numerator and the denominator by the complex conjugate, we obtain $A_r \approx -i$.

To understand the physical meaning of this result, we can also write $A_r = \exp(-i\pi/2)$, and consider the phase factor, namely, $\exp(-i\omega t)$. The effect of A_r is

$$A_r \exp(-i\omega t) = \exp\left(-i\frac{\pi}{2}\right)\exp(-i\omega t) = \exp\left[-i\left(\frac{\pi}{2} + \omega t\right)\right],$$

which is a phase advance at the turning point.

Closing Remarks

The reflection-angle and transmission-angle expressions derived in this chapter result from the continuity of phase across the interface. Analogous expressions, resulting from the conserved quantity associated with Fermat's principle of stationary traveltime, are discussed in Chapter 14.

Herein, the reflection-amplitude and transmission-amplitude expressions are derived for *SH* waves in transversely isotropic continua. This formulation provides a convenient illustration of the derivation process resulting from the boundary conditions that imply the equality of the sum of displacements and the equality of the traction components across the interface. Such a formulation can also be used in more general cases.

Note, however, that the illustration using *SH* waves does not address the fact that, in general, in anisotropic continua, displacement direction is neither parallel nor orthogonal to the wavefront normal, as discussed in Section 9.2.3. This property would introduce additional complications that are not addressed in this chapter.

Our formulation of the reflection and transmission amplitudes is based on the plane-wave assumption. Considering a point source, the plane-wave assumption provides a good approximation to a general formulation for distant sources. Moreover, other wavefront shapes can be considered as a composition of plane waves. In other words, any wavefront can be decomposed into plane waves.[8]

If we wish to derive a more general formulation, numerous assumptions must be investigated. For instance, considering ray methods in transversely isotropic continua, *SH* waves are decoupled from the *qP* and *qSV* waves. In general, in continua exhibiting different symmetries, all three waves are coupled. Also, for the interface considered in this chapter, the two transversely isotropic continua are oriented in such a way that their rotation-symmetry axes are normal to the interface. Furthermore, the boundary conditions used in this chapter are based on the assumption of the welded

[8] Readers interested in evaluation of the applicability of the plane-wave assumption and its extensions might refer to Grant, F.S., and West, G.F., (1965) Interpretation theory in applied geophysics: McGraw-Hill Book Co., Chapter 6.

contact at the interface. Many of the above concerns are addressed in the existing literature.[9]

10.3. Exercises

EXERCISE 10.1. [10]Show that expressions (10.1.6) and (10.1.11) are equivalent to one another.

SOLUTION 10.1. Consider expression (10.1.11). In view of the symmetry of an ellipse, we know that $V_x = v_x$ and $V_z = v_z$. Hence, we can write

$$\mathfrak{p} = \frac{1}{v_x\sqrt{\left(\frac{v_x}{v_z}\right)^2 \cot^2\theta + 1}}.$$

[9] Readers interested in the formulation of reflection and transmission coefficients for P, SV and SH waves at different boundary conditions might refer to Aki, K. and Richards, P.G., (2002) Quantitative seismology (2*nd* edition): University Science Books, pp. 128 – 149, and to Červený, V., (2001) Seismic ray theory: Cambridge University Press, pp. 477–505. The former reference also contains a convenient weak-inhomogeneity approximation.

Readers interested in a formulation involving qP and qSV waves in transversely isotropic continua might refer to Mavko, G., Mukerji, T., and Dvorkin, J., (1998) The rock physics handbook: Cambridge University Press, pp 65–70.

Readers interested in a formulation of reflection and transmission coefficients in anelastic continua might refer to Le, L.H.T., Krebes, E.S., and Quiroga-Goode, G.E., (1994) Synthetic seismograms for SH waves in anelastic transversely isotropic media: Geophys. J. Int., **116**, 598–604.

Readers interested in a formulation accounting for phenomena resulting from slip interfaces, including interfaces between two identical continua, might refer to Schoenberg, M., (1980) Elastic wave behaviour across linear slip interfaces: J. Acoust. Soc. Am., **68** (5), 1516 – 1521.

[10] Also see Section 14.6.

Recalling expression (8.7.12), we express the ray angle in terms of the phase angle to obtain

$$\mathfrak{p}=\frac{1}{v_x\sqrt{\left(\frac{v_x}{v_z}\right)^2\frac{1}{\tan^2\left\{\arctan\left[\left(\frac{v_x}{v_z}\right)^2\tan\vartheta\right]\right\}}+1}}.$$

Using trigonometric identities, we get

$$\mathfrak{p}=\frac{1}{v_x\sqrt{\left(\frac{v_x}{v_z}\right)^2\frac{1}{\left(\frac{v_x}{v_z}\right)^4\tan^2\vartheta}+1}}=\frac{1}{\sqrt{v_z^2\cot^2\vartheta+v_x^2}}.$$

Multiplying both numerator and denominator by $\sin\vartheta$, we obtain

$$\mathfrak{p}=\frac{\sin\vartheta}{\sqrt{v_z^2\cos^2\vartheta+v_x^2\sin^2\vartheta}}.$$

In view of expression (6.11.11), we can immediately write

$$\mathfrak{p}=\frac{\sin\vartheta}{v(\vartheta)},$$

which is expression (10.1.6), as required.

Exercise 10.2. Show that in isotropic continua, expressions (10.1.6) and (10.1.11) are identical to one another.

Solution 10.2. Consider expression (10.1.11). In isotropic continua, $V := V_x = V_z$. Hence, we can write

$$\mathfrak{p}=\frac{1}{V\sqrt{\cot^2\theta+1}}.$$

For isotropic continua, the magnitudes of the phase and ray velocities coincide, namely, $V=v$. Also, the phase and ray angles coincide, namely, $\theta=\vartheta$. Thus, invoking trigonometric identities, we obtain

$$\mathfrak{p}=\frac{\sin\vartheta}{v},$$

which is the isotropic form of expression (10.1.6), as required.

EXERCISE 10.3. Following expression (10.2.14), state the expressions for the reflection and transmission amplitudes for isotropic continua in terms of mass density, ρ , and velocity, v .

SOLUTION 10.3. In view of matrices (5.12.1) and (5.12.4), we let $\mu := C_{44} = C_{66}$ and write

$$A_r(\vartheta) = \frac{\sqrt{\rho_1\mu_1}\cos\vartheta_i - \sqrt{\rho_2\mu_2}\cos\vartheta_t}{\sqrt{\rho_1\mu_1}\cos\vartheta_i + \sqrt{\rho_2\mu_2}\cos\vartheta_t}$$

and

$$A_t(\vartheta) = \frac{2\sqrt{\rho_1\mu_1}\cos\vartheta_i}{\sqrt{\rho_1\mu_1}\cos\vartheta_i + \sqrt{\rho_2\mu_2}\cos\vartheta_t},$$

for the reflection and transmission amplitudes, respectively. In view of $v = \sqrt{\mu/\rho}$ and since both μ and ρ are positive real numbers, we can manipulate these expressions and restate them as

$$A_r(\vartheta) = \frac{\rho_1\sqrt{\dfrac{\mu_1}{\rho_1}}\cos\vartheta_i - \rho_2\sqrt{\dfrac{\mu_2}{\rho_2}}\cos\vartheta_t}{\rho_1\sqrt{\dfrac{\mu_1}{\rho_1}}\cos\vartheta_i + \rho_2\sqrt{\dfrac{\mu_2}{\rho_2}}\cos\vartheta_t} = \frac{\rho_1 v_1\cos\vartheta_i - \rho_2 v_2\cos\vartheta_t}{\rho_1 v_1\cos\vartheta_i + \rho_2 v_2\cos\vartheta_t} \tag{10.3.1}$$

and

$$A_t(\vartheta) = \frac{2\rho_1\sqrt{\dfrac{\mu_1}{\rho_1}}\cos\vartheta_i}{\rho_1\sqrt{\dfrac{\mu_1}{\rho_1}}\cos\vartheta_i + \rho_2\sqrt{\dfrac{\mu_2}{\rho_2}}\cos\vartheta_t} = \frac{2\rho_1 v_1\cos\vartheta_i}{\rho_1 v_1\cos\vartheta_i + \rho_2 v_2\cos\vartheta_t}. \tag{10.3.2}$$

Following Snell's law, namely, $\vartheta_t = \arcsin\left[(v_2/v_1)\sin\vartheta_i\right]$, we can express both A_r and A_t in terms of the angle of incidence, ϑ_i .

EXERCISE 10.4. Using expressions (10.2.14) and (10.2.15), state the expressions for the reflection and transmission amplitudes for normal incidence in terms of mass density, ρ , and velocity, $v_{SH}(0)$.

SOLUTION 10.4. Consider expressions (10.2.14) and (10.2.15). Letting $\vartheta_i = \vartheta_t = 0$, we obtain

$$A_r(0) = \frac{\sqrt{\rho_1 C_{44}^I} - \sqrt{\rho_2 C_{44}^{II}}}{\sqrt{\rho_1 C_{44}^I} + \sqrt{\rho_2 C_{44}^{II}}}$$

and

$$A_t(0) = \frac{2\sqrt{\rho_1 C_{44}^I}}{\sqrt{\rho_1 C_{44}^I} + \sqrt{\rho_2 C_{44}^{II}}},$$

for the reflection and transmission amplitudes, respectively. In view of expressions (9.2.18) and (9.2.22), we obtain $v := v_{SH}(0) = \sqrt{C_{44}/\rho}$, where v is a positive real number, to restate the above expressions as

$$A_r(0) = \frac{\rho_1\sqrt{\frac{C_{44}^I}{\rho_1}} - \rho_2\sqrt{\frac{C_{44}^{II}}{\rho_2}}}{\rho_1\sqrt{\frac{C_{44}^I}{\rho_1}} + \rho_2\sqrt{\frac{C_{44}^{II}}{\rho_2}}} = \frac{\rho_1 v_1 - \rho_2 v_2}{\rho_1 v_1 + \rho_2 v_2} \tag{10.3.3}$$

and

$$A_t(0) = \frac{2\rho_1\sqrt{\frac{C_{44}^I}{\rho_1}}}{\rho_1\sqrt{\frac{C_{44}^I}{\rho_1}} + \rho_2\sqrt{\frac{C_{44}^{II}}{\rho_2}}} = \frac{2\rho_1 v_1}{\rho_1 v_1 + \rho_2 v_2}, \tag{10.3.4}$$

where $v_1 \equiv v_{SH}(0)$ in the medium of incidence and $v_2 \equiv v_{SH}(0)$ in the medium of transmission.

EXERCISE 10.5. Expressions (10.3.3) and (10.3.4) are identical to expressions (6.8.13) and (6.8.14), respectively, which were formulated in a different way in Chapter 6. Comment on the formulation differences and, hence, the information lost and gained about incidence, reflection and transmission.

SOLUTION 10.5. Formulation of expressions (6.8.13) and (6.8.14) is rooted in general solution (6.8.6) of the wave equation in a single spatial dimension. Formulation of expressions (10.3.3) and (10.3.4) is rooted in trial solution (6.11.14) of the wave equation in three spatial dimensions.

These expressions are valid for any wave provided we are dealing with a single spatial dimension or a normal incidence in a laterally homogeneous continuum. The former requirement is due to restrictions in formulating d'Alembert's solution (6.8.6). The latter is due to the equality of the lateral components of the phase-slowness vectors that is used in formulating expressions (10.3.3) and (10.3.4). Our examination of these formulations provides us with further information.

In the context of expressions (6.8.13) and (6.8.14), we learn from solution (6.8.12) about the wavelengths of the reflected and transmitted waves. Wavelength information is lost in derivation of expressions (10.3.3) and (10.3.4) since the angular frequency, ω, from solution (6.11.14) disappears in simplification, as stated on page 450. However, independently of this derivation, we could invoke the standard relation among wavelength, λ, angular frequency and velocity, v, namely,

$$\lambda = 2\pi \frac{v}{\omega},$$

to relate the wavelengths of the incident and transmitted waves as $\lambda_i/\lambda_t = v_1/v_2$, which is consistent with the second expression of solution (6.8.12).

In the context of expressions (10.3.3) and (10.3.4), we learn from expressions (10.3.1) and (10.3.2) about the effect of the angle of incidence on the reflection and transmission amplitudes for any wave in a laterally homogeneous isotropic continuum—information that is not available if we consider only the single spatial dimension. Also, consideration of the single spatial dimension does not allow us to study the effects of anisotropy, which we do by examining expressions (10.2.14) and (10.2.15); these are valid for *SH* waves in transversely isotropic continua with the rotation-symmetry axes normal to the interface.

EXERCISE 10.6. Consider expression

$$\langle \mathfrak{E} \rangle = \frac{1}{2}\rho v \omega^2 A^2, \tag{10.3.5}$$

where $\langle \mathfrak{E} \rangle$ is the average energy density carried by the wave and ω is its angular frequency. Using the expressions for the normal-incidence reflection and transmission amplitudes, derived in Exercise 10.4, show that the energy is conserved.

SOLUTION 10.6. The balance of energy states that the energy carried by the incident wave must be equal to the sum of the energies carried by the reflected and transmitted waves, namely,

$$\langle \mathfrak{E}_i \rangle = \langle \mathfrak{E}_r \rangle + \langle \mathfrak{E}_t \rangle \,.$$

Considering monochromatic waves and normalizing incident-wave amplitude to unity, in accordance with expression (10.3.5), we obtain

$$\frac{1}{2}\rho_1 v_1 = \frac{1}{2}\rho_1 v_1 A_r^2 + \frac{1}{2}\rho_2 v_2 A_t^2 \,,$$

which can be rewritten as

$$1 = A_r^2 + \frac{\rho_2 v_2}{\rho_1 v_1} A_t^2 \,. \tag{10.3.6}$$

Inserting expressions (10.3.3) and (10.3.4) into expression (10.3.6), we get

$$\begin{aligned} 1 &= \left(\frac{\rho_1 v_1 - \rho_2 v_2}{\rho_1 v_1 + \rho_2 v_2}\right)^2 + \frac{\rho_2 v_2}{\rho_1 v_1}\left(\frac{2\rho_1 v_1}{\rho_1 v_1 + \rho_2 v_2}\right)^2 \\ &= \left(\frac{\rho_1 v_1 - \rho_2 v_2}{\rho_1 v_1 + \rho_2 v_2}\right)^2 + \frac{4\rho_1 v_1 \rho_2 v_2}{(\rho_1 v_1 + \rho_2 v_2)^2} = 1 \,, \end{aligned}$$

as required.

EXERCISE 10.7. [11]Show that if $\sin\vartheta_t > 1$, the magnitude of A_r is equal to unity.

SOLUTION 10.7. If $\sin\vartheta_t > 1$, then $\cos\vartheta_t$ is a pure imaginary number. In that case, expression (10.2.14) is of the form

$$A_r = \frac{a - bi}{a + bi} \,.$$

[11] See also Section 10.2.2.

The magnitude is given by

$$|A_r| = \sqrt{A_r A_r^*}\,,$$

where

$$A_r^* := \frac{a+bi}{a-bi}$$

is the complex conjugate. Therefore,

$$|A_r| = \sqrt{\frac{a-bi}{a+bi}\frac{a+bi}{a-bi}} = 1\,,$$

as required.

CHAPTER 11

Lagrange's Ray Equations

> The ancient Greeks had a hard time defining objects like "curves" and "surfaces" in a general way since their algebra was not well developed and always remained on a rather modest level. In fact, some historians think that the final stagnation of Greek mathematics was caused by the Greeks' failure to develop algebra and to apply it to geometry.
>
> *Stephan Hildebrandt and Anthony Tromba (1996)*
> *The parsimonious universe*

Preliminary Remarks

In Chapter 8, we obtained Hamilton's ray equations, which allow us to study seismic signals in an anisotropic inhomogeneous continuum. In a three-dimensional continuum, Hamilton's equations constitute a system of six first-order ordinary differential equations, which are expressed in terms of Hamiltonian $\mathscr{H}$ and exist in the $\mathbf{xp}$-space. This system can be also expressed as a system consisting of three ordinary second-order differential equations, which are expressed in terms of Lagrangian $\mathscr{L}$, where function $\mathscr{L}(\mathbf{x}, \dot{\mathbf{x}})$ is Legendre's transformation of function $\mathscr{H}(\mathbf{x}, \mathbf{p})$. These second-order differential equations are Lagrange's ray equations, which exist in the $\mathbf{x}\dot{\mathbf{x}}$-space.[1]

[1] In classical mechanics, the $\mathbf{x}\dot{\mathbf{x}}$-space corresponds to the velocity phase space. In this book, however, to avoid the confusion with the term "phase" that we use in the specific context of wave phenomena, we do not use this nomenclature.

In this chapter, we transform Hamilton's ray equations into Lagrange's ray equations. Lagrange's ray equations allow us to study ray theory in the realm of the calculus of variations. Thus, this chapter can be viewed as a transition between *Part 2* and *Part 3* of the book.

Readers who are not familiar with Legendre's transformation might find it useful to study this chapter together with Appendix B.

11.1. Legendre's Transformation of Hamiltonian

In view of Appendix B, where we discuss Legendre's transformation, we follow expression (B.3.8) to consider a new function given by

$$\mathscr{L}(\mathbf{x},\dot{\mathbf{x}}) = \sum_{j=1}^{3} p_j(\mathbf{x},\dot{\mathbf{x}})\dot{x}_j - \mathscr{H}(\mathbf{x},\mathbf{p}(\mathbf{x},\dot{\mathbf{x}})), \qquad (11.1.1)$$

where $\mathscr{L}$ is referred to as the ray-theory Lagrangian[2] corresponding to a given $\mathscr{H}$, and $p_j(\mathbf{x},\dot{\mathbf{x}})$ is a solution of

$$\dot{x}_j = \frac{\partial \mathscr{H}(\mathbf{x},\mathbf{p})}{\partial p_j}, \qquad j \in \{1,2,3\}, \qquad (11.1.2)$$

which is equation (B.3.7). Hence, in view of Appendix B, $\mathscr{L}$ is Legendre's transformation of $\mathscr{H}$. Now, we wish to rewrite Hamilton's ray equations (11.2.1) in terms of Lagrangian (11.1.1).

11.2. Formulation of Lagrange's Ray Equations

To obtain Lagrange's ray equations, recall Hamilton's ray equations (8.2.7), namely,

$$\left\{\begin{array}{l} \dot{x}_i = \dfrac{\partial \mathscr{H}}{\partial p_i} \\ \\ \dot{p}_i = -\dfrac{\partial \mathscr{H}}{\partial x_i} \end{array}\right., \qquad i \in \{1,2,3\}. \qquad (11.2.1)$$

[2] In this book we use four distinct Lagrangians, which are denoted by $\mathscr{L}$, $\mathscr{F}$, $\mathbb{L}$ and F. Consequently, in the text, we avoid a generic reference to "the Lagrangian", unless it is clear from the context which one among the four is considered.

Using expression (11.1.1), consider its derivative with respect to the first and second arguments, namely, x_i and $\dot{x}_i$, where $i \in \{1,2,3\}$. We obtain

$$\begin{aligned}\frac{\partial \mathscr{L}(\mathbf{x},\dot{\mathbf{x}})}{\partial x_i} &= \frac{\partial}{\partial x_i}\left[\sum_{j=1}^{3} p_j(\mathbf{x},\dot{\mathbf{x}})\dot{x}_j - \mathscr{H}(\mathbf{x},\mathbf{p}(\mathbf{x},\dot{\mathbf{x}}))\right] \\ &= \sum_{j=1}^{3}\frac{\partial p_j(\mathbf{x},\dot{\mathbf{x}})}{\partial x_i}\dot{x}_j - \frac{\partial \mathscr{H}(\mathbf{x},\mathbf{p}(\mathbf{x},\dot{\mathbf{x}}))}{\partial x_i} \\ &\quad - \sum_{j=1}^{3}\frac{\partial \mathscr{H}(\mathbf{x},\mathbf{p}(\mathbf{x},\dot{\mathbf{x}}))}{\partial p_j}\frac{\partial p_j(\mathbf{x},\dot{\mathbf{x}})}{\partial x_i}\end{aligned} \tag{11.2.2}$$

and

$$\begin{aligned}\frac{\partial \mathscr{L}(\mathbf{x},\dot{\mathbf{x}})}{\partial \dot{x}_i} &= \frac{\partial}{\partial \dot{x}_i}\left[\sum_{j=1}^{3} p_j(\mathbf{x},\dot{\mathbf{x}})\dot{x}_j - \mathscr{H}(\mathbf{x},\mathbf{p}(\mathbf{x},\dot{\mathbf{x}}))\right] \\ &= \sum_{j=1}^{3}\frac{\partial p_j(\mathbf{x},\dot{\mathbf{x}})}{\partial \dot{x}_i}\dot{x}_j + p_i(\mathbf{x},\dot{\mathbf{x}}) \\ &\quad - \sum_{j=1}^{3}\frac{\partial \mathscr{H}(\mathbf{x},\mathbf{p}(\mathbf{x},\dot{\mathbf{x}}))}{\partial p_j}\frac{\partial p_j(\mathbf{x},\dot{\mathbf{x}})}{\partial \dot{x}_i},\end{aligned} \tag{11.2.3}$$

respectively.

Using Hamilton's ray equations (11.2.1), we can restate expressions (11.2.2) and (11.2.3) as

$$\frac{\partial \mathscr{L}(\mathbf{x},\dot{\mathbf{x}})}{\partial x_i} = \sum_{j=1}^{3}\frac{\partial p_j(\mathbf{x},\dot{\mathbf{x}})}{\partial x_i}\dot{x}_j + \dot{p}_i(\mathbf{x},\dot{\mathbf{x}}) - \sum_{j=1}^{3}\dot{x}_j\frac{\partial p_j(\mathbf{x},\dot{\mathbf{x}})}{\partial x_i}, \quad i \in \{1,2,3\}, \tag{11.2.4}$$

and

$$\frac{\partial \mathscr{L}(\mathbf{x},\dot{\mathbf{x}})}{\partial \dot{x}_i} = \sum_{j=1}^{3}\frac{\partial p_j(\mathbf{x},\dot{\mathbf{x}})}{\partial \dot{x}_i}\dot{x}_j + p_i(\mathbf{x},\dot{\mathbf{x}}) - \sum_{j=1}^{3}\dot{x}_j\frac{\partial p_j(\mathbf{x},\dot{\mathbf{x}})}{\partial \dot{x}_i}, \quad i \in \{1,2,3\}, \tag{11.2.5}$$

respectively.

Examining expressions (11.2.4) and (11.2.5), we see that the first and the third terms on the right-hand sides cancel each other. Hence, $\partial \mathscr{L}/\partial x_i =$

$\dot{p}_i$ and $\partial\mathscr{L}/\partial\dot{x}_i = p_i$. Thus, we conclude that

$$\frac{\partial\mathscr{L}(\mathbf{x},\dot{\mathbf{x}})}{\partial x_i} = \frac{\mathrm{d}}{\mathrm{d}t}\left(\frac{\partial\mathscr{L}(\mathbf{x},\dot{\mathbf{x}})}{\partial\dot{x}_i}\right), \qquad i \in \{1,2,3\},$$

which we can rewrite as

$$\frac{\partial\mathscr{L}}{\partial x_i} - \frac{\mathrm{d}}{\mathrm{d}t}\frac{\partial\mathscr{L}}{\partial\dot{x}_i} = 0, \qquad i \in \{1,2,3\}, \tag{11.2.6}$$

where $\mathscr{L}$ is given in expression (11.1.1). System (11.2.1) is composed of six first-order ordinary differential equations in t to be solved for $\mathbf{x}(t)$ and $\mathbf{p}(t)$. As a result of Legendre's transformation, we expressed this system as three second-order ordinary differential equations (11.2.6) in t. These equations constitute a system of three second-order ordinary differential equations to be solved for $\mathbf{x}(t)$, which is the curve corresponding to the ray. We refer to these equations as Lagrange's ray equations.

Note that when we introduce $\mathscr{L}(\mathbf{x},\dot{\mathbf{x}})$ in expression (11.1.1), $\dot{\mathbf{x}}$ denoted a new variable, which, *a priori*, had no association with $\mathbf{x}$. However, if we consider the solution of system (11.2.1), which is given by $(\mathbf{x}(t),\mathbf{p}(t))$, then, in view of $\mathbf{p}(t) = \mathbf{p}(\mathbf{x}(t),\dot{\mathbf{x}}(t))$, we also have the corresponding solution $(\mathbf{x}(t),\dot{\mathbf{x}}(t))$. By examining equation (11.1.2) together with the first equation of system (11.2.1), we see that $\mathrm{d}\mathbf{x}(t)/\mathrm{d}t = \dot{\mathbf{x}}(t)$. Consequently, at the end, our initial abuse of notation did no harm and, rather, might be viewed as insightful. In other words, depending on the context, $\dot{\mathbf{x}}$ can be viewed as an independent variable or as a function of t.

In view of this derivation, system (11.2.6) is equivalent to system (11.2.1). Herein, we have obtained Lagrange's ray equations from Hamilton's ray equations. The duality of Legendre's transformation is such that we can also obtain Hamilton's ray equations from Lagrange's ray equations, as shown in Exercise 11.2 and in Appendix B. This leads to the following proposition.

PROPOSITION 11.2.1. *Rays, parametrized by time, can be obtained either by solving Hamilton's ray equations (11.2.1) or by solving Lagrange's ray equations (11.2.6).*

We note that, in view of Legendre's transformation, the derivation of Lagrange's ray equations from Hamilton's ray equations requires regularity of Hamiltonian $\mathscr{H}$, namely,

$$\det\left[\frac{\partial^2\mathscr{H}}{\partial p_i \partial p_j}\right] \neq 0\,, \qquad i,j \in \{1,2,3\}\,,$$

which is a necessary condition for Legendre's transformation to be a local diffeomorphism. This limitation is discussed in Section 13.1.2.

11.3. Beltrami's Identity

For our subsequent work, we notice that we can write all the equations of system (11.2.6) as a single equation, namely,

$$\frac{\partial\mathscr{L}}{\partial t} + \frac{\mathrm{d}}{\mathrm{d}t}\left(\sum_{i=1}^{3}\dot{x}_i\frac{\partial\mathscr{L}}{\partial\dot{x}_i} - \mathscr{L}\right) = 0\,. \tag{11.3.1}$$

Equation (11.3.1) is also valid for an n-dimensional case, where $i \in \{1,\ldots,n\}$. The verification of this expression, for the two-dimensional case, is shown in Exercise 11.1.

We refer to equation (11.3.1) as Beltrami's identity, since this expression was formulated in 1868 by Eugenio Beltrami. Beltrami's identity plays an important role in our raytracing methods, as illustrated in Section 12.3 and in Chapter 14. This importance results from the fact that if $\mathscr{L}$ does not explicitly depend on t , the first term on the left-hand side in equation (11.3.1) vanishes and, hence, the term in parentheses is equal to a constant. Furthermore, if $\mathscr{L}$ is homogeneous in the $\dot{x}_i$, the Lagrangian is conserved along the solution, $\mathbf{x}(t)$, as shown in Exercise 13.2.

Closing Remarks

To describe rays in anisotropic inhomogeneous continua, we can use either Hamilton's ray equations or Lagrange's ray equations. Herein, Hamilton's ray equations are directly rooted in fundamental physical principles discussed in Chapter 2, while Lagrange's ray equations are based on these principles via Legendre's transformation, which links the two systems. Thus, Lagrange's ray equations are subject to the singularities of this transformation.

However, we can also treat Lagrange's ray equations in their own right without invoking Hamilton's ray equations. Lagrange's ray equations belong to the realm of variational methods. Hence, in *Part 3*, we introduce the tools of the calculus of variations, which allow us to base these equations on Fermat's variational principle.

11.4. Exercises

EXERCISE 11.1. [3]Considering a two-dimensional continuum, verify that, given Lagrangian $\mathscr{L}$ that satisfies Lagrange's ray equations (11.2.6), Beltrami's identity (11.3.1) is also satisfied.

SOLUTION 11.1. For a two-dimensional continuum, let $x := x_1$ and $z := x_2$. Then, we can write $\mathscr{L} = \mathscr{L}(x, z, \dot{x}, \dot{z}, t)$. Consequently, Beltrami's identity (11.3.1) can be written as

$$\frac{\partial \mathscr{L}}{\partial t} + \frac{\mathrm{d}}{\mathrm{d}t}\left(\dot{x}\frac{\partial \mathscr{L}}{\partial \dot{x}} + \dot{z}\frac{\partial \mathscr{L}}{\partial \dot{z}} - \mathscr{L}\right) = 0. \tag{11.4.1}$$

Differentiating the left-hand side of equation (11.4.1), we obtain

$$\begin{aligned}
&\frac{\partial \mathscr{L}}{\partial t} + \frac{\mathrm{d}}{\mathrm{d}t}\left(\dot{x}\frac{\partial \mathscr{L}}{\partial \dot{x}} + \dot{z}\frac{\partial \mathscr{L}}{\partial \dot{z}} - \mathscr{L}\right) \\
&\quad = \frac{\partial \mathscr{L}}{\partial t} + \frac{\mathrm{d}}{\mathrm{d}t}\left(\dot{x}\frac{\partial \mathscr{L}}{\partial \dot{x}}\right) + \frac{\mathrm{d}}{\mathrm{d}t}\left(\dot{z}\frac{\partial \mathscr{L}}{\partial \dot{z}}\right) - \frac{\mathrm{d}\mathscr{L}}{\mathrm{d}t} \\
&\quad = \frac{\partial \mathscr{L}}{\partial t} + \ddot{x}\frac{\partial \mathscr{L}}{\partial \dot{x}} + \dot{x}\frac{\mathrm{d}}{\mathrm{d}t}\left(\frac{\partial \mathscr{L}}{\partial \dot{x}}\right) + \ddot{z}\frac{\partial \mathscr{L}}{\partial \dot{z}} + \dot{z}\frac{\mathrm{d}}{\mathrm{d}t}\left(\frac{\partial \mathscr{L}}{\partial \dot{z}}\right) \\
&\qquad - \left[\frac{\partial \mathscr{L}}{\partial x}\dot{x} + \frac{\partial \mathscr{L}}{\partial z}\dot{z} + \frac{\partial \mathscr{L}}{\partial \dot{x}}\ddot{x} + \frac{\partial \mathscr{L}}{\partial \dot{z}}\ddot{z} + \frac{\partial \mathscr{L}}{\partial t}\right] \\
&\quad = \dot{x}\left[\frac{\mathrm{d}}{\mathrm{d}t}\left(\frac{\partial \mathscr{L}}{\partial \dot{x}}\right) - \frac{\partial \mathscr{L}}{\partial x}\right] + \dot{z}\left[\frac{\mathrm{d}}{\mathrm{d}t}\left(\frac{\partial \mathscr{L}}{\partial \dot{z}}\right) - \frac{\partial \mathscr{L}}{\partial z}\right] = 0,
\end{aligned}$$

[3] See also Section 14.5.

which agrees with the right-hand side, as required. Note that the vanishing of the left-hand side results from the fact that each expression in brackets corresponds to a ray equation from system (11.2.6), namely,

$$\begin{cases} \dfrac{\partial \mathscr{L}}{\partial x} - \dfrac{\mathrm{d}}{\mathrm{d}t}\left(\dfrac{\partial \mathscr{L}}{\partial \dot{x}}\right) = 0 \\ \\ \dfrac{\partial \mathscr{L}}{\partial z} - \dfrac{\mathrm{d}}{\mathrm{d}t}\left(\dfrac{\partial \mathscr{L}}{\partial \dot{z}}\right) = 0 \end{cases}.$$

EXERCISE 11.2. Assuming that Hamiltonian $\mathscr{H}$ and Lagrangian $\mathscr{L}$ do not explicitly depend on t, following expression (11.1.1) and using equations (11.2.6), derive Hamilton's ray equations (8.2.7).

SOLUTION 11.2. Consider $\mathscr{H}(x,p)$ and $\mathscr{L}(x,\dot{x})$. Following expression (11.1.1), the differential of $\mathscr{H}$ becomes

$$\mathrm{d}\mathscr{H} = \sum_{i=1}^{3} \mathrm{d}p_i\dot{x}_i + \sum_{i=1}^{3} p_i\mathrm{d}\dot{x}_i - \sum_{i=1}^{3} \frac{\partial \mathscr{L}}{\partial x_i}\mathrm{d}x_i - \sum_{i=1}^{3} \frac{\partial \mathscr{L}}{\partial \dot{x}_i}\mathrm{d}\dot{x}_i .$$

In view of expression (B.3.2), we can write $p_i = \partial \mathscr{L}/\partial \dot{x}_i$. Hence, the second and last summations on the right-hand side cancel one another, and we obtain

$$\mathrm{d}\mathscr{H} = \sum_{i=1}^{3} \dot{x}_i\mathrm{d}p_i - \sum_{i=1}^{3} \frac{\partial \mathscr{L}}{\partial x_i}\mathrm{d}x_i . \qquad (11.4.2)$$

Also, the differential of $\mathscr{H}$, can be formally written as

$$\mathrm{d}\mathscr{H} = \sum_{i=1}^{3} \frac{\partial \mathscr{H}}{\partial p_i}\mathrm{d}p_i + \sum_{i=1}^{3} \frac{\partial \mathscr{H}}{\partial x_i}\mathrm{d}x_i . \qquad (11.4.3)$$

Equating the corresponding terms of expression (11.4.2) and its formal statement (11.4.3), we can write

$$\begin{cases} \dot{x}_i = \dfrac{\partial \mathscr{H}}{\partial p_i} \\ \\ \dfrac{\partial \mathscr{L}}{\partial x_i} = -\dfrac{\partial \mathscr{H}}{\partial x_i} \end{cases}, \qquad i \in \{1,2,3\} .$$

Invoking Lagrange's ray equation (11.2.6) and recalling expression (B.3.2), we can write

$$\frac{\partial \mathscr{L}}{\partial x_i} - \frac{\mathrm{d}p_i}{\mathrm{d}t} \equiv \frac{\partial \mathscr{L}}{\partial x_i} - \dot{p}_i = 0, \qquad i \in \{1,2,3\} .$$

Hence, we obtain

$$\dot{p}_i = \frac{\partial \mathscr{L}}{\partial x_i}, \qquad i \in \{1,2,3\} .$$

Thus, we can write

$$\begin{cases} \dot{x}_i = \dfrac{\partial \mathscr{H}}{\partial p_i} \\ \\ \dot{p}_i = -\dfrac{\partial \mathscr{H}}{\partial x_i} \end{cases}, \qquad i \in \{1,2,3\} ,$$

which are Hamilton's ray equations (8.2.7), as required.

EXERCISE 11.3. [4]In view of the polar reciprocity of the phase-slowness curve and the ray-velocity curve, derive the equation that, while analogous to expression (8.4.12), relates phase angle to both ray velocities and ray angles, namely,

$$\tan \vartheta = \frac{\tan \theta - \dfrac{1}{V(\theta)} \dfrac{\partial V}{\partial \theta}}{1 + \dfrac{\tan \theta}{V(\theta)} \dfrac{\partial V}{\partial \theta}} .$$

SOLUTION 11.3. Phase angle is given by

$$\tan \vartheta = \frac{p_1}{p_3},$$

where, following Legendre's transformation, p_i is the phase-slowness component given by

$$p_i = \frac{\partial \mathscr{L}}{\partial \dot{x}_i},$$

[4] See also Section 10.1.2.

and $\mathscr{L}$ is the ray-theory Lagrangian. Considering a two-dimensional medium and following the definition of the Lagrangian, we can write

$$\mathscr{L} = \frac{\dot{x}_1^2 + \dot{x}_3^2}{[V(\theta)]^2},$$

where $\theta = \arctan(\dot{x}_1/\dot{x}_3)$. Consider differential operator $\partial/\partial\dot{x}_i$. To express the differential operator in terms of the ray angle, we can write

$$\frac{\partial}{\partial\dot{x}_1} = \frac{\partial\theta}{\partial\dot{x}_1}\frac{\partial}{\partial\theta} = \frac{\partial\arctan\dfrac{\dot{x}_1}{\dot{x}_3}}{\partial\dot{x}_1}\frac{\partial}{\partial\theta} = \frac{\dfrac{1}{\dot{x}_3}}{1+\left(\dfrac{\dot{x}_1}{\dot{x}_3}\right)^2}\frac{\partial}{\partial\theta} = \frac{\dot{x}_3}{V^2}\frac{\partial}{\partial\theta},$$

and

$$\frac{\partial}{\partial\dot{x}_3} = \frac{\partial\theta}{\partial\dot{x}_3}\frac{\partial}{\partial\theta} = \frac{\partial\arctan\dfrac{\dot{x}_1}{\dot{x}_3}}{\partial\dot{x}_3}\frac{\partial}{\partial\theta} = -\frac{\dfrac{\dot{x}_1}{\dot{x}_3^2}}{1+\left(\dfrac{\dot{x}_1}{\dot{x}_3}\right)^2}\frac{\partial}{\partial\theta} = -\frac{\dot{x}_1}{V^2}\frac{\partial}{\partial\theta}.$$

Consider the expression for the phase-slowness components and for the ray-theory Lagrangian. Using the quotient rule, we can write

$$p_1 = \frac{\partial\mathscr{L}}{\partial\dot{x}_1} = \frac{2\dot{x}_1V^2 - 2\left(\dot{x}_1^2+\dot{x}_3^2\right)V\dfrac{\partial V}{\partial\dot{x}_1}}{V^4} = \frac{2\dot{x}_1 - 2V\dfrac{\partial V}{\partial\dot{x}_1}}{V^2},$$

where we used the fact that the expression in parentheses is equal to the square of the magnitude of the ray velocity, namely, V^2. Using, for $\partial/\partial\dot{x}_1$, the differential operator derived above, we obtain

$$p_1 = 2\frac{\dot{x}_1 - \dfrac{\dot{x}_3}{V}\dfrac{\partial V}{\partial\theta}}{V^2}.$$

Similarly, we get

$$p_3 = 2\frac{\dot{x}_3 + \dfrac{\dot{x}_1}{V}\dfrac{\partial V}{\partial\theta}}{V^2}.$$

Thus,

$$\tan\vartheta = \frac{p_1}{p_3} = \frac{\dot{x}_1 - \frac{\dot{x}_3}{V}\frac{\partial V}{\partial\theta}}{\dot{x}_3 + \frac{\dot{x}_1}{V}\frac{\partial V}{\partial\theta}} = \frac{\frac{\dot{x}_1}{\dot{x}_3} - \frac{1}{V}\frac{\partial V}{\partial\theta}}{1 + \frac{\frac{\dot{x}_1}{\dot{x}_3}}{V}\frac{\partial V}{\partial\theta}} = \frac{\tan\theta - \frac{1}{V}\frac{\partial V}{\partial\theta}}{1 + \frac{\tan\theta}{V}\frac{\partial V}{\partial\theta}}, \tag{11.4.4}$$

as required, which shows that $1/v(\vartheta)$ and $V(\theta)$ are polar reciprocals of one another.

REMARK 11.4.1. Expression (11.4.4) requires a closed form expression for the ray velocity as a function of the ray angle, $V(\theta)$. Such an expression can be formulated only for elliptical velocity dependence. In such a case the ray-velocity curve is an ellipse.

EXERCISE 11.4. Using expressions (8.4.12) and (11.4.4) and following standard trigonometric identities, show that

$$\frac{\partial}{\partial\vartheta}\ln v = \frac{\partial}{\partial\theta}\ln V.$$

SOLUTION 11.4. Note that expression (8.4.12) can be written as

$$\tan\theta = \frac{\tan\vartheta + \frac{\partial}{\partial\vartheta}\ln v}{1 - \tan\vartheta\frac{\partial}{\partial\vartheta}\ln v} = \frac{\tan\vartheta + \tan\left[\arctan\left(\frac{\partial}{\partial\vartheta}\ln v\right)\right]}{1 - \tan\vartheta\tan\left[\arctan\left(\frac{\partial}{\partial\vartheta}\ln v\right)\right]},$$

which, following trigonometric identities, we can write as

$$\theta = \vartheta + \arctan\left(\frac{\partial}{\partial\vartheta}\ln v\right). \tag{11.4.5}$$

Similarly, expression (11.4.4) can be written as

$$\vartheta = \theta - \arctan\left(\frac{\partial}{\partial\theta}\ln V\right). \tag{11.4.6}$$

Solving expression (11.4.6) for θ and equating it to expression (11.4.5), we obtain

$$\frac{\partial}{\partial\vartheta}\ln v = \frac{\partial}{\partial\theta}\ln V,$$

as required.

Part 3

Variational formulation of rays

Introduction to Part 3

> What you do is to invent various curves, and calculate on each curve a certain quantity. If you calculate this quantity for one route, and then for another, you will get a different number for each route. There is one route which gives the least possible number, however, and that is the route that the particle in nature actually takes. We are now describing the actual motion by saying something about the whole curve. We have lost the idea of causality, that the particle feels the pull and moves in accordance with it. Instead of that, in some grand fashion it smells all the curves, all the possibilities, and decides which one to take by choosing that for which our quantity is least.[5]
>
> *Richard Feynman (1967)*
> *The Character of Physical Law*

The fundamental formulation of ray theory was presented in *Part 2*. This theory is based on Cauchy's equations of motion in anisotropic inhomogeneous continua and results in Hamilton's ray equations. Also, in *Part 2*,

[5] Readers interested in the philosophical aspects of this statement, in the context of analytical mechanics, might refer to Toretti, R., (1999) The philosophy of physics: Cambridge University Press, p. 92, and to Johns, R., (2002) A theory of physical probability: University of Toronto Press, pp. 66–69.

we used Legendre's transformation of the ray-theory Hamiltonian to obtain Lagrange's ray equations. Thus, within the limitations of this transformation, we have two equivalent forms of the ray equations.

In *Part 3*, we study ray theory in the context of Lagrange's ray equations. We show that they are the stationarity conditions of the calculus of variations. Hence, we show that rays, wavefronts and traveltimes can be studied by invoking the concept of stationary traveltime.

In search of the stationarity condition for a definite integral that describes the traveltime of the signal between a source and a receiver, we use the calculus of variations. Since, in the variational approach to ray theory, either time or distance constitutes the single variable, the stationarity conditions are a system of ordinary differential equations. Consequently, the variational formulation is an elegant method to describe rays, wavefronts and traveltimes. Also, an intuitive concept of stationarity is a fruitful starting point for many investigations.

The first scientific statement of a variational principle was formulated in optics by Pierre de Fermat in 1657.[6] In its original formulation, this principle was referred to as the principle of least time.[7] Following Fermat's principle, the principle of least action in mechanics was proposed in the first half of the eighteenth century by Pierre-Louis Moreau de Maupertuis and, then, rigorously stated by William Rowan Hamilton in 1835.[8]

The theory of the calculus of variations originated with the statement of Johannes Bernoulli, who, in 1696, posed the problem to determine the shape of a wire along which a bead might slide in the shortest possible time. While this problem might have initially appeared quite particular, it led to an important general theory. In 1900, David Hilbert delivered a talk on "Mathematical Problems" during which he made the following statement.

---

[6] Interested readers might refer to Born, M., and Wolf, E., (1999) Principles of optics (*7th* edition): Cambridge University Press, p. xxvi.

[7] Readers interested in discussion of Fermat's principle in the context of quantum electrodynamics might refer to Feynman, R.P., (1985/2006) QED: The strange theory of light and matter: Princeton University Press, pp. 38–45 and pp. 50 – 52.

[8] Readers interested in formal relations between the classical-mechanics principle of stationary action and the ray-theory principle of stationary traveltime as well as their relation to quantum mechanics might refer to Goldstein, H., (1950/1980) Classical mechanics: Addison-Wesley Publishing Co., pp. 365–371 and pp. 484–492.

The mathematicians of past centuries were accustomed to devote themselves to the solution of difficult individual problems with passionate zeal. They knew the value of difficult problems. I remind you only of the 'problem of the line of quickest descent', proposed by Johannes Bernoulli. [...] It is an error to believe that rigour in the proof is the enemy of simplicity. On the contrary, we find it confirmed by numerous examples that the rigorous method is at the same time simpler and the more easily comprehended. [...] the most striking example of my statement is the calculus of variations.

CHAPTER 12

Euler's Equations

> For since the shape of the whole universe is most perfect and, in fact, designed by the wisest creator, nothing at all takes place in the universe in which a rule of maximum or minimum does not appear.[1]
>
> *Leonhard Euler (1744)*
> *Methodus inveniendi lineas curvas maximi minimive proprietate gaudentes, sive solutio problematis isoperimetrici latissimo sensu accepti*[2]

Preliminary Remarks

In Chapter 11, we derived Lagrange's ray equations. These equations are variational equations and, hence, allow us to consider ray theory in the context of the calculus of variations.

We begin this chapter with a brief discussion of stationarity of a definite integral and the derivation of the stationarity condition of the calculus of

[1] Readers interested in the modern perspective on this statement might refer to Brown, J.R., (2001) Who rules in science: An opinionated guide to the wars: Harvard University Press, pp. 107–108:

> During the heydey of fruitful interaction of science and religion, Kepler, Newton, Leibniz, and many others brought religious considerations to bear on scientific theorizing. [...] Their religious beliefs were not mere acts of faith, but the consequence of rational considerations. [...] It is no longer rational to set religious constraints on theorizing in biology or astronomy or anywhere else.

[2] Method of finding curved lines enjoying the maximum and minimum property; or the solution of the isoperimetric problem understood in the broadest sense.

variations, namely, Euler's equation. This is followed by formulations of the generalized and special forms of Euler's equations, which are again used in Chapters 13 and 14. We conclude this chapter by relating Euler's equations to Lagrange's ray equations.

This chapter is intended to give a brief introduction to the calculus of variations for readers who are not familiar with this subject. Otherwise, it can be omitted without affecting the study of subsequent chapters.

12.1. Mathematical Background

The calculus of variations is the study of methods to obtain stationary values of definite integrals. These values depend on functions that compose a given integrand. In other words, the domain of a definite integral is a set of functions. An integral operates on a set of functions and we seek a particular function that gives a stationary value of this integral. Analogously, in differential calculus, a function operates on a set of points and we seek a particular point that gives a stationary value of this function.

In differential calculus, the condition for stationarity of a function is the vanishing of its first derivative. We wish to formulate an analogous condition for stationarity of a definite integral.[3]

Herein, we focus our study on two-dimensional problems that are contained in the xz-plane. In this study, we require stationary values of an integral expressed as

$$I = \int_{a}^{b} F\left(z(x), z'(x); x\right) \mathrm{d}x. \tag{12.1.1}$$

Thus, we seek function $z(x)$ that makes integral (12.1.1) stationary. Assuming that $z(x)$ is continuous and smooth, we can view it as a curve in the xz-plane.

Integrand F contains three arguments, namely, $z(x)$, $z'(x) \equiv \mathrm{d}z/\mathrm{d}x$ and x. In formulating the condition of stationarity, we consider these three arguments as independent.

[3] Readers interested in a definition of stationarity of a definite integral might refer to Arnold, V.I., (1989) Mathematical methods of classical mechanics (2*nd* edition): Springer-Verlag, p. 57.

Note that to avoid any confusion, we could choose to write

$$F\left(z(x), z'(x); x\right) \equiv F\left(\xi_1, \xi_2, \xi_3\right). \tag{12.1.2}$$

However, we do not introduce these additional symbols.

We need, however, a new operator symbol. In the search for stationarity, Lagrange introduced a special symbol denoted by δ, which refers to the variations of curve $z(x)$. In other words, among all the variations of $z(x)$ between the fixed end-points a and b, we look for a curve that renders the value of a given integral stationary. This curve is a solution of the variational problem. Hence, the problem of looking for such a curve is symbolically stated as $\delta \int_a^b F \mathrm{d}x = 0$. Note the distinction between the variational and differential operators; symbol $\delta z(x)$ refers to a variation from curve to curve for a given x, whereas symbol $\mathrm{d}z(x)$ refers to a differential change along a given curve for a change in x.[4]

In this chapter and in Chapter 14, we restrict our study to curves in the form $z = z(x)$, rather than in the parametric form, $\mathbf{x}(t)$, used to formulate Hamilton's and Lagrange's ray equations in Chapters 8 and 11, respectively.[5]

The condition of stationarity of integral (12.1.1) was derived by Euler in 1744.[6] This condition is discussed in the next section.

12.2. Formulation of Euler's Equation

In this section, we derive the stationarity condition for integral (12.1.1). In other words, among all continuously differentiable functions $z(x)$ that satisfy the boundary conditions at $z(a)$ and $z(b)$, we establish the condition to

---

[4] Readers interested in the δ operator might refer to Lanczos, C., (1949/1986) The variational principles of mechanics: Dover, pp. 38–40, and to Ewing, M.G., (1969/1985) Calculus of variations with applications: Dover, pp. 86–88.

[5] Readers interested in the relation between the explicit and parametric formulations of the Euler equations might refer to Ewing, M.G., (1969/1985) Calculus of variations with applications: Dover, pp. 140–141, to Gelfand, I.M., and Fomin, S.V., (1963/2000) Calculus of variations: Dover, pp 38 – 42, to Sagan, H., (1969/1992) Introduction to the calculus of variations: Dover, pp. 197–202, or to Weinstock, R., (1952/1974) Calculus of variations with applications to physics and engineering: Dover, pp. 34–36.

[6] It is also common to refer to this equation as the Euler-Lagrange equation. Readers interested in the history of this equation might refer to Marsden, J.E., and Ratiu, T.S., (1999) Introduction to mechanics and symmetry: A basic exposition of classical mechanical systems (*2nd* edition): Springer-Verlag, pp. 231–234.

choose a function that renders integral (12.1.1) stationary. This stationarity condition is stated by the following theorem.

THEOREM 12.2.1. *Function $z(x)$ with the continuous first derivative on interval $[a,b]$ yields a stationary value of integral (12.1.1), namely,*

$$I = \int_a^b F\left(z(x), z'(x); x\right) \mathrm{d}x, \tag{12.2.1}$$

in the class of functions with boundary conditions $z(a) = z_a$ and $z(b) = z_b$, if equation

$$\frac{\partial F}{\partial z} - \frac{\mathrm{d}}{\mathrm{d}x}\left(\frac{\partial F}{\partial z'}\right) = 0 \tag{12.2.2}$$

is satisfied.[7]

We refer to equation (12.2.2) as Euler's equation. Euler's equation (12.2.2) is a second-order ordinary differential equation.[8]

To see the connection between integral (12.2.1) and its stationarity condition, given by Euler's equation (12.2.2), consider the following heuristic argument.

Replace the integral by a finite sum of subdivisions given by $x_0, x_1, \ldots, x_{n-1}, x_n$, where the interval of integration $[a,b]$ is $[x_0, x_n]$. The subdivisions are assumed to be equally spaced and we denote this spacing by $\Delta x = (b-a)/n$. A discrete expression approximating integral (12.2.1) can be written as

$$S_n = \sum_{i=0}^{n-1} F\left(z_{i+1}, z'_{i+1}; x_{i+1}\right) \Delta x,$$

where

$$z'_{i+1} := \frac{z_{i+1} - z_i}{\Delta x}. \tag{12.2.3}$$

Here S_n is viewed as a function of the $n-1$ variables, $z_1, \ldots, z_{n-1}$.

---

[7] Readers interested in a rigorous proof of Theorem 12.2.1 might refer to Arnold, V.I., (1989) Mathematical methods of classical mechanics (2*nd* edition): Springer-Verlag, pp. 57–58.

[8] Readers interested in a thorough study of Euler's equations might refer to Courant, R., and Hilbert, D., (1924/1989) Methods of mathematical physics: John Wiley & Sons, Vol. I, pp. 183–206, and to Morse P.M., and Feshbach H., (1953) Methods of theoretical physics: McGraw-Hill, Inc., Part I, pp. 276–280.

Note that z_0 and z_n are not included as variables because they are fixed by the boundary conditions, namely, $z_0 = z_a$ and $z_n = z_b$.

To find the stationary value of S_n , we find the stationary points for $n-1$ variables. This is equivalent to setting to zero all partial derivatives of S_n with respect to z_i . In other words, the stationarity condition is

$$\frac{\partial S_n}{\partial z_i} = 0, \qquad i \in \{1,\ldots,n-1\}. \tag{12.2.4}$$

In view of expression (12.2.3), in the sum S_n , for any given $i \in (1,\ldots,n-1)$, there are only two consecutive terms that explicitly contain a given z_i , namely,

$$F\left(z_i, z_i'; x_i\right)\Delta x + F\left(z_{i+1}, z_{i+1}'; x_{i+1}\right)\Delta x. \tag{12.2.5}$$

Applying stationarity condition (12.2.4), we take the derivative of expression (12.2.5) with respect to z_i and obtain

$$\left[\frac{\partial F}{\partial z}\left(z_i, z_i'; x_i\right) + \frac{\partial F}{\partial z'}\left(z_i, z_i'; x_i\right)\frac{\partial z_i'}{\partial z_i}\right]\Delta x$$
$$+ \left[\frac{\partial F}{\partial z'}\left(z_{i+1}, z_{i+1}'; x_{i+1}\right)\frac{\partial z_{i+1}'}{\partial z_i}\right]\Delta x = 0, \tag{12.2.6}$$

where $i \in \{1,\ldots,n-1\}$.

Note that an analogous approach can be followed by viewing S_n as a function of x_i and, hence, by setting all partial derivatives with respect to x_i to zero. As shown in Exercise 12.2, by following this approach, we obtain Beltrami's identity (12.3.1).

In view of equation (12.2.6) and recalling expression (12.2.3), we have

$$\frac{\partial z_i'}{\partial z_i} = \frac{\partial}{\partial z_i}\frac{z_i - z_{i-1}}{\Delta x} = \frac{1}{\Delta x},$$

and

$$\frac{\partial z_{i+1}'}{\partial z_i} = \frac{\partial}{\partial z_i}\frac{z_{i+1} - z_i}{\Delta x} = -\frac{1}{\Delta x}.$$

Hence, equation (12.2.6) becomes

$$\left[\frac{\partial F}{\partial z}\left(z_i, z_i'; x_i\right) + \frac{\partial F}{\partial z'}\left(z_i, z_i'; x_i\right)\frac{1}{\Delta x}\right]\Delta x - \left[\frac{\partial F}{\partial z'}\left(z_{i+1}, z_{i+1}'; x_{i+1}\right)\frac{1}{\Delta x}\right]\Delta x = 0,$$

where $i \in \{1,\ldots,n-1\}$. This equation can be rearranged to give

$$\frac{\partial F}{\partial z}\left(z_i, z_i'; x_i\right) - \frac{1}{\Delta x}\left[\frac{\partial F}{\partial z'}\left(z_{i+1}, z_{i+1}'; x_{i+1}\right) - \frac{\partial F}{\partial z'}\left(z_i, z_i'; x_i\right)\right] = 0, \tag{12.2.7}$$

where $i \in \{1,\ldots,n-1\}$.

We now assume that as $\Delta x \to 0$ and $x_i \to x \in [a,b]$, z_i approaches $z(x)$ and $z_i' = (z_i - z_{i-1})/\Delta x$ approaches $z'(x)$. Then, equation (12.2.7) becomes

$$\frac{\partial F}{\partial z}\left(z(x), z'(x); x\right) - \frac{\mathrm{d}}{\mathrm{d}x}\left[\frac{\partial F}{\partial z'}\left(z(x), z'(x); x\right)\right] = 0,$$

which is Euler's equation (12.2.2), as required.

12.3. Beltrami's Identity

A convenient form of Euler's equation (12.2.2) is Beltrami's identity, discussed in Section 11.3. In the two-dimensional case, where we look for the $z(x)$ that is a solution of Euler's equation (12.2.2), Beltrami's identity is equivalent to that equation. Hence, we can write

$$\frac{\partial F}{\partial z} - \frac{\mathrm{d}}{\mathrm{d}x}\left(\frac{\partial F}{\partial z'}\right) = 0 = \frac{\partial F}{\partial x} + \frac{\mathrm{d}}{\mathrm{d}x}\left(z'\frac{\partial F}{\partial z'} - F\right). \tag{12.3.1}$$

In general, Beltrami's identity is not equivalent to the corresponding Euler's equations or Lagrange's ray equations. In Chapter 11, for instance, a single expression of Beltrami's identity (11.3.1) refers to three equations of system (11.2.6) and, hence, by itself, cannot give a unique solution of system (11.2.6). A verification of Beltrami's identity (12.3.1) and its derivation are shown in Exercises 12.1 and 12.2, respectively.

Beltrami's identity is particularly useful if the integrand does not explicitly depend on x, namely, $F = F(z, z')$. In such a case, the first term on the right-hand side of equation (12.3.1) vanishes. Important consequences of this simplification are discussed in Chapter 14 in the context of ray parameters.

12.4. Generalizations of Euler's Equation

12.4.1. Introductory comments

Integral (12.1.1) depends on a single variable, x, on a single function, $z(x)$, and on its first derivative, $z'(x)$. In mathematical considerations of physically motivated problems, a given integral whose stationary value we seek can also depend on several variables, on several functions and on higher-order derivatives. Such formulations result in stationarity conditions that are second-order partial differential equations, systems of second-order ordinary differential equations and higher-order ordinary differential equations, respectively.

Although a given problem can depend on all of the above quantities, each of the three cases is described separately below.

12.4.2. Case of several variables

Let us consider an integral that contains a single function of several variables. To begin, we consider an integral whose integrand contains a function of two variables, namely,

$$I = \int_{a_y}^{b_y}\int_{a_x}^{b_x} F\left(z(x,y), z_x, z_y; x, y\right) \mathrm{d}x\,\mathrm{d}y,$$

where $z_x := \partial z/\partial x$ and $z_y := \partial z/\partial y$. Thus, within given constraints on a boundary, we look for a smooth surface, $z(x,y)$, that renders I stationary. In this case, Euler's equation becomes

$$\frac{\partial F}{\partial z} - \left[\frac{\partial}{\partial x}\left(\frac{\partial F}{\partial z_x}\right) + \frac{\partial}{\partial y}\left(\frac{\partial F}{\partial z_y}\right)\right] = 0, \tag{12.4.1}$$

which is a second-order partial differential equation. The generalization for n variables follows the same pattern, thereby giving

$$\frac{\partial F}{\partial z} - \sum_{i=1}^{n} \frac{\partial}{\partial x_i}\left(\frac{\partial F}{\partial z_{x_i}}\right) = 0, \tag{12.4.2}$$

where $z = z(x_1, \ldots, x_n)$ and $z_{x_i} := \partial z/\partial x_i$, with $i \in \{1, \ldots, n\}$.

Problems involving multiple integrals were considered by Lagrange in his papers dating from 1760–1762. A physical example of a double integral is discussed in Section 13.2.4.

12.4.3. Case of several functions

Let us consider an integral that contains several single-variable functions and their first derivatives. To begin, we consider an integral whose integrand contains two functions, namely,

$$I = \int_a^b F\left(y(x), y'(x), z(x), z'(x); x\right) \mathrm{d}x .$$

Thus, we look for smooth curves $y(x)$ and $z(x)$ that render I stationary, subject to constraints

$$\begin{cases} y(a) = a_1 \\ z(a) = a_2 \\ y(b) = b_1 \\ z(b) = b_2 \end{cases}, \tag{12.4.3}$$

where a_i and b_i are constants. In this case, Euler's equations become a system of second-order ordinary differential equations

$$\begin{cases} \dfrac{\partial F}{\partial y} - \dfrac{\mathrm{d}}{\mathrm{d}x}\left(\dfrac{\partial F}{\partial y'}\right) = 0 \\[2ex] \dfrac{\partial F}{\partial z} - \dfrac{\mathrm{d}}{\mathrm{d}x}\left(\dfrac{\partial F}{\partial z'}\right) = 0 \end{cases}.$$

The generalization for n functions follows the same pattern thereby giving a system of n equations,

$$\frac{\partial F}{\partial \zeta_i} - \frac{\mathrm{d}}{\mathrm{d}x}\left(\frac{\partial F}{\partial \zeta_i'}\right) = 0, \qquad i \in \{1, \ldots, n\}, \tag{12.4.4}$$

where $\zeta_i = \zeta_i(x)$ and $\zeta_i' = \mathrm{d}\zeta_i(x)/\mathrm{d}x$.

12.4.4. Higher-order derivatives

Let us consider an integral whose integrand contains higher-order derivatives of a single-variable function. To begin, we consider an integral whose integrand contains both the first and second derivatives, namely,

$$I = \int_a^b F\left(z(x), z'(x), z''(x)\,; x\right) \mathrm{d}x .$$

Thus, we look for a smooth curve $z(x)$ that renders I stationary, subject to constraints

$$\begin{cases} z(a) = a_1 \\ z'(a) = a_2 \\ z(b) = b_1 \\ z'(b) = b_2 \end{cases}, \qquad (12.4.5)$$

where a_i and b_i are constants. In this case, Euler's equation becomes

$$\frac{\partial F}{\partial z} - \frac{\mathrm{d}}{\mathrm{d}x}\left(\frac{\partial F}{\partial z'}\right) + \frac{\mathrm{d}^2}{\mathrm{d}x^2}\left(\frac{\partial F}{\partial z''}\right) = 0 .$$

This is a fourth-order ordinary differential equation. The generalization for nth-order derivatives follows the same pattern to yield

$$\frac{\partial F}{\partial z} - \frac{\mathrm{d}}{\mathrm{d}x}\left(\frac{\partial F}{\partial z'}\right) + \frac{\mathrm{d}^2}{\mathrm{d}x^2}\left(\frac{\partial F}{\partial z''}\right) + \ldots + (-1)^n \frac{\mathrm{d}^n}{\mathrm{d}x^n}\left(\frac{\partial F}{\partial z^{(n)}}\right) = 0 ,$$

which is an ordinary differential equation of order $2n$.

12.5. Special Cases of Euler's Equation

12.5.1. Introductory comments

There are cases where, due to the explicit absence of certain arguments or to the particular form of integral (12.1.1), Euler's equation (12.2.2) becomes a simpler equation.

Note that in evaluating partial derivatives, only explicit appearances of the variable of differentiation are taken into account. For instance, if we differentiate $F(z(x))$ with respect to z, namely, $\partial F/\partial z$, no allowance is made for the fact that a change in x also results in a change of z. Following expression (12.1.2), we could choose to write such a differentiation as $\partial F(\xi_1)/\partial\xi_1$ and, thus, at the expense of introducing an additional symbol, avoid any confusion.

12.5.2. Independence of z

Let us consider an integrand that is explicitly independent of z, namely, $F = F(z';x)$. We see that Euler's equation (12.2.2) is reduced to

$$\frac{\mathrm{d}}{\mathrm{d}x}\left(\frac{\partial F}{\partial z'}\right) = 0 .$$

The vanishing of the total derivative implies that

$$\frac{\partial F}{\partial z'} = C_1 , \qquad (12.5.1)$$

where C_1 denotes a constant. Thus, $z(x)$ is obtained as a solution of first-order ordinary differential equation (12.5.1).

12.5.3. Independence of x and z

Let us consider an integrand that is explicitly independent of both x and z. In other words, it is only dependent on z', namely, $F = F(z')$. Since z' is the only variable, we can immediately rewrite equation (12.5.1) as

$$\frac{\mathrm{d}F(z')}{\mathrm{d}z'} = C_1 . \qquad (12.5.2)$$

Denoting $\mathrm{d}F(z')/\mathrm{d}z'$ as $\mathrm{f}(z')$, we can write equation (12.5.2) as

$$\mathrm{f}(z') = C_1 .$$

Assuming that $\mathrm{d}\mathrm{f}/\mathrm{d}z' \neq 0$, we can consider inverse function f^{-1}. Thus, we can write

$$z' = \mathrm{f}^{-1}(C_1) .$$

Recalling that $z' \equiv \mathrm{d}z/\mathrm{d}x$ and denoting $\mathrm{f}^{-1}(C_1) = C_2$, we can write

$$\frac{\mathrm{d}z}{\mathrm{d}x} = C_2 . \qquad (12.5.3)$$

This is a first-order ordinary differential equation, whose solution,

$$z = C_2 x + C_3 ,$$

is obtained directly by integration.

Thus, finding the curve which gives a stationary value of $\int_a^b F(z')\mathrm{d}x$ consists of writing the equation of a straight line passing through points $[a, z(a)]$ and $[b, z(b)]$.

In a seismological context, this implies that in homogeneous continua, whether the continua be isotropic or anisotropic, if the properties do not depend on position, rays are straight.

12.5.4. Independence of x

Let us consider an integrand that is explicitly independent of x, namely, $F = F(z, z')$. Using Beltrami's identity (12.3.1), we obtain

$$\frac{\mathrm{d}}{\mathrm{d}x}\left(z'\frac{\partial F}{\partial z'} - F\right) = 0 .$$

The vanishing of the total derivative implies that

$$z'\frac{\partial}{\partial z'} - F = C , \tag{12.5.4}$$

where C denotes a constant. Thus, $z(x)$ is obtained as a solution of first-order ordinary differential equation (12.5.4).

In a seismological context, the case where the traveltime integral is independent of x implies that the continuum is homogeneous along the x-axis—a case commonly encountered in layered media. In such media, the constant in expression (12.5.4) is a ray parameter, discussed in Chapter 14.

12.5.5. Total derivative

Let integrand $F(x, z, z')$ be a total derivative of function $f(x, z)$ with respect to x, namely,

$$F(z, z'; x) = \frac{\mathrm{d}f(x, z)}{\mathrm{d}x} = \frac{\partial f}{\partial x} + \frac{\partial f}{\partial z}z' . \tag{12.5.5}$$

Consider the left-hand side of Euler's equation (12.2.2). Inserting function (12.5.5), we obtain

$$\frac{\partial F}{\partial z}-\frac{\mathrm{d}}{\mathrm{d}x}\left(\frac{\partial F}{\partial z'}\right)=\frac{\partial^2 f}{\partial x\partial z}+\frac{\partial^2 f}{\partial z^2}z'-\frac{\mathrm{d}}{\mathrm{d}x}\left(\frac{\partial f}{\partial z}\right)$$
$$=\frac{\partial^2 f}{\partial x\partial z}+\frac{\partial^2 f}{\partial z^2}z'-\left(\frac{\partial^2 f}{\partial x\partial z}+\frac{\partial^2 f}{\partial z^2}z'\right)=0.$$

Thus, equation (12.2.2) is identically satisfied. Consequently, if F is a total derivative, Euler's equation (12.2.2) is satisfied by any $z(x)$. In other words, if a variational problem involves the integral of a total differential, namely,

$$\delta\int_a^b \mathrm{d}f(x,z)=0,$$

the value of the integral is independent of the integration path and depends only on the limits of integration.

Note that, considering a fixed-ends variational problem, we can add to the integrand a term that is a total derivative without changing the solution of Euler's equations, as shown in Exercise 12.3.[9] Considering such cases, we note that, although a solution curve is not affected by this addition, the value of the integral is changed. For instance, identical rays can result in distinct traveltimes, depending on the properties of the continuum.

12.5.6. Function of x and z

Euler's equation

In physically motivated problems, we often encounter an integral given by $\int_a^b h(\mathbf{x})\mathrm{d}s$, which is an integral of function h, whose value depends on position $\mathbf{x}$ along the arclength element $\mathrm{d}s$. Such an integral represents a certain quantity measured along a trajectory that connects points a and b. Considering the two-dimensional case and assuming that the trajectory

[9] In electromagnetic theory, this property is associated with the gauge invariance. Interested readers might refer to Morse P.M., and Feshbach H., (1953) Methods of theoretical physics: McGraw-Hill, Inc., pp. 210 –212.

can be expressed as $z = z(x)$, we can write such an integral as

$$\int_{a_x}^{b_x} h(x,z)\sqrt{1+\left(\frac{\mathrm{d}z}{\mathrm{d}x}\right)^2}\,\mathrm{d}x. \tag{12.5.6}$$

Thus, Euler's equation (12.2.2) becomes

$$\frac{\partial}{\partial z}\left[h(x,z)\sqrt{1+(z')^2}\right] - \frac{\mathrm{d}}{\mathrm{d}x}\left[\frac{\partial}{\partial z'}\left(h(x,z)\sqrt{1+(z')^2}\right)\right] = 0,$$

where $z' \equiv \mathrm{d}z/\mathrm{d}x$. Performing partial derivatives, we obtain

$$\frac{\partial h(x,z)}{\partial z}\sqrt{1+(z')^2} - \frac{\mathrm{d}}{\mathrm{d}x}\left[h(x,z)\frac{z'}{\sqrt{1+(z')^2}}\right] = 0.$$

Then, by the product rule, we get

$$\frac{\partial h(x,z)}{\partial z}\sqrt{1+(z')^2} - \left[\frac{\mathrm{d}h(x,z)}{\mathrm{d}x}\frac{z'}{\sqrt{1+(z')^2}} + h(x,z)\frac{\mathrm{d}}{\mathrm{d}x}\frac{z'}{\sqrt{1+(z')^2}}\right] = 0.$$

Letting $h := h(x,z)$ and using the quotient and chain rules, we obtain

$$\frac{\partial h}{\partial z}\sqrt{1+(z')^2} - \left[\left(\frac{\partial h}{\partial x} + \frac{\partial h}{\partial z}z'\right)\frac{z'}{\sqrt{1+(z')^2}} + h\frac{z''\sqrt{1+(z')^2} - z'\dfrac{z'z''}{\sqrt{1+(z')^2}}}{1+(z')^2}\right] = 0.$$

An algebraic simplification leads to

$$\frac{\partial h}{\partial z}\sqrt{1+(z')^2} - \frac{\partial h}{\partial x}\frac{z'}{\sqrt{1+(z')^2}} - \frac{\partial h}{\partial z}\frac{(z')^2}{\sqrt{1+(z')^2}} - h\frac{z''}{\left[1+(z')^2\right]^{\frac{3}{2}}} = 0.$$

Rearranging the common factor, we obtain

$$\frac{1}{\sqrt{1+(z')^2}}\left[\frac{\partial h}{\partial z}+\frac{\partial h}{\partial z}\left(z'\right)^2-\frac{\partial h}{\partial x}z'-\frac{\partial h}{\partial z}\left(z'\right)^2-h\frac{z''}{1+\left(z'\right)^2}\right]=0\,.$$

The cancellation of identical terms results in

$$\frac{1}{\sqrt{1+(z')^2}}\left[\frac{\partial h}{\partial z}-\frac{\partial h}{\partial x}z'-h\frac{z''}{1+\left(z'\right)^2}\right]=0\,.$$

Since the factor in front of the brackets is never zero, Euler's equation becomes

$$\frac{\partial h}{\partial z}-\frac{\partial h}{\partial x}\frac{\mathrm{d}z}{\mathrm{d}x}-h\frac{\dfrac{\mathrm{d}^2 z}{\mathrm{d}x^2}}{1+\left(\dfrac{\mathrm{d}z}{\mathrm{d}x}\right)^2}=0\,. \tag{12.5.7}$$

To study equation (12.5.7) in the context of ray theory, let function $h(x,z)$ describe slowness in an isotropic inhomogeneous continuum. Hence, letting the velocity function be $v(x,z)=1/h(x,z)$ and rearranging equation (12.5.7), we obtain

$$v\frac{\mathrm{d}^2 z}{\mathrm{d}x^2}-\frac{\partial v}{\partial x}\left(\frac{\mathrm{d}z}{\mathrm{d}x}\right)^3+\frac{\partial v}{\partial z}\left(\frac{\mathrm{d}z}{\mathrm{d}x}\right)^2-\frac{\partial v}{\partial x}\frac{\mathrm{d}z}{\mathrm{d}x}+\frac{\partial v}{\partial z}=0\,, \tag{12.5.8}$$

where we assume $v(x,z)\neq 0$. In such a case, integral (12.5.6), which can be rewritten as

$$\int\limits_{a_x}^{b_x}\frac{\sqrt{1+(z')^2}}{v(x,z)}\,\mathrm{d}x\,, \tag{12.5.9}$$

represents the traveltime between two points. Thus, a solution of equation (12.5.8) is a ray, $z(x)$, satisfying Fermat's principle of stationary traveltime.

If $v(x,z)$ is given by a constant, integral (12.5.9) is explicitly independent of x and z, and it corresponds to the case discussed in Section 12.5.3. In such a case, equation (12.5.8) reduces to $\mathrm{d}^2z/\mathrm{d}x^2=0$, whose solutions are $z=C_2x+C_3$, where C_2 and C_3 are constants that depend on the limits of integration in integral (12.5.9).

Geometrical interpretation and physical meaning

Integral (12.5.6) has a simple geometrical interpretation. Let function h be a smooth and continuous function whose values are positive. Consider an orthonormal coordinate system, where $h(x,z)$ can be represented as a surface above the xz-plane. Let $z(x)$ be a smooth and continuous curve in the xz-plane that connects points a and b. Integral (12.5.6) is the surface area of a strip that is orthogonal to the xz-plane and whose edges are given by $z(x)$ and the corresponding values of $h(x,z)$. This strip can be viewed as a fence that follows curve $z(x)$, and whose height, at any point, is determined by function $h(x,z)$.

A solution of equation (12.5.7), namely, $z = z(x)$, is the curve along which the area of the corresponding strip is stationary. Herein, given the geometry of the variational problem, the area of the strip that results from equation (12.5.7) is minimum, as illustrated in Exercise 12.4.

Traveltime is the product of slowness and distance travelled. If function h represents the slowness in an isotropic inhomogeneous continuum, the area of the strip represents the traveltime between the two points, which are given by the limits of integration. Hence, a solution of equation (12.5.8) is a trajectory along which the traveltime is stationary.

12.6. First Integrals

Special cases of Euler's equation, which result from the absence of particular arguments in the integrand function, are called first integrals. This name originates in the period of mathematical history when many differential equations were solved by integration. The description shown in Section 12.5.3, where the integrand is explicitly independent of both x and z, exemplifies such an approach.

The term "first integral" implies that the order of the differential equation has been reduced by one, which is equivalent to the integration process. Formally, the meaning of first integral is described in the following definition.

DEFINITION 12.6.1. If an nth-order differential equation

$$f\left(x,z,z',\ldots,z^{(n)}\right)=0, \tag{12.6.1}$$

can be transformed to the equivalent form

$$\frac{\mathrm{d}}{\mathrm{d}x}g\left(x,z,z',\ldots,z^{(n-1)}\right)=0,$$

we see that

$$g\left(x,z,z',\ldots,z^{(n-1)}\right)=C, \tag{12.6.2}$$

where C is a constant. Expression (12.6.2) is a "first integral" of equation (12.6.1).

Note that the fact that the integrand of a variational problem does not explicitly depend on a particular argument is equivalent to saying that this problem is invariant with respect to that argument. This invariance and the associated first integral are contained in Noether's theorem, published by Emmy Noether in Göttingen in 1918 in her paper entitled "Invariante Variationsprobleme". We can state this theorem in the following way.[10]

THEOREM 12.6.2. *To every change of dependent or independent variables that leaves the integral of the Lagrangian invariant, there corresponds a conservation law.*

In the context of ray theory, we use the property that a first integral of a differential equation is a function that has a constant value along a solution curve. This conserved quantity is a ray parameter, which we discuss in Chapter 14.

12.7. Lagrange's Ray Equations as Euler's Equations

To use the calculus of variations in the study of ray theory, we wish to show that Lagrange's ray equations (11.2.6) belong to the realm of Euler's equations. The parametric form of Euler's equation (12.2.2) corresponds

---

[10]Readers interested in rigorous derivations and proofs might refer to Gelfand, I.M., and Fomin, S.V., (1963/2000) Calculus of variations: Dover, pp. 79–83 and pp. 176–179, and to Goldstein, H., (1950/1980) Classical mechanics: Addison-Wesley Publishing Co., pp. 588–596.

Readers interested in variational aspects of Noether's theorem might refer to Lanczos, C., (1949/1986) The variational principles of mechanics: Dover, pp. 401–405. Readers interested in foundational aspects of Noether's theorem might refer to Anastopoulos, C., (2008) Particle or wave: The evolution of the concept of matter in modern physics: Princeton University Press, pp. 132–134.

to a system of two Euler's equations, namely,

$$\begin{cases} \dfrac{\partial G}{\partial x} - \dfrac{\mathrm{d}}{\mathrm{d}t}\left(\dfrac{\partial G}{\partial \dot{x}}\right) = 0 \\ \\ \dfrac{\partial G}{\partial z} - \dfrac{\mathrm{d}}{\mathrm{d}t}\left(\dfrac{\partial G}{\partial \dot{z}}\right) = 0 \end{cases}, \tag{12.7.1}$$

where $G = G(x, z, \dot{x}, \dot{z})$ with $\dot{x} := \mathrm{d}x/\mathrm{d}t$ and $\dot{z} := \mathrm{d}z/\mathrm{d}t$. A solution of system (12.7.1) is a curve in the xz-plane given by $[x(t), z(t)]$ that corresponds to variational problem $\delta \int G\,\mathrm{d}t = 0$. To see the relation between G and F, which is stated in integral (12.1.1), we can write $\mathrm{d}t = \mathrm{d}x/\dot{x}$ and $z' := \mathrm{d}z/\mathrm{d}x$. Hence, $G(x, z, \dot{x}, \dot{z}) = F(z, \dot{z}/\dot{x}, x)\,\dot{x}$.

This parametric formulation allows us to use Euler's equations for an n-dimensional space. In general, we can write a system of n Euler's equations, namely,

$$\frac{\partial G}{\partial x_i} - \frac{\mathrm{d}}{\mathrm{d}t}\frac{\partial G}{\partial \dot{x}_i} = 0, \qquad i \in \{1, \ldots, n\}, \tag{12.7.2}$$

whose solution is a curve in the $\mathbf{x}$-space given by $\mathbf{x}(t)$. Examining systems (11.2.6) and (12.7.2), we conclude that Lagrange's ray equations (11.2.6) possess the form of Euler's equations (12.7.2).

The fact that Euler's and Lagrange's equations have equivalent forms is the reason why equations of the form (11.2.6) and (12.7.2) are often referred to as the Euler-Lagrange equations. In this book, we use the term Euler's equations to refer to the mathematical condition of stationarity while we reserve the term Lagrange's equations to refer to those among Euler's equations that are endowed with physical meaning associated with ray theory or classical mechanics.

Closing Remarks

The fact that Lagrange's ray equations are also Euler's equations implies that rays can be obtained as solutions of a variational problem. This fact allows us to use the tools of the calculus of variations in our investigations of ray theory.

In the calculus of variations, a stationary curve is given by Euler's equation. The conditions to specify that this curve results in a minimum or a maximum value of a given integral are difficult to formulate mathematically and are not addressed in this book.[11] Yet, in physically motivated problems the minimum or maximum nature of the stationary curve is often obvious from the physical context.

In Chapter 13, we study Fermat's variational principle of stationary traveltime. We show that the search for a ray is equivalent to the search for a curve along which the traveltime is stationary. Also in Chapter 13, we discuss Hamilton's variational principle of stationary action, which we use to derive the wave equation. In Chapter 14, we show that first integrals, which correspond to conserved quantities along the rays, can be used in raytracing techniques.

12.8. Exercises

EXERCISE 12.1. In view of Euler's equation (12.2.2), verify Beltrami's identity (12.3.1).

SOLUTION 12.1. Consider $F = F(x,z,z')$ and Beltrami's identity (12.3.1). We can write

$$\begin{aligned}
&\frac{\partial F}{\partial x} + \frac{\mathrm{d}}{\mathrm{d}x}\left(z'\frac{\partial F}{\partial z'} - F\right) \\
&\quad = \frac{\partial F(x,z,z')}{\partial x} + \frac{\mathrm{d}}{\mathrm{d}x}\left(z'\frac{\partial F(x,z,z')}{\partial z'}\right) - \frac{\mathrm{d}F(x,z,z')}{\mathrm{d}x} \\
&\quad = \frac{\partial F}{\partial x} + z''\frac{\partial F}{\partial z'} + z'\frac{\mathrm{d}}{\mathrm{d}x}\left(\frac{\partial F}{\partial z'}\right) - \left(\frac{\partial F}{\partial x} + \frac{\partial F}{\partial z}z' + \frac{\partial F}{\partial z'}z''\right) \\
&\quad = z'\frac{\mathrm{d}}{\mathrm{d}x}\left(\frac{\partial F}{\partial z'}\right) - \frac{\partial F}{\partial z}z' = -z'\left[\frac{\partial F}{\partial z} - \frac{\mathrm{d}}{\mathrm{d}x}\left(\frac{\partial F}{\partial z'}\right)\right],
\end{aligned}$$

[11] Readers interested in geodesic fields and its implication to minima and maxima might refer to Kreyszig, E., (1959/1991) Differential geometry: Dover, pp. 162–168.

where the terms in brackets is Euler's equation (12.2.2). Thus,

$$\frac{\partial F}{\partial x}+\frac{\mathrm{d}}{\mathrm{d}x}\left(z'\frac{\partial F}{\partial z'}-F\right)=0,$$

as required.

EXERCISE 12.2. Following the argument used to justify Theorem 12.2.1, derive the explicit form of Beltrami's identity (12.3.1).

SOLUTION 12.2. To obtain Beltrami's identity, consider term (12.2.5), namely,

$$F\left(z_i,z_i';x_i\right)\Delta x+F\left(z_{i+1},z_{i+1}';x_{i+1}\right)\Delta x,$$

where $i\in\{1,\ldots,n-1\}$. Differentiating with respect to x, we obtain

$$\left[\frac{\partial F}{\partial x}\left(z_i,z_i';x_i\right)+\frac{\partial F}{\partial z'}\left(z_i,z_i';x_i\right)\frac{\partial z_i'}{\partial x_i}\right]\Delta x+F\left(z_i,z_i';x_i\right)\frac{\partial \Delta x}{\partial x_i}$$
$$+\frac{\partial F}{\partial z'}\left(z_{i+1},z_{i+1}';x_{i+1}\right)\frac{\partial z_{i+1}'}{\partial x_i}\Delta x+F\left(z_{i+1},z_{i+1}';x_{i+1}\right)\frac{\partial \Delta x}{\partial x_i}. \tag{12.8.1}$$

Recalling expression (12.2.3) and the appropriate expression for Δx, we can write

$$\frac{\partial z_i'}{\partial x_i}=\frac{\partial}{\partial x_i}\left(\frac{z_i-z_{i-1}}{x_i-x_{i-1}}\right)=-\frac{z_i-z_{i-1}}{\left(x_i-x_{i-1}\right)^2}=-\frac{z_i-z_{i-1}}{(\Delta x)^2},$$
$$\frac{\partial \Delta x}{\partial x_i}\left(z_i,z_i';x_i\right)=\frac{\partial}{\partial x_i}\left(x_i-x_{i-1}\right)=1,$$
$$\frac{\partial z_{i+1}'}{\partial x_i}=\frac{\partial}{\partial x_i}\left(\frac{z_{i+1}-z_i}{x_{i+1}-x_i}\right)=\frac{z_{i+1}-z_i}{\left(x_{i+1}-x_i\right)^2}=\frac{z_{i+1}-z_i}{(\Delta x)^2},$$
$$\frac{\partial \Delta x}{\partial x_i}\left(z_{i+1},z_{i+1}';x_{i+1}\right)=\frac{\partial}{\partial x_i}\left(x_{i+1}-x_i\right)=-1.$$

Hence, expression (12.8.1) becomes

$$\left[\frac{\partial F}{\partial x}\left(z_i,z_i';x_i\right)-\frac{\partial F}{\partial z'}\left(z_i,z_i';x_i\right)\frac{z_i-z_{i-1}}{(\Delta x)^2}\right]\Delta x+F\left(z_i,z_i';x_i\right)$$
$$+\frac{\partial F}{\partial z'}\left(z_{i+1},z_{i+1}';x_{i+1}\right)\frac{z_{i+1}-z_i}{(\Delta x)^2}\Delta x-F\left(z_{i+1},z_{i+1}';x_{i+1}\right).$$

As in stationarity condition (12.2.4), we have a system of $n-1$ equations, namely,

$$\begin{aligned}\frac{\partial S_n}{\partial x_i} &= 0 \\ &= F\left(z_i, z_i'; x_i\right) - F\left(z_{i+1}, z_{i+1}'; x_{i+1}\right) \\ &\quad - \frac{\partial F}{\partial z'}\left(z_i, z_i'; x_i\right)\frac{z_i - z_{i-1}}{\Delta x} + \frac{\partial F}{\partial z'}\left(z_{i+1}, z_{i+1}'; x_{i+1}\right)\frac{z_{i+1} - z_i}{\Delta x} \\ &\quad + \frac{\partial F}{\partial x}\left(z_i, z_i'; x_i\right)\Delta x,\end{aligned}$$

where $i \in \{1, \ldots, n-1\}$. Dividing both sides of each equation by Δx and using the appropriate definition of Δx, we can write

$$\begin{aligned}0 = &\frac{F\left(z_i, z_i'; x_i\right) - F\left(z_{i+1}, z_{i+1}'; x_{i+1}\right)}{\Delta x} \\ &+ \frac{\frac{\partial F}{\partial z'}\left(z_{i+1}, z_{i+1}'; x_{i+1}\right)\frac{z_{i+1} - z_i}{\Delta x} - \frac{\partial F}{\partial z'}\left(z_i, z_i'; x_i\right)\frac{z_i - z_{i-1}}{\Delta x}}{\Delta x} \\ &+ \frac{\partial F}{\partial x}\left(z_i, z_i'; x_i\right),\end{aligned}$$

where $i \in \{1, \ldots, n-1\}$. Letting $\Delta x \to 0$, we see that z_i approaches $z(x)$ so that $z_i' = (z_i - z_{i-1})/\Delta x$ approaches $z'(x)$. Recognizing in the resulting statement the definitions of the derivatives, we obtain a single equation

$$0 = -\frac{\mathrm{d}F}{\mathrm{d}x} + \frac{\mathrm{d}}{\mathrm{d}x}\left(\frac{\partial F}{\partial z'}z'\right) + \frac{\partial F}{\partial x}.$$

Rearranging and using the linearity of the differential operator, we get

$$\frac{\partial F}{\partial x} + \frac{\mathrm{d}}{\mathrm{d}x}\left[z'\frac{\partial F}{\partial z'} - F\right] = 0,$$

which is Beltrami's identity (12.3.1), as required.

EXERCISE 12.3. [12]Prove the following lemma.

[12]See also Section 12.5.5.

NOTATION 12.8.1. To state Lemma 12.8.2, below, we use the parametric form of the variational problems, rather than the explicit form used in this chapter.

LEMMA 12.8.2. *The two variational problems given by*

$$\delta \int F(\mathbf{x}, \dot{\mathbf{x}}) \, \mathrm{d}t = 0 \tag{12.8.2}$$

and

$$\delta \int \left[cF(\mathbf{x}, \dot{\mathbf{x}}) + \sum_{j=1}^{n} f_j(\mathbf{x}) \dot{x}_j \right] \mathrm{d}t = 0 \tag{12.8.3}$$

have the same function $\mathbf{x}(t)$ that renders the corresponding integrals stationary, if $f_j(\mathbf{x})$ are the components of a gradient of a function and c is a nonzero constant.

SOLUTION 12.3. We show two different proofs of Lemma 12.8.2. Proof A invokes the properties of a variational fixed-ends problem, while Proof B utilizes standard properties of differential calculus in the context of Euler's equations.

PROOF. [Proof A] To prove that $\mathbf{x}(t)$ is the same for variational problems (12.8.2) and (12.8.3), we reduce problem (12.8.3) to problem (12.8.2). Consider the integral of variational problem (12.8.3). In view of the linearity of the integral operator, we can write

$$\begin{aligned} \int \left[cF(\mathbf{x}, \dot{\mathbf{x}}) + \sum_{j=1}^{n} f_j(\mathbf{x}) \dot{x}_j \right] \mathrm{d}t &= c \int F(\mathbf{x}, \dot{\mathbf{x}}) \, \mathrm{d}t + \int \sum_{j=1}^{n} f_j(\mathbf{x}) \dot{x}_j \mathrm{d}t \\ &= c \int F(\mathbf{x}, \dot{\mathbf{x}}) \, \mathrm{d}t + \int \sum_{j=1}^{n} f_j(\mathbf{x}) \, \mathrm{d}x_j . \end{aligned}$$

Consider the integral that involves the summation. Since the $f_j(\mathbf{x})$ are the components of ∇g, for some function $g(\mathbf{x})$, we can restate this integral as

$$\int \sum_{j=1}^{n} f_j(\mathbf{x}) \, \mathrm{d}x_j = \int \sum_{j=1}^{n} \frac{\partial g(\mathbf{x})}{\partial x_j} \mathrm{d}x_j . \tag{12.8.4}$$

Since integral (12.8.4) is the integral of total differential

$$\frac{\partial g(\mathbf{x})}{\partial x_1}\mathrm{d}x_1 + \ldots + \frac{\partial g(\mathbf{x})}{\partial x_n}\mathrm{d}x_n = \mathrm{d}g(\mathbf{x}),$$

the value of integral (12.8.4) is independent of the integration path. Hence, term $\sum f_j(\mathbf{x})\dot{x}_j$ has no effect on the choice of function $\mathbf{x}(t)$. Recalling that $c \neq 0$, we have reduced variational problem (12.8.3) to variational problem (12.8.2) and, hence, the proof is complete.

[Proof B] Consider variational problem (12.8.2). The corresponding Euler's equations are

$$\frac{\partial F}{\partial x_i} - \frac{\mathrm{d}}{\mathrm{d}t}\left(\frac{\partial F}{\partial \dot{x}_i}\right) = 0, \qquad i \in \{1,\ldots,n\}. \tag{12.8.5}$$

Now, consider variational problem (12.8.3). The corresponding Euler's equations are

$$\frac{\partial \left[cF + \sum\limits_{j=1}^{n} f_j(\mathbf{x})\dot{x}_j\right]}{\partial x_i} - \frac{\mathrm{d}}{\mathrm{d}t}\left(\frac{\partial \left[cF + \sum\limits_{j=1}^{n} f_j(\mathbf{x})\dot{x}_j\right]}{\partial \dot{x}_i}\right) = 0,$$

$$i \in \{1,\ldots,n\},$$

which we can write as

$$c\left[\frac{\partial F}{\partial x_i} - \frac{\mathrm{d}}{\mathrm{d}t}\left(\frac{\partial F}{\partial \dot{x}_i}\right)\right] + \frac{\partial \left[\sum\limits_{j=1}^{n} f_j(\mathbf{x})\dot{x}_j\right]}{\partial x_i} - \frac{\mathrm{d}}{\mathrm{d}t}\left(\frac{\partial \left[\sum\limits_{j=1}^{n} f_j(\mathbf{x})\dot{x}_j\right]}{\partial \dot{x}_i}\right) = 0, \tag{12.8.6}$$

where $i \in \{1,\ldots,n\}$. To prove that solution $\mathbf{x}(t)$ is the same for equations (12.8.5) and (12.8.6), we prove that these two systems of equations are equivalent to one another. Recalling that $c \neq 0$, to prove that equations (12.8.5) and (12.8.6) are equivalent, we need to show that

$$\frac{\partial\left[\sum_{j=1}^{n} f_j(\mathbf{x})\dot{x}_j\right]}{\partial x_i} = \frac{\mathrm{d}}{\mathrm{d}t}\left(\frac{\partial\left[\sum_{j=1}^{n} f_j(\mathbf{x})\dot{x}_j\right]}{\partial \dot{x}_i}\right), \qquad i \in \{1,\ldots,n\}. \tag{12.8.7}$$

Consider the left-hand side of equation (12.8.7). Using the linearity of the differential operator, we can write

$$\frac{\partial\left[\sum_{j=1}^{n} f_j(\mathbf{x})\dot{x}_j\right]}{\partial x_i} = \sum_{j=1}^{n} \frac{\partial f_j(\mathbf{x})}{\partial x_i}\dot{x}_j, \qquad i \in \{1,\ldots,n\}. \tag{12.8.8}$$

Consider the right-hand side of equation (12.8.7). Using the linearity of the differential operator and taking into account the fact that the only term of $\sum f_j(\mathbf{x})\dot{x}_j$ that is dependent on $\dot{x}_i$ is the term where $j = i$, we obtain

$$\frac{\mathrm{d}}{\mathrm{d}t}\left(\frac{\partial\left[\sum_{j=1}^{n} f_j(\mathbf{x})\dot{x}_j\right]}{\partial \dot{x}_i}\right) = \frac{\mathrm{d}f_i(\mathbf{x})}{\mathrm{d}t} = \sum_{j=1}^{n} \frac{\partial f_i(\mathbf{x})}{\partial x_j}\dot{x}_j, \qquad i \in \{1,\ldots,n\}. \tag{12.8.9}$$

Examining the coefficients of $\dot{x}_j$ in expressions (12.8.8) and (12.8.9), we see that we need to show the equality given by

$$\frac{\partial f_j(\mathbf{x})}{\partial x_i} = \frac{\partial f_i(\mathbf{x})}{\partial x_j}, \qquad i,j \in \{1,\ldots,n\}. \tag{12.8.10}$$

Recall that $[f_1(\mathbf{x}),\ldots,f_n(\mathbf{x})]$ are the components of ∇g, for some function $g(\mathbf{x})$, namely, $[\partial g/\partial x_1,\ldots,\partial g/\partial x_n]$. Thus, we can write the left-hand side of equation (12.8.10) as

$$\frac{\partial f_j(\mathbf{x})}{\partial x_i} = \frac{\partial}{\partial x_i}\left[\frac{\partial g(\mathbf{x})}{\partial x_j}\right] = \frac{\partial^2 g(\mathbf{x})}{\partial x_i \partial x_j}, \qquad i,j \in \{1,\ldots,n\}.$$

Analogously, we can write the right-hand side of equation (12.8.10) as

$$\frac{\partial f_i(\mathbf{x})}{\partial x_j} = \frac{\partial}{\partial x_j}\left[\frac{\partial g(\mathbf{x})}{\partial x_i}\right] = \frac{\partial^2 g(\mathbf{x})}{\partial x_j \partial x_i}, \qquad i,j \in \{1,\ldots,n\}.$$

Hence, due to the equality of mixed partial derivatives—which we can write as $\partial^2 g/\partial x_i \partial x_j = \partial^2 g/\partial x_j \partial x_i$—the proof is complete. □

EXERCISE 12.4. Consider a variational problem given by integral (12.5.6). Let $f(x,z) = 1$, and let the endpoints be $(0,0)$ and $(1,1)$. Find function $z(x)$ that renders this integral stationary and calculate the value of the integral along this function. Choose another function that connects the endpoints and show that the resulting value of the integral is greater than the one corresponding to the extremizing function. In view of Section 12.5.6, provide a geometrical explanation.

SOLUTION 12.4. The variational problem in question is

$$\delta \int_0^1 \sqrt{1 + \left(\frac{\mathrm{d}z}{\mathrm{d}x}\right)^2}\, \mathrm{d}x = 0. \tag{12.8.11}$$

Since integral (12.8.11) depends only on z', in view of Section 12.5.3, the extremizing function is a straight line given by $z(x) = x$. Inserting $z = x$ into integral (12.8.11), we obtain the distance along the extremizing function, namely,

$$\int_0^1 \sqrt{2}\,\mathrm{d}x = \sqrt{2},$$

as also expected from Pythagoras' theorem. Now, let us calculate the distance along another curve, for instance, $z(x) = x^2$. Integral (12.8.11) becomes

$$\int_0^1 \sqrt{1 + (2x)^2}\,\mathrm{d}x = \frac{1}{4}\left[2x\sqrt{1+(2x)^2} + \operatorname{Arc\,sinh}(2x)\right]_0^1 \approx 1.48,$$

which is greater than $\sqrt{2}$, as expected.

In view of Section 12.5.6, integral (12.8.11) is the surface area of a strip whose width is equal to unity, due to $f(x,z) = 1$, and whose length corresponds to the curve $z(x)$, between $x=0$ and $x=1$. Since the width of the strip is constant, the least surface area corresponds to the shortest curve connecting the two points. Hence, the extremizing function is $z(x)=x$, which is a straight line.

EXERCISE 12.5. Express Euler's equation (12.2.2) as the corresponding Hamilton's equations in $\mathrm{d}z/\mathrm{d}x$ and $\mathrm{d}p/\mathrm{d}x$.

SOLUTION 12.5. Consider integrand (12.1.1), namely, $F(z,z';x)$. In view of Legendre's transformation, discussed in Appendix B, let the variable of transformation be denoted by

$$p := \frac{\partial F}{\partial z'}, \tag{12.8.12}$$

and the new function be

$$H(z,p;x) = pz' - F\left(z,z';x\right), \tag{12.8.13}$$

which is the Hamiltonian corresponding to F. Hence, by the duality of Legendre's transformation, we can write

$$z' = \frac{\partial H}{\partial p}. \tag{12.8.14}$$

Invoking Euler's equation (12.2.2) and in view of expression (12.8.12), we can write

$$\frac{\partial F}{\partial z} = \frac{\mathrm{d}}{\mathrm{d}x}\left(\frac{\partial F}{\partial z'}\right) = \frac{\mathrm{d}p}{\mathrm{d}x} \equiv p'. \tag{12.8.15}$$

Hence, using expression (12.8.13) to express function F in terms of function H, we obtain

$$p' = \frac{\partial F}{\partial z} = \frac{\partial}{\partial z}\left[pz' - H(z,p;x)\right] = -\frac{\partial H}{\partial z}. \tag{12.8.16}$$

Thus, using equations (12.8.14) and (12.8.16), we can write a system of first-order ordinary differential equations in $\mathrm{d}z/\mathrm{d}x$ and $\mathrm{d}p/\mathrm{d}x$, namely,

$$\left\{\begin{aligned} z' &= \frac{\partial H}{\partial p} \\ p' &= -\frac{\partial H}{\partial z} \end{aligned}\right., \tag{12.8.17}$$

which are the required Hamilton's equations.

EXERCISE 12.6. In view of Exercise 12.5, prove the following theorem.

THEOREM 12.8.3. *For an integral given by expression (12.1.1), namely,*

$$\int_a^b F\left[z(x), z'(x); x\right] \mathrm{d}x,$$

if F does not explicitly depend on x, the corresponding Hamiltonian, H, is the first integral of equation (12.2.2).

SOLUTION 12.6. Let us proceed with the following proof.

PROOF. We can formally write

$$\frac{\mathrm{d}H(p,z;x)}{\mathrm{d}x} = \frac{\partial H}{\partial p}\frac{\mathrm{d}p}{\mathrm{d}x} + \frac{\partial H}{\partial z}\frac{\mathrm{d}z}{\mathrm{d}x} + \frac{\partial H}{\partial x}. \tag{12.8.18}$$

Invoking equations (12.8.17), the first two terms on the right-hand side of equation (12.8.18) vanish and, hence, this equation is reduced to

$$\frac{\mathrm{d}H(p,z;x)}{\mathrm{d}x} = \frac{\partial H}{\partial x}.$$

In view of expression (12.8.13), H does not depend on x explicitly if and only if F does not depend on x explicitly. In such a case, we obtain

$$\frac{\mathrm{d}H}{\mathrm{d}x} = 0,$$

and, hence, H is constant. Using expressions (12.8.12) and (12.8.13), we can write

$$H = pz' - F = \frac{\partial F}{\partial z'}z' - F = C,$$

where C denotes a constant. This is the first integral of equation (12.2.2) that is given by expression (12.5.4). □

CHAPTER 13

Variational Principles

> There is hardly any other branch of mathematical sciences in which abstract mathematical speculations and concrete physical evidence go so beautifully together and complement each other so perfectly. [...] In spite of all differences in the interpretation, the variational principles of mechanics continue to hold their ground in the description of all the phenomena of nature.
>
> *Cornelius Lanczos (1949)*
> *The variational principles of mechanics*

Preliminary Remarks

In Chapter 11, we derived Lagrange's ray equations, which, as shown in Chapter 12, are the stationarity conditions for a definite integral. In this chapter, we show that this definite integral corresponds to the traveltime of a signal between two points in an anisotropic inhomogeneous continuum. Consequently, we can study ray theory in terms of Fermat's variational principle of stationary traveltime.

In general, physical applications of the calculus of variations are based on the fact that the behaviours of physical systems appear to coincide with the extremals of certain integrals. For instance, while in ray theory this integral corresponds to the traveltime, in classical mechanics this integral is given in terms of the kinetic and potential energies.

We begin this chapter with the statement of Fermat's principle as a theorem dealing with rays. Proof of this theorem is rooted in Hamilton's ray equations, where the mathematical concept of a ray originates. Hence, we investigate several properties of the ray-theory Hamiltonian and the resulting Lagrangian and, using these properties, obtain a proof of this theorem. We also discuss another variational principle that is pertinent to our studies; namely, Hamilton's principle of stationary action.

13.1. Fermat's Principle

NOTATION 13.1.1. In this section, to show the generality of the formulation, all expressions are derived for an n-dimensional space.

13.1.1. Statement of Fermat's principle

[1]In 1657, Pierre de Fermat formulated his variational principle for the propagation of light. He stated that light travels along a curve that renders the traveltime minimum. In modern notation, a generic form of this principle, to which we refer as the principle of stationary traveltime, can be restated by the following theorem.

THEOREM 13.1.2. *Rays are the solutions of the variational problem*

$$\delta \int_{A}^{B} \frac{\mathrm{d}s}{V(\mathbf{x},\mathbf{n})} = 0, \tag{13.1.1}$$

where $\mathrm{d}s$ *is an arclength element and* $V(\mathbf{x},\mathbf{n})$ *is the ray velocity in direction* $\mathbf{n} = \mathrm{d}\mathbf{x}/\mathrm{d}s$ *at point* $\mathbf{x}$. *A and B are the fixed endpoints of this variational problem.*

Note that, in expression (13.1.1) and throughout Section 13.1, $\mathbf{n}$ denotes a vector tangent to the ray and not a vector normal to the wavefront, as is the case in other sections of this book.

[1] This section is based on the work that was published by Bóna, A. and Slawinski, M.A., (2003) Fermat's principle for seismic rays in elastic media. Journal of Applied Geophysics **54**, 445–451.

13.1.2. Properties of Hamiltonian $\mathcal{H}$

To prove Theorem 13.1.2, we must show that the solution of variational problem (13.1.1) is equivalent to the solution of Hamilton's ray equations (8.2.7); namely,

$$\begin{cases} \dot{x}_i = \dfrac{\partial \mathcal{H}}{\partial p_i} \\ \\ \dot{p}_i = -\dfrac{\partial \mathcal{H}}{\partial x_i} \end{cases}, \qquad i \in \{1, \ldots, n\} . \tag{13.1.2}$$

Let us investigate the properties of the Hamiltonian that is given by expression (8.2.8); namely,

$$\mathcal{H}(\mathbf{x}, \mathbf{p}) = \frac{1}{2} p^2 v^2 (\mathbf{x}, \mathbf{p}) ,$$

and which, in view of v being homogeneous of degree 0 in the p_i, can also be stated as

$$\mathcal{H}(\mathbf{x}, \mathbf{p}) = \frac{1}{2} p^2 v^2 \left(\mathbf{x}, \frac{\mathbf{p}}{|\mathbf{p}|} \right) , \tag{13.1.3}$$

where $|\mathbf{p}|$ is the magnitude of the phase-slowness vector.

By examining expression (13.1.3), we note the following properties of this Hamiltonian. $\mathcal{H}$ is homogeneous of degree 2 in the p_i. Also, since $\mathcal{H}$ does not explicitly depend on time, its value is conserved along the ray. The latter property can be stated by the following lemma.

LEMMA 13.1.3. *Hamiltonian $\mathcal{H}(\mathbf{x}, \mathbf{p})$, given by expression (13.1.3), is conserved along the ray.*

PROOF. Differentiating $\mathcal{H}(\mathbf{x}, \mathbf{p})$ with respect to t, we get

$$\frac{\mathrm{d}\mathcal{H}}{\mathrm{d}t} = \sum_{i=1}^{n} \frac{\partial \mathcal{H}}{\partial x_i} \dot{x}_i + \sum_{i=1}^{n} \frac{\partial \mathcal{H}}{\partial p_i} \dot{p}_i + \frac{\partial \mathcal{H}}{\partial t} .$$

Since $\mathcal{H}$ does not explicitly depend on time, we write

$$\frac{\mathrm{d}\mathcal{H}}{\mathrm{d}t} = \sum_{i=1}^{n} \frac{\partial \mathcal{H}}{\partial x_i} \dot{x}_i + \sum_{i=1}^{n} \frac{\partial \mathcal{H}}{\partial p_i} \dot{p}_i .$$

Using system (13.1.2), we obtain

$$\frac{\mathrm{d}\mathscr{H}}{\mathrm{d}t} = -\sum_{i=1}^{n} \dot{p}_i \dot{x}_i + \sum_{i=1}^{n} \dot{x}_i \dot{p}_i = 0,$$

as required. □

Moreover, the value of the Hamiltonian, which is conserved along the ray, is equal to $1/2$. This results from the fact that the eikonal equation, which is shown in equation (7.3.8), must be satisfied along the rays. Hence, in view of this equation, which states that $p^2 v^2 = 1$, and expression (8.2.8), we require that

$$\mathscr{H}(\mathbf{x}, \mathbf{p}) = \frac{1}{2} \tag{13.1.4}$$

along a ray.

13.1.3. Variational equivalent of Hamilton's ray equations

To show that rays obtained from Hamilton's ray equations (13.1.2) are solutions of variational problem (13.1.1), we express these equations in the context of the calculus of variations. As stated in Section 12.7, Lagrange's ray equations (11.2.6); namely,

$$\frac{\partial \mathscr{L}}{\partial x_i} - \frac{\mathrm{d}}{\mathrm{d}t}\frac{\partial \mathscr{L}}{\partial \dot{x}_i} = 0, \qquad i \in \{1, \ldots, n\}, \tag{13.1.5}$$

possess the form of Euler's equations. Consequently, in view of Chapter 12, we can state the following proposition.

PROPOSITION 13.1.4. *Rays are the solutions of the variational problem*

$$\delta \int \mathscr{L} \mathrm{d}t = 0, \tag{13.1.6}$$

where the ray-theory Lagrangian $\mathscr{L}$ *is given by expression (11.1.1), namely*

$$\mathscr{L}(\mathbf{x}, \dot{\mathbf{x}}) = \sum_{j=1}^{n} p_j(\mathbf{x}, \dot{\mathbf{x}}) \dot{x}_j - \mathscr{H}(\mathbf{x}, \mathbf{p}). \tag{13.1.7}$$

13.1.4. Properties of Lagrangian $\mathscr{L}$

To examine variational formulations (13.1.1) and (13.1.6), we must study the properties of Lagrangian $\mathscr{L}$, given by expression (13.1.7), in terms of the corresponding Hamiltonian, $\mathscr{H}$. We begin by stating the following lemma.

LEMMA 13.1.5. *If $\mathscr{H}(\mathbf{x},\mathbf{p})$ is homogeneous of degree 2 in the p_i, then*

$$\mathscr{L}(\mathbf{x},\dot{\mathbf{x}}(\mathbf{x},\mathbf{p})) = \mathscr{H}(\mathbf{x},\mathbf{p}),$$

where, by Legendre's transformation, $\dot{x}_i = \partial\mathscr{H}/\partial p_i$.

PROOF. Consider Lagrangian

$$\mathscr{L}(\mathbf{x},\dot{\mathbf{x}}(\mathbf{x},\mathbf{p})) = \sum_{i=1}^{n} p_i\dot{x}_i - \mathscr{H}.$$

In view of Legendre's transformation, we can write

$$\mathscr{L}(\mathbf{x}(t),\dot{\mathbf{x}}(\mathbf{x}(t),\mathbf{p}(t))) = \sum_{i=1}^{n} p_i\frac{\partial\mathscr{H}}{\partial p_i} - \mathscr{H}.$$

If $\mathscr{H}$ is homogeneous of degree 2 in the p_i, by Theorem A.2.1, stated in Appendix A, we obtain

$$\mathscr{L}(\mathbf{x}(t),\dot{\mathbf{x}}(\mathbf{x}(t),\mathbf{p}(t))) = 2\mathscr{H} - \mathscr{H} = \mathscr{H},$$

which completes the proof. □

In view of the conserved value of Hamiltonian $\mathscr{H}$, as shown in Lemma 13.1.3, and following expression (13.1.4), we obtain the following corollary of Lemma 13.1.5.

COROLLARY 13.1.6. *Along each ray, Lagrangian $\mathscr{L}$ is equal to $1/2$.*

In view of $\mathscr{H}$ being homogeneous of degree 2 in the p_i, the analogous property of $\mathscr{L}$ is shown in the following lemma.

LEMMA 13.1.7. *If Hamiltonian $\mathscr{H}(\mathbf{x},\mathbf{p})$ is homogeneous of degree 2 in the p_i, then Lagrangian $\mathscr{L}(\mathbf{x},\dot{\mathbf{x}})$ is homogeneous of degree 2 in the $\dot{x}_i$.*

PROOF. By Lemma 13.1.5, $\mathscr{H}(\mathbf{x},\mathbf{p}) = \mathscr{L}(\mathbf{x},\dot{\mathbf{x}}(\mathbf{x},\mathbf{p}))$, where $\dot{\mathbf{x}}$ and $\mathbf{p}$ are related by Legendre's transformation $\dot{x}_i = \partial\mathscr{H}/\partial p_i$. Let $\mathbf{p}' := a\mathbf{p}$, where a is a constant. The corresponding Hamilton's equations are

$$\dot{x}_i' = \frac{\partial\mathscr{H}(\mathbf{x},\mathbf{p}')}{\partial p_i'} = \frac{\partial\mathscr{H}(\mathbf{x},a\mathbf{p})}{\partial(ap_i)}, \qquad i\in\{1,\ldots,n\}.$$

By the homogeneity of $\mathscr{H}$ and the property of the differential operator, we can write

$$\dot{x}_i' = \frac{\partial \mathscr{H}(\mathbf{x}, a\mathbf{p})}{\partial (ap_i)} = \frac{a^2 \dfrac{\partial \mathscr{H}(\mathbf{x},\mathbf{p})}{\partial p_i}}{\dfrac{\partial (ap_i)}{\partial p_i}} = \frac{a^2 \dfrac{\partial \mathscr{H}(\mathbf{x},\mathbf{p})}{\partial p_i}}{a}, \qquad i \in \{1,\ldots,n\}.$$

Hence,

$$\dot{x}_i' = a \frac{\partial \mathscr{H}(\mathbf{x},\mathbf{p})}{\partial p_i}, \qquad i \in \{1,\ldots,n\},$$

which, in view of Hamilton's ray equations, given by system (13.1.2), can be stated as

$$\dot{x}_i' = a\dot{x}_i, \qquad i \in \{1,\ldots,n\}.$$

Consequently, we can write

$$\mathscr{L}(\mathbf{x}, a\dot{\mathbf{x}}) = \mathscr{L}(\mathbf{x}, \dot{\mathbf{x}}'),$$

which, by Lemma 13.1.5, yields

$$\begin{aligned} \mathscr{L}(\mathbf{x}, \mathbf{a}\dot{\mathbf{x}}) &= \mathscr{L}(\mathbf{x}, \dot{\mathbf{x}}') = \mathscr{H}(\mathbf{x}, \mathbf{p}') = \mathscr{H}(\mathbf{x}, a\mathbf{p}) \\ &= a^2 \mathscr{H}(\mathbf{x}, \mathbf{p}) = a^2 \mathscr{L}(\mathbf{x}, \dot{\mathbf{x}}), \end{aligned}$$

where the expression in the middle results from the homogeneity of $\mathscr{H}$. This means that Lagrangian $\mathscr{L}(\mathbf{x}, \dot{\mathbf{x}})$ is homogeneous of degree 2 in the $\dot{x}_i$. □

Lemma 13.1.7 implies that variational problem (13.1.6) has a fixed parametrization since $\mathscr{L}$ is homogeneous of degree 2 in the $\dot{x}_i$. For a variational problem to be independent of parametrization, the integrand must be homogeneous of degree 1 in the $\dot{x}_i$, as shown in Exercise 13.1.

Note that, as shown in Section 8.2, the solutions of Hamilton's ray equations (13.1.2) are parametrized by time; hence, the solutions of system (13.1.5) are also parametrized by time. Also note that, in view of the homogeneity of the Lagrangian and its not being explicitly dependent on t, Beltrami's identity together with Euler's homogeneous-function theorem imply that $\mathscr{L}$ is conserved along any ray, as shown in Exercise 13.2. As expected, this result is consistent with Corollary 13.1.6.

13.1.5. Parameter-independent Lagrange's ray equations

Parametrization independence is necessary to state Fermat's principle since its generic form, shown in expression (13.1.1), is parametrization independent. This results from the fact that the integrand in expression (13.1.1) is homogeneous of degree 1 in the $\dot{x}_i$.

Let us consider a Lagrangian given by

$$\mathscr{F} = \sqrt{2\mathscr{L}}, \tag{13.1.8}$$

where $\mathscr{L}$ is given by expression (13.1.7). Note that, following Definition A.1.1, stated in Appendix A, $\mathscr{F}$ is absolute-value homogeneous of degree 1 in the $\dot{x}_i$. Under certain conditions, which are satisfied in our case, the solutions of Lagrange's ray equations (13.1.5) are also the solutions of the equations given by

$$\frac{\partial \mathscr{F}}{\partial x_i} - \frac{\mathrm{d}}{\mathrm{d}t}\left(\frac{\partial \mathscr{F}}{\partial \dot{x}_i}\right) = 0, \qquad i \in \{1, \ldots, n\}. \tag{13.1.9}$$

This is stated by the following lemma.

LEMMA 13.1.8. *A solution of equations (13.1.5) that satisfies the condition given in Corollary 13.1.6, where $\mathscr{L}$ is given by expression (13.1.7), is also a solution of equations (13.1.9), where $\mathscr{F} = \sqrt{2\mathscr{L}}$.*

PROOF. Inserting $\mathscr{L} = \mathscr{F}^2/2$ into equations (13.1.5), we obtain

$$\begin{aligned}\frac{\partial}{\partial x_i}\left(\frac{\mathscr{F}^2}{2}\right) - \frac{\mathrm{d}}{\mathrm{d}t}\left[\frac{\partial}{\partial \dot{x}_i}\left(\frac{\mathscr{F}^2}{2}\right)\right] &= \mathscr{F}\frac{\partial \mathscr{F}}{\partial x_i} - \frac{\mathrm{d}}{\mathrm{d}t}\left[\mathscr{F}\frac{\partial \mathscr{F}}{\partial \dot{x}_i}\right] \\ &= \mathscr{F}\left[\frac{\partial \mathscr{F}}{\partial x_i} - \frac{\mathrm{d}}{\mathrm{d}t}\left(\frac{\partial \mathscr{F}}{\partial \dot{x}_i}\right)\right] - \frac{\mathrm{d}\mathscr{F}}{\mathrm{d}t}\frac{\partial \mathscr{F}}{\partial \dot{x}_i} \\ &= 0, \qquad i \in \{1, \ldots, n\}.\end{aligned}$$

Since $\mathscr{L} = 1/2$ along a ray, as shown in Corollary 13.1.6, then $\mathscr{F} = 1$ and, hence, $\mathrm{d}\mathscr{F}/\mathrm{d}t = 0$ along the solutions of equations (13.1.5). Thus, equations (13.1.5) become equations (13.1.9), as required. □

Consequently, equations (13.1.9) can also be viewed as Lagrange's ray equations.

If we can show that

$$\mathscr{F} = \frac{|\dot{\mathbf{x}}|}{V\left(\mathbf{x}, \dfrac{\dot{\mathbf{x}}}{|\dot{\mathbf{x}}|}\right)}, \tag{13.1.10}$$

where $|\dot{\mathbf{x}}| = \mathrm{d}s/\mathrm{d}t$ and $\dot{\mathbf{x}}/|\dot{\mathbf{x}}| = \mathbf{n}$, then we prove Theorem 13.1.2, since the right-hand side of equation (13.1.10) is the integrand of equation (13.1.1).

13.1.6. Ray velocity

To show that the right-hand side of equation (13.1.10) is the integrand of equation (13.1.1), we must formulate ray velocity in a variational context. Since, as shown in Lemma 13.1.7, Lagrangian $\mathscr{L}$ is homogeneous of degree 2 in the $\dot{x}_i$, we can write

$$\mathscr{L}(\mathbf{x}, \dot{\mathbf{x}}) = \mathscr{L}(\mathbf{x}, |\dot{\mathbf{x}}|\,\mathbf{n}) = |\dot{\mathbf{x}}|^2 \mathscr{L}(\mathbf{x}, \mathbf{n}),$$

where $\mathbf{n} = \dot{\mathbf{x}}/|\dot{\mathbf{x}}|$ is a unit vector tangent to the ray. Since, as stated in Corollary 13.1.6, the value of Lagrangian $\mathscr{L}$ along a ray is $1/2$, we can write

$$\frac{1}{2} = |\dot{\mathbf{x}}|^2 \mathscr{L}(\mathbf{x}, \mathbf{n}).$$

Since this expression is valid along any ray, the ray velocity V, given by $|\dot{\mathbf{x}}|$, can be expressed as

$$V(\mathbf{x}, \mathbf{n}) := |\dot{\mathbf{x}}| = \frac{1}{\sqrt{2\mathscr{L}(\mathbf{x}, \mathbf{n})}}, \tag{13.1.11}$$

which is consistent with expression (8.4.3).

Now, we are ready to complete the proof of Theorem 13.1.2.

13.1.7. Proof of Fermat's principle

PROOF. By Lemma 13.1.8, rays are the solutions of Euler's equations stated in system (13.1.9). Consequently, rays are the solutions of variational problem

$$\delta \int \mathscr{F}\, \mathrm{d}t = 0. \tag{13.1.12}$$

In view of expression (13.1.8), we can restate this variational problem as

$$\delta \int \mathscr{F}(\mathbf{x}, \dot{\mathbf{x}})\, \mathrm{d}t = \delta \int \sqrt{2\mathscr{L}(\mathbf{x}, \dot{\mathbf{x}})}\, \mathrm{d}t = 0.$$

Since, as stated in Lemma 13.1.7, $\mathscr{L}$ is homogeneous of degree 2 in the $\dot{x}_i$, we can write

$$\delta\int\mathscr{F}(\mathbf{x},\dot{\mathbf{x}})\,\mathrm{d}t=\delta\int\sqrt{2\left|\dot{\mathbf{x}}\right|^2\mathscr{L}(\mathbf{x},\mathbf{n})}\,\mathrm{d}t=\delta\int\left|\dot{\mathbf{x}}\right|\sqrt{2\mathscr{L}(\mathbf{x},\mathbf{n})}\,\mathrm{d}t=0\,.$$

In view of expressions (13.1.11) and since $|\dot{\mathbf{x}}|\;\mathrm{d}t=\mathrm{d}s$, we conclude that

$$\delta\int\mathscr{F}(\mathbf{x},\dot{\mathbf{x}})\,\mathrm{d}t=\delta\int\frac{\mathrm{d}s}{V(\mathbf{x},\mathbf{n})}=0\,.$$

Hence, the solutions of Hamilton's ray equations that correspond to rays are the solutions of variational problem (13.1.1). □

Theorem 13.1.2 states that seismic rays in anisotropic inhomogeneous continua obey Fermat's principle of stationary traveltime. Since our proof relies on Legendre's transformation, discussed in Appendix B, it is valid only if the Hamiltonian, $\mathscr{H}$, is regular; namely,

$$\det\left[\frac{\partial^2\mathscr{H}}{\partial p_i\partial p_j}\right]\neq 0\,,\qquad i,j\in\{1,\ldots,n\}\,. \tag{13.1.13}$$

In other words, we are unable to prove Theorem 13.1.2 at the inflection points of the phase-slowness surface. As stated in Theorem 9.3.1, for an elastic continuum defined by constitutive equations (3.2.1), the innermost phase-slowness surface is always convex and, hence, the Hamiltonian associated with the fastest wave is always regular. For the slower waves, however, there are points where the Hamiltonian is irregular. This does not mean that Fermat's principle does not hold in general; however, the proof of Theorem 13.1.2 in the context of a phase-velocity function giving an irregular Hamiltonian remains an open problem.

Heuristically, the principle of stationary traveltime can be justified by the fact that among all signals of finite duration, the signals arriving at the receiver at the same instant constructively interfere and, consequently, contribute to the recorded observation, while the contribution of a multitude of signals arriving at different times is negligible.

13.2. Hamilton's Principle: Example

13.2.1. Introductory comments

Born and Wolf, in their classic book entitled "Principles of optics", make the following statement.

> Variational considerations are of considerable importance as they often reveal analogies between different branches of physics. In particular there is a close analogy between geometrical optics and the mechanics of a moving particle; this was brought out very clearly by the celebrated investigations of Sir W.R. Hamilton, whose approach became of great value in modern physics, especially in applications to de Broglie's wave mechanics.

In this book, we focus on the variational formulation of geometrical optics without explicitly studying the analogies among different branches of physics.[2] In this section, however, we illustrate the analogy with classical mechanics by deriving the wave equation using Hamilton's variational principle.

13.2.2. Action

Fermat's principle, discussed in Section 13.1, plays an important role in ray theory. Another variational principle, which is pertinent to wave phenomena in elastic continua, is that of Hamilton. As stated by Arnold, in "Mathematical methods of classical mechanics",

> the fundamental notions of classical mechanics arose by the transforming of several very simple and natural notions of geometrical optics, guided by a particular variational principle—that of Fermat, into general variational principles.

[2] Interested readers might refer to Basdevant, J-L., (2007) Variational principles in physics: Springer-Verlag, where the author discusses these principle in such branches of physics as thermodynamics and quantum mechanics.

In this section, we illustrate Hamilton's principle in a simple context where the resulting Lagrange's equations of motion can be viewed as a restatement of Newton's second law of motion. Consequently, using the particular case of Hamilton's variational principle, we derive the one-dimensional wave equation, which corresponds to homogeneous continua.

While Newton proposed to measure motion by the rate of change of momentum, Leibniz suggested another quantity, the *vis viva*[3]. In the standard formulation of classical mechanics, *vis viva*, which underlies the concept of action, can be viewed as twice the kinetic energy.

The commonly accepted definition of action is

$$\mathbb{A} := \int_{t_1}^{t_2} \mathbb{L}\, \mathrm{d}t\,, \tag{13.2.1}$$

where $\mathbb{L}$ is the classical-mechanics Lagrangian that is defined by

$$\mathbb{L} := T - U\,, \tag{13.2.2}$$

with T and U denoting the kinetic energy and the potential energy, respectively.[4]

In classical mechanics, the principle of least action was proposed by de Maupertuis who, in 1744 in a document appropriately entitled *Accord des différentes lois de la nature qui avait jusqu'ici paru incompatibles*[5], stated that

> l'action est proportionnelle au produit de la masse par la vitesse et par l'espace. Maintenant, voici ce principe si

---

[3] *Living force.* Readers interested in the origin of this entity might refer to Toretti, R., (1999) The philosophy of physics: Cambridge University Press, pp. 33–36. Readers interested in the principle of *vis viva* as seen by early mechanicians might refer to Kuhn, T.S., (1996) The structure of scientific revolutions (3*rd* edition): The University of Chicago Press, pp. 189–191.

[4] Readers interested in developments of the definition of action might refer to Ekeland, I., (2000) Le meilleur des mondes possibles: Mathématiques et destinée: Seuil, pp. 57–98, or—for the English version—to Ekeland, I., (2006) The best of all possible worlds: Mathematics and destiny, pp. 44–78.

[5] Agreement of various laws of nature which until now appeared incompatible

> digne de l'Être suprême: Lorsqu'il arrive quelque changement dans la Nature, la quantité d'action employée pour ce changement est toujours la plus petite qu'il soit possible.[6]

However, careful analysis of the variational methods led to the formulation of the principle of stationary action rather than the principle of least action. The stationary-action principle was rigorously stated by Hamilton who wrote that

> although the law of least action has thus attained a rank among the highest theorems of physics, yet its pretensions to a cosmological necessity, on the grounds of economy in the universe, are now generally rejected. And the rejection appears just, for this, among other reasons, that the quantity pretended to be economized is in fact often lavishly expended.[7]

In other words, action may be either a minimum or maximum. As a result, in classical mechanics, the principle of stationary action proposed by Hamilton states that

> if the positions of a conservative system are given at two instants, t_1 and t_2, the value of the time integral of Lagrangian $\mathbb{L}$ is stationary for the path actually described by this system, as compared to any other path that connects the two positions and obeys the constraints of the system.

In other words, in view of definition (13.2.1), finding a stationary value of action is equivalent to variational problem

$$\delta \mathbb{A} = \delta \int_{t_1}^{t_2} \mathbb{L} \mathrm{d}t = 0 . \tag{13.2.3}$$

---

[6] Action is proportional to the product of mass, velocity, and displacement. Consequently, the principle so worthy of the Supreme Being: When there is a change in Nature, the value of action used for this change is the smallest possible.

[7] Hamilton, W.R., (1833) On a general method of expressing the paths of light, and of the planets, by the coefficients of a characteristic function: Dublin University Review.

From the variational principle of action, it is possible to derive many equations of mathematical physics. In particular, a variational derivation of the wave equation is shown in Section 13.2.4. In the context of this illustration, the potential energy is assumed to be a function of position alone, while the kinetic energy is assumed to be a function of velocity alone. In other words, for this illustration of Hamilton's principle, we confine our interests to homogeneous continua.

13.2.3. Lagrange's equations of motion

In this section, we introduce Lagrange's equations of motion using the concepts of particle mechanics in order to familiarize the reader with this classical formulation. In the context of seismic wave propagation, the reader can omit this section and proceed directly to Section 13.2.4.

Considering Hamilton's principle, stated in equation (13.2.3), and in view of the stationarity conditions, discussed in Chapter 12, the motion of a particle must satisfy Euler's equations. The parametric form of Euler's equations can be written as

$$\frac{\partial \mathbb{L}}{\partial x_i} - \frac{\mathrm{d}}{\mathrm{d}t}\left(\frac{\partial \mathbb{L}}{\partial \dot{x}_i}\right) = 0, \qquad i \in \{1,2,3\}, \tag{13.2.4}$$

where t denotes time, x_i is the position coordinate and, hence, $\dot{x}_i$ is a component of the velocity vector tangent to the trajectory of this particle. Equations (13.2.4) are Lagrange's equations of motion. In the context of this section, since the kinetic energy does not depend on position, Lagrange's equations of motion (13.2.4) can be viewed as a restatement of Newton's second law of motion.[8] To justify this view, consider the following description.

Considering the first term of Lagrange's equations of motion (13.2.4) and recalling that T is assumed to be a function of velocity alone, we obtain

$$\frac{\partial \mathbb{L}}{\partial x_i} = \frac{\partial \left[T\left(\dot{\mathbf{x}}\right) - U\left(\mathbf{x}\right)\right]}{\partial x_i} = -\frac{\partial U}{\partial x_i} =: F_i, \qquad i \in \{1,2,3\},$$

[8] Readers interested in important philosophical subtleties of the relation between Lagrange's equations of motion and Newton's second law of motion might refer to Johns, R., (2002) A theory of physical probability: University of Toronto Press, pp. 67–69.

which is the expression for a component of force in a conservative field. Considering the expression in parentheses in the second term of Lagrange's equations of motion (13.2.4) and recalling that U is assumed to be a function of position alone, we obtain

$$\frac{\partial \mathbb{L}}{\partial \dot{x}_i} = \frac{\partial \left[T(\dot{\mathbf{x}}) - U(\mathbf{x})\right]}{\partial \dot{x}_i} = \frac{\partial T}{\partial \dot{x}_i} =: p_i , \qquad i \in \{1,2,3\} , \tag{13.2.5}$$

which is the expression for a component of momentum.

Since the first term of equations (13.2.4) is the component of force, while the second term is the rate of change of the corresponding component of momentum, Lagrange's equations of motion (13.2.4) are equivalent to Newton's second law of motion; namely,

$$F_i - \frac{\mathrm{d}p_i}{\mathrm{d}t} = 0, \qquad i \in \{1,2,3\} .$$

Also, as shown in Exercises 13.6 and 13.7, we can derive Hamilton's equations of motion from Newton's laws of motion.

To gain further insight into Lagrange's equations of motion, let us invoke Beltrami's identity (11.3.1) to write

$$\frac{\partial \mathbb{L}}{\partial t} + \frac{\mathrm{d}}{\mathrm{d}t}\left(\sum_{i=1}^{n} \dot{x}_i \frac{\partial \mathbb{L}}{\partial \dot{x}_i} - \mathbb{L}\right) = 0.$$

If $\mathbb{L}$ does not depend explicitly on time—a concept used in Exercise 13.2, below, in the context of ray theory—this equation becomes

$$\frac{\mathrm{d}}{\mathrm{d}t}\left(\sum_{i=1}^{n} \dot{x}_i \frac{\partial \mathbb{L}}{\partial \dot{x}_i} - \mathbb{L}\right) = 0,$$

which implies that the term in parentheses is constant, C.[9] If $\mathbb{L}$ is homogeneous of degree 2 in the $\dot{x}_i$, in view of Theorem A.2.1 from Appendix A, we obtain $2\mathbb{L} - \mathbb{L} = C$. Thus, Lagrangian $\mathbb{L}$ is conserved along trajectory $\mathbf{x}(t)$. This constant of motion can be viewed also as the conserved energy of the system. To see that—in view of expression (13.2.5), and expression

[9] Readers interested in a different derivation of this conserved quantity might refer to Basdevant, J-L., (2007) Variational principles in physics: Springer-Verlag, pp. 54–55.

(B.3.1) from Appendix B—we write the conserved energy as the classical-mechanics Hamiltonian, namely,

$$\mathbb{H} = \sum_{i=1}^{n} \dot{x}_i p_i - \mathbb{L}.$$

Herein, $\mathbb{H} = T + U$: the sum of kinetic and potential energies. We illustrate the relation between this expression and the definition of $\mathbb{L}$ stated by expression (13.2.2) in Exercise 13.5.

Lagrange's equations of motion (13.2.4) apply to discrete systems, where the Lagrangian depends on the position of each particle. However, as shown in the following section, we can use the principle of stationary action in the context of continua, where the motion is defined by coordinates that are functions of both time and position variables.[10]

13.2.4. Wave equation

Continuous systems and Lagrangian density

A seismological application of stationary-action principle (13.2.3) and, consequently, of Lagrange's equations of motion for elastic continua, is exemplified by the derivation of the wave equation.[11] The coordinates of a three-dimensional continuous system are given by three position variables, x_1 , x_2 , x_3 , and the time variable, t . Consequently, the displacement is given as a function of four independent variables; namely, $u = u(x_1, x_2, x_3, t)$. Hence, for a three-dimensional continuum, Lagrangian $\mathbb{L}$ is associated with an element of volume and is given by

$$\mathbb{L} = \iiint \mathfrak{L} \, \mathrm{d}x_1 \, \mathrm{d}x_2 \, \mathrm{d}x_3 , \qquad (13.2.6)$$

where $\mathfrak{L}$ is the Lagrangian density

$$\mathfrak{L} = \mathfrak{L}\left(u, \frac{\partial u}{\partial x_i}, \frac{\partial u}{\partial t}, x_i, t\right), \qquad i = \{1, 2, 3\} .$$

[10] Readers interested in the energy propagation in the seismological context of the continuum using the Lagrange equations of motion and the generalized coordinates, might refer to Udías, A., (1999) Principles of seismology: Cambridge University Press, pp. 36–38.

[11] Readers interested in further descriptions of the Lagrangian formulation for continuous systems might refer to Goldstein, H., (1980) Classical mechanics: Addison-Wesley Publishing Co., pp. 548–555.

Variational derivation of wave equations

Consider oscillations of a finite-length string with fixed ends. Let the string itself be massless, have a length ℓ, and contain n equal masses, m_i, spaced at equal intervals, Δx. Let the longitudinal displacements of masses be $u_0, \ldots, u_{n+1}$, with $u_0 = u_{n+1} = 0$ being the boundary conditions corresponding to fixed ends. Assume the force, F, required to stretch a length Δx of the string by amount u, to be

$$F = \frac{k}{\Delta x} u, \tag{13.2.7}$$

where k denotes a constant.

Note that the term $k/\Delta x$ has the units of $[N/m]$ and expression (13.2.7) can be viewed as a one-dimensional statement of Hooke's law.

The potential energy, U, is associated with the elasticity of the string and is given by the strain-energy function, discussed in Chapter 4. Following equation (4.5.1), we write the potential energy of a segment of the string as

$$U_i = \int_0^{\Delta u} F \mathrm{d}u = \frac{k}{\Delta x} \int_0^{\Delta u} u \mathrm{d}u = \frac{1}{2} \frac{k}{\Delta x} (\Delta u)^2 ,$$

where $\Delta u \equiv u_i - u_{i-1}$. Summing all the segments, the potential energy along the entire string containing n discrete mass points is

$$U = \sum_{i=1}^{n} U_i = \frac{1}{2} k \sum_{i=1}^{n} \frac{(u_i - u_{i-1})^2}{\Delta x} . \tag{13.2.8}$$

If $n \to \infty$ and $\Delta x \to 0$ in such a way that $(n+1)\Delta x = l$, the potential energy, U, can be written as

$$U = \frac{1}{2} k \sum_{i=1}^{\infty} \left[\frac{u(x_i, t) - u(x_{i-1}, t)}{x_i - x_{i-1}} \right]^2 (x_i - x_{i-1}),$$

where $x_i - x_{i-1} \equiv \Delta x$. Thus, in the limit, the term in brackets represents a partial derivative with respect to x, while the summation results in integration. Hence, we can write

$$U = \frac{1}{2} k \int_0^{l} \left[\frac{\partial u(x,t)}{\partial x} \right]^2 \mathrm{d}x . \tag{13.2.9}$$

The kinetic energy, T, for the entire string containing n discrete mass points, each of which has a mass m, is

$$T = \frac{1}{2} m \sum_{i=1}^{n} \left[\frac{\partial u(x_i, t)}{\partial t} \right]^2 . \tag{13.2.10}$$

Rearranging and using the limit, we can write

$$T = \frac{1}{2} \frac{m}{\Delta x} \sum_{i=1}^{\infty} \left[\frac{\partial u(x_i, t)}{\partial t} \right]^2 \Delta x = \frac{1}{2} \rho \int_0^l \left[\frac{\partial u(x, t)}{\partial t} \right]^2 \mathrm{d}x, \tag{13.2.11}$$

where $\rho := \lim_{\Delta x \to 0} m/\Delta x$, is the mass density of the one-dimensional continuum.

Since the kinetic energy, given in expression (13.2.11), is not a function of position, we can invoke the classical-mechanics Lagrangian, given by expression (13.2.2). Thus, using expressions (13.2.9) and (13.2.11), we can write

$$\begin{aligned} \mathbb{L}(x, t) &= T(\dot{x}) - U(x) \\ &= \frac{1}{2} \rho \int_0^l \left[\frac{\partial u(x, t)}{\partial t} \right]^2 \mathrm{d}x - \frac{1}{2} k \int_0^l \left[\frac{\partial u(x, t)}{\partial x} \right]^2 \mathrm{d}x \\ &= \int_0^l \left\{ \frac{\rho}{2} \left[\frac{\partial u(x, t)}{\partial t} \right]^2 - \frac{k}{2} \left[\frac{\partial u(x, t)}{\partial x} \right]^2 \right\} \mathrm{d}x. \end{aligned} \tag{13.2.12}$$

Since we are presently dealing with a one-dimensional continuum, considering expression (13.2.6), we can write

$$\mathbb{L} = \int_0^l \mathfrak{L} \, \mathrm{d}x, \tag{13.2.13}$$

where, in view of integral (13.2.12), $\mathfrak{L}$ is the Lagrangian density given by

$$\mathfrak{L} \equiv \frac{\rho}{2} \left(\frac{\partial u}{\partial t} \right)^2 - \frac{k}{2} \left(\frac{\partial u}{\partial x} \right)^2 . \tag{13.2.14}$$

To invoke a variational formulation, in view of expression (13.2.13) and following equation (13.2.3), we can write

$$\delta\int_0^t \mathbb{L}\,\mathrm{d}t = \delta\int_0^t\int_0^l \mathfrak{L}\,\mathrm{d}x\,\mathrm{d}t = 0 .$$

Thus, we seek the stationary value of a definite integral that depends on two variables. In view of the corresponding Euler's equation, namely, equation (12.4.1), we can write the stationarity condition as

$$\frac{\partial\mathfrak{L}}{\partial u} - \left[\frac{\partial}{\partial t}\left(\frac{\partial\mathfrak{L}}{\partial u_t}\right) + \frac{\partial}{\partial x}\left(\frac{\partial\mathfrak{L}}{\partial u_x}\right)\right] = 0, \qquad (13.2.15)$$

where $u_t := \partial u/\partial t$ and $u_x := \partial u/\partial x$.

Equation (13.2.15) is Lagrange's equation of motion for a one-dimensional continuum.

Inserting the Lagrangian density, stated in expression (13.2.14), into equation (13.2.15) and considering ρ and k as constants, we obtain

$$-\left[\frac{\partial}{\partial t}\left(\rho\frac{\partial u}{\partial t}\right) - \frac{\partial}{\partial x}\left(k\frac{\partial u}{\partial x}\right)\right] = -\rho\frac{\partial^2 u}{\partial t^2} + k\frac{\partial^2 u}{\partial x^2} = 0 .$$

Rearranging, we can write

$$\frac{\partial^2 u}{\partial x^2} = \frac{1}{\frac{k}{\rho}}\frac{\partial^2 u}{\partial t^2},$$

which is a one-dimensional wave equation for longitudinal waves in elastic continua, where $\sqrt{k/\rho}$ denotes the speed of propagation with the units of speed resulting from $[k] = [kg\,m/s^2]$ and $[\rho] = [kg/m]$.

Note that the solution of the one-dimensional wave equation is surface $u(x,t)$—in the xt-space—that renders $\iint \mathfrak{L}\mathrm{d}x\mathrm{d}t$ stationary. This illustrates the fact that a solution of Euler's equation involving two variables is a surface, as stated in Section 12.4.2.

The variational approach to the one-dimensional wave equation for transverse waves is shown in Exercise 13.9.

Closing Remarks

As shown in this chapter, rays—originally formulated in terms of Hamilton's ray equations (8.2.7)—coincide with the curves exhibiting stationary traveltime. This property allows us to invoke Fermat's principle and, hence, to study ray theory using the tools of the calculus of variations. In Chapter 14, we use the stationarity of traveltime to study raytracing techniques.

Variational formulations are equivalent to Hamilton's ray equations provided we can, using Legendre's transformation, write a given ray-theory Hamiltonian as the corresponding ray-theory Lagrangian. This requirement is satisfied for all convex phase-slowness surfaces. As stated in Theorem 9.3.1, the phase-slowness surface of the fastest wave is convex. Consequently, we can always use Fermat's principle to study the qP wave. When dealing with the qS wave, we must be aware of the inflection points of its phase-slowness surface. The study of such points belongs to the realm of singularity theory, which is not considered in this book.[12]

13.3. Exercises

EXERCISE 13.1. Consider a traveltime integral in an anisotropic inhomogeneous continuum; namely,

$$\check{C} = \int_a^b \mathscr{F}(\mathbf{x}, \dot{\mathbf{x}})\,\mathrm{d}t\,. \tag{13.3.1}$$

Show that if $\mathscr{F}(\mathbf{x}, \dot{\mathbf{x}})$ is homogeneous of degree 1 in $\dot{\mathbf{x}}$, the integral is independent of parametrization.

[12] Interested readers might refer to Hanyga, A. and Slawinski, M.A., (2001) Caustics in qSV rayfields of transversely isotropic and vertically inhomogeneous media: Anisotropy 2000: Fractures, converted waves, and case studies: SEG (Special Issue), 409–418.

SOLUTION 13.1. Let $s = f(t)$ be an arbitrary parametrization. Hence,

$$\mathrm{d}s = \frac{\mathrm{d}f}{\mathrm{d}t}\mathrm{d}t =: \dot{f}\mathrm{d}t\,,$$

and

$$\dot{f} = \frac{\mathrm{d}s}{\mathrm{d}t}\,. \tag{13.3.2}$$

Consider $\mathscr{F}(\mathbf{x}, \dot{\mathbf{x}})$, where

$$\dot{\mathbf{x}} = \frac{\mathrm{d}\mathbf{x}}{\mathrm{d}s}\frac{\mathrm{d}s}{\mathrm{d}t}\,,$$

which, in view of expression (13.3.2), can be written as

$$\dot{\mathbf{x}} = \frac{\mathrm{d}\mathbf{x}}{\mathrm{d}s}\frac{\mathrm{d}f}{\mathrm{d}t} =: \mathbf{x}'\dot{f}\,.$$

For the value of the integral (13.3.1) to be independent of parametrization, we require

$$\mathscr{F}(\mathbf{x}, \dot{\mathbf{x}})\,\mathrm{d}t = \mathscr{F}\left(\mathbf{x}, \mathbf{x}'\right)\mathrm{d}s\,. \tag{13.3.3}$$

Consider the left-hand side of equation (13.3.3). Since $\dot{\mathbf{x}} = \mathbf{x}'\dot{f}$ and $\mathrm{d}t = \mathrm{d}s/\dot{f}$, we can write it as

$$\mathscr{F}(\mathbf{x}, \dot{\mathbf{x}})\,\mathrm{d}t = \mathscr{F}\left(\mathbf{x}, \mathbf{x}'\dot{f}\right)\frac{\mathrm{d}s}{\dot{f}}\,.$$

If $\mathscr{F}$ is homogeneous of degree 1 in $\dot{\mathbf{x}}$, following Definition A.1.1, stated in Appendix A, we obtain

$$\mathscr{F}(\mathbf{x}, \dot{\mathbf{x}})\,\mathrm{d}t = \dot{f}\mathscr{F}\left(\mathbf{x}, \mathbf{x}'\right)\frac{\mathrm{d}s}{\dot{f}} = \mathscr{F}\left(\mathbf{x}, \mathbf{x}'\right)\mathrm{d}s\,,$$

which is equation (13.3.3), as required.

REMARK 13.3.1. Exercise 13.1 shows that the general statement of Fermat's principle, namely,

$$\delta\int\mathscr{F}(\mathbf{x}, \dot{\mathbf{x}})\,\mathrm{d}t = 0,$$

is independent of parametrization. This is the case since $\mathscr{F}(\mathbf{x}, \dot{\mathbf{x}})\,\mathrm{d}t = \mathrm{d}s/V$ is homogeneous of degree 1 in $\dot{\mathbf{x}}$. Note that $\mathrm{d}s$ is homogeneous of degree 1 in $\dot{\mathbf{x}}$, while V is homogeneous of degree 0 in $\dot{\mathbf{x}}$.[13]

---

[13] See also Section 11.3.

EXERCISE 13.2. In view of Lemma 13.1.7, use Beltrami's identity (11.3.1), namely,

$$\frac{\partial \mathscr{L}}{\partial t}+\frac{\mathrm{d}}{\mathrm{d}t}\left(\sum_{i=1}^{n}\dot{x}_i\frac{\partial \mathscr{L}}{\partial \dot{x}_i}-\mathscr{L}\right)=0, \tag{13.3.4}$$

to show that Lagrangian $\mathscr{L}$ is conserved along the ray.

SOLUTION 13.2. Since $\mathscr{L}$, given by expression (13.1.7), does not explicitly depend on t, equation (13.3.4) becomes $\mathrm{d}\left(\sum_{i=1}^{n}\dot{x}_i\partial\mathscr{L}/\partial\dot{x}_i-\mathscr{L}\right)/\mathrm{d}t=0$, which implies that $\sum_{i=1}^{n}\dot{x}_i\partial\mathscr{L}/\partial\dot{x}_i-\mathscr{L}=C$, where C denotes a constant. Since, by Lemma 13.1.7, $\mathscr{L}$ is homogeneous of degree 2 in the $\dot{x}_i$, in view of Theorem A.2.1, we obtain $2\mathscr{L}-\mathscr{L}=C$. Thus, Lagrangian $\mathscr{L}$ is equal to a constant and, hence, it is conserved along the ray.

EXERCISE 13.3. Consider the system of six characteristic equations for an isotropic inhomogeneous continuum, derived in Exercises 8.8 and 8.9; namely,

$$\left\{\begin{array}{l}\dfrac{\mathrm{d}x_i}{\mathrm{d}t}=v^2p_i\\[3em]\dfrac{\mathrm{d}p_i}{\mathrm{d}t}=-\dfrac{\partial\ln v}{\partial x_i}\end{array}\right., \qquad i\in\{1,2,3\}. \tag{13.3.5}$$

Express system (13.3.5) as three second-order equations.

SOLUTION 13.3. Solving the first equation of system (13.3.5) for the components of the phase-slowness vector, we get

$$p_i=\frac{1}{v^2(\mathbf{x})}\frac{\mathrm{d}x_i}{\mathrm{d}t}, \qquad i\in\{1,2,3\}.$$

Differentiating with respect to t, we can write

$$\frac{\mathrm{d}p_i}{\mathrm{d}t}=\frac{\mathrm{d}}{\mathrm{d}t}\left[\frac{1}{v^2(\mathbf{x})}\frac{\mathrm{d}x_i}{\mathrm{d}t}\right], \qquad i\in\{1,2,3\},$$

which we can equate to the second equation of set (13.3.5) to obtain

$$\frac{\partial\ln v}{\partial x_i}+\frac{\mathrm{d}}{\mathrm{d}t}\left[\frac{1}{v^2(\mathbf{x})}\frac{\mathrm{d}x_i}{\mathrm{d}t}\right]=0, \qquad i\in\{1,2,3\}, \tag{13.3.6}$$

as required.

EXERCISE 13.4. Consider the traveltime integral in an isotropic inhomogeneous continuum. Show that equations (13.3.6) are equivalent to a parametric form of Euler's equations.

SOLUTION 13.4. Let the integrand of the traveltime integral in an isotropic inhomogeneous continuum be written as

$$\mathscr{F}(\mathbf{x},\dot{\mathbf{x}};t)=\frac{\sqrt{\sum_{i=1}^{3}\dot{x}_i\dot{x}_i}}{V(\mathbf{x})},$$

where $\dot{\mathbf{x}}\equiv \mathrm{d}\mathbf{x}/\mathrm{d}t$. We invoke equations (13.1.9), namely,

$$\frac{\partial\mathscr{F}}{\partial x_i}-\frac{\mathrm{d}}{\mathrm{d}t}\left(\frac{\partial\mathscr{F}}{\partial\dot{x}_i}\right)=0,\qquad i\in\{1,2,3\}, \tag{13.3.7}$$

which are a parametric form of Euler's equations. Considering integrand $\mathscr{F}$ and the first term of equations (13.3.7), we obtain

$$\frac{\partial\mathscr{F}}{\partial x_i}=-\frac{1}{V^2(\mathbf{x})}\frac{\partial V(\mathbf{x})}{\partial x_i}\sqrt{\sum_{i=1}^{3}\dot{x}_i\dot{x}_i},\qquad i\in\{1,2,3\}.$$

Using the fact that t denotes time, and, hence, as shown in expression (8.4.3),

$$\sqrt{\sum_{i=1}^{3}\dot{x}_i\dot{x}_i}=:V(\mathbf{x}), \tag{13.3.8}$$

where V is the magnitude of ray velocity, we can write

$$\frac{\partial\mathscr{F}}{\partial x_i}=-\frac{1}{V(\mathbf{x})}\frac{\partial V(\mathbf{x})}{\partial x_i},\qquad i\in\{1,2,3\}.$$

Using the chain rule, we can rewrite this expression as

$$\frac{\partial\mathscr{F}}{\partial x_i}=-\frac{\partial}{\partial x_i}\ln V(\mathbf{x}),\qquad i\in\{1,2,3\}. \tag{13.3.9}$$

Considering integrand $\mathscr{F}$ and the second term of equations (13.3.7), we obtain

$$\frac{\partial\mathscr{F}}{\partial\dot{x}_i}=\frac{1}{V(\mathbf{x})}\frac{\dot{x}_i}{\sqrt{\sum_{i=1}^{3}\dot{x}_i\dot{x}_i}},\qquad i\in\{1,2,3\},$$

which, using expression (13.3.8), we can write as

$$\frac{\partial \mathscr{F}}{\partial \dot{x}_i}=\frac{1}{V^2(\mathbf{x})}\frac{\mathrm{d}x_i}{\mathrm{d}t}, \qquad i\in\{1,2,3\}. \tag{13.3.10}$$

Consequently, using expressions (13.3.9) and (13.3.10), we can write Euler's equations (13.3.7) as

$$\frac{\partial \ln V}{\partial x_i}+\frac{\mathrm{d}}{\mathrm{d}t}\left[\frac{1}{V^2(\mathbf{x})}\frac{\mathrm{d}x_i}{\mathrm{d}t}\right]=0, \qquad i\in\{1,2,3\}. \tag{13.3.11}$$

Since in isotropic continua, phase and ray velocities coincide, namely, $V\equiv v$, equations (13.3.11) are equivalent to equations (13.3.6), as required.

REMARK 13.3.2. Exercises 13.3 and 13.4 show that the characteristic equations that are the solutions of the eikonal equation in isotropic inhomogeneous continua are tantamount to Euler's equations that provide the stationarity condition for the traveltime of the signal in such continua. In other words, these exercises verify Fermat's principle in isotropic inhomogeneous continua.[14]

EXERCISE 13.5. Recall the classical-mechanics Lagrangian given in expression (13.2.2), namely,

$$\mathbb{L}:=T-U, \tag{13.3.12}$$

where T and U are the kinetic and potential energies. Let the classical-mechanics Hamiltonian be

$$\mathbb{H}:=T+U. \tag{13.3.13}$$

Using the standard expression for kinetic energy and letting p_i be a component of linear momentum, verify Legendre's transformation between $\mathbb{L}$ and $\mathbb{H}$.

SOLUTION 13.5. In view of Legendre's transformation, discussed in Appendix B, we can write

$$\mathbb{H}=\sum_{i=1}^{n}p_i\dot{x}_i-\mathbb{L},$$

[14]Readers interested in a formulation linking rays and Fermat's principle in isotropic inhomogeneous continua might also refer to Elmore, W.C., and Heald, M.A., (1969/1985) Physics of waves: Dover, pp. 320–322.

where $p_i = mv_i$, with v_i being a component of velocity given by $v_i = \mathrm{d}x_i/\mathrm{d}t \equiv \dot{x}_i$. Hence, we can write

$$\mathbb{H} = m\sum_{i=1}^{n}\dot{x}_i^2 - \mathbb{L} = mv^2 - \mathbb{L}, \qquad (13.3.14)$$

where v stands for the magnitude of velocity. Recalling definitions (13.3.12) and (13.3.13), we can write expression (13.3.14) as

$$T + U = mv^2 - (T - U),$$

where T and U are the kinetic and potential energies, respectively. Simplifying, we obtain

$$T = \frac{1}{2}mv^2,$$

which is the standard expression for kinetic energy.

EXERCISE 13.6. Given Newton's second law of motion, stated as a single second-order ordinary differential equation, namely,

$$m\frac{\mathrm{d}^2x_i}{\mathrm{d}t^2} = -\frac{\partial U(\mathbf{x})}{\partial x_i}, \qquad i \in \{1,2,3\}, \qquad (13.3.15)$$

where $U(\mathbf{x})$ denotes the scalar potential, write the corresponding two first-order ordinary differential equations in t to be solved for the x_i and the p_i, where p_i is a component of the linear momentum.

SOLUTION 13.6. We can denote the components of the momentum vector as

$$p_i := m\frac{\mathrm{d}x_i}{\mathrm{d}t}, \qquad i \in \{1,2,3\}. \qquad (13.3.16)$$

Differentiating both sides of equations (13.3.16) with respect to t, we obtain

$$\frac{\mathrm{d}p_i}{\mathrm{d}t} = m\frac{\mathrm{d}^2x_i}{\mathrm{d}t^2}, \qquad i \in \{1,2,3\},$$

which are equations (13.3.15). Hence, Newton's second law of motion can be written as a set of two first-order differential equations,

$$\left\{\begin{aligned}\frac{\mathrm{d}x_i}{\mathrm{d}t}&=\frac{p_i}{m}\\ \frac{\mathrm{d}p_i}{\mathrm{d}t}&=-\frac{\partial U}{\partial x_i}\end{aligned}\right., \qquad i\in\{1,2,3\}. \tag{13.3.17}$$

EXERCISE 13.7. [15]Using expression (13.3.13), show that equations (13.3.17), obtained in Exercise 13.6, correspond to Hamilton's equations of motion that are given by

$$\left\{\begin{aligned}\frac{\mathrm{d}x_i}{\mathrm{d}t}&=\frac{\partial \mathbb{H}}{\partial p_i}\\ \frac{\mathrm{d}p_i}{\mathrm{d}t}&=-\frac{\partial \mathbb{H}}{\partial x_i}\end{aligned}\right., \qquad i\in\{1,2,3\}.$$

SOLUTION 13.7. Consider expression (13.3.13). Using the standard expression for the kinetic energy, as well as the definition of linear momentum, we can write this expression as

$$\mathbb{H}=T+U=\frac{1}{2}m\left(\frac{\mathrm{d}x_i}{\mathrm{d}t}\right)^2+U=\frac{1}{2m}p_i^2+U, \qquad i\in\{1,2,3\}. \tag{13.3.18}$$

Differentiating equations (13.3.18) with respect to both the p_i and the x_i, we obtain

$$\left\{\begin{aligned}\frac{\partial \mathbb{H}}{\partial p_i}&=\frac{p_i}{m}\\ \frac{\partial \mathbb{H}}{\partial x_i}&=\frac{\partial U}{\partial x_i}\end{aligned}\right., \qquad i\in\{1,2,3\}.$$

[15] See also Section 13.2.3.

Using Newton's second law of motion, which is stated in expression (13.3.17), we obtain

$$\left\{\begin{aligned} \frac{\mathrm{d}x_i}{\mathrm{d}t} &= \frac{\partial \mathbb{H}}{\partial p_i} \\ \frac{\mathrm{d}p_i}{\mathrm{d}t} &= -\frac{\partial \mathbb{H}}{\partial x_i} \end{aligned}\right., \qquad i \in \{1,2,3\},$$

which are Hamilton's equations of motion, as required.

EXERCISE 13.8. Considering a free-falling body in the vacuum, show that Hamilton's principle is consistent with Newton's concept of acceleration due to gravity.

SOLUTION 13.8. Let $T = mv^2/2$ and $U = mgz$, where m is mass, v is velocity, g is acceleration due to gravity and z denotes height. Since $v = \mathrm{d}z/\mathrm{d}t$, we can write the classical-mechanics Lagrangian as

$$\mathbb{L} = T - U = \frac{1}{2}m\left(\frac{\mathrm{d}z}{\mathrm{d}t}\right)^2 - mgz.$$

Consider the action integral

$$\mathbb{A} = \int_{t_0}^{t_1} \mathbb{L}\,\mathrm{d}t.$$

Hamilton's principle implies

$$\delta \int_{t_0}^{t_1} (T - U)\,\mathrm{d}t = \delta \int_{t_0}^{t_1} \left(\frac{1}{2}m\,(\dot{z})^2 - mgz\right)\mathrm{d}t = 0,$$

where $\dot{z} := \mathrm{d}z/\mathrm{d}t$. Invoking Euler's equation, which corresponds to Lagrange's equations of motion (13.2.4), we obtain

$$\frac{\partial \mathbb{L}}{\partial z} - \frac{\mathrm{d}}{\mathrm{d}t}\left(\frac{\partial \mathbb{L}}{\partial \dot{z}}\right) = -mg - \frac{\mathrm{d}}{\mathrm{d}t}(m\dot{z}) = -mg - m\ddot{z} = 0,$$

which can be written as

$$\frac{\mathrm{d}^2 z(t)}{\mathrm{d}t^2} = -g,$$

where g is the free-fall acceleration, as required.

EXERCISE 13.9. Following Section 13.2.4, derive a one-dimensional wave equation for transverse waves.

SOLUTION 13.9. Let the transverse displacements of masses be $u_0, \ldots, u_{n+1}$, with $u_0 = u_{n+1} = 0$, which are *boundary conditions* corresponding to fixed ends. The potential energy, U, is associated with the tension, μ, of the string. The potential energy per segment is

$$dU = \mu \left[\sqrt{(\Delta x)^2 + (u_i - u_{i-1})^2} - \Delta x \right], \tag{13.3.19}$$

where the term in parentheses constitutes the extension of the segment Δx, which is the difference between its original length, Δx, and its strained length, $\sqrt{(\Delta x)^2 + (u_i - u_{i-1})^2}$. We can rewrite expression (13.3.19) as

$$dU = \mu \Delta x \left[\sqrt{1 + \left(\frac{u_i - u_{i-1}}{\Delta x} \right)^2} - 1 \right].$$

Expanding the square root as a power series gives us

$$dU = \mu \Delta x \left[\frac{1}{2} \left(\frac{u_i - u_{i-1}}{\Delta x} \right)^2 - \frac{1}{8} \left(\frac{u_i - u_{i-1}}{\Delta x} \right)^4 + \ldots \right].$$

Assuming that the term in parentheses is much smaller than unity, which implies that the transverse displacement is much smaller than the length of a segment, we obtain

$$dU \approx \frac{\mu}{2} \Delta x \left(\frac{u_i - u_{i-1}}{\Delta x} \right)^2.$$

Thus, the potential energy for the entire string is

$$U \approx \frac{1}{2} \mu \sum_{i=1}^{n} \left(\frac{u_i - u_{i-1}}{\Delta x} \right)^2 \Delta x.$$

Letting $n \longrightarrow \infty$ and $\Delta x \longrightarrow 0$ in such a way that $n\Delta x = l$, where ℓ is the length of the string, and noticing that the term in parentheses represents a partial derivative with respect to x, we can write

$$U = \frac{1}{2}\mu \sum_{i=1}^{\infty} \left[\frac{\partial u(x_i,t)}{\partial x}\right]^2 \Delta x .$$

Thus, in the limit, we obtain

$$U = \frac{1}{2}\mu \int_0^l \left[\frac{\partial u(x,t)}{\partial x}\right]^2 \mathrm{d}x .$$

The kinetic energy is given by expression (13.2.11); namely,

$$T = \frac{1}{2}\rho \int_0^l \left[\frac{\partial u(x,t)}{\partial t}\right]^2 \mathrm{d}x .$$

Thus, using $\mathbb{L}$, given in expression (13.2.2), we can write

$$\mathbb{L} = T - U = \int_0^l \left[\frac{\rho}{2}\left(\frac{\partial u}{\partial t}\right)^2 - \frac{\mu}{2}\left(\frac{\partial u}{\partial x}\right)^2\right] \mathrm{d}x .$$

Invoking Hamilton's principle, stated in expression (13.2.3), we obtain

$$\delta \int_0^t \mathbb{L}\,\mathrm{d}t = \delta \int_0^t \int_0^l \left[\frac{\rho}{2}\left(\frac{\partial u}{\partial t}\right)^2 - \frac{\mu}{2}\left(\frac{\partial u}{\partial x}\right)^2\right] \mathrm{d}x\,\mathrm{d}t = 0 .$$

Using the corresponding Euler's equation (12.4.1), we get

$$\frac{\partial^2 u}{\partial x^2} = \frac{1}{\frac{\mu}{\rho}} \frac{\partial^2 u}{\partial t^2} . \tag{13.3.20}$$

Equation (13.3.20) is a one-dimensional wave equation for transverse waves where the transverse displacement, u, is assumed to be much smaller than the length of the string, ℓ.

CHAPTER 14

Ray Parameters

> En général la conservation des forces vives donne toujours une intégrale première des différentes équations différentielles de chaque problème; ce qui est d'une grande utilité dans plusieurs occasions.[1]
>
> *Joseph-Louis Lagrange (1788)*
> *Mécanique Analytique*

Preliminary Remarks

In the context of ray theory, the trajectories of seismic signals as well as their traveltimes can be obtained by solving Hamilton's ray equations or Lagrange's ray equations, discussed in Chapters 8 and 11, respectively. In certain cases, particular properties of the continuum result in simplifications of these equations, thereby allowing us to obtain their solutions more easily, as well as to gain further insight into these solutions.

In Chapter 13, we showed that rays are the solutions of the variational problem of stationary traveltime and, hence, they are the solutions of the corresponding Euler's equations. For the continua that exhibit particular homogeneities, Euler's equations can be simplified by obtaining the corresponding first integrals, which were introduced in Section 12.6. First

[1] In general, the conservation of living forces yields always a first integral of various differential equations of each problem; this is of great utility on numerous occasions.

integrals are the conserved quantities. In ray theory, these quantities, which are constant along a given ray, are called ray parameters.

We begin this chapter, in which we study only two-dimensional continua, with the formulation of the ray parameter for an anisotropic continuum that is homogeneous along one axis. By integrating the ray-parameter expression, we obtain the expression for the ray. Also, using the ray parameter, we obtain the expression for the traveltime. Then we briefly discuss a case in which ray equations do not possess corresponding ray parameters. We conclude this chapter by discussing the conserved quantities in the context of Hamilton's ray equations.

14.1. Traveltime Integrals

Let us consider a two-dimensional continuum that is contained in the xz-plane. The traveltime between two points A and B within this continuum can be stated as

$$\check{C} = \int_A^B \frac{\sqrt{1+(z')^2}}{V(x,z,z')}\,\mathrm{d}x =: \int_A^B \mathrm{F}\,\mathrm{d}x, \tag{14.1.1}$$

where $z' := \mathrm{d}z/\mathrm{d}x$. Since $\mathrm{d}z/\mathrm{d}x = \cot\theta$, where θ is the ray angle, we see that the ray velocity, V, is a function of position, (x,z), and direction, z'. In other words, integral (14.1.1) allows us to study traveltimes in anisotropic inhomogeneous continua.

Also, let us view the x-axis and the z-axis as the horizontal and vertical axes, respectively, where the vertical axis corresponds to depth within a geological model.

In view of Fermat's principle, discussed in Chapter 13, rays correspond to curves along which the traveltime is stationary. Since integral (14.1.1) is of the type given by integral (12.1.1), in general, we can obtain such a curve using Euler's equation (12.2.2). Consequently, F is a ray-theory Lagrangian.

As discussed in Section 12.5, a particular form of the integral, whose stationary value we seek, may result in simplifications of Euler's equation. Herein, we wish to study special cases of traveltime integral (14.1.1) that are pertinent to seismic investigations.

14.2. Ray Parameters as First Integrals[2]

In this section, we study horizontally layered media. In such a case, where the ray velocity, V, may vary with depth, z, and direction, z', traveltime integral (14.1.1) becomes

$$\check{C}=\int_A^B \frac{\sqrt{1+(z')^2}}{V(z,z')}\,\mathrm{d}x \equiv \int_A^B \mathrm{F}(z,z')\,\mathrm{d}x. \tag{14.2.1}$$

Since traveltime integral (14.2.1) does not exhibit an explicit dependence on x, to obtain the ray, we use Beltrami's identity (12.3.1), namely,

$$\frac{\partial \mathrm{F}}{\partial x}+\frac{\mathrm{d}}{\mathrm{d}x}\left(z'\frac{\partial \mathrm{F}}{\partial z'}-\mathrm{F}\right)=0,$$

which immediately leads to

$$z'\frac{\partial \mathrm{F}}{\partial z'}-\mathrm{F}=C, \tag{14.2.2}$$

where C is a constant. Expression (14.2.2) is first integral (12.5.4) and C is a conserved quantity along the ray.

We wish to study this conserved quantity. Inserting integrand F, given in integral (14.2.1), into expression (14.2.2), we obtain

$$z'\frac{\partial\left(\frac{\sqrt{1+(z')^2}}{V(z,z')}\right)}{\partial z'}-\frac{\sqrt{1+(z')^2}}{V(z,z')}=-\frac{V'z'\sqrt{1+(z')^2}}{V^2}-\frac{1}{V\sqrt{1+(z')^2}}=C, \tag{14.2.3}$$

where, for convenience, we denote $V:=V(z,z')$ and $V':=\partial V/\partial z'$. The chain rule implies $-V'/V^2=\partial\left[1/V(z,z')\right]/\partial z'$ and, hence, we get

$$\frac{\partial}{\partial z'}\left(\frac{1}{V}\right)z'\sqrt{1+(z')^2}-\frac{1}{V\sqrt{1+(z')^2}}=C. \tag{14.2.4}$$

---

[2] This section is based on Slawinski, M.A., and Webster, P.S., (1999) On generalized ray parameters for vertically inhomogeneous and anisotropic media, Canadian Journal of Exploration Geophysics, **35**, No. 1/2, 28–31.

Expression (14.2.4) is a first integral of Euler's equation (12.2.2) for traveltime integral (14.2.1).

To express the first integral in terms of the ray angle, we use $z' \equiv \mathrm{d}z/\mathrm{d}x = \cot\theta$, where θ is the ray angle measured from the z-axis. Also, the differential operator in expression (14.2.4) can be restated as $\partial/\partial z' = (\partial\theta/\partial z')\,\partial/\partial\theta$. Hence, using trigonometric identities, we obtain another form of expression (14.2.4) given by

$$\mathfrak{p} = \cos\theta \frac{\partial}{\partial\theta}\left[\frac{1}{V(z,\theta)}\right] + \frac{\sin\theta}{V(z,\theta)}, \tag{14.2.5}$$

where $\mathfrak{p} = -C$ and where V and θ denote ray velocity and ray angle, respectively. Since $\mathfrak{p}$ is conserved along a given ray, $z(x)$, we refer to this conserved quantity as ray parameter. Expression (14.2.5) is the ray parameter for anisotropic vertically inhomogeneous continua.

For expression (14.2.5) to be valid, the ray velocity may vary along the z-axis but not along the x-axis. The directional dependence of velocity, however, need not exhibit any particular symmetry. In other words, the angular velocity dependence is arbitrary.

Note that in the context of elasticity theory, the availability of exact and explicit ray-velocity expressions $V(\theta)$ is limited due to the requirements of Legendre's transformation. An explicit, closed-form expression for $V(\theta)$ is only possible for the case of elliptical velocity dependence.

14.3. Example: Elliptical Anisotropy and Linear Inhomogeneity

14.3.1. Introductory comments

In this section, we study a particular case of wave propagation that is associated with both an elliptical velocity dependence with direction and a linear velocity dependence with depth. This assumption allows us to obtain analytic expressions for rays and traveltimes. Also, the same case was treated in Section 8.5 in the context of Hamilton's ray equations. Thus, our examination of Sections 8.5 and 14.3 allows us to investigate the same physical problem using the two different approaches that are available to study seismic ray theory.

Since Euler's equation (12.2.2), or its Beltrami's identity (12.3.1), is a second-order ordinary differential equation, in view of Definition 12.6.1, first integral (14.2.4) and ray parameter (14.2.5) are first-order ordinary differential equations. If the integration of the ray parameter is possible, this integration results in a solution of Euler's equation and its Beltrami's identity, which can be given by $z(x)$ or $x(z)$. In other words, the expressions for ray velocity, V, that result in integrable expression (14.2.5) allow us to obtain rays by integration.[4] Ray velocity that results in a conveniently integrable ray parameter is provided by the case of elliptical velocity dependence with direction and linear velocity dependence with depth.

14.3.2. Rays

Derivation

To obtain an analytic expression for a ray, we wish to use an exact ray-velocity expression to be inserted into expression (14.2.5). For this purpose, we consider expression (8.7.13), namely,

$$V(\theta) = V_z \sqrt{\frac{1+\tan^2\theta}{1+\left(\frac{V_z}{V_x}\right)^2 \tan^2\theta}}, \tag{14.3.1}$$

which gives the magnitude of the ray velocity as a function of the ray angle for the case of elliptical velocity dependence. For convenience, let the measure of ellipticity be given by

$$\chi := \frac{V_x^2 - V_z^2}{2V_z^2}, \tag{14.3.2}$$

where V_x and V_z stand for the magnitude of the horizontal and the vertical ray velocities, respectively. This definition of χ is consistent with definition (8.5.1) since—for elliptical velocity dependence—the ray and phase velocities are equal to one another along the axes of the ellipse.

[4] For certain cases with applications to continua exhibiting folded layers, readers might refer to Epstein, M. and Slawinski, M.A., (1999) On rays and ray parameters in inhomogeneous isotropic media: Canadian Journal of Exploration Geophysics. **35**, No. 1/2, 7–19.

Using χ, we can write expression (14.3.1) as

$$V(\theta) = V_z \sqrt{\frac{1+2\chi}{1+2\chi\cos^2\theta}}.$$

Let us assume that the ray velocity varies along the z-axis in such a way that χ remains constant. This implies that the ratio of magnitudes of horizontal and vertical ray velocities remains constant. In such a case, we can write

$$V(\theta, z) = V_z(z) \sqrt{\frac{1+2\chi}{1+2\chi\cos^2\theta}}. \qquad (14.3.3)$$

Furthermore, we assume that the magnitude of the ray velocity increases linearly along the z-axis.[5] In such a case, we can write expression (14.3.3) as

$$V(\theta, z) = (a+bz) \sqrt{\frac{1+2\chi}{1+2\chi\cos^2\theta}}, \qquad (14.3.4)$$

where a and b are positive constants.

Inserting expression (14.3.4) into expression (14.2.5), we obtain

$$\begin{aligned}\mathfrak{p} &= \cos\theta \frac{\partial}{\partial\theta}\left[\frac{1}{(a+bz)\sqrt{\frac{1+2\chi}{1+2\chi\cos^2\theta}}}\right] + \frac{\sin\theta}{(a+bz)\sqrt{\frac{1+2\chi}{1+2\chi\cos^2\theta}}} \\ &= \frac{1}{(a+bz)\sqrt{1+2\chi}}\left(\cos\theta\frac{\partial}{\partial\theta}\sqrt{1+2\chi\cos^2\theta} + \sin\theta\sqrt{1+2\chi\cos^2\theta}\right) \\ &= \frac{\sin\theta}{(a+bz)\sqrt{1+2\chi}\sqrt{1+2\chi\cos^2\theta}}. \qquad (14.3.5)\end{aligned}$$

To obtain an expression for a ray, we wish to state ray parameter (14.3.5) in terms of position variables x and z. Dividing both the numerator and the

[5] Readers interested in a seismological formulation of linearly increasing velocity might refer to Epstein, M. and Slawinski, M.A., (1999) On raytracing in constant velocity-gradient media: Geometrical approach, Canadian Journal of Exploration Geophysics. **35**, No. 1/2, 1–6, and to Slawinski, R.A. and Slawinski, M.A., (1999) On raytracing in constant velocity-gradient media: Calculus approach, Canadian Journal of Exploration Geophysics. **35**, No. 1/2, 24–27.

denominator by $\sin\theta$, we rewrite expression (14.3.5) as

$$\mathfrak{p}=\frac{1}{(a+bz)\sqrt{1+2\chi}\sqrt{1+\dfrac{2\chi+1}{\tan^2\theta}}}.$$

Squaring both sides and rearranging, we obtain

$$\frac{1}{\tan^2\theta}=\frac{1-\mathfrak{p}^2(a+bz)^2(1+2\chi)}{\mathfrak{p}^2(a+bz)^2(1+2\chi)^2}. \tag{14.3.6}$$

Since $1/\tan^2\theta=(\mathrm{d}z/\mathrm{d}x)^2$, we have a first-order ordinary differential equation, which is a special case of first integral (12.5.4). We can write equation (14.3.6) as

$$\frac{\mathrm{d}z}{\mathrm{d}x}=\frac{\sqrt{1-\mathfrak{p}^2(a+bz)^2(1+2\chi)}}{\mathfrak{p}(a+bz)(1+2\chi)}, \tag{14.3.7}$$

which can be restated as

$$\mathrm{d}x=\frac{\mathfrak{p}(a+bz)(1+2\chi)}{\sqrt{1-\mathfrak{p}^2(a+bz)^2(1+2\chi)}}\mathrm{d}z. \tag{14.3.8}$$

To integrate equation (14.3.8), we set the initial conditions in such a way that $z(0)=0$. In other words, the source is located at the origin of the coordinate system. Hence, integrating both sides, namely,

$$\int_0^x \mathrm{d}\xi=\int_0^z \frac{\mathfrak{p}(a+b\zeta)(1+2\chi)}{\sqrt{1-\mathfrak{p}^2(a+b\zeta)^2(1+2\chi)}}\mathrm{d}\zeta,$$

where ξ and ζ are the integration variables, we obtain

$$x=\frac{1}{\mathfrak{p}b}\left[\sqrt{1-\mathfrak{p}^2a^2(1+2\chi)}-\sqrt{1-\mathfrak{p}^2(a+bz)^2(1+2\chi)}\right], \tag{14.3.9}$$

which describes the ray given by $x(z)$ for elliptical velocity dependence with direction and a linear velocity dependence with depth.

Interpretation

To obtain a geometrical interpretation of equation (14.3.9), we rearrange it and write

$$\frac{\left(x-\frac{\sqrt{1-\mathfrak{p}^2a^2(1+2\chi)}}{\mathfrak{p}b}\right)^2}{\left(\frac{1}{\mathfrak{p}b}\right)^2}+\frac{\left(z+\frac{a}{b}\right)^2}{\left(\frac{1}{\mathfrak{p}b\sqrt{1+2\chi}}\right)^2}=1. \tag{14.3.10}$$

This is the equation of an ellipse whose axes are parallel to the axes of the coordinate system with the origin of this system located at the source. The centre of the ellipse is located at

$$\left[\frac{\sqrt{1-\mathfrak{p}^2a^2(1+2\chi)}}{\mathfrak{p}b},-\frac{a}{b}\right]. \tag{14.3.11}$$

Equation (14.3.10) is identical to equation (8.5.17), which we obtained in Section 8.5.6 using Hamilton's ray equations.

In a seismological notation, with the z-axis being vertical and pointing downwards, the centre of the ellipse is located on the horizontal line positioned a/b units above the x-axis. In view of $v(z)=a+bz$, this line corresponds to the level where the velocity vanishes. Ellipse (14.3.10) passes through the origin, as can be verified by setting $z=0$ in equation (14.3.9). The segment of the ellipse that is below the x-axis corresponds to the ray. This interpretation is consistent with with our discussion in Section 8.5, as expected.

The greater the distance between the source and the centre of the ellipse, the smaller the curvature of the ray. For constant-velocity fields, where $b=0$, the centre of the ellipse is located infinitely far from the source. In such a case, the ray is a straight line, as also shown in Exercise 14.3. For a signal propagating along the z-axis, $\theta=0$ and, following expression (14.3.5), $\mathfrak{p}=0$. Hence, in view of expression (14.3.11), the x-coordinate of the centre of the ellipse is located infinitely far from the source. In such a case, the ray is a vertical straight line. For the isotropic case, where $\chi=0$, equation (14.3.10) reduces to the expression for a circle. In such a case, the rays are circular arcs.

14.3.3. Traveltimes

We can use ray parameter (14.2.5) to obtain the traveltime along the corresponding ray. For this purpose, we wish to rewrite integral (14.2.1) to include the ray parameter for a given source-receiver pair.

Integral (14.2.1) can be viewed as $\int \mathrm{d}s/V$, where $\mathrm{d}s$ is the arclength element along the ray. In the xz-plane, the arclength element can be written as $\mathrm{d}s = \mathrm{d}x/\sin\theta$ or as $\mathrm{d}s = \mathrm{d}z/\cos\theta$, where θ is the ray angle. We choose the former case, since along the x-axis a ray is expressed as a single-valued function; this is not the case along the z-axis due to the downgoing and upgoing signal. Thus, traveltime integral (14.2.1) between the source at $(0,0)$ and the receiver at (X,Z) can be written as

$$\check{C} = \int_0^X \frac{\mathrm{d}x}{V(z,\theta)\sin\theta}, \qquad (14.3.12)$$

where $V(z,\theta)$ is given by expression (14.3.4).

To integrate, we must express $\sin\theta$ in terms of constants a, b, χ, $\mathfrak{p}$, and integration variable x. Combining expressions (14.3.4) and (14.3.5), we write

$$\begin{aligned} &V(z,\theta)\sin\theta \\ &\quad = (a+bz)\sqrt{\frac{1+2\chi}{1+2\chi\cos^2\theta}}\,\mathfrak{p}\,(a+bz)\sqrt{1+2\chi}\sqrt{1+2\chi\cos^2\theta} \\ &\quad = \mathfrak{p}\,(a+bz)^2(1+2\chi), \end{aligned}$$

which is the denominator in integral (14.3.12). To proceed with integration, we must express z in terms of x. To do so, we write expression (14.3.9) as

$$\left(\mathfrak{p}bx - \sqrt{1-\mathfrak{p}^2a^2(1+2\chi)}\right)^2 = 1-\mathfrak{p}^2(a+bz)^2(1+2\chi).$$

Combining the last two equations, we get

$$\left(\mathfrak{p}bx - \sqrt{1-\mathfrak{p}^2a^2(1+2\chi)}\right)^2 = 1-\mathfrak{p}V(z,\theta)\sin\theta.$$

Solving for $V \sin\theta$, we obtain

$$V(z,\theta)\sin\theta = \frac{1-\left(\mathfrak{p}bx-\sqrt{1-\mathfrak{p}^2a^2(1+2\chi)}\right)^2}{\mathfrak{p}}, \tag{14.3.13}$$

which is the denominator in integral (14.3.12) expressed in terms of variable x. Inserting expression (14.3.13) into integral (14.3.12) and using the fact that $\mathfrak{p}$ is constant for a given source-receiver pair, we write

$$\check{C} = \mathfrak{p}\int_0^X \frac{\mathrm{d}x}{1-\left(\mathfrak{p}bx-\sqrt{1-\mathfrak{p}^2a^2(1+2\chi)}\right)^2}. \tag{14.3.14}$$

Integrating between $x=0$ and $x=X$, while treating $\mathfrak{p}$ as a constant, we obtain—as shown in Exercise 14.5—the expression for the value of the traveltime; namely,

$$\check{C} = \frac{\tanh^{-1}\left[\mathfrak{p}bX-\sqrt{1-\mathfrak{p}^2a^2(1+2\chi)}\right]+\tanh^{-1}\sqrt{1-\mathfrak{p}^2a^2(1+2\chi)}}{b}, \tag{14.3.15}$$

which is expression (8.5.18), as could be expected. We treat $\mathfrak{p}$ as a constant since, for a given source-receiver pair in a laterally homogeneous continuum, $\mathfrak{p}$ is a conserved quantity along the ray.

To find the expression for $\mathfrak{p}$ that corresponds to the source at $(0,0)$ and the receiver at (X,Z), we can write expression (14.3.9) as

$$X = \frac{1}{\mathfrak{p}b}\left[\sqrt{1-\mathfrak{p}^2a^2(1+2\chi)}-\sqrt{1-\mathfrak{p}^2(a+bZ)^2(1+2\chi)}\right]. \tag{14.3.16}$$

Solving for $\mathfrak{p}$, we obtain

$$\mathfrak{p} = \frac{2X}{\sqrt{\left[X^2+(1+2\chi)Z^2\right]\left[(2a+bZ)^2(1+2\chi)+b^2X^2\right]}}. \tag{14.3.17}$$

Expression (14.3.17) is identical to expression (8.5.19), which we obtained in Section 8.5.7 using Hamilton's ray equations. Herein, however, $\mathfrak{p}$ is treated as a constant for a particular choice of X and Z.

REMARK 14.3.1. Expression (14.3.15) has a singularity at $X=0$; the inverse hyperbolic function tends to infinity.

Studying the properties of the continuum in terms of a, b and χ, we can use expression (14.3.15) with $\mathfrak{p}$ given by expression (14.3.17) to obtain the traveltime between the source and the receiver. These expressions are convenient to use for inverse problems that are based on traveltime measurements.[6] For such a study, we might wish to know if the receiver has been reached by a downgoing or an upgoing signal travelling along an elliptical arc. As shown in Exercise 14.7, the subsurface receiver at (X,Z) is reached by the downgoing signal from the source at $(0,0)$ if

$$X < \sqrt{\frac{1+2\chi}{b}\,(2a+bZ)\,Z}.$$

For the value of X equal to the right-hand side, the signal is at its deepest point when it reaches the receiver. Also as shown in Exercise 14.7, the curve distinguishing between the downgoing and upgoing arrivals is a hyperbola whose asymptote is

$$x = \sqrt{1+2\chi}\left(z+\frac{a}{b}\right).$$

As shown in Section 6.11.3, by using a linear transformation of coordinates we can treat elliptical velocity dependence as an isotropic case. Consequently, we can also obtain the traveltime and ray-parameter expressions by such a transformation, as illustrated in Exercise 14.8.

14.4. Rays in Isotropic Continua

In Sections 14.2 and 14.3, we studied ray equations in two-dimensional anisotropic continua and obtained analytic expressions for rays and traveltimes. The availability of analytic expressions resulted from the assumption of homogeneity along the x-axis and, hence, from the existence of a first integral. In this section, to emphasize the convenience of first integrals, we look briefly at ray equations in a two-dimensional isotropic continuum that is contained in the xz-plane. The traveltime between two points A and

[6] Interested readers might refer to Slawinski, M.A., Wheaton, C.J., and Powojowski, M. (2004) VSP traveltime inversion for linear inhomogeneity and elliptical anisotropy: Geophysics, **69**(2), 373–377. However, expression (11), therein—which, herein, is expression (8.5.20), on page 375—is valid only for the downgoing signal.

B within this continuum can be stated as

$$\check{C} = \int\limits_{A}^{B} \frac{\sqrt{1 + (z')^2}}{V(x,z)} \, \mathrm{d}x . \tag{14.4.1}$$

Since the continuum is isotropic, V is not a function of z'. However, the integrand is an explicit function of x, z and z', and, hence, the corresponding Euler's equation does not have a first integral.

In view of the stationarity of traveltime and Section 12.5.6, the corresponding ray equation, which results from Euler's equation (12.2.2), is given by equation (12.5.8), namely,

$$V \frac{\mathrm{d}^2 z}{\mathrm{d}x^2} - \frac{\partial V}{\partial x} \left(\frac{\mathrm{d}z}{\mathrm{d}x} \right)^3 + \frac{\partial V}{\partial z} \left(\frac{\mathrm{d}z}{\mathrm{d}x} \right)^2 - \frac{\partial V}{\partial x} \frac{\mathrm{d}z}{\mathrm{d}x} + \frac{\partial V}{\partial z} = 0, \tag{14.4.2}$$

where, due to the isotropy of the continuum, phase and ray velocities coincide, namely, $v \equiv V$. Equation (14.4.2) is a nonlinear ordinary differential equation, which requires numerical methods to obtain rays and corresponding traveltimes.

14.5. Lagrange's Ray Equations in xz-Plane

In this chapter, as well as in Chapter 12, Euler's equations and Lagrange's ray equations are formulated in the context of explicit functions. Such a formulation is convenient for many raytracing applications. It rules out, however, complicated rays that are given by multiple-valued functions. To generalize the formulation so as to allow such rays, we can formulate our problem in a parametric form.

Consider traveltime integral (14.1.1). An analogous parametric representation can be given in terms of $x(t)$, $z(t)$, $\dot{x} := \mathrm{d}x/\mathrm{d}t$ and $\dot{z} := \mathrm{d}z/\mathrm{d}t$. Then, the traveltime integral is

$$\check{C} = \int \frac{\mathrm{d}s}{V} = \int \frac{\sqrt{\dot{x}^2 + \dot{z}^2}}{V(x,z,\dot{x},\dot{z})} \, \mathrm{d}t := \int \mathscr{F} \, \mathrm{d}t , \tag{14.5.1}$$

where $\mathscr{F}$ is a two-dimensional form of expression (13.1.10).

In view of the principle of stationary traveltime, we can use Lagrange's ray equations (13.1.9). In the two-dimensional case, discussed herein, these equations constitute the system

$$\left\{\begin{aligned}\frac{\partial\mathscr{F}}{\partial x}-\frac{\mathrm{d}}{\mathrm{d}t}\left(\frac{\partial\mathscr{F}}{\partial\dot{x}}\right)&=0\\ \frac{\partial\mathscr{F}}{\partial z}-\frac{\mathrm{d}}{\mathrm{d}t}\left(\frac{\partial\mathscr{F}}{\partial\dot{z}}\right)&=0\end{aligned}\right., \tag{14.5.2}$$

where $\mathscr{F}$ denotes the integrand of the traveltime integral.

Also, the equations of system (14.5.2) are related by Beltrami's identity; namely,

$$\frac{\partial\mathscr{F}}{\partial t}+\frac{\mathrm{d}}{\mathrm{d}t}\left(\dot{x}\frac{\partial\mathscr{F}}{\partial\dot{x}}+\dot{z}\frac{\partial\mathscr{F}}{\partial\dot{z}}-\mathscr{F}\right)=0. \tag{14.5.3}$$

The justification for this form of Beltrami's identity is shown in Exercise 11.1.

In view of Theorem A.2.1, stated in Appendix A, $\mathscr{F}$ cannot depend explicitly on parameter t. Mathematically, we can justify this corollary in the following way.

Since ray-velocity function, V, is homogeneous of degree 0 in the variables $\dot{x}$ and $\dot{z}$ and $\sqrt{\dot{x}^2+\dot{z}^2}$ is absolute-value homogeneous of degree 1 in the same variables, the integrand of the traveltime integral is absolute-value homogeneous of degree 1 in these variables. Thus, since $\mathscr{F}$ is absolute-value homogeneous of degree 1, it follows from Theorem A.2.1 that

$$\mathscr{F}=\dot{x}\frac{\partial\mathscr{F}}{\partial\dot{x}}+\dot{z}\frac{\partial\mathscr{F}}{\partial\dot{z}}. \tag{14.5.4}$$

Consequently, the term in parentheses of Beltrami's identity (14.5.3) vanishes and equation (14.5.3) implies that $\mathscr{F}$ cannot depend explicitly on t, and, hence, V does not explicitly depend on t, which justifies our corollary.

Physically, this independence means that the ray-velocity function does not change with time. In other words, the properties of the continuum are time-invariant.

Also, the parametric formulation of the traveltime integral conveniently allows us to obtain ray parameters. Consider system (14.5.2). If $\mathscr{F}$ is not explicitly dependent on x, the first equation becomes $\partial\mathscr{F}/\partial\dot{x}=\mathfrak{p}$, where $\mathfrak{p}$ is a conserved quantity. This conserved quantity is equivalent to ray-parameter expression (14.2.5), as shown in Exercise 14.4.

14.6. Conserved Quantities and Hamilton's Ray Equations

In this chapter, we study the conserved quantities along the ray in the context of the calculus of variations. In other words, we study these quantities using the Lagrangian formulation of the ray theory. In view of the fact that we can study ray theory in terms of both the Hamiltonian and Lagrangian formulations, let us briefly look at the conserved quantities in terms of Hamilton's ray equations (8.2.7); namely,

$$\left\{\begin{array}{l}\dot{x}_i=\dfrac{\partial\mathscr{H}}{\partial p_i}\\[2ex] \dot{p}_i=-\dfrac{\partial\mathscr{H}}{\partial x_i}\end{array}\right., \qquad i\in\{1,2,3\}.$$

Before discussing ray parameters, to gain more insight into Hamilton's ray equations let us consider $\mathscr{H}$ itself. If $\mathscr{H}$ does not explicitly depend on t, it is conserved along the ray, as stated by Lemma 13.1.3. Also, in view of expression (13.3.13), we can infer that in classical mechanics the Hamiltonian that is independent of time implies the conservation of energy.

Let us return to ray parameters. Examining the second equation, we see that if $\mathscr{H}$ does not explicitly depend on the x_i, the corresponding p_i is constant along solution curve $\mathbf{x}(t)$, since $\mathrm{d}p_i/\mathrm{d}t=0$. To elucidate the consequences of this statement, recall expression (8.2.8); namely,

$$\mathscr{H}=\frac{1}{2}p^2v^2(\mathbf{x},\mathbf{p}).$$

We see that $\mathscr{H}$ does not explicitly depend on the x_i if and only if function v does not depend on the x_i coordinate. Since phase velocity, v, is a function of the properties of the continuum, we conclude that $\mathscr{H}$ does not depend on the x_i if and only if the continuum is homogeneous along the x_i-axis.

Also, in view of Lagrange's ray equations (11.2.6), namely,

$$\frac{\partial\mathscr{L}}{\partial x_i}-\frac{\mathrm{d}}{\mathrm{d}t}\left(\frac{\partial\mathscr{L}}{\partial\dot{x}_i}\right)=0,\qquad i\in\{1,2,3\}, \tag{14.6.1}$$

if $\mathscr{L}$ does not explicitly depend on the x_i, the equation of system (14.6.1) that corresponds to the given subscript i is reduced to

$$\frac{\mathrm{d}}{\mathrm{d}t}\left(\frac{\partial \mathscr{L}}{\partial \dot{x}_i}\right) = 0, \tag{14.6.2}$$

which implies that the term in parentheses of expression (14.6.2) is constant. In view of Legendre's transformation, following expressions (B.3.2), shown in Appendix B, we can write

$$p_i = \frac{\partial \mathscr{L}}{\partial \dot{x}_i}, \tag{14.6.3}$$

where $\mathfrak{p} = p_i$ is the conserved quantity along the solution curve $\mathbf{x}(t)$.

Expressions given by $\mathrm{d}p_i/\mathrm{d}t = 0$ and $\mathrm{d}(\partial \mathscr{L}/\partial \dot{x}_i)/\mathrm{d}t = 0$, formulated in terms of Hamilton's and Lagrange's ray equations, respectively, result from the homogeneity of the continuum along the x_i-axis. These are different formulations of the same conserved quantity. Fundamentally, this quantity results from Noether's theorem, which relates the conserved quantities to the symmetries.

The fact that the same ray parameter can be obtained from both Hamilton's ray equations and Lagrange's ray equations allows us to use both phase velocities, v, and ray velocities, V, as well as phase angles, ϑ, and ray angles, θ, to study rayfields in the context of conserved quantities. For instance, considering anisotropic vertically inhomogeneous continua, we can write

$$\mathfrak{p} = \frac{\sin\vartheta}{v(z,\vartheta)} = \cos\theta \frac{\partial}{\partial\theta}\left[\frac{1}{V(z,\theta)}\right] + \frac{\sin\theta}{V(z,\theta)},$$

where the relation between the magnitudes of phase and ray velocities is given by expressions (8.4.9), while the relation between the phase and ray angles is given by expression (8.4.12). An example illustrating this equivalence is shown in Exercise 10.1.

Closing Remarks

In this chapter, we used ray parameters, which are first integrals of ray equations, to obtain expressions for rays and traveltimes. In a general inhomogeneous continuum, there are no ray parameters since the integrand of the traveltime integral is an explicit function of all the coordinates. In other words, the inhomogeneity of the continuum does not possess any convenient symmetry that would allow us to formulate expressions for conserved quantities. In such cases, we can still solve Hamilton's or Lagrange's ray equations to obtain rays, even though these equations may be analytically and numerically involved.

14.7. Exercises

EXERCISE 14.1. Using polar coordinates, formulate the conserved quantity for radially inhomogeneous continua, where the traveltime integral is given by

$$\int_a^b \frac{\sqrt{r^2+(r')^2}}{V(r)}\,\mathrm{d}\xi\,.$$

Explain the physical context of the conserved quantity.

SOLUTION 14.1. Consider Beltrami's identity given by expression

$$\frac{\partial \mathrm{F}}{\partial \xi}+\frac{\mathrm{d}}{\mathrm{d}\xi}\left(r'\frac{\partial \mathrm{F}}{\partial r'}-\mathrm{F}\right)=0\,,$$

where F denotes the integrand of the traveltime integral and $r' := \mathrm{d}r/\,\mathrm{d}\xi$. Due to the explicit absence of the latitude angle, ξ, we obtain

$$r'\frac{\partial \mathrm{F}}{\partial r'}-\mathrm{F}=C\,,$$

where C is a constant. Thus, performing the partial differentiation, we obtain

$$C=-\frac{r^2}{V(r)\sqrt{r^2+(r')^2}}\,,$$

which is the expression for the conserved quantity. The conserved quantity results from the traveltime integral's invariance to the latitude angle. In other words, the velocity field consists of concentric circles.

REMARK 14.7.1. Noticing that

$$\frac{r}{\sqrt{r^2+(r')^2}}=\sin\theta\,,$$

where θ is the ray angle measured between the ray and the radial direction, we can write

$$C=-\frac{r\sin\theta}{V(r)}\,, \tag{14.7.1}$$

which is a standard form of the ray parameter for radially inhomogeneous continua.[7] Note that ray parameter (14.7.1) has different units than ray parameter (14.2.5).

EXERCISE 14.2. Given expression (14.3.5), examine how the value of the anisotropy parameter χ affects the maximum depth of a ray.

SOLUTION 14.2. Solving equation (14.3.5) for z, we obtain

$$z=\frac{1}{\mathfrak{p}b}\left(\frac{\sin\theta}{\sqrt{1+2\chi}\sqrt{1+2\chi\cos^2\theta}}-\mathfrak{p}a\right).$$

The maximum depth is given by setting $\theta=\pi/2$. Thus, we get

$$z_{\max}=\frac{1}{\mathfrak{p}b}\left(\frac{1}{\sqrt{1+2\chi}}-\mathfrak{p}a\right). \tag{14.7.2}$$

To state expression (14.7.2) in terms of the take-off ray angle, we set $z=0$ in expression (14.3.5), and denote the corresponding θ as θ_0. Hence, expression (14.3.5) becomes

$$\mathfrak{p}=\frac{\sin\theta_0}{a\sqrt{1+2\chi}\sqrt{1+2\chi\cos^2\theta_0}}. \tag{14.7.3}$$

Inserting expression (14.7.3) into expression (14.7.2), we obtain

$$z_{\max}=\frac{a}{b}\left(\frac{\sqrt{1+2\chi\cos^2\theta_0}}{\sin\theta_0}-1\right),$$

[7] Readers interested in traveltime expressions for rays whose ray parameter is given by expression (14.7.1) might refer to Kennett, B.L.N., (2001) The seismic wavefield, Vol. I: Introduction and theoretical development: Cambridge University Press, pp. 171–174.

which gives the maximum depth for a given take-off angle. This expression shows that for $\chi \in (-0.5, 0)$, the maximum depth reached is less than that for the isotropic case, $\chi = 0$. Conversely, for $\chi \in (0, \infty)$, the maximum depth reached is greater than that for the isotropic case, $\chi = 0$. In other words, negative values of parameter χ increase the curvature of the ray while positive values decrease it.

REMARK 14.7.2. In most seismological studies of sedimentary layers, χ is positive. Hence, as shown in Exercise 14.2, the presence of anisotropy in a vertically inhomogeneous medium tends to straighten the rays and, hence, increase the maximum depth they reach.

EXERCISE 14.3. Using expression (14.3.9) show that in homogeneous continua rays are straight lines.

SOLUTION 14.3. Consider equation (14.3.9); namely,

$$x(z; a, b) = \frac{\sqrt{1 - \mathfrak{p}^2 a^2 (1 + 2\chi)} - \sqrt{1 - \mathfrak{p}^2 (a + bz)^2 (1 + 2\chi)}}{\mathfrak{p} b}.$$

For a homogeneous continuum, $b = 0$; hence, we can write

$$\lim_{b \to 0} x(z; a, b) = \lim_{b \to 0} \frac{\sqrt{1 - \mathfrak{p}^2 a^2 (1 + 2\chi)} - \sqrt{1 - \mathfrak{p}^2 (a + bz)^2 (1 + 2\chi)}}{\mathfrak{p} b}.$$

Since both the numerator and the denominator vanish, we invoke French mathematician Guillaume de l'Hôpital's rule to write

$$\lim_{b \to 0} x(z; a, b) = \lim_{b \to 0} \frac{\dfrac{\partial \left[\sqrt{1 - \mathfrak{p}^2 a^2 (1 + 2\chi)} - \sqrt{1 - \mathfrak{p}^2 (a + bz)^2 (1 + 2\chi)}\right]}{\partial b}}{\dfrac{\partial (\mathfrak{p} b)}{\partial b}}.$$

Performing the differentiation, we get

$$\lim_{b \to 0} x(z; a, b) = \lim_{b \to 0} \frac{\mathfrak{p} (a + bz)(1 + 2\chi) z}{\sqrt{1 - \mathfrak{p}^2 (a + bz)^2 (1 + 2\chi)}}.$$

Taking the limit, we obtain

$$\lim_{b\to 0} x(z;a,b) = \frac{\mathfrak{p}a(1+2\chi)}{\sqrt{1-\mathfrak{p}^2a^2(1+2\chi)}}z.$$

Since the term given by the fraction is constant and there is no additional term, this is the equation of a straight line going through the origin, $(x,z) = (0,0)$.

To interpret this result, let us rewrite the above expression as

$$z(x) = \frac{\sqrt{1-\mathfrak{p}^2a^2(1+2\chi)}}{\mathfrak{p}a(1+2\chi)}x,$$

which describes the ray in an elliptically anisotropic homogeneous continuum. Let us also recall expression (14.3.5) and consider it for $b = 0$. Thus we get

$$\mathfrak{p} = \frac{\sin\theta}{a\sqrt{1+2\chi}\sqrt{1+2\chi\cos^2\theta}}.$$

Combining these two expressions, we get

$$z(x) = \frac{\sqrt{1-\left(\dfrac{\sin\theta}{\sqrt{1+2\chi\cos^2\theta}}\right)^2}}{\dfrac{\sin\theta\sqrt{1+2\chi}}{\sqrt{1+2\chi\cos^2\theta}}}x.$$

Simplifying, we obtain

$$z(x) = \cot\theta\, x,$$

where $\cot\theta$ is the slope of the line. This is true for both isotropic homogeneous and anisotropic homogeneous cases; χ does not appear in the final result.

EXERCISE 14.4. Show that the parametric form of the ray parameter, given by $\mathfrak{p} = \partial\mathscr{F}/\partial\dot{x}$ and discussed in Section 14.5, is equivalent to ray parameter (14.2.5).

SOLUTION 14.4. Using the first equation of system (14.5.2) and considering the case where x is not explicitly present in the integrand, $\mathscr{F}$, we obtain the conserved quantity given by

$$\mathfrak{p} = \frac{\partial \mathscr{F}}{\partial \dot{x}}. \tag{14.7.4}$$

In view of $\mathscr{F}$ given in expression (14.5.1), we obtain

$$\mathfrak{p} = \frac{1}{V}\frac{\dot{x}}{\sqrt{\dot{x}^2+\dot{z}^2}} + \sqrt{\dot{x}^2+\dot{z}^2}\frac{\partial}{\partial \dot{x}}\left(\frac{1}{V}\right). \tag{14.7.5}$$

To state expression (14.7.5) in terms of the ray angle, θ, we can write the differential operator as

$$\frac{\partial}{\partial \dot{x}} = \frac{\partial \theta}{\partial \dot{x}}\frac{\partial}{\partial \theta} + \frac{\partial \dot{z}}{\partial \dot{x}}\frac{\partial}{\partial \dot{z}} = \frac{\partial \theta}{\partial \dot{x}}\frac{\partial}{\partial \theta} = \frac{1}{\frac{\partial \dot{x}}{\partial \theta}}\frac{\partial}{\partial \theta}.$$

Since $\dot{x} = \dot{z}\tan\theta$, we obtain

$$\frac{\partial}{\partial \dot{x}} = \frac{1}{\dot{z}\frac{1}{\cos^2\theta}}\frac{\partial}{\partial \theta} = \frac{\cos^2\theta}{\dot{z}}\frac{\partial}{\partial \theta}.$$

Thus, returning to expression (14.7.5), we can write

$$\begin{aligned}
\mathfrak{p} &= \frac{\sin\theta}{V} + \sqrt{\dot{x}^2+\dot{z}^2}\frac{\cos^2\theta}{\dot{z}}\frac{\partial}{\partial \theta}\left(\frac{1}{V}\right) \\
&= \frac{\sin\theta}{V} + \sqrt{\tan^2\theta+1}\cos^2\theta\frac{\partial}{\partial \theta}\left(\frac{1}{V}\right) \\
&= \frac{\sin\theta}{V} + \cos\theta\frac{\partial}{\partial \theta}\left(\frac{1}{V}\right),
\end{aligned}$$

which is identical to expression (14.2.5), as required.

EXERCISE 14.5. Integrate traveltime expression (14.3.14), namely,

$$\mathfrak{p}\int_0^X \frac{\mathrm{d}x}{1-\left(\mathfrak{p}bx-\sqrt{1-\mathfrak{p}^2a^2(1+2\chi)}\right)^2}, \tag{14.7.6}$$

using the fact that $\mathfrak{p}$ is constant.

SOLUTION 14.5. Let us make the following substitution.

$$\zeta := \mathfrak{p}bx - \sqrt{1-\mathfrak{p}^2a^2\left(1+2\chi\right)}.$$

Thus, we rewrite integral (14.7.6) as

$$\frac{1}{b}\int\limits_{-\sqrt{1-\mathfrak{p}^2a^2(1+2\chi)}}^{\mathfrak{p}bX-\sqrt{1-\mathfrak{p}^2a^2(1+2\chi)}}\frac{\mathrm{d}\xi}{1-\zeta^2}.$$

Using partial fractions, we write

$$\frac{1}{2b}\int\limits_{-\sqrt{1-\mathfrak{p}^2a^2(1+2\chi)}}^{\mathfrak{p}bX-\sqrt{1-\mathfrak{p}^2a^2(1+2\chi)}}\left(\frac{1}{1-\zeta}+\frac{1}{1+\zeta}\right)\mathrm{d}\zeta.$$

Integrating, we get

$$\frac{1}{2b}\ln\frac{1+\zeta}{1-\zeta}\Bigg|_{-\sqrt{1-\mathfrak{p}^2a^2(1+2\chi)}}^{\mathfrak{p}bX-\sqrt{1-\mathfrak{p}^2a^2(1+2\chi)}}.$$

Evaluating, we obtain

$$\frac{1}{2b}\left(\ln\frac{1-\sqrt{1-\mathfrak{p}^2a^2\left(1+2\chi\right)}+\mathfrak{p}bX}{1+\sqrt{1-\mathfrak{p}^2a^2\left(1+2\chi\right)}-\mathfrak{p}bX}-\ln\frac{1-\sqrt{1-\mathfrak{p}^2a^2\left(1+2\chi\right)}}{1+\sqrt{1-\mathfrak{p}^2a^2\left(1+2\chi\right)}}\right).$$

Using an identity for hyperbolic functions, namely,

$$\tanh^{-1}\zeta = \frac{1}{2}\ln\frac{1+\zeta}{1-\zeta},$$

we can also write

$$\frac{\tanh^{-1}\left[\mathfrak{p}bX-\sqrt{1-\mathfrak{p}^2a^2\left(1+2\chi\right)}\right]+\tanh^{-1}\sqrt{1-\mathfrak{p}^2a^2\left(1+2\chi\right)}}{b}.$$

EXERCISE 14.6. In a manner analogous to the one described in Section 14.3.3, derive the traveltime expression by integrating traveltime integral (14.2.1) along the z-axis. Discuss the validity of the resulting expression.

SOLUTION 14.6. In the xz-plane, the arclength element can be written as $\mathrm{d}s = \mathrm{d}z/\cos\theta$, where θ is the ray angle. Hence, traveltime integral (14.2.1) between the source at $(0,0)$ and the receiver at (X,Z) is

$$\check{C} = \int_0^Z \frac{\mathrm{d}z}{V(z,\theta)\cos\theta}, \tag{14.7.7}$$

where $V(z,\theta)$ is given by expression (14.3.4). Hence, we can explicitly write

$$\check{C} = \int_0^Z \frac{\mathrm{d}z}{(a+bz)\sqrt{\frac{1+2\chi}{1+2\chi\cos^2\theta}}\cos\theta}. \tag{14.7.8}$$

To integrate, we must express $\cos\theta$ in terms of constants a, b, χ, $\mathfrak{p}$, and integration variable z. Using expression (14.3.5) and trigonometric identities, we obtain

$$\cos\theta = \sqrt{\frac{1-\mathfrak{p}^2(a+bz)^2(1+2\chi)}{1+2\chi\mathfrak{p}^2(a+bz)^2(1+2\chi)}}. \tag{14.7.9}$$

Inserting expression (14.7.9) into integral (14.7.8), after algebraic manipulation, we obtain

$$\check{C} = \int_0^Z \frac{\mathrm{d}z}{(a+bz)\sqrt{1-\mathfrak{p}^2(a+bz)^2(1+2\chi)}}. \tag{14.7.10}$$

Integrating between $z=0$ and $z=Z$, while treating $\mathfrak{p}$ as a constant, we obtain the expression for the value of the traveltime; namely,

$$\check{C} = \frac{1}{b}\ln\left[\frac{a+bZ}{a}\frac{1+\sqrt{1-\mathfrak{p}^2a^2(1+2\chi)}}{1+\sqrt{1-\mathfrak{p}^2(a+bZ)^2(1+2\chi)}}\right], \tag{14.7.11}$$

with $\mathfrak{p}$ given by expression (14.3.17).

REMARK 14.7.3. Expression (14.7.11) is valid for the downward rays only, which could be illustrated by plotting this expression versus x for a set value of Z; such a plot would exhibit a cusp at x that corresponds to the case of the signal reaching the receiver, (X,Z), at its deepest point—just prior to its beginning an upward path. Between the source and the cusp, the receiver is reached by a downgoing signal travelling along an elliptical trajectory. Beyond the cusp, the receiver is reached by an upgoing signal travelling along this trajectory. The cusp in such a plot is the result of the traveltime beyond the cusp being calculated with the opposite sign than the traveltime before the cusp, due to the change of the direction of integration that is set by $\mathrm{d}z$.

EXERCISE 14.7. In view of Remark 14.7.3 and considering a source at $(0,0)$, derive an expression for the curve in the xz-plane that separates the receiver locations, (X,Z), reached by the downgoing signal from those reached by the upgoing one.

SOLUTION 14.7. At a given depth (set value of Z), the receivers at different horizontal locations (varying values of X) are reached either by a downgoing or upgoing signal. The receiver at $(0,Z)$, which is directly below the source, is reached by the downgoing signal whose takeoff ray angle is $\theta=0$. As X grows, so does θ, until the takeoff ray angle corresponds to the ray that horizontally grazes the receiver at (X,Z); the takeoff angle is at its maximum. From that point on, the receiver is reached by the upgoing signal travelling along an elliptical arc. Consequently, the ray takeoff angle begins to decrease, and the receiver at an infinite horizontal distance would be reached by a ray whose takeoff angle is nearly zero—the ray is nearly vertical at takeoff. In view of expression (8.5.12), we see that the increase or decrease in the ray angle corresponds to the increase or decrease of the phase angle, since $\chi>-1/2$, as we can deduce from definition (8.5.1). Also, in view of expression (8.5.13), we see that the increase or decrease of the phase takeoff angle corresponds to the increase or decrease of ray parameter. Hence, to find point x at depth Z at which the ray angle reaches its largest value, let us write expression (14.3.17) as

$$p(x;Z)=\frac{2x}{\sqrt{\left[x^2+(1+2\chi)Z^2\right]\left[(2a+bZ)^2(1+2\chi)+b^2x^2\right]}},$$

and consider its derivative, namely,

$$\frac{\mathrm{d}p}{\mathrm{d}x}=\frac{8a^2Z^2(1+2\chi)+8abZ^3(1+2\chi)+2b^2\left(Z^4(1+2\chi)^2-x^4\right)}{\left(\left(x^2+(1+2\chi)^2Z^2\right)\left(b^2x^2+(1+2\chi)(2a+bZ)^2\right)\right)^{\frac{3}{2}}}.$$

To find x that corresponds to the maximum of the takeoff angle, we set the numerator to zero and proceed to rearrange the resulting equation using the fact that a, b, χ, Z are real and a, b, Z are positive. Thus we obtain the required solution, namely,

$$x=\sqrt{\frac{1+2\chi}{b}(2a+bZ)Z}.$$

We can rewrite this solution as

$$\frac{x^2}{\left(\frac{a}{b}\right)^2(1+2\chi)}-\frac{\left(z+\frac{a}{b}\right)^2}{\left(\frac{a}{b}\right)^2}=1,$$

which is an expression for hyperbola. To find its asymptote, we set the right-hand side to zero, and solve for x to obtain

$$x=\sqrt{1+2\chi}\left(z+\frac{a}{b}\right).$$

EXERCISE 14.8. Consider the traveltime expression for the downgoing signal between the source at $(0,0)$ and the receiver at (X,Ξ) in an isotropic and linearly inhomogeneous continuum in the $x\zeta$-plane, namely,[8]

$$\check{C}=\frac{1}{b}\ln\left[\frac{\alpha+b\Xi}{\alpha}\frac{1+\sqrt{1-\mathfrak{p}^2\alpha^2}}{1+\sqrt{1-\mathfrak{p}^2(\alpha+b\Xi)^2}}\right], \tag{14.7.12}$$

where

$$\mathfrak{p}=\frac{2X}{\sqrt{(X^2+\Xi^2)\left[(2\alpha+b\Xi)^2+b^2X^2\right]}}. \tag{14.7.13}$$

[8] Readers interested in the derivation of expressions (14.7.12) and (14.7.13), might refer to Slotnick, M.M., (1959) Lessons in seismic computing: Society of Exploration Geophysicists, Lesson 37.

Using a transformation of coordinates, derive equations (14.7.11) and (14.3.17).

SOLUTION 14.8. Since equations (14.7.11) and (14.3.17) deal with an elliptical velocity dependence, consider the magnitudes of the horizontal and vertical velocities given by

$$v_x = a\sqrt{1+2\chi}\,, \tag{14.7.14}$$

with χ given by expression (14.3.2), and by

$$v_z = a\,, \tag{14.7.15}$$

respectively.

Note that, since v_x and v_z are the magnitudes of velocities along the symmetry axes, expressions (14.7.14) and (14.7.15) are the same for both phase and ray velocities.

Regardless of the inhomogeneity of the model, infinitesimal wavefronts resulting from any point source within the medium are ellipses with axes $(\mathrm{d}t)\,v_x$ and $(\mathrm{d}t)\,v_z$, where $\mathrm{d}t$ is the traveltime increment and v_x and v_z are the magnitudes of velocity at a given point. We can write such a wavefront as

$$\frac{(\mathrm{d}x)^2}{(\mathrm{d}t)^2\,v_x^2} + \frac{(\mathrm{d}z)^2}{(\mathrm{d}t)^2\,v_z^2} = 1\,,$$

which, using expressions (14.7.14) and (14.7.15), we can rewrite as

$$\frac{(\mathrm{d}x)^2}{1+2\chi} + (\mathrm{d}z)^2 = (\mathrm{d}t)^2\,a^2\,. \tag{14.7.16}$$

Since v_x and v_z are the magnitudes of velocities along the x-axis and the z-axis, respectively, we can scale the z-axis by a factor of $\sqrt{1+2\chi}$ to obtain circular wavefronts, which correspond to an isotropic case. In other words, we transform the xz-plane into the $x\zeta$-plane, where

$$\zeta = z\sqrt{1+2\chi}\,. \tag{14.7.17}$$

Thus, in view of expression (14.7.17), we let $z = \zeta/\sqrt{1+2\chi}$ in expression (14.7.16) to write it as

$$(\mathrm{d}x)^2 + (\mathrm{d}\zeta)^2 = (\mathrm{d}t)^2\,\alpha^2\,, \tag{14.7.18}$$

where

$$\alpha = a\sqrt{1+2\chi} \tag{14.7.19}$$

is the velocity in the $x\zeta$-plane. Equation (14.7.18) describes infinitesimal circular wavefronts in the $x\zeta$-plane.

Since equations (14.7.11) and (14.3.17) deal with the magnitude of the velocity that increases linearly, let us set $v(\zeta) = \alpha + b\zeta$. Note that the units of b are $[1/s]$. Consequently, its value does not depend on the scaling of position coordinates. Substituting expression (14.7.19) into expressions (14.7.12) and (14.7.13), as well as—in view of expression (14.7.17)—letting $\Xi = Z\sqrt{1+2\chi}$, we obtain expressions (14.7.11) and (14.3.17), as required.

EXERCISE 14.9. In view of Lemma 12.8.2, show that if $V(z,\theta) = A(z)$ $B(\theta)$, where $B(\theta) = 1/(1+C\cos\theta)$, then the ray in an anisotropic inhomogeneous continuum, $V(z,\theta)$ is the same as the ray in an isotropic inhomogeneous continuum, $A(z)$.

SOLUTION 14.9. To express $B(\theta)$, where θ is measured from the z-axis, in terms of z', we invoke trigonometric identity $\cos\theta = \cot\theta/\sqrt{1+\cot^2\theta}$. Noting that $\cot\theta = \mathrm{d}z/\mathrm{d}x := z'$, we obtain $\cos\theta = z'/\sqrt{1+(z')^2}$. Consequently,

$$B(z') = \frac{1}{1+C\dfrac{z'}{\sqrt{1+(z')^2}}},$$

and,

$$V(z,z') = A(z)B(z') = \frac{A(z)}{1+C\dfrac{z'}{\sqrt{1+(z')^2}}}. \tag{14.7.20}$$

Consider traveltime integral

$$\check{C} = \int_{x_1}^{x_2} \frac{\sqrt{1+(z')^2}}{V(z,z')}\mathrm{d}x = \int_{x_1}^{x_2} \frac{\sqrt{1+(z')^2}}{\dfrac{A(z)}{1+C\dfrac{z'}{\sqrt{1+(z')^2}}}}\mathrm{d}x.$$

Upon algebraic manipulations, we obtain

$$\check{C}=\int_{x_1}^{x_2}\frac{\sqrt{1+(z')^2}+Cz'}{A(z)}\,\mathrm{d}x\equiv\int_{x_1}^{x_2}\mathrm{F}\,\mathrm{d}x\,.$$

To find the ray, we invoke Euler's equation (12.2.2) to obtain

$$\frac{\partial \mathrm{F}}{\partial z}-\frac{\mathrm{d}}{\mathrm{d}x}\left(\frac{\partial \mathrm{F}}{\partial z'}\right)$$
$$=-\left(\sqrt{1+(z')^2}+Cz'\right)\frac{\frac{\partial A}{\partial z}}{A^2}-\frac{\mathrm{d}}{\mathrm{d}x}\left[\frac{1}{A}\left(\frac{z'}{\sqrt{1+(z')^2}}+C\right)\right]=0\,. \tag{14.7.21}$$

Considering only factors that contain C, we obtain

$$-Cz'\frac{\frac{\partial A}{\partial z}}{A^2}+Cz'\frac{\frac{\partial A}{\partial z}}{A^2}=0\,.$$

Thus, Euler's equation is independent of C. In view of expression (14.7.20), the ray resulting from equation (14.7.21) is the same for both $V(z,z')$ and $A(z)$.

Part 4

Appendices

Introduction to Part 4

> Physics is the science upon which all other sciences rest, since it attempts to explain the nature of the universe of things. [...] Mathematics is a language, which enables us to express certain kinds of ideas (e.g., order, patterns) much more precisely than whatever everyday language we speak. [...] Mathematical physics can therefore be regarded as the 'dialect' of mathematics spoken by physicists when they wish to express and use the 'laws' or theories of physics clearly and unambiguously.
>
> *Michael Grant Rochester (1997)*
> *Lecture notes on mathematical physics*

In the presentation of this book, we assume that the reader is familiar with several mathematical subjects typically taught in undergraduate studies in the faculty of science. These subjects consist of linear algebra, differential and integral calculus, vector and tensor calculus, as well as ordinary and partial differential equations. Another subject that plays an important role in this book—but is not commonly included in an undergraduate curriculum—consists of the calculus of variations. Chapter 12 is devoted to the aspects of this subject that are pertinent to this book.

In *Part 4*, we describe two additional mathematical concepts that are used in the book and with which the reader might not be familiar; namely, Euler's homogeneous-function theorem and Legendre's transformations. Notably, in the context of this book, the applications of these two concepts are often associated with one another. In view of Euler's theorem, different degrees of homogeneity exhibited by several functions formulated in this book give us insight into their physical meanings and allow us to manipulate them. Legendre's transformation is the tool that allows us to transform Hamilton's ray equations into Lagrange's ray equations. Consequently, this transformation links the concepts discussed in *Part 2* with those discussed in *Part 3*.

Throughout the book, the meaning of a given symbol used in an equation is stated in the proximity of the pertinent equation to avoid ambiguity among several meanings that can be associated with the same symbol. To facilitate clarity, certain symbols are uniquely associated with a particular mathematical or physical meaning. These symbols, together with their meanings, are listed in *List of symbols*.

APPENDIX A

Euler's Homogeneous-Function Theorem

> Mathematicians can pursue many conflicting directions to derive new results. In the absence of internal criteria that favour or justify one direction rather than another, a choice must be based on external considerations. Of these, certainly the most important is the traditional and still most justifiable reason for the creation and development of mathematics, its value to the sciences.
>
> *Morris Kline (1980)*
> *Mathematics: The loss of certainty*

Preliminary Remarks

In this book, seismological quantities are expressed in terms of mathematical entities. In accordance with physical principles, we require that these entities possess certain mathematical properties. Using these properties, we can study these mathematical formulations to obtain further insight into their physical meaning. The homogeneity of a function and Euler's homogeneous-function theorem are of particular use in our work.

We begin this appendix by stating the definition of a homogeneous function. Then, we state and prove Euler's homogeneous-function theorem.

A.1. Homogeneous Functions

Several functions that play an important role in this book are homogeneous. Notably, the Hamiltonian, stated in expression (8.2.8); namely,

$$\mathscr{H}(\mathbf{x},\mathbf{p}) = \frac{1}{2}p^2 v^2(\mathbf{x},\mathbf{p}), \tag{A.1.1}$$

where $p^2 \equiv \mathbf{p}\cdot\mathbf{p}$, is homogeneous of degree 2 in $\mathbf{p}$. To see this property, consider Definition A.1.1.

DEFINITION A.1.1. A real function $f(x_1,\ldots,x_n)$ is homogeneous of degree r in the variables $x_1,\ldots,x_n$ if

$$f(cx_1,\ldots,cx_n) = c^r f(x_1,\ldots,x_n),$$

for every real number c. If $f(cx_i) = |c|^r f(x_i)$, where $i\in\{1,\ldots,n\}$, we say that f is absolute-value homogeneous of degree r in the x_i.

REMARK A.1.2. Both terms "degree" and "order" are commonly used to describe the homogeneity of a function. In this book, we use the former term since it refers to the value of the exponent and, hence, is consistent with other uses of this term, such as "degree of a polynomial".

Now, consider the fact that v is the phase-velocity function that depends on position $\mathbf{x}$ and direction, which is given by the vector normal to the wavefront; namely, $\mathbf{p}$. Since the orientation of the wavefront, indicated by $\mathbf{p}$, does not depend on the magnitude of $\mathbf{p}$, we can rewrite expression (A.1.1) as

$$\mathscr{H}(\mathbf{x},\mathbf{p}) = \frac{1}{2}p^2 v^2\left(\mathbf{x},\frac{\mathbf{p}}{|\mathbf{p}|}\right), \tag{A.1.2}$$

where $\mathbf{p}/|\mathbf{p}|$ is a unit vector normal to the wavefront. Hence, we see that v is homogeneous of degree 0 in $\mathbf{p}$. In other words, we can multiply $\mathbf{p}$ by any number and v remains unchanged. In view of Definition A.1.1, we can write

$$v\left(\mathbf{x},\frac{c\mathbf{p}}{|c\mathbf{p}|}\right) = c^0 v\left(\mathbf{x},\frac{\mathbf{p}}{|\mathbf{p}|}\right) = v\left(\mathbf{x},\frac{\mathbf{p}}{|\mathbf{p}|}\right).$$

This immediately implies that function (A.1.2) is homogeneous of degree 2 in $\mathbf{p}$, since

$$\begin{aligned}\mathscr{H}(\mathbf{x},c\mathbf{p}) &= \frac{1}{2}[(c\mathbf{p})\cdot(c\mathbf{p})]\,v^2\left(\mathbf{x},\frac{c\mathbf{p}}{|c\mathbf{p}|}\right) = \frac{1}{2}c^2p^2v^2\left(\mathbf{x},\frac{c\mathbf{p}}{|c\mathbf{p}|}\right)\\ &= \frac{c^2}{2}p^2v^2\left(\mathbf{x},\frac{\mathbf{p}}{|\mathbf{p}|}\right) = c^2\mathscr{H}(\mathbf{x},\mathbf{p})\,. \qquad \text{(A.1.3)}\end{aligned}$$

We can also illustrate Definition A.1.1 by the following straightforward example.

EXAMPLE A.1.3. Consider function

$$f(x_1,x_2,x_3) = x_1x_2x_3 + x_1x_2^2 + x_2x_3^2\,. \qquad \text{(A.1.4)}$$

Let

$$\begin{aligned}f(cx_1,cx_2,cx_3) &= cx_1cx_2cx_3 + cx_1(cx_2)^2 + cx_2(cx_3)^2\\ &= c^3\left(x_1x_2x_3 + x_1x_2^2 + x_2x_3^2\right)\\ &= c^3 f(x_1,x_2,x_3)\,.\end{aligned}$$

Thus, in view of Definition A.1.1, f is homogeneous of degree 3 in the x_i.

Homogeneity of a function allows us to use Euler's homogeneous-function theorem, stated in Theorem A.2.1. This theorem plays an important role in the formulations described in this book. It allows us to simplify numerous expressions and gain insight into their physical meaning.

A.2. Homogeneous-Function Theorem

[1]Euler's homogeneous-function theorem can be stated in the following way.

THEOREM A.2.1. *If function $f(x_1,\ldots,x_n)$ is homogeneous of degree r in $x_1,\ldots,x_n$, then*

$$\sum_{i=1}^{n}\frac{\partial f}{\partial x_i}(x_1,\ldots,x_n)\,x_i = rf(x_1,\ldots,x_n)\,. \qquad \text{(A.2.1)}$$

[1] Interested readers might refer to Olmsted, J.M.H., (1961) Advanced calculus: Prentice-Hall, Inc., p. 272.

PROOF. In view of Definition A.1.1, we can write

$$f(cx_1, \ldots, cx_n) = c^r f(x_1, \ldots, x_n) \,. \tag{A.2.2}$$

Differentiating both sides of equation (A.2.2) with respect to c, we obtain

$$\sum_{i=1}^{n} f_i(cx_1, \ldots, cx_n) \frac{\partial (cx_i)}{\partial c} = rc^{r-1} f(x_1, \ldots, x_n) \,, \tag{A.2.3}$$

where f_i denotes the derivative of function f with respect to its ith argument. To obtain the expression stated in Theorem A.2.1, we consider a particular case where $c = 1$. Letting $c = 1$, we can rewrite equation (A.2.3) as

$$\sum_{i=1}^{n} \frac{\partial f}{\partial x_i}(x_1, \ldots, x_n)\, x_i = rf(x_1, \ldots, x_n) \,, \tag{A.2.4}$$

which is equation (A.2.1), as required. □

To illustrate Theorem A.2.1, we can study function (A.1.4), as shown in the following example.

EXAMPLE A.2.2. Using function (A.1.4), namely,

$$f(x_1, x_2, x_3) = x_1 x_2 x_3 + x_1 x_2^2 + x_2 x_3^2 \,,$$

we can write the left-hand side of equation (A.2.1) as

$$\begin{aligned} \sum_{i=1}^{3} \frac{\partial f}{\partial x_i} x_i &= \left(x_2 x_3 + x_2^2\right) x_1 + \left(x_1 x_3 + 2x_1 x_2 + x_3^2\right) x_2 + \left(x_1 x_2 + 2x_2 x_3\right) x_3 \\ &= 3\left(x_1 x_2 x_3 + x_1 x_2^2 + x_2 x_3^2\right) \\ &= 3f(x_1, x_2, x_3) \,. \end{aligned} \tag{A.2.5}$$

Expression (A.2.5) is the right-hand side of equation (A.2.1) for a function that is homogeneous of degree 3 in the x_i, as expected from Theorem A.2.1.

Equation (A.2.1) is often invoked in this book. For instance, in the proof of Lemma 13.1.5—knowing that $\mathscr{H}$ is homogeneous of degree 2 in $\mathbf{p}$, as shown in expression (A.1.3)—we can write

$$\sum_{i=1}^{n} \frac{\partial \mathscr{H}}{\partial p_i} p_i = 2\mathscr{H} \,,$$

which allows us to complete that proof.

The following example illustrates equation (A.2.1) in the context of physics.

EXAMPLE A.2.3. Following the standard classical-mechanics formulation, let the kinetic energy be

$$T(v) = \frac{1}{2} m v^2 ,$$

where m and v denote mass and velocity, respectively. In view of Definition A.1.1, T is homogeneous of degree 2 in v since

$$T(cv) = \frac{1}{2} m (cv)^2 = \frac{c^2}{2} m v^2 = c^2 T(v) ,$$

where c denotes a constant. Thus, following Theorem A.2.1, we can write

$$\frac{\partial T}{\partial v} v = 2T .$$

We can directly verify this result; namely,

$$\frac{\partial T}{\partial v} v = \left[\frac{\partial}{\partial v} \left(\frac{m v^2}{2} \right) \right] v = m v^2 = 2T .$$

Closing Remarks

Note that a multivariable function can be homogeneous in a particular set of variables. In this book, $\mathscr{H}(\mathbf{x}, \mathbf{p})$, given in expression (8.2.8), is homogeneous of degree 2 in $\mathbf{p}$. $\mathscr{L}(\mathbf{x}, \dot{\mathbf{x}})$, given in expression (11.1.1), is homogeneous of degree 2 in $\dot{\mathbf{x}}$. $\mathscr{F}(\mathbf{x}, \dot{\mathbf{x}})$, given in expression (13.1.8), is absolute-value homogeneous of degree 1 in $\dot{\mathbf{x}}$. None of these functions is homogeneous in $\mathbf{x}$. The properties of homogeneity of these functions allow us to prove Theorem 13.1.2, which is the statement of Fermat's principle.

Certain functions used in our studies exhibit no homogeneity. For instance, traveltime integrand $\mathrm{F}(z, z')$, given in expression (14.2.1), is not homogeneous in either variable.

Euler's homogeneous-function theorem is explicitly used in Chapters 8, 11 and 13.

APPENDIX B

Legendre's Transformation

> To penetrate into symplectic geometry while bypassing the long historical route, it is simplest to use the axiomatic method, which has, as Bertrand Russell observed, many advantages, similar to the advantages of stealing over honest work.
>
> *Vladimir Igorevitch Arnold (1992)*
> *Catastrophe theory*

Preliminary Remarks

Legendre's transformation is a transformation in which we replace a function by a new function that depends on partial derivatives of the original function with respect to original independent variables. In the context of this book, we replace the ray-theory Hamiltonian, $\mathscr{H}(\mathbf{x},\mathbf{p})$, by the ray-theory Lagrangian, $\mathscr{L}(\mathbf{x},\dot{\mathbf{x}})$, which depends on the $\dot{x}_i = \partial\mathscr{H}/\partial p_i$, where $i \in \{1,2,3\}$.

We begin this appendix with the derivation of Legendre's transformation in a geometrical context, where we consider functions of single variables. Then we proceed to multivariable functions and formulate Legendre's transformation between $\mathscr{H}(\mathbf{x},\mathbf{p})$ and $\mathscr{L}(\mathbf{x},\dot{\mathbf{x}})$. We conclude by using Legendre's transformations of these functions to derive the corresponding ray equations.

B.1. Geometrical Context

B.1.1. Surface and its tangent planes[1]

Legendre's transformation can be illustrated in a geometrical context. Let an n-dimensional surface in the $(n+1)$-dimensional space be given by equation

$$y = f(x_1, \ldots, x_n) . \tag{B.1.1}$$

Consider the set of all possible n-dimensional planes that are tangent to this surface. The envelope of these planes is the original surface. We wish to derive the equation that describes these tangent planes.

A general form of the equation of an n-dimensional plane is $y = u_1 x_1 + \ldots + u_n x_n - v$, where $u_1, \ldots, u_n$ and v are real numbers that define the plane uniquely and, hence, can be viewed as coordinates of the plane. The transformation from the equation of a surface, given by equation (B.1.1), to the equation that describes all of its tangent planes, given by

$$v = g(u_1, \ldots, u_n) ,$$

is Legendre's transformation.

In general, Legendre's transformation is possible if the surface is differentiable and if there are no tangent planes to this surface that are parallel to each other. Otherwise, for the same values of $u_1, \ldots, u_n$, we have different values of v. In other words, v is not a single-value function of $u_1, \ldots, u_n$. In certain cases, however, the issue of nonuniqueness resulting from a multivalue function can be overcome, as discussed in Example B.2.2.

B.1.2. Single-variable case

To illustrate the geometrical context, consider a smooth curve in the xy-plane. We can describe this curve as a set of points in the plane, where the y-coordinate is determined by the function of one variable; namely, $y = f(x)$. Also, this curve can be regarded as the envelope of its tangent lines. We wish to derive equation $v = g(u)$ that describes all the lines $y = ux - v$, in the xy-plane, that are tangent to the original curve.

[1] Readers interested in the geometrical motivation of the Legendre transformation might refer to Courant, R., and Hilbert, D., (1924/1989) Methods of mathematical physics: John Wiley & Sons., Vol. II, pp. 32–39.

The line $y = ux - v$ is tangent to the curve $y = f(x)$, at point x, if and only if the line passes through the point $(x, f(x))$ and has the same slope as the curve at this point. In other words,

$$v = ux - f(x), \tag{B.1.2}$$

and

$$u = \frac{\mathrm{d}f}{\mathrm{d}x}, \tag{B.1.3}$$

respectively.

To derive function $g(u)$, we would like to express x in terms of u. In general, this is not possible if we cannot solve uniquely equation (B.1.3) for x. Thus, our ability to express x in terms of u depends on the form of function $f(x)$.

Assuming that we can obtain $x = x(u)$, the set of all tangent lines is described by $v = g(u)$, where

$$g(u) = ux(u) - f(x(u)). \tag{B.1.4}$$

Function $g(u)$ is Legendre's transformation of $f(x)$, where u and x are the transformation variables related by equation (B.1.3). This construction is illustrated by the following example.

EXAMPLE B.1.1. Let $f(x) = x^2$. Following equation (B.1.3), we can write

$$u = \frac{\mathrm{d}f}{\mathrm{d}x} = 2x. \tag{B.1.5}$$

Solving for x, we obtain

$$x = \frac{u}{2}.$$

In accordance with equation (B.1.4), we write

$$g(u) = ux(u) - f(x(u)) = \frac{u^2}{2} - \left(\frac{u}{2}\right)^2 = \frac{u^2}{4}.$$

Therefore, $g(u) = u^2/4$ is Legendre's transformation of $f(x) = x^2$, where u and x are the transformation variables related by equation (B.1.5).

We can also view Legendre's transformation in a different way. Consider a curve $y = f(x)$ and a straight line $y = ux$, where u is a real number. For a given x-coordinate, we can view $h(x) = ux - f(x)$ as the distance between a point on the curve and a point on the straight line. We wish to find point $x(u)$ that maximizes that distance. Therefore, we set

$$\frac{\mathrm{d}h}{\mathrm{d}x} = u - \frac{\mathrm{d}f}{\mathrm{d}x} = 0,$$

which gives

$$u = \frac{\mathrm{d}f}{\mathrm{d}x}.$$

If we can solve this equation, which is equation (B.1.3), for x, namely, $x = x(u)$, then $g(u) = h(x(u))$ is Legendre's transformation of $f(x)$.

REMARK B.1.2. In Example B.1.1, the slope of a tangent to $f(x) = x^2$ is $y = 2x$. Since the relation between the function and its slope is one-to-one for all $x \in \mathbb{R}$, Legendre's transformation of x^2 is a single-value function and, hence, possesses an inverse, as discussed in Appendix B.2. Similarly, for $f(x) = x^4$ and its tangent, $y = 4x^3$, the relation is one-to-one and, again, Legendre's transformation of x^4 possesses an inverse. For $f(x) = x^3$, however, the slope of the tangent is $y = 3x^2$, which is not a one-to-one relation between the function and the slope of its tangent line. This function possesses Legendre's transformation, but, as discussed in Example (B.2.1), the result of this transformation has to be transformed piecewise to obtain the inverse.

In general, the invertibility of Legendre's transformation requires a one-to-one relation between a curve and its tangent line or between a surface and its tangent plane, as stated, for a closed surface, by condition (13.1.13) on page 515, which therein is tantamount to limiting the presented proof to surfaces without inflection points. Herein, to discuss issues of invertibility of Legendre's transformation, we invoke a necessary condition for the existence of an inflection point, x_0, which is $f''(x)|_{x=x_0} = 0$, where $'$ denotes a derivative, and a sufficient condition, which requires $f''(x)$ to have opposite signs in the neighborhood of x_0.

Neither condition is satisfied for x^2, since its second derivative is equal to 2. Hence, Legendre's transformation of x^2 is invertible for all $x \in \mathbb{R}$, as expected. The necessary condition, on the other hand, is satisfied, at $x_0 = 0$, by both x^3 and x^4. The sufficient condition, however, is satisfied only by the former. Thus, Legendre's transformation of x^4 is invertible for all $x \in \mathbb{R}$, as expected, but x^3 is only piecewise invertible.

Let us examine $f(x) = x^3$. Solving equation (B.1.3), we obtain $x = \pm\sqrt{u/3}$. Using equation (B.1.4), we write Legendre's transformation

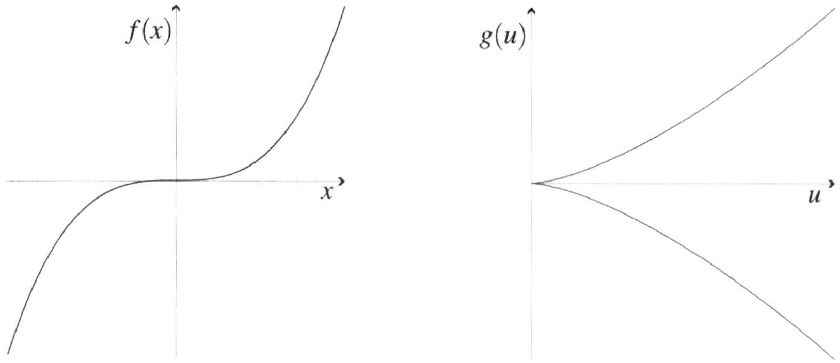

Fig. B.1.1 Function $f(x) = x^3$ and its Legendre's transformation, given in expressions (B.1.7) and (B.1.8); the upper branch corresponds to the transformation for $x \geqslant 0$ and the lower branch for $x < 0$.

of $f(x) = x^3$ as

$$g(u) = \pm u\sqrt{\frac{u}{3}} - \left(\pm\sqrt{\frac{u}{3}}\right)^3 ; \tag{B.1.6}$$

this results in

$$g(u) = -\frac{2}{3\sqrt{3}}u^{\frac{3}{2}}, \qquad u \geqslant 0, \tag{B.1.7}$$

which corresponds to the negative square root in expression (B.1.6), $x < 0$, and

$$g(u) = \frac{2}{3\sqrt{3}}u^{\frac{3}{2}}, \qquad u \geqslant 0, \tag{B.1.8}$$

which corresponds to the positive square root in expression (B.1.6), $x > 0$.

In contrast to Example B.1.1—where Legendre's transformation of x^2 is $u^2/4$—Legendre's transformation of x^3, illustrated in Figure B.1.1, results in two values of $g(u)$ corresponding to each value of $u \geqslant 0$. This ambiguity requires a piecewise Legendre's transformation of $g(u)$ to obtain $f(x)$.

An analogous case appears in wave propagation in anisotropic Hookean solids, where the inflection points on a slowness surface result in three values of ray velocity corresponding to a single value of the ray angle. In such a case, in a manner analogous to duplication in Figure B.1.1, there is a triplication of a wavefront.[2] Also, in a manner analogous to Figure B.1.1, the slowness surfaces are smooth and—if they have inflection points—the corresponding wavefronts, which are their Legendre's transformations, have cusps.

B.2. Duality of Transformation

Legendre's transformation is referred to as a dual transformation since if transformation of f leads to g, then, under certain restrictions, transformation of g leads to f.[3] We can illustrate this property by inverting the transformation shown in Example B.1.1. Also, we can illustrate certain restrictions by inverting the transformation of $f(x) = x^3$, discussed in Remark B.1.2.

EXAMPLE B.2.1. Let $g(u) = u^2/4$. In view of equation (B.1.4), consider a new function given by

$$\begin{aligned} f(x) &= ux - g(u) \\ &= ux - \frac{u^2}{4}, \end{aligned} \tag{B.2.1}$$

where the new independent variable is

$$x = \frac{\mathrm{d}g}{\mathrm{d}u} = \frac{u}{2}. \tag{B.2.2}$$

[2] Interested readers might refer to Figures 1 and 4 in Hanyga, A. and Slawinski, M.A., (2001) Caustics in qSV rayfields of transversely isotropic and vertically inhomogeneous media: Anisotropy 2000: Fractures, converted waves, and case studies: SEG (Special Issue), 409–418.

[3] For a proof of this duality, readers might refer to Arnold, V.I., (1989) Mathematical methods of classical mechanics (2nd edition): Springer-Verlag, p. 63, where the theorem is stated as follows.

> The Legendre transformation is involutive, i.e., its square is the identity: if under the Legendre transformation f is taken to g, then the Legendre transform [*sic*] of g will again be f.

Herein, we can uniquely express u in terms of x; namely, $u = 2x$. Hence, we can write function (B.2.1) as

$$f(x) = 2x^2 - \frac{(2x)^2}{4} = x^2,$$

as expected from Example B.1.1. Therefore, $f(x) = x^2$ is Legendre's transformation of $g(u) = u^2/4$, where x and u are the transformation variables related by equation (B.2.2).

If a transformed function has inflection points, as is the case illustrated in Figure B.1.1, there is a constrained duality, as illustrated in Example B.2.2. Also, such is the case of slowness surfaces and wavefronts. If the slowness surface is convex, they are duals of one another for all angles, since the corresponding wavefront is also convex. If the slowness surface is not convex, the duality between this surface and the corresponding wavefront is constrained to specific values of angles.

EXAMPLE B.2.2. Following Remark B.1.2 and proceeding in a manner analogous to Example B.2.1, we let

$$g(u) = -\frac{2}{3\sqrt{3}} u^{\frac{3}{2}}, \qquad u \geqslant 0,$$

which is expression (B.1.7); it corresponds to $x < 0$, as imposed by the forward transformation. In view of equation (B.1.4),

$$f(x) = ux - g(u) = ux + \frac{2}{3\sqrt{3}} u^{\frac{3}{2}},$$

where the new independent variable is

$$x = \frac{\mathrm{d}g}{\mathrm{d}u} = -\sqrt{\frac{u}{3}}.$$

To solve for u, we square both sides to write

$$x^2 = \frac{u}{3}.$$

To include the restriction of $x < 0$, we take the square root of both sides and—using the fact that, in general, $\sqrt{x^2} = |x|$, where $|\cdot|$ denotes the absolute value—we obtain

$$|x| = \sqrt{\frac{u}{3}}.$$

To solve for u, we square both sides to write

$$|x|^2 = \frac{u}{3},$$

which means that $u = 3|x|^2$. Thus,

$$f(x) = 3|x|^2 x + 2|x|^3.$$

Since $x < 0$, removing the absolute value, we write $|x| = -x$; hence,

$$f(x) = 3x^3 - 2x^3 = x^3,$$

which is $f(x)$, as expected in view of the original function. Also, let us consider

$$g(u) = \frac{2}{3\sqrt{3}} u^{\frac{3}{2}}, \qquad u \geqslant 0,$$

which is expression (B.1.8); it corresponds to $x \geqslant 0$. In view of equation (B.1.4),

$$f(x) = ux - \frac{2}{3\sqrt{3}} u^{\frac{3}{2}},$$

where the new independent variable is

$$|x| = \sqrt{\frac{u}{3}},$$

which means that, again, $u = 3|x|^2$, and

$$f(x) = 3|x|^2 x - 2|x|^3.$$

Since $x \geqslant 0$, upon removal of the absolute value, $|x| = x$, and, hence, we obtain $f(x) = 3x^3 - 2x^3 = x^3$, as expected. Thus, we obtain the original function, $f(x) = x^3$, where $x \in \mathbb{R}$, by two piecewise Legendre's transformations.

Similarly, a concave slowness surface can be obtained by a piecewise transformation of a triple wavefront. Commonly, it is achieved pointwise by using equation (8.4.12) on page 360.

B.3. Transformation between Lagrangian $\mathscr{L}$ and Hamiltonian $\mathscr{H}$

NOTATION B.3.1. In this appendix, to familiarize the reader with the fact that the phase slowness is a covector, $\mathbf{p}$, while the ray velocity is a vector, $\dot{\mathbf{x}}$, following the standard convention, their components appear as subscripts and superscripts, respectively. This distinction is not used in the text of the book.

REMARK B.3.2. *Throughout this book, we formulate our expressions in terms of orthonormal coordinates. The distinction between vectors and covectors becomes important if curvilinear coordinates are used.*

NOTATION B.3.3. In this appendix, to show the generality of the formulation, all expressions are derived for an n-dimensional space.[4]

In the context of this book, Legendre's transformation relates the ray-theory Lagrangian, $\mathscr{L}$, to the ray-theory Hamiltonian, $\mathscr{H}$. The transformation between functions $\mathscr{L}(\mathbf{x},\cdot)$ and $\mathscr{H}(\mathbf{x},\cdot)$ is analogous to the transformation between functions $f(\cdot)$ and $g(\cdot)$, discussed above, where $\cdot$ stands for the variables of transformation. Note that $\mathbf{x}$, while specifying the point in the continuum where the transformation is performed, plays no role in this transformation. At point $\mathbf{x}$, variables $\dot{\mathbf{x}}$ and $\mathbf{p}$ are the active variables of transformation.

Let $\mathscr{L}=\mathscr{L}(\mathbf{x},\dot{\mathbf{x}})$. In view of expression (B.1.4), consider a new function given by

$$\mathscr{H}(\mathbf{x},\mathbf{p})=\sum_{i=1}^{n}\dot{x}^{i}p_{i}-\mathscr{L}(\mathbf{x},\dot{\mathbf{x}}), \tag{B.3.1}$$

where, in view of expression (B.1.3), the new variables are

$$p_{i}=\frac{\partial\mathscr{L}}{\partial\dot{x}^{i}},\qquad i\in\{1,\ldots,n\}. \tag{B.3.2}$$

[4] Readers interested in an insightful description of Legendre's transformation, including the duality of the transformation and the application of the transformation to $\mathscr{L}$ and $\mathscr{H}$, might refer to Lanczos, C., (1949/1986) The variational principles in mechanics: Dover, pp. 161–172.

Following expression (B.3.1), we can write the differential of $\mathscr{H}(\mathbf{x},\mathbf{p})$ as

$$\begin{aligned}\mathrm{d}\mathscr{H} &= \sum_{i=1}^{n}\left(p_i\mathrm{d}\dot{x}^i + \dot{x}^i\mathrm{d}p_i\right) - \sum_{i=1}^{n}\left(\frac{\partial\mathscr{L}}{\partial x^i}\mathrm{d}x^i + \frac{\partial\mathscr{L}}{\partial\dot{x}^i}\mathrm{d}\dot{x}^i\right)\\ &= \sum_{i=1}^{n}\left(p_i\mathrm{d}\dot{x}^i + \dot{x}^i\mathrm{d}p_i - \frac{\partial\mathscr{L}}{\partial x^i}\mathrm{d}x^i - \frac{\partial\mathscr{L}}{\partial\dot{x}^i}\mathrm{d}\dot{x}^i\right). \qquad \text{(B.3.3)}\end{aligned}$$

In view of expression (B.3.2), the first and the last term in expression (B.3.3) cancel one another. Thus, we obtain

$$\mathrm{d}\mathscr{H} = \sum_{i=1}^{n}\left(\dot{x}^i\mathrm{d}p_i - \frac{\partial\mathscr{L}}{\partial x^i}\mathrm{d}x^i\right). \qquad \text{(B.3.4)}$$

Also, we can formally write the differential of $\mathscr{H}(\mathbf{x},\mathbf{p})$ as

$$\mathrm{d}\mathscr{H} = \sum_{i=1}^{n}\left(\frac{\partial\mathscr{H}}{\partial x^i}\mathrm{d}x^i + \frac{\partial\mathscr{H}}{\partial p_i}\mathrm{d}p_i\right), \qquad \text{(B.3.5)}$$

which is a statement of the chain rule.

Equating the right-hand sides of equations (B.3.4) and (B.3.5), we can write

$$\sum_{i=1}^{n}\left(\dot{x}^i\mathrm{d}p_i - \frac{\partial\mathscr{L}}{\partial x^i}\mathrm{d}x^i\right) = \sum_{i=1}^{n}\left(\frac{\partial\mathscr{H}}{\partial x^i}\mathrm{d}x^i + \frac{\partial\mathscr{H}}{\partial p_i}\mathrm{d}p_i\right). \qquad \text{(B.3.6)}$$

By examining equation (B.3.6), we conclude that

$$\dot{x}^i = \frac{\partial\mathscr{H}}{\partial p_i}, \qquad i\in\{1,\ldots,n\}. \qquad \text{(B.3.7)}$$

Examining expressions (B.3.2) and (B.3.7), we recognize the duality of these expressions. Thus, we can write the counterpart of expression (B.3.1), namely,

$$\mathscr{L}(\mathbf{x},\dot{\mathbf{x}}) = \sum_{i=1}^{n}\dot{x}^i p_i - \mathscr{H}(\mathbf{x},\mathbf{p}), \qquad \text{(B.3.8)}$$

where the active variables are given by expression (B.3.7).

B.4. Transformation and Ray Equations

Knowing that $\mathscr{H}(\mathbf{x},\mathbf{p})$ is Legendre's transformation of $\mathscr{L}(\mathbf{x},\dot{\mathbf{x}})$, and vice versa, we wish to consider the effect of this transformation on the corresponding ray equations. Herein, in view of the duality of the transformation, we derive Hamilton's ray equations from Lagrange's ray equations. This process is the inverse of the transformation used in Chapter 11.

Recall Lagrange's ray equations (11.2.6), which, in general, can be written as

$$\frac{\partial \mathcal{L}}{\partial x^i} = \frac{\mathrm{d}}{\mathrm{d}t}\frac{\partial \mathcal{L}}{\partial \dot{x}^i}, \qquad i \in \{1, \ldots, n\}. \tag{B.4.1}$$

By examining equation (B.3.6), we conclude that

$$-\frac{\partial \mathcal{L}(\mathbf{x}, \dot{\mathbf{x}})}{\partial x^i} = \frac{\partial \mathcal{H}(\mathbf{x}, \mathbf{p})}{\partial x^i}, \qquad i \in \{1, \ldots, n\}, \tag{B.4.2}$$

where relations between $\dot{\mathbf{x}}$ and $\mathbf{p}$ are given by expressions (B.3.2) and (B.3.7). Hence, using expressions (B.3.2) and (B.4.2), we can write equation (B.4.1) as

$$-\frac{\partial \mathcal{H}}{\partial x^i} = \frac{\mathrm{d}p_i}{\mathrm{d}t}, \qquad i \in \{1, \ldots, n\},$$

which can be immediately restated as

$$\dot{p}_i = -\frac{\partial \mathcal{H}}{\partial x^i}, \qquad i \in \{1, \ldots, n\}. \tag{B.4.3}$$

Thus, we conclude that using the new function, given in expression (B.3.1), and the new variables, given in expression (B.3.2), we obtain expression (B.3.7), while, invoking Lagrange's ray equations (11.2.6), we obtain expression (B.4.3). We notice that the system composed of equations (B.3.7) and (B.4.3) are Hamilton's ray equations (8.2.7); namely,

$$\left\{\begin{aligned} \dot{x}^i &= \frac{\partial \mathcal{H}}{\partial p_i} \\ \dot{p}_i &= -\frac{\partial \mathcal{H}}{\partial x^i} \end{aligned}\right., \qquad i \in \{1, \ldots, n\}. \tag{B.4.4}$$

Hence, Legendre's transformation of $\mathcal{L}$, which leads to $\mathcal{H}$, allows us to derive Hamilton's ray equations from Lagrange's ray equations. In view of this derivation, we recognize that the first equation of system (B.4.4) is the definition of a variable for Legendre's transformation, while the second equation is endowed with the physical content since it results from Lagrange's ray equations.

Closing Remarks

In the context of our work, the fundamental physical principles are directly contained in Hamilton's ray equations, which originate in Cauchy's equations of motion. The fundamental justification of Lagrange's ray equations relies on Legendre's transformation and, hence, it is subject to the singularities of this transformation.[5] Furthermore, if we wish to express the governing equations explicitly in terms of Lagrangian $\mathscr{L}$, we need to solve equations (B.3.7) for the p_i, in a closed form, which is not always possible.

In the context of elastic continua, the desired transformation is possible for any convex phase-slowness surface. Furthermore, if $\mathscr{H}$ is a quadratic function in the p_i—in other words, the phase-slowness surface is elliptical—we can always obtain explicit, closed-form expressions for the ray velocity and the ray angle.

Legendre's transformation links the Hamiltonian formulation discussed in *Part 2* with the Lagrangian one discussed in *Part 3*. The transformation is explicitly used in Chapters 11 and 13.

[5] In general, depending on the context, we can view either the Hamiltonian or the Lagrangian formulation as being more fundamental. Readers interested in this question might refer to Marsden, J.E., and Ratiu, T.S., (1999) Introduction to mechanics and symmetry: A basic exposition of classical mechanical systems (2*nd* edition): Springer-Verlag, pp. 1–6.

APPENDIX C

List of Symbols

> Our symbolic mechanism is eminently useful and powerful, but the danger is ever-present that we become drowned in a language which has its well-defined grammatical rules but evidently loses all content and becomes a nebulous sham.
>
> *Cornelius Lanczos (1961)*
> *Linear differential operators*

REMARK C.0.1. Symbols listed herein correspond to the given meaning throughout the entire book.

C.1. Mathematical Relations and Operations

$=$: equality
$\approx$: approximation
$\equiv$: identity
$:=$: "defines/denotes"
$=:$: "is defined/denoted by"
$\sim$: asymptotic relation
$\circ$: orthogonal-transformation operator
$\cdot$: scalar product
$\times$: vector product
∇ : gradient
$\nabla\cdot$: divergence
$\nabla\times$: curl
d : total derivative
∂ : partial derivative
D : material time derivative
δ : variation

δ_{ij} : Kronecker's delta
$\delta(t)$: Dirac's delta
ε_{ijk} : permutation
$\in$: "belongs to a set"
$\to$: "maps to" or "tends to"
Θ : coordinate-rotation angle
J : Jacobian
$\mathbb{R}$: real numbers
$\mathbb{R}^n$: n-dimensional space of real numbers
$\mathbf{e}_i$: unit vector along the x_i-axis
$f(\cdot)|_a$: function $f(\cdot)$ evaluated at $\cdot = a$

C.2. Physical Quantities

C.2.1. Greek letters

$\varepsilon_{k\ell}$: strain tensor—see expression (1.4.6)
ϑ : phase angle—see expressions (9.2.22) and (8.4.5)
θ : ray angle—see expressions (8.7.4) and (8.4.12)
κ : compressibility—see expressions (5.12.11)
λ : Lamé's parameter—see expressions (5.12.2)
μ : Lamé's parameter, also known as rigidity modulus—see expressions (5.12.2)
ν : Poisson's ratio—see expressions (5.14.35) and (5.14.38)
ξ_{ij} : rotation tensor—see expression (1.5.1)
Ψ : rotation vector—see expression (1.5.2)
ρ : mass density—see expression (2.1.1)
σ_{ij} : stress tensor—see expression (2.5.2) and Figure 2.5.1
φ : dilatation—see expression (1.4.18)
ϕ : displacement angle—see expression (9.2.30)
ω : angular frequency—see expression (6.10.2)

C.2.2. Roman letters

$c_{ijk\ell}$: elasticity tensor—see expression (3.2.1)
C_{mn} : elasticity-matrix entries, also known as elasticity parameters—see expression (4.2.8)

E : Young's modulus—see Remark 5.14.7

$\mathscr{F}$: ray-theory Lagrangian, absolute-value homogeneous of degree 1 in the x_i—see expression (13.1.8)

F : ray-theory Lagrangian—see expression (14.1.1)

$\mathscr{H}$: ray-theory Hamiltonian, homogeneous of degree 2 in the p_i—see expression (8.2.8)

$\mathbb{H}$: classical-mechanics Hamiltonian—see Exercise 13.5

K : bulk modulus—see Section 5.12.4

$\mathscr{L}$: ray-theory Lagrangian, homogeneous of degree 2 in the $\dot{x}_i$—see expression (11.1.1)

$\mathbb{L}$: classical-mechanics Lagrangian— see expression (13.2.2)

W : strain energy—see expressions (4.1.3) and (4.1.1)

Bibliography

This list contains works that served the author as a source of information and inspiration in writing this book. Many concepts discussed in this book result from the author's studying the works included in this list.

This bibliography contains numerous republications of previously published works. In certain cases, there are two dates that refer to the original publication and to the edition, translation or facsimile used herein.

Abbott, M.B., (1966) An introduction to the method of characteristics: Elsevier

Achenbach, J.D., (1973) Wave propagation in elastic solids: North Holland

Achenbach, J.D., (2003) Reciprocity in elastodynamics: Cambridge University Press

Achenbach, J.D., Gautesen, A.K., and McMaken, H., (1982) Ray methods for waves in elastic solids: with applications to scattering by cracks: Pitman Publishing Inc.

Ait-Haddou, R. (1996) Courbes à hodographe Pythagorien en géométrie de Minkowski et modélisation géométric (Ph.D. thesis): Université Joseph Fourier–Grenoble I

Ait-Haddou, R., Biard, L., Slawinski, M.A., (2000) Minkowski isoperimetric hodograph curves. Comp. Aided Geom. Design. **17**, Issue 9, 835–861.

Aki, K., and Richards, P.G., (1980) Quantitative seismology: Theory and methods: W.H. Freeman & Co.

Aki, K., and Richards, P.G., (2002) Quantitative seismology (2*nd* edition): University Science Books

Akivis, M.A., and Goldberg, V.V., (1972) An introduction to linear algebra and tensors: Dover

Aoki, H., Syono, Y., and Hemley, R.J., (editors), (2000) Physics meets mineralogy: Condensed-matter physics in geosciences: Cambridge University Press

Aleksandrov, A.D., Kolmogorov, A.N., and Lavrentev, M.A., (editors), (1969/1999) Mathematics: Its content, methods and meaning: Dover

Anastopoulos, C., (2008) Particle or wave: The evolution of the concept of matter in modern physics: Princeton University Press

Antman, S.S., (1995) Nonlinear problems of elasticity: Springer-Verlag

Anton, H., (1973) Elementary linear algebra: John Wiley & Sons

Anton, H., (1984) Calculus with analytic geometry: John Wiley & Sons

Antonelli, P.L., Ingarden, R.S., and Matsumoto, M., (1993) The theory of sprays and Finsler spaces with applications in physics and biology: Kluwer

Appel, W., (2005) Mathématiques pour la physique et les physiciens (3*rd* edition): H & K Editions

Appell, P., (1905) Cours de mécanique à l'usage des élèves de la classe de mathématique spéciale: Gauthier-Villars, Paris

Arianrhod, R., (2005) Einstein's heroes: Imagining the world through the language of mathematics: Oxford University Press

Armstrong, M.A., (1988) Groups and symmetry: Springer-Verlag

Arfken, G.B., (1985) Mathematical methods for physicists (3*rd* edition): Academic Press, Inc.

Arfken, G.B, and Weber, H.J., (2001) Mathematical methods for physicists (5*th* edition): Harcourt/Academic Press

Arnold, V.I. (1984) Ordinary differential equations (3*rd* edition): Springer-Verlag

Arnold, V.I., (1989) Mathematical methods of classical mechanics (2*nd* edition): Springer-Verlag

Arnold, V.I., (1991) The theory of singularities and its applications: Cambridge University Press

Arnold, V.I., (1992) Catastrophe theory (*3rd* edition): Springer-Verlag

Arnold, V.I., (2004) Lectures on partial differential equations: Springer-Verlag

Aster, C.R., Borchers, B., and Thurber, C.H., (2005) Parameter estimation and inverse problems: Elsevier

Auld, B.A., (1973) Acoustic fields and waves in solids: John Wiley and Sons

Ayres, F., (1962) Matrices: Schaum's Outlines, McGraw-Hill, Inc.

Babich, V.M., (1961) Ray method of calculating the intensity of wavefronts in the case of a heterogeneous, anisotropic, elastic medium: In *Problems of the dynamic theory of propagation of seismic waves* (in Russian), **5**, ed. G.I. Petrashen, 36–46: Leningrad University Press. (English translation: Geophys. J. Int., 1994, **118**, 379–383.)

Babich, V.M., and Buldyrev, V.S., (1972/1991) Short-wavelength diffraction theory: Asymptotic methods: Springer-Verlag

Backus, G., (1970) A geometrical picture of anisotropic elastic tensors: Rev. Geophys., **8** (3), 633–671.

Bao, D., Chern, S.-S., and Shen, Z., (Editors), (1996) Finsler geometry: American Mathematical Society

Bao, D., Chern, S.-S., and Shen, Z., (2000) An introduction to Riemann-Finsler geometry: Springer-Verlag

Barton, G., (1989) Elements of Green's functions and propagation: Potentials, diffusion and waves: Oxford Science Publications

Båth, M., and Berkhout, A.J., (1984) Mathematical aspects of seismology: Geophysical Press

Basdevant, J-L., (2007) Variational principles in physics: Springer-Verlag

Basdevant, J-L., (2005) Principes variationneles et dynamique: Vuibert

Batterman, R.W., (2002) The devil in the details: Asymptotic reasoning in explanation, reduction, and emergence: Oxford University Press

Bellman, R., (1966/2003) Perturbation techniques in mathematics, engineering and physics: Dover

Bender, C.M., and Orszag, S.A., (1978/1999) Advanced mathematical methods for scientists and engineers: Asymptotic methods and perturbation theory: Springer-Verlag

Ben-Menahem, A., and Singh, S.J., (1981/2000) Seismic waves and sources: Dover

Benson, D.J., (2007) Music: A mathematical offering: Cambridge University Press

Bérest, P., (1997) Calcul des variations, Application à la mécanique et à la physique: Ellipses

Beyer, W.H., (editor), (1984) CRC Standard mathematical tables ($27th$ edition): CRC Press, Inc.

Bishop, R.L., and Goldberg, S.I., (1968/1980) Tensor analysis on manifolds: Dover

Bleistein, N., (1984) Mathematical methods for wave phenomena: Academic Press

Bleistein, N., Cohen, J.K., and Stockwell, J.W., (2001) Mathematics of multidimensional seismic imaging, migration, and inversion: Springer-Verlag

Bleistein, N., and Handelsman, R.A., (1975/1986) Asymptotic expansions of integrals: Dover

Boccara, N., (1997) Distributions: Ellipses

Bóna, A. and Slawinski, M.A., (2021) Wavefronts and rays as characteristics and asymptotics ($3rd$ edition): World Scientific

Bóna A., Bucataru, I., and Slawinski, M.A., (2008) Space of $SO(3)$-orbits of elasticity tensors: Archives of Mechanics **60**(2), 121–136.

Bóna, A., Bucataru, I., and Slawinski, M.A., (2007) Coordinate-free classification of elasticity tensor: Journal of Elasticity **87**(2-3), 109–132.

Bóna, A., Bucataru, I., and Slawinski, M.A., (2005) Characterization of elasticity-tensor symmetries using SU(2): Journal of Elasticity **75**(3), 267–289.

Bóna, A., Bucataru, I., and Slawinski, M.A., (2004) Material symmetries of elasticity tensor: The Quarterly Journal of Mechanics and Applied Mathematics **57**(4), 583–598.

Bóna, A. and Slawinski, M.A., (2003) Fermat's principle for seismic rays in elastic media: Journal of Applied Geophysics **54**, 445–451.

Bond, W.L., (1943) The mathematics of the physical properties of crystals: The Bell System Technical Journal, **22**, No. 1, 1–72.

Born, M. and Wolf, E., (1999) Principles of optics (*7th* edition): Cambridge University Press

Borcherdt, R.D., (2009) Viscoelastic waves in layered media: Cambridge University Press

Bos, L., Slawinski, M.A., Stanoev, T., Vianello, M., (2020) On orthogonal transformations of Christoffel equations: International Journal on Geomathematics

Bos, L., Slawinski, M.A., Stanoev, T., Vianello, M., (2019) On orthogonal transformations of Christoffel equations: arXiv 1901.03926 [physics.geo-ph].

Bos, L. and Slawinski, M.A., (2010) Elastodynamic equations: Characteristics, wavefronts and rays: The Quarterly Journal of Mechanics and Applied Mathematics **63**(1), 23–37.

Bos, L.P., Gibson, P.C., Kotchetov, M., Slawinski, M.A., (2004) Classes of anisotropic media: a tutorial: Studia geophysica and geodætica, **48**, 265–287.

Brekhovskikh, L.M., and Goncharov, V., (1982/1994) Mechanics of continua and wave dynamics: Springer-Verlag

Bridgman, P.W., (1927) The logic of modern physics: The MacMillan Co.

Broglie, L. de, (1941) Continu et discontinu en physique moderne: Albin Michel

Bronson, R., (1973) Modern introductory differential equations: Schaum's Outlines: McGraw-Hill, Inc.

Brown, J.R., (2001) Who rules in science: An opinionated guide to the wars: Harvard University Press

Brown, J.R., (1994) Smoke and mirrors: How science reflects reality: Routledge

Brown, J.R., (1991) The laboratory of the mind: Thought experiments in the natural sciences: Routledge

Brown, J.R. and Slawinski, M.A., (2017) On foundations of seismology: Bringing idealizations down to Earth: World Scientific

Bucataru, I. and Slawinski, M.A., (2009) On convexity and detachment of innermost wavefront-slowness surface: Geophysics **74**(5), WB63-WB66.

Bucataru, I. and Slawinski, M.A., (2005) Generalized orthogonality of rays and wavefronts in anisotropic inhomogeneous media. Nonlinear analysis: Series B: Real-world applications **6**, 111–121.

Buchanan, G.R., (1995) Finite element analysis: Schaum's Outlines: McGraw-Hill, Inc.

Buchdahl, H.A., (1970/1993) An introduction to Hamiltonian optics: Dover

Bullen, K.E., and Bolt, B.A., (1947/1985) An introduction to the theory of seismology: Cambridge University Press

Bunge, M., (2006) Chasing reality: Strife over realism: University of Toronto Press

Bunge, M., (2003) Emergence and convergence: Qualitative novelty and the unity of knowledge: University of Toronto Press

Bunge, M. A., (1998) Philosophy of Science, Vol. I: From problem to theory (Revised edition): Transaction Publishers

Bunge, M. A., (1998) Philosophy of Science, Vol. II: From explanation to justification (Revised edition): Transaction Publishers

Bunge, M., (1973) Philosophy of physics: D. Reidel Publishing Co.

Bunge, M., (1973) Method, model and matter: D. Reidel Publishing Co.

Bunge, M., (1967) Foundations of physics: Springer-Verlag

Bunge, M., (editor), (1964) Critical approaches to science and philosophy: Transaction Publishers

Carcione, J.M., (2001) Wave fields in real media: wave propagation in anisotropic, anelastic and porous media: Pergamon

Cauchy, A. L., (1827) De la pression ou tension dans un corps solide: Ex. de Math, **2**, 42–56.

Červený, V., (2002) Fermat's variational principle for anisotropic inhomogeneous media: Studia geophysica and geodætica, **46**, 567–588.

Červený, V., (2001) Seismic ray theory: Cambridge University Press

Červený, V., and Ravindra, R., (1971) Theory of seismic head waves: University of Toronto Press

Červený, V., Molotkov, I.A., and Pšenčík, I., (1977) Ray method in seismology: Univerzita Karlova

Chadwick, P., (1976/1999) Continuum mechanics: Concise theory and problems: Dover

Chapman, C., (2004) Fundamentals of seismic wave propagation: Cambridge University Press

Cheney, W., (2001) Analysis for applied mathematics: Springer-Verlag

Chew, W.C., (1990) Waves and fields in inhomogeneous media: Van Nostrand Reinhold

Chou, P.C., and Pagano, N.J., (1967/1992) Elasticity: tensor, dyadic, and engineering approaches: Dover

Close, F., (2009) Nothing: A Very Short Introduction: Oxford University Press

Cox, D., Little, J., and O'Shea, D., (1992) Ideals, varieties and algorithms (2*nd* edition): Springer-Verlag

Coirier, J., (1997) Mécanique des milieux continues: Concepts de base: Cours et exercices corrigés: Dunod

Coirier, J., (2001) Mécanique des milieux continues: Aide-mémoire: Dunod

Coleman, B.D., and Noll, W., (1961) Foundations of linear viscoelasticity: Revue of modern physics, **33**, 239–249.

Collins, R.E., (1968/1999) Mathematical methods for physicists and engineers: Dover

Corben, H.C., and Stehle, P., (1950/1994) Classical mechanics: Dover

Courant, R., and Hilbert, D., (1924/1989) Methods of mathematical physics: John Wiley & Sons

Courant, R., and John, F., (1974/1989) Introduction to calculus and analysis: Springer-Verlag

Courant, R., and Robbins, H., (1941/1996) What is mathematics?: Oxford University Press

Dahlen, F.A., and Tromp, J., (1998) Theoretical global seismology: Princeton University Press

Danek, T., Kochetov, M., Slawinski, M.A., (2013) Uncertainty analysis of effective elasticity tensors using quaternion-based global optimization and Monte-Carlo method. The Quarterly Journal of Mechanics and Applied Mathematics **66**(2), 253–272.

Das, A., (2007) Tensors: The mathematics of relativity theory and continuum mechanics: Springer-Verlag

d'Alembert, J., (1747) Recherches sur la courbe que forme une corde tendue mise en vibration: Hist. Acad. Sci. Berlin **3**, 214 – 219.

de Bruijn, N.G., (1958/1981) Asymptotic methods in analysis: Dover

de Hoop, A. T., (1995) Textbook of radiation and scattering of waves: Academic Press, Inc.

Demidov, A.S., (2001) Generalized functions in mathematical physics: Main ideas and concepts: Nova Science Publishers, Inc.

Dennery, P., and Krzywicki, A., (1967) Mathematics for physicists: Harper and Row

Dingle, R.B., (1973) Asymptotic expansions: Their derivation and interpretation: Academic Press

Dix, C.H., (1955) Seismic velocities from surface measurements: Geophysics **20**, 60–86.

do Carmo, M.P., (1971/1994) Differential forms and applications: Springer-Verlag

do Carmo, M.P., (1976) Differential geometry of curves and surfaces: Prentice-Hall, Inc.

Doran, C., and Lasenby, A., (2003) Geometric algebra for physicists, Cambridge University Press

Dugas, R., (1950/1996) Histoire de la mécanique: Éditions Jacques Gabay

Duistermaat, J.J., (1996) Fourier Integral operators: Birkhäuser

Einstein, A., and Infeld, L., (1938) Evolution of physics from early concepts to relativity and quanta: Simon & Schuster

Eisenhart, L. P., (1925/1997) Riemannian Geometry: Princeton University Press

Ekeland, I., (2000) Le meilleur des mondes possibles: Mathématiques et destinée: Seuil

Elmore, W.C., and Heald, M.A., (1969/1985) Physics of waves: Dover

Elsgolc, L.E., (1962) Calculus of Variations: Pergamon

Epstein, M., and Elżanowski, M., (2007) Material inhomogeneities and their evolution: A geometric approach: Springer-Verlag

Epstein, M. and Slawinski, M.A., (1999) On raytracing in constant velocity-gradient media: Geometrical approach, Canadian Journal of Exploration Geophysics, **35**, No. 1/2, 1–6.

Epstein, M. and Slawinski, M.A., (1999) On rays and ray parameters in inhomogeneous isotropic media: Canadian Journal of Exploration Geophysics, **35**, No. 1/2, 7–19.

Epstein, M. and Slawinski, M.A., (1998) On some aspects of the continuum-mechanics context: Revue de l'Institut Français du Pétrole, **53**, No. 5, 669–677.

Epstein, M., and Śniatycki, J., (1992) Fermat's principle in elastodynamics: Journal of Elasticity, **27**, 45–56.

Eringen, A.C., (1967) Mechanics of Continua: John Wiley & Sons

Eringen, A.C., (1999) Microcontinuum field theories I: Foundations and solids: Springer-Verlag

Evans, L.C., (1998) Partial differential equations: AMS

Ewing, M.G., (1969/1985) Calculus of variations with applications: Dover

Ewing, W.M., Jardetzky, W.S., and Press, F., (1957) Elastic waves in layered media: McGraw-Hill, Inc.

Faraut, J., (2008) Analysis on Lie groups: An introduction: Cambridge University Press

Farlow, S.J., (1982) Partial differential equations for scientists and engineers: Dover

Fedorov, F.I., (1968) Theory of elastic waves in crystals: Plenum Press

Feynman, R.P., (1985/2006) QED: The strange theory of light and matter: Princeton University Press

Feynman, R.P., (1967) The character of physical law: The MIT Press

Feynman, R.P., Leighton, R.B., and Sands, M., (1963/1989) Feynman's lectures on physics: Addison-Wesley Publishing Co.

Flanders, H., (1963/1989) Differential forms with applications to the physical sciences: Dover

Flügge, S., (1960) Handbuch der Physik: Springer-Verlag

Folland, G.B., (1995) Introduction to partial differential equations (2*nd* edition): Princeton University Press

Forest, S., (2006) Milieux continus généralisés et matériaux hétérogènes: Mines Paris

Frankel, T., (1997) The geometry of physics: An introduction: Cambridge University Press

Franklin, J.N., (1968/2000) Matrix theory: Dover

Freegarde, T., (2013) Introduction to the physics of waves: Cambridge University Press

Fung, Y.C., (1965) Foundations of solid mechanics: Prentice-Hall, Inc.

Fung, Y.C., (1977) A first course in continuum mechanics: Prentice-Hall, Inc.

Garrity, T.A., (2001) All the mathematics you missed [but need to know for graduate school]: Cambridge University Press

Gelfand, I.M., and Fomin, S.V., (1963/2000) Calculus of variations: Dover

Gibson, C.G., (2001) Elementary geometry of differentiable curves: An undergraduate introduction: Cambridge University Press

Goldberg, S.I., (1962/1998) Curvature and homology: Dover

Goldstein, H., (1950/1980) Classical mechanics: Addison-Wesley Publishing Co.

Graff, K.F., (1975/1991) Wave motion in elastic solids: Dover

Grant, F.S., and West, G.F., (1965) Interpretation theory in applied geophysics: McGraw-Hill, Inc.

Gray, J., (2008) Plato's ghost: The modernist transformation of mathematics: Princeton University Press

Greenspan, D., (1961/2000) Introduction to partial differential equations: Dover

Greenwood, D.T., (1977/1997) Classical dynamics: Dover

Guggenheimer, H.W., (1963/1977) Differential geometry: Dover

Guillemin, V., and Sternberg, S., (1977) Geometric asymptotics: American Mathematical Society

Gurtin, M. E., (1981/2003) An introduction to continuum mechanics: Academic Press

Hadsell, F., (1995) Tensors of geophysics for mavericks and mongrels: SEG

Hadsell, F., and Hansen, R., (1999) Tensors of geophysics: Generalized functions and curvilinear coordinates: SEG

Hankins, T.L., (1980) Sir William Rowan Hamilton: The Johns Hopkins University Press

Hanna, J.R., (1982) Fourier series and integrals of boundary value problems: John Wiley and Sons

Hanyga, A., (1985) Seismic wave propagation in the Earth: Elsevier

Hanyga, A., (1985) Mathematical theory of non-linear elasticity: PWN—Polish Scientific Publishers

Hanyga, A., (2001) Fermat's principle in anisotropic elastic media: Advances in anisotropy: Selected theory, modeling, and case studies: Society of Exploration Geophysicists (Special Issue), 271–288.

Hanyga, A. and Slawinski, M.A., (2001), Caustics in qSV rayfields of transversely isotropic and vertically inhomogeneous media: Anisotropy 2000: Fractures, converted waves, and case studies: Society of Exploration Geophysicists (Special Issue), 409–418.

Hardy, G.H., (1940/1996) A mathematician's apology: Cambridge University Press

Harper, C., (1976) Introduction to mathematical physics: Prentice-Hall, Inc.

Hausner, M., (1965/1998) A vector space approach to geometry: Dover

Helbig, K., (1994) Foundations of anisotropy for exploration seismics: Pergamon

Herman, B., (1945) Some theorems of the theory of anisotropic media: Comptes Rendus (Doklady) de l'Académie des Sciences de l'URSS, **48** (2), 89–92.

Hesse, M.B., (1962/2005) Forces and fields: The concept of action at a distance in the history of physics: Dover

Hildebrand, F.B., (1948) Advanced calculus for applications: Prentice-Hall, Inc.

Hildebrandt, S. and Tromba, A., (1996) The parsimonious universe: shape and form in the natural world: Springer-Verlag

Holm, D.D., (2008) Geometric mechanics: Imperial College Press

Ihlenburg, F., (1998) Finite element analysis of acoustic scattering: Springer-Verlag

Isakov, V., (1998) Inverse problems for partial differential equations: Springer-Verlag

Ivanov, Y., and Stovas, A. (2019) 3D mapping of kinematic attributes in anisotropic media: Geophysics **84**(3), C159–C170.

Jaynes, E.T. (2003) Probability theory: The logic of science: Cambridge University Press

Jeans, J., (1943/1981) Physics and philosophy: Dover

Jeffreys, H., and Jeffreys, B., (1946/1999) Methods of Mathematical Physics: Cambridge University Press

John, F., (1982) Partial differential equations ($4th$ edition): Springer-Verlag

John, F., (1955/2004) Plane waves and spherical means applied to partial differential equations: Dover

Johns, R., (2002) A theory of physical probability: University of Toronto Press

Kac, M., and Ulam, S.M., (1968/1992) Mathematics and logic: Dover

Karal, F.C., and Keller, J.B., (1959) Elastic wave propagation in homogeneous and inhomogeneous media: J. Acoust. Soc. Am., **31** (6), 694–705.

Kennett, B.L.N., (2001) The seismic wavefield, Vol. I: Introduction and theoretical development: Cambridge University Press

Kennett, B.L.N., (2002) The seismic wavefield, Vol. II: Interpretation of seismograms on regional and global scales: Cambridge University Press

Kennett, B.L.N., and Bunge, H.-P., (2008) Geophysical continua: Deformation in the Earth's interior: Cambridge University Press

Kilmister, C.W., and Reeve, J.E., (1966) Rational mechanics: Longmans

Kittel, C. (2005) Introduction to solid state physics ($8th$ edition): John Wiley and Sons

Klein, E., (1991) Conversations avec le Sphinx: Les paradoxes en physique: Editions Albin Michel

Kline, M., (1959/1981) Mathematics and the physical world: Dover

Kline, M., (1972) Mathematical thought from ancient to modern times: Oxford University Press

Kline, M., (1980) Mathematics: The loss of certainty: Oxford University Press

Knobel, R., (2000) An introduction to the mathematical theory of waves: American Mathematical Society, Institute for Advance Studies

Kochetov, M. and Slawinski, M.A. (2009) On obtaining effective transversely isotropic elasticity tensors: Journal of Elasticity **94**(1), 1–13.

Koks, D., (2006) Explorations in mathematical physics: The concepts behind an elegant language: Springer-Verlag

Kolmogorov, A.N., and Fomin, S.V., (1954–1960/1999) Elements of the theory of functions and functional analysis: Dover

Kolsky, H., (1953/1963) Stress waves in solids: Dover

Kravtsov, Y.A., (2005) Geometrical optics in engineering physics: Alpha Science International Ltd.

Kravtsov, Y.A., and Orlov, Y.I., (1990) Geometrical optics of inhomogeneous media: Springer-Verlag

Kravtsov, Y.A., and Orlov, Y.I., (1999) Caustics, catastrophes and wave fields: Springer-Verlag

Krebes, E.S., (2004) Seismic theory and methods (Lecture notes): University of Calgary

Kreyszig, E., (1959/1991) Differential geometry: Dover

Krysicki, W., and Włodarski, L., (1980) Analiza matematyczna w zadaniach: PWN—Polish Scientific Publishers

Kuhn, T.S., (1996) The structure of scientific revolutions (3*rd* edition): The University of Chicago Press

Lafontaine, J., (1996) Introduction aux variétés différentielles: Presses Universitaires de Grenoble

Lagrange, J-L., (1788/1989) Mécanique analytique: Éditions Jacques Gabay

Lancaster, P., and Tismenetsky, M., (1985) The theory of matrices with applications (2*nd* edition): Academic Press

Lancaster, P., and Šalkauskas, K., (1996) Transform methods in applied mathematics: An introduction: John Wiley and Sons

Lanczos, C., (1949/1986) The variational principles of mechanics: Dover

Lanczos, C., (1956/1988) Applied analysis: Dover

Lanczos, C., (1961/1997) Linear differential operators: Dover

Landau, L.D., and Lifshitz, E.M., (1976) Course of theoretical physics, Vol. 1: Mechanics (3*rd* edition): Elsevier

Landau, L.D., and Lifshitz, E.M., (1975) Course of theoretical physics, Vol. 2: The classical theory of fields (4*th* edition): Elsevier

Landau, L.D., and Lifshitz, E.M., (1986) Course of theoretical physics, Vol. 7: Theory of elasticity (3*rd* edition): Elsevier

Lang, S., (1973) Calculus of several variables: Addison-Wesley Publishing Co.

Lang, S., (1995) Differential and Riemannian manifolds: Springer-Verlag

Le, L.H.T., Krebes, E.S., and Quiroga-Goode, G.E., (1994) Synthetic seismograms for *SH* waves in anelastic transversely isotropic media: Geophys. J. Int., **116**, 598–604.

Lee, J.M., (2000) Introduction to topological manifolds: Springer-Verlag

Lipschutz, M.M., (1969) Theory and problems of differential geometry: Schaum's Outline Series, McGraw-Hill, Inc.

Love, A.E.H., (1892/1944) A treatise on the mathematical theory of elasticity: Dover

Long, C., (1967) On the completeness of the Lamé potentials: Acta Mechanica, **3**, 371–375.

Lovelock, D., and Rund, H., (1975/1989) Tensors, differential forms, and variational principles: Dover

MacBeth, C., (2002) Multi-component VSP analysis for applied seismic anisotropy: Pergamon

Macelwane, J.B., and Sohon, F.W., (1936) Introduction to theoretical seismology, Part I: Geodynamics: John Wiley and Sons

Mach, E., (1893/1989) The science of mechanics: A critical and historical account of its development: Open Court Publishing Co.

Malvern, L.E., (1969) Introduction to the mechanics of a continuous medium: Prentice-Hall

Marion, J.B., and Thornton, S.T., (1970/1988) Classical dynamics of particles and systems: Harcourt Brace Jovanovich

Marsden, J.E., and Hughes, T.J.R., (1983/1994) Mathematical foundations of elasticity: Dover

Marsden, J.E., and Ratiu, T.S., (1999) Introduction to mechanics and symmetry: A basic exposition of classical mechanical systems (2*nd* edition): Springer-Verlag

Marsden, J.E., and Tromba, A., (2012) Vector calculus (6*th* edition): W.H. Freeman & Co.

Marsden, J.E., and Tromba, A., (1981) Vector calculus (2*nd* edition): W.H. Freeman & Co.

Martin, D., (2002) Manifold theory: An introduction for mathematical physicists: Horwood Publishing

Mase, G.E., (1970) Theory and problems of continuum mechanics: Schaum's Outlines, McGraw-Hill, Inc.

Mathai, A.M., and Haubold, H.J., (2008) Special functions for applied scientists: Springer-Verlag

Matthews, P.C., (1998) Vector calculus: Springer-Verlag

Mavko, G., Mukerji, T., and Dvorkin, J., (1998) The rock physics book: Cambridge University Press

McCuskey, S.W., (1959) An introduction to advanced dynamics: Addison-Wesley Publishing Co.

McOwen, R.C., (1996) Partial differential equations: Methods and applications: Prentice-Hall, Inc.

Meirovitch, L., (1970) Methods of analytical dynamics: McGraw-Hill, Inc.

Menzel, D.H., (1947/1961) Mathematical physics: Dover

Menzel, D.H., (1955/1960) Fundamental formulas of physics: Dover

Morse P.M., and Feshbach H., (1953) Methods of theoretical physics: McGraw-Hill, Inc.

Musgrave, M.J.P., (1970) Crystal acoustics: Introduction to the study of elastic waves and vibrations in crystals: Holden-Day

Muskhelishvili, N.I., (1933/1975) Some basic problems of the mathematical theory of elasticity: Noordhoff International Publishing

Nahin, P.J., (2006) Dr. Euler's fabulous formula: Cures many mathematical ills: Princeton University Press

Neusel, M.D., and Smith, L., (2002) Invariant theory of finite groups: American Mathematical Society

Newnham, R.E., (2005) Properties of materials: Anisotropy, symmetry, structure: Oxford University Press

Nolet, G., (2008) A breviary of seismic tomography: Imaging the interior of the Earth and Sun: Cambridge University Press

Noll, W., (1974) The foundations of mechanics and thermodynamics: Springer-Verlag

Novozhilov, V.V., (1948/1999) Foundations of the nonlinear theory of elasticity: Dover

Nye, J.F., (1957/1987) Physical properties of crystals: their representation by tensors and matrices: Oxford University Press

Nye, J.F., (1999) Natural focusing and fine structure of light: Caustics and wave dislocations: Institute of Physics Publishing

Officer, C.B., (1974) Introduction to theoretical geophysics: Springer-Verlag

Okasha, S., (2002) Philosophy of science: A very short introduction: Oxford University Press

Olmsted, J.M.H., (1961) Advanced calculus: Prentice-Hall, Inc.

Olver, P.J., (1993) Applications of Lie groups to differential equations (2*nd* edition): Springer-Verlag

Pesic, P., (2003) Abel's proof: An essay on the sources and meaning of mathematical unsolvability: The MIT Press

Porteous, I.R., (1994) Geometric differentiation for the intelligence of curves and surfaces: Cambridge University Press

Pitt, V.H., (1977) The Penguin's dictionary of physics: Penguin books

Polya, G., (1954) Mathematics and plausible reasoning, Vol. I: Induction and analogy in mathematics: Princeton University Press

Polya, G., (1968) Mathematics and plausible reasoning, Vol. II: Patterns of plausible inference: Princeton University Press

Potter, M.C., (1978) Mathematical methods in the physical sciences: Prentice-Hall, Inc.

Poincaré, H., (1902/1968) La science et l'hypothèse: Flammarion

Prager, W., (1961) Introduction to mechanics of continua: Ginn and Company

Pšenčík, I., (1994) Introduction to seismic methods (Lecture notes): PPPG/UFBa, Salvador

Pujol, J., (2003) Elastic wave propagation and generation in seismology: Cambridge University Press

Radin, S.H., and Folk, R.T., (1982) Physics for scientists and engineers: Prentice-Hall, Inc.

Rasband, S. N., (1990) Chaotic dynamics of nonlinear systems: John Wiley & Sons

Rauch, J., (1991) Partial differential equations: Springer-Verlag

Rayleigh, J.W.S., (1877/1945) The theory of sound: Dover

Renardy, M., and Rogers, R.C., (1993) An introduction to partial differential equations: Springer-Verlag

Riley, K.F., (1974) Mathematical methods for the physical sciences: An informal treatment for students of physics and engineering: Cambridge University Press

Rochester, M.G., (2010) Note on the necessary conditions for P and S wave propagation in a homogeneous isotropic elastic solid: Journal of Elasticity, **98**(1), 111–114.

Rochester, M.G., (1986) Lecture notes on continuum mechanics: Memorial University of Newfoundland

Rochester, M.G., (1997) Lecture notes on mathematical physics: Memorial University of Newfoundland

Rogister, Y., (2000) Etude théorique des deformations globales de la terre (Ph.D. thesis): Université de Liège

Rogister, Y. and Slawinski, M.A., (2005) Analytic solution of ray-tracing equations for a linearly inhomogeneous and elliptically anisotropic velocity model: Geophysics, **70**, D37–D41.

Ross, S.L., (1989) Introduction to ordinary differential equations (4*th* edition): John Wiley & Sons

Ruggeri, T., (2007) Introduzione alla termomeccanica dei continui: Monduzzi Editore

Rund, H., (1959) The differential geometry of Finsler spaces: Springer-Verlag

Rund, H., (1966) The Hamilton-Jacobi theory in the calculus of variations: D. Van Nostrand

Rund, H., (1971) Invariant theory of variational problems on subspaces of a Riemannian manifold: Vandenhoeck & Ruprecht

Rychlewski, J., (1985) On Hooke's law: Prikl. Matem. Mekhan., **48** (3), 420–435.

Rymarz, C., (1993) Mechanika ośrodkow ciągłych: PWN—Polish Scientific Publishers

Sagan, H., (1969/1992) Introduction to the calculus of variations: Dover

Schey, H.M., (2005) div, grad, curl, and all that: An informal text on vector calculus (4*th* edition): W.W. Norton & Co.

Schoenberg, M., (1980) Elastic wave behaviour across linear slip interfaces: J. Acoust. Soc. Am., **68** (5), 1516–1521.

Schouten, J.A., (1951/1989) Tensor analysis for physicists: Dover

Schrödinger, E., (1996) Nature and the Greeks: Cambridge University Press

Schutz, B., (2003) Gravity from the ground up: Cambridge University Press

Schutz, B., (1980) Geometrical methods of mathematical physics: Cambridge University Press

Schwartz, L., (1966) Mathematics for the physical sciences: Addison-Wesley Publishing Co.

Seaborn, J.B., (2002) Mathematics for the physical sciences: Springer-Verlag

Sedov, L.I., (1971) A course in continuum mechanics: Wolters-Noordhoff Publishing

Segel, L.A., (1977/1987) Mathematics applied to continuum mechanics: Dover

Segev, R. and Rodnay, G., (1999) Cauchy's theorem on manifolds: Journal of Elasticity, **56**, 129–144.

Semay, C., and Silvestre-Brac, B., (2007) Introduction au calcul tensoriel: Applications à la physique: Dunod

Seto, W.W., (1971) Acoustics: Schaum's Outlines, McGraw-Hill, Inc.

Shearer, P.M., (1999) Introduction to seismology: Cambridge University Press

Shen, Z., (2001) Differential geometry of spray and Finsler spaces: Kluwer Academic Publishers

Sheriff, R.E., (2002) Encyclopedic dictionary of applied geophysics (*4th* edition): Society of Exploration Geophysicists

Sheriff, R.E., and Geldart, L.P., (1982) Exploration seismology: Cambridge University Press

Shilov, G.E., (1974/1996) Elementary functional analysis: Dover

Simmonds, J.G., (1994) A brief on tensor analysis: Springer-Verlag

Slater, J.C., and Frank, N.H., (1947) Mechanics: McGraw-Hill, Inc.

Slawinski, M.A. (2021) Waves and rays in seismology: Answers to unasked questions (*3rd* edition): World Scientific

Slawinski, M.A., Wheaton, C.J., and Powojowski, M. (2004) VSP traveltime inversion for linear inhomogeneity and elliptical anisotropy: Geophysics, **69**(2), 373–377.

Slawinski, M.A., and Webster, P.S., (1999) On generalized ray parameters for vertically inhomogeneous and anisotropic media: Canadian Journal of Exploration Geophysics, **35**, No. 1/2, 28–31.

Slawinski, M.A., Slawinski, R.A., Brown, R.J., and Parkin, J.M., (2000) A generalized form of Snell's law in anisotropic media: Geophysics, **65**, No. 2, 632–637.

Slawinski, R.A. and Slawinski, M.A. (1999) On raytracing in constant velocity-gradient media: Calculus approach: Canadian Journal of Exploration Geophysics, **35**, No. 1/2, 24–27.

Slotnick, M.M., (1959) Lessons in seismic computing: Society of Exploration Geophysicists

Smith, D.R., (1974/1998) Variational methods in optimization: Dover

Snieder, R., (2004) A guided tour of mathematical methods for the physical sciences: Cambridge University Press

Soper, D.E, (1976/2008) Classical field theory: Dover

Spiegel, M.R., (1967) Theory and problems of theoretical mechanics with an introduction to Lagrange's equations and Hamiltonian theory: Schaum's Outlines, McGraw-Hill, Inc.

Spivak, M., (1970/1999) A comprehensive introduction to differential geometry: Publish or Perish, Inc.

Stebbing L.S., (1937/1958) Philosophy and the physicists: Dover

Steiner, M., (1998) The applicability of mathematics as a philosophical problem: Harvard University Press

Stewart, J., (1995) Multivariable calculus (3*rd* edition): Brooks/Cole Publishing Co.

Stolk, C.C., (2000) On the modeling and inversion of seismic data (Ph.D. thesis): Universiteit Utrecht

Strauss, W.A., (2008) Partial differential equations: An introduction (2*nd* edition): John Wiley and Sons

Strauss, W.A., (2008) Partial differential equations: An introduction (2*nd* edition): Solutions manual: John Wiley and Sons

Struik, D.J., (1950/1988) Lectures on classical differential geometry: Dover

Synge, J.L., (1937/1962) Geometrical optics: An introduction to Hamilton's methods: Cambridge University Press

Synge, J.L., and Griffith, B.A., (1949) Principles of mechanics: McGraw-Hill, Inc.

Synge, J.L., and Schild, A., (1949/1978) Tensor calculus: Dover

Szekeres, P., (2004) A course in modern mathematical physics: Groups, Hilbert space and differential geometry: Cambridge University Press

Talpaert, Y., (2001) Differential geometry with applications to mechanics and physics: Marcel Dekker, Inc.

Tarantola, A., (2006) Elements of physics: Quantities, qualities, and intrinsic theories: Springer-Verlag

Taylor, M.E., (1996) Partial differential equations; Basic theory: Springer-Verlag

Telford, W.M., Geldart, L.P., Sheriff, R.E., and Keys, D.A., (1976) Applied geophysics: Cambridge University Press

Ter Haar, D., (1961) Elements of Hamiltonian mechanics: North Holland

Thomsen, L., (1986) Weak elastic anisotropy: Geophysics, **51**, 1954–1966.

Thomson, W. (Lord Kelvin), (1890) Mathematical and physical papers, Vol. III, elasticity, heat, electromagnetism: Cambridge University Press

Timoshenko, S.P., and Goodier, J.N., (1934/1987) Theory of elasticity: McGraw-Hill, Inc.

Ting, T.C.T., (1996) Anisotropic elasticity: Theory and applications: Oxford University Press

Toretti, R., (1999) The philosophy of physics: Cambridge University Press

Trèves, F., (1975/2006) Basic linear partial differential equations: Dover

Truesdell, C.A., (1966) Six lectures on modern natural philosophy: Springer-Verlag

Tsvankin, I., (2001) Seismic signatures and analysis of reflection data in anisotropic media: Pergamon

Tsvankin, I., and Grechka V., (2011) Seismology of azimuthally anisotropic media and seismic fracture characterization: Society of Exploration Geophysicists

Tymoczko, T., (editor) (1986/1998) New directions in the philosophy of mathematics: Princeton University Press

Udías, A., (1999) Principles of seismology: Cambridge University Press

van Brunt, B., (2006) The calculus of variations: Springer-Verlag

Voigt, W., (1910) Lehrbuch der Kristallphysics: Teubner

Wang, C.C., and Truesdell, C., (1973) Introduction to rational elasticity: Noordhoff International Publishing

Weinert, F., (2005) The scientist as philosopher: Springer-Verlag

Weinstock, R., (1952/1974) Calculus of variations with applications to physics and engineering: Dover

Weisstein, E.W., (1999) CRC concise encyclopedia of mathematics: CRC Press

Weyl, H., (1921/1952) Space-time-matter: Dover

Weyl, H., (1939/1946) The classical groups: Their invariants and representations: Princeton University Press

Weyl, H., (1952) Symmetry: Princeton University Press

Wilcox, H.J., and Myers, D.L., (1978/1994) An introduction to Lebesgue integration and Fourier series: Dover

Wilkinson, J.H., (1965/2004) The algebraic eigenvalue problem: Oxford Science Publications

Winterstein, D.F., (1990) Velocity anisotropy terminology for geophysicists: Geophysics, **55**, 1070–1088.

Woodhouse, N.M.J., (1992) Geometric quantization (2*nd* edition): Oxford Science Publications

Woodhouse, N.M.J., (1987) Introduction to analytical dynamics: Oxford Science Publications

Zemanian, A.H., (1965/1987) Distribution theory and transform analysis: An introduction to generalized functions, with applications: Dover

Zill, D.G., and Cullen, M.R., (1997) Differential equations with boundary-value problems (4*th* edition): Brooks/Cole Publishing Co.

Zwillinger D., (1992) Book of differential equations (2*nd* edition): Academic Press

Index

About the Author

Michael A. Slawinski is a professor in the Department of Earth Sciences at Memorial University in St. John's, Newfoundland, and an adjunct professor in the Department of Mathematics and Statistics at the University of Calgary.

photo by Massimo Cervetti

He studied at the University of Warsaw, University of Paris and, subsequently, at the University of Calgary. There, in 1988, he obtained his M.Sc. with a thesis entitled "Investigation of inhomogeneous body waves in an elastic/anelastic medium." In 1996, he obtained his Ph.D. from the University of Calgary with a thesis entitled "On elastic-wave propagation in anisotropic media: Reflection/refraction laws, raytracing and traveltime inversion."

Between 1986 and 1997, Slawinski worked for several years as a geophysicist in the petroleum industry. In 1998, he joined the faculty of the Department of Mechanical Engineering at the University of Calgary. In 2001, he assumed the position of Research Chair in Applied Seismology at Memorial University.

Since 1998, Slawinski has been the director of The Geomechanics Project, a theoretical-research group whose research focuses on seismic ray theory in the context of differential geometry and continuum mechanics.

www.ingramcontent.com/pod-product-compliance
Ingram Content Group UK Ltd.
Pitfield, Milton Keynes, MK11 3LW, UK
UKHW020405150926
8624IPUK00004B/359

9 789811 226403